# FORAMINIFERA AND THEIR APPLICATIONS

The abundance and diversity of Foraminifera ('forams') make them uniquely useful in studies of modern marine environments and the ancient rock record, and for key applications in palaeoecology and biostratigraphy for the oil industry. In a one-stop resource, this book provides a state-of-the-art overview of all aspects of pure and applied foram studies.

Building from introductory chapters on the history of foraminiferal research, and research methods, the book then takes the reader through biology, ecology, palaeoecology, biostratigraphy and sequence stratigraphy. This is followed by key chapters detailing practical applications of forams in petroleum geology, mineral geology, engineering geology, environmental science and archaeology. All applications are fully supported by numerous case studies selected from around the world, providing a wealth of real-world data. The book also combines lavish illustrations, including over 70 stunning original picture-diagrams of Foraminifera, with comprehensive references for further reading, and online data tables providing additional information on hundreds of foram families and species.

Accessible and practical, this is a vital resource for graduate students, academic micropalaeontologists, and professionals across all disciplines and industry settings that make use of foram studies.

ROBERT WYNN JONES has 30 years' experience working as a foraminiferal micropalaeontologist and biostratigrapher in the oil industry, from gaining his Ph.D. in 1982, until his recent retirement from BG Group PLC. Throughout his career, he also maintained an active interest in academic research, producing over one hundred publications, which include seven books, among them *Applied Palaeontology* (Cambridge, 2006) and *Applications in Palaeontology: Techniques and Case Studies* (Cambridge, 2011). Dr Jones is also the author of a book on the history of London, and is an Honorary Scientific Associate of the Natural History Museum in London.

# FORAMINIFERA AND THEIR APPLICATIONS

## ROBERT WYNN JONES
*The Natural History Museum, London*

CAMBRIDGE
UNIVERSITY PRESS

Shaftesbury Road, Cambridge CB2 8EA, United Kingdom

One Liberty Plaza, 20th Floor, New York, NY 10006, USA

477 Williamstown Road, Port Melbourne, VIC 3207, Australia

314–321, 3rd Floor, Plot 3, Splendor Forum, Jasola District Centre, New Delhi – 110025, India

103 Penang Road, #05–06/07, Visioncrest Commercial, Singapore 238467

Cambridge University Press is part of Cambridge University Press & Assessment, a department of the University of Cambridge.

We share the University's mission to contribute to society through the pursuit of education, learning and research at the highest international levels of excellence.

www.cambridge.org
Information on this title: www.cambridge.org/9781107036406

First published 2014

*A catalogue record for this publication is available from the British Library*

ISBN    978-1-107-03640-6    Hardback

Additional resources for this publication at www.cambridge.org/foraminifera

Dedicated to John Haynes and John Whittaker,
and to the memory of Fred Banner

# Contents

# Preface

## What, if anything, are Foraminifera?

This is a question that I am often asked. I tend to try to keep my answer simple, so as to be as comprehensible as possible to the layman. I say that they are single-celled organisms similar to *Amoebae*, but differing in possessing shells.

## Why should I care?

This is another question that I am often asked in one form or another (such as 'So what?'), usually immediately after I have given my answer to the previous one. I say: on account of their numerical importance in modern environments and in the ancient rock record; and of their practical importance to Science and to Human-kind, in developing an understanding of modern environments and of the ancient rock record.

## Importance of Foraminifera

The number of nominal living species of Foraminifera, including synonyms, has been estimated to be at least 6000 (Jones, 1994, based on a count of those in the Ellis & Messina Catalogue), and could be considerably more (Pawlowski *et al.*, 2003a; Murray, 2007; Lipps & Finger, 2010; Pawlowski *et al.*, 2010; Pignatti *et al.*, 2010). The total number of fossil and living species, again including synonyms, has been estimated to be at least 38 000. Foraminifera have been estimated to constitute approximately 2% of all the animals, or animal-like organisms, known from the Cambrian–Recent, and 38% of the animal-like protists or 'protozoans' (Boltovskoy & Wright, 1976).

As individuals as opposed to species, Foraminifera are common in all modern and ancient marine environments, such that their study has been, and continues to be, central to a number of branches of natural science, including biology, ecology, oceanography, geology and palaeontology. It is perhaps as fossils that they are best known, and that they have been, and continue to be, most useful to Science and to Humankind. Fossil Foraminifera are sufficiently common throughout the rock record as to have proved particularly invaluable in the science of stratigraphic palaeontology, or biostratigraphy, and in palaeoenvironmental interpretation, both in academia and in industry, for example in petroleum, mineral and engineering geology. Some, so-called larger benthic Foraminifera or LBFs, are sufficiently large as well as common in parts of the rock record as to be rock-forming, and even, locally, reservoir-forming!

## Books on Foraminifera

There have been a large number of previous books on Foraminifera, including, for example, those of Loeblich & Tappan (1964), Murray (1973), Boltovskoy & Wright (1976), Haynes (1981), Bolli *et al.* (1985), van Morkhoven *et al.* (1986), Loeblich & Tappan (1987), Murray (1991), Sen Gupta (1999) and Murray (2006). These books each cover one or two of the key aspects of biology, morphology and classification; ecology; palaeobiology or palaeoecology; biostratigraphy; sequence stratigraphy; and applications. This book covers all aspects, although it focuses on applications, and contains numerous case studies of applications, in petroleum geology, mineral geology, engineering geology, environmental science and archaeology.

# Acknowledgements

My sincere thanks are due to Cambridge University Press, the Geological Society, and the Grzybowski Foundation, for providing permission to reproduce copyrighted figures.

To Laura Clark, Abigail Jones, and their colleagues at CUP, and the copy-editor, Zoë Lewin, for seeing the project through to publication.

To Hildegard von Bingen, Antonella de Caserta, Pablo Bruna, Pietro Nardini, Yevstigney Fomin, Benjamin Carr, Ion Ivanovici, Ricardo Castro, Ture Rangstrom, Harry Partch, Daniel Jones, Avner Dorman and Yanni, for their musical inspiration – and to my prospective daughter-in-law Jess for putting me on to them!

And last, but not least, to my wife Heather and sons Wynn and Gethin, for putting up with me for so long!

# 1

## Past, present and future foraminiferal research

This chapter deals with past, present and future foraminiferal research.

### 1.1 Past research (retrospective)

Although they are classed as microfossils, at least certain Foraminifera, the so-called larger benthic Foraminifera, or LBFs, are large enough to be visible to the naked eye. The group has therefore been known to Humankind since early antiquity. The first written reference to Foraminifera is by Strabo, who wrote, of his observations of what we now know to be the LBF *Nummulites gizehensis*: 'There are heaps of stone chips lying in front of the pyramids and among them are found chips that are like lentils both in form and size ... They say that what was left of the food of the workmen has petrified and this is not improbable'.

There may be said to have been two, partially overlapping, phases of past research on the Foraminifera, namely, the descriptive and the interpretive. The emphasis shifted between the two phases, from pure to applied research.

The descriptive phase began with the first formal descriptions of species of Foraminifera dating back to the late eighteenth to nineteenth centuries. Those undertaken by the so-called 'Continental School', personified by the great French naturalist Alcide Dessalines d'Orbigny, embodied a narrower, or 'splitting', species concept than would be widely accepted today, albeit possibly a more accurate one (le Calvez, in Hedley & Adams, 1974); while those undertaken by the 'English School', personified by Henry Bowman Brady, embodied a wider, or 'lumping' concept (Jones, 1990, 1994; Jones, in Lightman, 2004; Jones, in Matthew, 2004; Jones, 2007; Jones, in Bowden *et al.*, in press; see also Box 1 below).

The earliest classification schemes were undertaken by the likes of d'Orbigny and Brady in the nineteenth century, and by the American Joseph Augustine Cushman in the early twentieth (for a fuller review, see Cifelli & Richardson, 1990; see also Section 3.3.2). More modern, later twentieth-century schemes

Box 1
## Henry Bowman Brady (1835–91): The Man, the Scientist, and the Legacy

**Brady the Man**

Henry Bowman Brady was born in 1835 in Gateshead in the north-east of England, the son of Henry Senior and Hannah, *nee* Bowman (Adams, in Hedley & Adams, 1978; Jones, 1990, 1994; Jones, in Lightman, 2004; Jones, in Matthew, 2004; Jones, 2007; Jones, in Bowden *et al.*, in press). His older brother was George Stewardson Brady, a noted ostracodologist. His younger brother Thomas's descendants survive to this day. (Thomas's great-granddaughter Pippa Senior recently worked as a Press Officer for the Royal Society.)

Brady received his education at two Quaker schools, Ackworth and Tulketh Hall, leaving in 1850 to serve as an apprentice to a chemist in Leeds. After going on to study pharmacy in Newcastle, he began a pharmaceutical career also in Newcastle, in 1855, and prospered from the start, eventually diversifying into the sale of scientific instruments, and thereby establishing contacts with a number of natural scientists. While still pursuing his career, Brady became an enthusiastic member of the Tyneside Naturalists' Field Club, and of the Northumberland, Durham and Newcastle-upon-Tyne Natural History Society, and wrote his first papers on the Foraminifera in the Transactions of the aforementioned societies, and in the Reports of the British Association for the Advancement of Science, in the 1860s. He was sufficiently successful in his career as to be able to retire, and to devote the remainder of his life to the full-time study of the Foraminifera, in 1876.

Brady died in early 1891 in Bournemouth in the south of England, where he had gone to attempt to recuperate from an illness contracted on his travels to the Upper Nile late the previous year. His obituary read 'Science has lost a steady and fruitful worker, and many men of science have lost a friend … whose place … no-one else can fill. His wide knowledge of many branches of scientific enquiry … made the hours spent with him always profitable; his sympathy with art and literature, and that special knowledge of men and things that belong only to the travelled man made him welcome also where science was unknown; while the brave patience with which he bore … enfeebled health … and a sense of humour which, when needed, led him to desert his usual staid demeanour for the merriment of the moment, endeared him to all his friends.'

**Brady the Scientist**

Brady ultimately produced over 30 important publications on the Foraminifera, ranging in age from Silurian to Recent, including some co-authored with other leading contemporary 'English school' foraminiferologists. Unfortunately, he died without achieving his final ambition – alluded to in a letter – on monographing the British Recent Foraminifera.

Box 1 (cont.)

In recognition of his services to foraminiferology, Brady was elected Fellow of the Geological Society in 1864, Fellow of the Royal Society in 1874, Corresponding Member of the Imperial Geological Museum of Vienna, and Honorary Member of the Royal Bohemian Museum, Prague, and was awarded an Honorary Doctorate from the University of Aberdeen, and a gold medal inscribed 'Insignia of the Royal and Imperial Austro-Hungarian Empire for Art and Science' from Emperor Franz Joseph I.

*Report on the Foraminifera dredged by HMS* Challenger ...
The pinnacle of Brady's achievements as a foraminiferologist was undoubtedly the publication of the *Report on the Foraminifera dredged by H.M.S.* Challenger ... , generally referred to simply as *The* Challenger *Report*, published, after six years work, in 1884, and which remains an indispensable reference even to this day (Jones, 1990, 1994; Henderson & Jones, 2006).

*The* Challenger *Report* describes, figures and includes distribution data on 915 species (15% of the total number of extant species), belonging to 368 genera (44% of the total number of extant genera), including the type-species of 284 (34%). It contains 814 pages of text, written in a delightfully discursive style, and 116 magnificent colour plates produced, under Brady's supervision, by A. T. Hollick (a deaf and dumb artist and lithographer, of whom it was said that 'these terrible disadvantages have been overcome by natural genius'). The quality of the plates was not matched until the advent of digital image capture technology in the 1990s, and has arguably never been bettered. It is perhaps Brady's most enduring legacy.

**Brady's Legacy**

*The Literal Legacy*
Brady left us a literal legacy of a library of books and miscellaneous papers relating to the Protozoa, which he bequeathed to the Royal Society in his will, together with a sum of money for the maintenance and augmentation of the same. The papers include three bound foolscap volumes of distribution data on *Challenger* Foraminifera, some of it not in the *The* Challenger *Report*. They include also letters to Brady from fellow natural scientists G. S. Brady, Carter, Guppy, Halkyard, Hantken, Howchin, Jukes-Browne, Millett, Murray, Robertson, Schwager and Sherborn. Incidentally, the letter from Jukes-Browne to Brady, written in 1889, enquires as to whether he had been able to interpret the palaeobathymetry of samples sent to him from the Oceanic Deposits of Barbados; and is accompanied by a scribbled note evidently written by Brady in preparation for his formal response, and referring to palaeobathymetries of '500 to 1000 fathoms' (see below).

Most of the *Challenger* Foraminifera are housed in the Heron-Allen Library in the Natural History Museum in London, including all of those figured by Brady in *The* Challenger *Report* (Adams, 1960; Adams *et al.*, 1980; Jones, in Lightman, 2004;

*Continued*

## Box 1 (cont.)

Jones, in Matthew, 2004). The *Challenger* collection is the most important, most cited and most consulted collection of Foraminifera in the Natural History Museum.

There is also some 'Bradyana' in the Local Studies Department in the Central Library in Newcastle, including photographs of the Brady family and friends, Brady's fellow foraminiferologists W. B. Carpenter, T. R. Jones, W. K. Parker, C. G. Ehrenberg, F. Karrer and A. E. Reuss, and other contemporary figures such as one might expect to have been admired by someone with Quaker sensibilities, such as the carer and social reformer Florence Nightingale and the abolitionist Abraham Lincoln (but also, bizarrely, one of 'Crockett the Lion Tamer', pictured in what one can only describe as a leopard-skin leotard).

Interestingly, there are two letters from Brady to Charles Darwin in the Darwin Archive in Cambridge, regarding observations on rattle-snake behaviour in relation to evolutionary theory, one dated 18 October, 1871 and the other 22 October, 1871. The tone of the latter indicates that Darwin may have written back to Brady regarding the former, although even if this were indeed the case, the whereabouts of Darwin's letter is not known.

### The Philosophical Legacy

Brady may also be said to have left us a philosophical legacy, in the form of a way of looking at Foraminifera or, better, a way of seeing them rather than simply looking at them; and of interpreting them rather than simply analysing them. His publications, in particular the *Monograph on Carboniferous and Permian Foraminifera* ... (Brady, 1876) and *The* Challenger *Report* (Brady, 1884), may certainly be said to have set modern standards of accuracy of observation; and also of presentation of data. Brady's data continue to have important applications both in biostratigraphy and in biology and palaeobiology. Importantly, he himself was apparently the first to apply – his own – bathymetric data in absolute palaeobathymetric interpretation, using uniformitarian principles, and in this respect was decades ahead of his time (see below).

*Taxonomy. The* Challenger *Report* (Brady, 1884) was conceived in the pervasive intellectual atmosphere of the 'English school' of the latter part of the nineteenth century, such that its taxonomy requires revision to enable it to be used in the type of synoptic and interpretive work being undertaken today (Jones, 1994). Brady's comment on p. vi, that 'the progress of knowledge will eventually break down all sharp demarcations and substitute series for divisions', indicates that his species concept, which is considerably broader than that acceptable today, was influenced by Thomas Henry Huxley ('Darwin's bulldog'). His decision on p. 55 to base his 'primary division' of the Foraminifera on the perforation rather than structural composition of the wall indicates that his suprageneric classification scheme, which would otherwise be widely acceptable today, was influenced by Carpenter.

## Box 1 (cont.)

*Biostratigraphy.* The *Monograph on Carboniferous and Permian Foraminifera ...* (Brady, 1876) set new standards of documentation of collecting localities and incorporation of stratigraphic control. An indication of the consequent lasting value of this work is that it was reprinted as recently as 1970.

*Biology and palaeobiology: biogeography and palaeobiogeography.* The Challenger *Report* (Brady, 1884) documents the distributions of Foraminifera dredged during the voyage of the *Challenger* (1872–1876), and also those dredged on the North Atlantic voyages of the *Lightning* (1868), *Porcupine* (1869) and *Knight Errant* (1879), and the Arctic voyages of the Austro-Hungarian and British North Polar Expeditions (1872–1874 and 1875–1876, respectively). The contained data enable the recognition of five foraminiferal biogeographic provinces in the North Atlantic and Arctic, namely: the Arctic; Subarctic; northern Cool-Temperate; southern Cool-Temperate; and Warm-Temperate provinces (Jones, 2006; Jones & Whittaker, in Whittaker & Hart, 2010). The modern foraminiferal biogeographic data derived from *The* Challenger *Report* has been used as a proxy for interpreting palaeobiogeography, for example in the Pleistocene–Holocene of the British Isles (Jones, 2006; Jones & Whittaker, in Whittaker & Hart, 2010; see also Section 13.2.1).

*Bathymetry and Palaeobathymetry.* The contained data in *The* Challenger *Report* (Brady, 1884) also enables the recognition of seven foraminiferal bathymetric zones, namely: the inner, middle and outer shelf; the upper, middle and lower slope; and the abyssal plain, bathymetric zones. The modern foraminiferal bathymetric data derived from *The* Challenger *Report* has been used as a proxy for interpreting palaeobathymetry, for example in the Palaeogene of the North Sea (Charnock & Jones, in Hemleben *et al.*, 1990; Jones, 1996; Jones, in Jones & Simmons, 1999; Jones, 2006; see also Sections 9.3.3 and 9.5.4). It has also been used as a proxy for interpreting the palaeobathymetric evolution or uplift history of Barbados (Jones, 2009).

As noted above, Brady himself was apparently the first to apply bathymetric data in absolute palaeobathymetric interpretation.

A paper he wrote on the so-called soapstone of Fiji (Brady, 1888) includes a palaeobathymetric interpretation, as follows: 'The depth at which the deposit may originally have taken place can ... be determined approximately. Comparing the list of species with similar lists compiled from material collected on the *Challenger* Expedition at various Pacific stations within the tropics, it is found to include several forms not recorded from depths of less than 129 fathoms, and certain others of which the minimum depth is about 150 fathoms; besides a few ... which are best known from much deeper water. ... [J]udging from its general facies, the Rhizopod-fauna is one that I should expect to find in a deposit forming at from 150 to 200 fathoms (more rather than less) in the neighbourhood of any of the volcanic islands of the Pacific.'

*Continued*

---

Box 1 (cont.)

A later paper on the geology of Barbados (Jukes-Browne & Harrison, 1892) includes a 'Report by the [then] late H. B. Brady', which in turn also includes a palaeobathymetric interpretation, as follows: 'I have made a preliminary examination, in respect of the Foraminifera they contain, of most of the samples you sent me. The results, though far from complete, possess considerable interest, and as I am unable at present [presumably, at this time, through ill health] to continue the investigation I send them to you as they stand. ... The aspect of the rhizopod fauna ... is not inconsistent with the idea of a sea-bottom of considerable depth, perhaps from 500 to 1000 fathoms.'

---

include those of Haynes, Saidova, Loeblich & Tappan, Lee and Mikhalevich (again, see Cifelli & Richardson, 1990; see also Section 3.3.2).

The interpretive phase began with the first use of Foraminifera in biostratigraphy, by the Pole, Josef Grzybowski, in the oilfield area of the Polish Carpathians, in the late nineteenth century (Charnock & Jones, in Hemleben *et al.*, 1990; Kaminski *et al.*, 1993). It continued with further applications in the petroleum industry, in areas as diverse as California, the US Gulf Coast, Iran, Nigeria, Papua New Guinea and Sarawak, in the early twentieth century. Applications in academia began with the establishment of regional larger benthic and, importantly, global planktic foraminiferal biostratigraphic zonation schemes in the late twentieth century (Bolli *et al.*, 1985); accompanied by improvements in the understanding of foraminiferal ecology, oceanography, palaeoecology, palaeoceanography and palaeoclimatology, and of biogeochemical proxies (Scott & Medioli, 1980; Vincent *et al.*, 1981; Lutze & Coulbourn, 1984; Corliss, 1985; Delaney & Boyle, 1987; Gooday & Lambshead, 1989; Mix, in Berger *et al.*, 1989; Herguera & Berger, 1991; Kaiho, 1994; Jorissen, in Sen Gupta, 1999; Pearson & Palmer, 2000). Significant advances on all fronts accompanied the 'big science' initiatives of the time, including the Deep-Sea Drilling Project or DSDP and succeeding Ocean Drilling Program or ODP, and CLIMAP (CLIMAP Project Members, 1981; Imbrie *et al.*, in Berger *et al.*, 1984; Shackleton *et al.*, 1990).

## 1.2 Present research (perspective)

Probably the most important advances in foraminiferology in recent years have been in the fields of molecular biology (Gregory *et al.*, 2006; Murray, 2006; see also Section 3.3.2). Advances have also been made in the fields of imaging technology (Briguglio & Hohenegger, 2010; Mucadam, 2010; Speijer *et al.*, 2010b; Szinger *et al.*, 2010; Gorog *et al.*, 2012); and of automated species identification (see, for example, Shan Yu *et al.*, 1996; Ranaweera *et al.*, 2009a; b; see also Section 2.3). Innovative uses have been made of Foraminifera in the fields of mineral and engineering geology (Hart *et al.*, in Jenkins, 1993; see also, respectively, Chapters 10 and 11); environmental science (Martin, 2000; see also Chapter 12); and archaeology (Whittaker, in Whittaker *et al.*, 2003; Jones & Whittaker, in Whittaker & Hart, 2010; see also Chapter 13). And if only for its massive economic impact, or what I once referred to as 'bang for your bug', then surely well-site operations in petroleum geology, and especially 'biosteering' must merit at least a mention (Jones *et al.*, 1999; Jones *et al.*, in Koutsoukos, 2005).

## 1.3 Future research (prospective)

I personally would like to see more of the same, please, in petroleum geology, especially as regards reservoir exploitation; in mineral and engineering geology and in archaeology; and, especially, in environmental science (at least this last is well underway in preparation for the implementation of the European Water Framework Directive or ENFD in 2015 and European Marine Strategy Framework Directive or EMSFD in 2020 (see, for example, Barras *et al.*, 2010; Jorissen, 2010; Schonfeld *et al.*, 2012)). And maybe some strategic research into how to explore for and exploit the world's remaining, in many cases stratigraphically rather than structurally trapped, petroleum reserves most efficiently and, at the same time, with the least environmental impact.

# 2

# Research methods

This chapter deals with research methods. It contains sections on work-flow; on sample acquisition and processing; on analytical data acquisition; on interpretation; and on integration. The section on sample acquisition and processing includes separate sub-sections on sample acquisition and on sample processing.

## 2.1 Work-flow

A generic work-flow for applied (micro)palaeontology is shown in Fig. 2.1. It will be seen that the key constituent elements are sample acquisition and processing (also analysis), analytical data acquisition, interpretation and integration. Each of these is discussed in turn below.

## 2.2 Sample acquisition and processing

### 2.2.1 Sample acquisition

*Field acquisition of samples of live Foraminifera*

The acquisition of samples of live Foraminifera, for laboratory culture and/or analysis, including sea-bed samples for benthics and sea-surface samples for planktics, is discussed by, among others, Murray (1973), Haynes (1981), Green (2001), Schonfeld (2012) and Schonfeld *et al.* (2012) (see also the Scientific Committee on Oceanic Research (SCOR) website). The best sea-bed samples for benthics are those acquired through the use of multiple or box corers, which preserve the sediment–water interface.

*Field acquisition of samples of fossil Foraminifera*

The field acquisition of samples of fossil Foraminifera for laboratory analysis is discussed by Green (2001), Jones (2006) and Jones (2011a) (see also Coe, 2010).

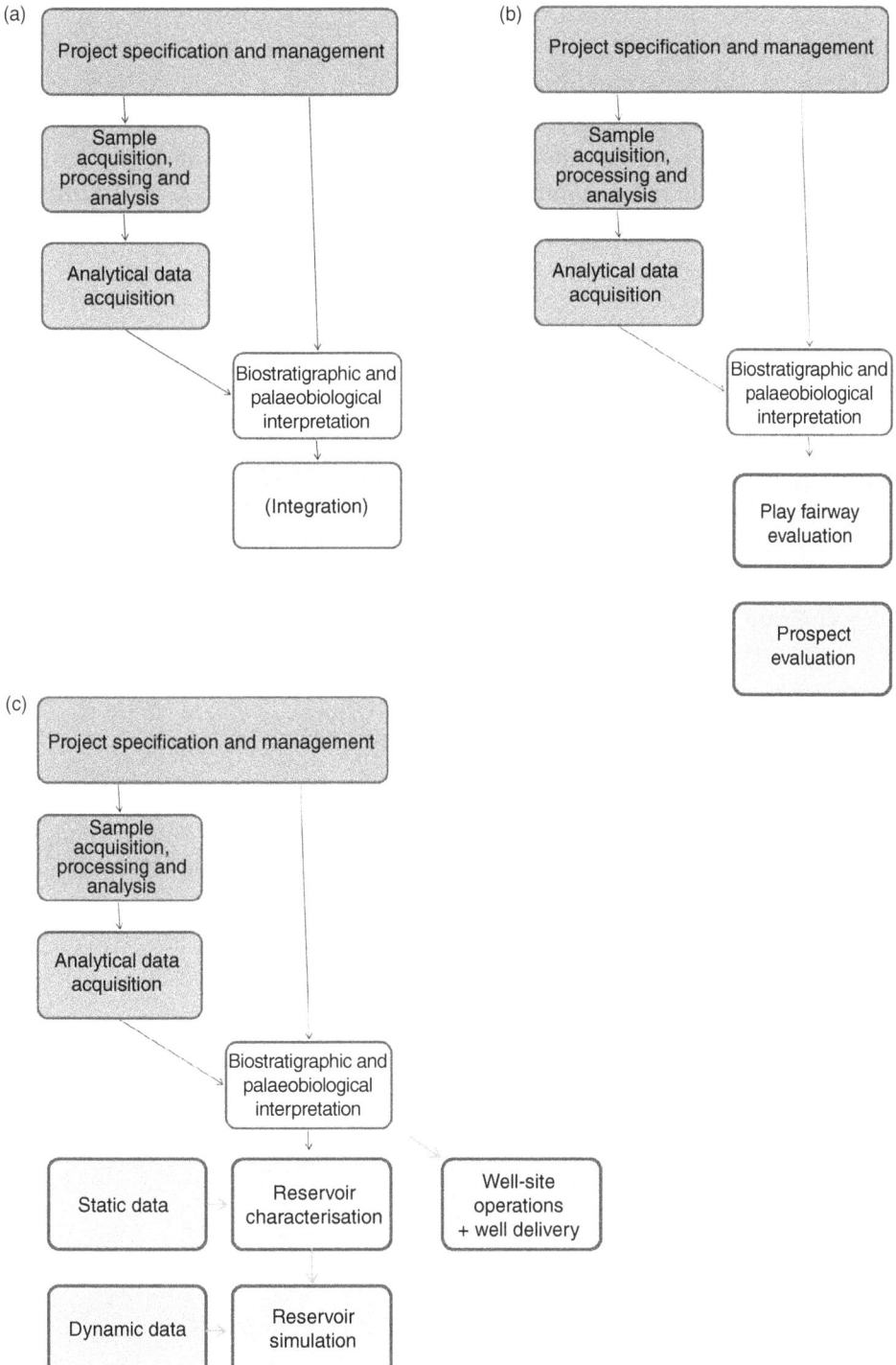

Fig. 2.1 **Work-flow for (applied) micropalaeontology.** (a) General; (b) petroleum exploration; (c) reservoir exploitation; from Jones (2011).

Sample acquisition is necessary for field mapping and laboratory research purposes, but it should still be undertaken responsibly and sustainably, so as to
conserve or preserve a finite natural resource for future generations. Note in this
context that sample acquisition is restricted in Sites of Special Scientific Interest,
or SSSIs, in the United Kingdom, and indeed is restricted by nature conservation
and by national monument protection legislation in 'geotopes' ('parts of the
geosphere ... clearly distinguishable from their surroundings in a geoscientific
fashion') in Germany.

The overall objectives of the field-work should be considered when determining
the sampling strategy. For example, if the objective is reconnaissance mapping, spot
sampling might be all that is required, whereas if the objective is detailed logging,
closer sampling will be required. As a general comment, the biostratigraphic or
palaeoenvironmental resolution of the analytical results will depend as much on the
sampling density as on the fossils themselves. Partly on account of this, and partly
on account of the logistical effort and financial cost of mobilising field parties, it is
always advisable to collect what might be thought of as too many rather than too
few samples. However, any restrictions should be respected (see above).

*Lithology*   Foraminifera are common in essentially all at least marginally marine
mudstones, marls and limestones. Fresh rather than weathered samples should be
acquired, through digging, augering or trenching, if necessary.

*Size of sample*   A 'Standard British Handful' is generally sufficient to ensure
recovery of Foraminifera. However, larger samples are required in areas characterised by high sediment accumulation rate, which trends to dilute the fossil content,
such as parts of the Polish Carpathians.

### *Well-site acquisition of samples of fossil Foraminifera*
### *(in the petroleum industry)*

The well-site acquisition of samples of fossil Foraminifera for laboratory analysis
is discussed by Jones (2006) and Jones (2011a). The most useful are conventional-
or side-wall- core samples, and the least useful are ditch cuttings (see also Section
9.1.2). This is because ditch-cuttings samples are prone to contamination not only
from caved material, but also on occasion from the drilling mud.

As above, the overall objectives of the well-site work should be considered
when determining the appropriate strategy, ideally well in advance. For example, if
the objective is routine monitoring in the exploration phase, coarsely-spaced ditch-
cuttings sampling might be all that is required, whereas if the objective is integrated reservoir characterisation in the exploitation phase, closely-spaced cuttings,
or conventional- or side-wall- core sampling might be required. Also as above,

larger samples are required in areas characterised by high sediment accumulation rate, such as the Columbus basin, offshore Trinidad.

Note that the drilling as well as the depositional environment, and also the bit type, need to be considered when selecting samples for foraminiferal analysis. 'Turbo-drilling', can have a particularly destructive effect on samples, effectively metamorphosing them and rendering them useless for analytical purposes. Coiled tubing, slim-hole and underbalance technologies have, though, recently been successfully employed in combination in the drilling of the Sajaa field in Sharjah in the United Arab Emirates, without unduly adversely affecting the quality of the samples used in foraminiferal analysis and in 'biosteering' (see Section 9.7.1).

### *2.2.2 Sample processing*

The processing of samples for foraminiferal analysis is discussed by Green (2001), Jones (2006, 2011a) and Schonfeld (2012) (see also Semensatto & Dias-Brito, 2007; de Moura *et al.*, 1999; Remin *et al.*, 2012). Unfortunately, processing methods are not standardised, leading to sometimes serious issues of comparability from one project to another (Schonfeld, 2012; Schonfeld *et al.*, 2012).

Most samples can simply be disaggregated by washing in water, with or without the addition of heat, and/or of chemicals such as acetone ($C_3H_6OH$), Alconox or other anionic detergents, ethanol ($C_2H_5OH$), hydrogen peroxide ($H_2O_2$), sodium carbonate ($Na_2CO_3$), sodium hypochlorite (NaOCl), sodium hydroxide (NaOH), or white spirit ($C_{10}H_{22}$); whereafter individual Foraminifera can be picked out of the sieved dried residues with the use of a fine paint-brush, and placed into a gridded slide. Note that the use of hydrogen peroxide is not recommended, as it can lead to over-vigorous oxidation and destruction of specimens containing or preserved in pyrite. Note also that the use of any chemical brings associated health and safety risks, which need to be mitigated. Carbon tetrachloride, or tetrachloromethane ($CCl_4$), once used for heavy liquid flotation, is no longer used, as it has been discovered to be a carcinogen.

Particularly indurated samples may need to be thin-sectioned. Non-calcareous, agglutinating Foraminifera can be extracted from indurated carbonates by acid treatment.

### 2.3 Analytical data acquisition

Once they have been picked into a gridded slide (see above), Foraminifera can be sorted into species, and species placed in separate individual grids for identification, and, if required, as in semi- or fully-quantitative, as opposed to qualitative, studies, for counting. Identification is generally undertaken manually, with reference to standard published works, although it has just started to be automated (see, for example, Shan Yu *et al.*, 1996; Ranaweera *et al.*, 2009a, b). Unfortunately, again, data acquisition methods are not standardised, sometimes leading to lack of comparability of results from one project to another (Schonfeld, 2012; Schonfeld *et al.*, 2012).

Once acquired, data can be displayed, generally in the form of a species distribution or range chart. Range charts are generated either manually or automatically. The petroleum industry standard software for the automation of data display and manipulation is StrataBugsTM (see www.stratadata.co.uk).

## 2.4  Interpretation

The environmental, palaeoenvironmental, biostratigraphic and sequence stratigraphic interpretation of foraminiferal data is discussed by Jones (1996), Jones (2006) and Jones (2011a), and, respectively, in Chapters 4, 5, 7 and 8. The use of quantitative techniques in interpretation in ecology, palaeoecology and biostratigraphy is discussed by Jones (1996), Jones (2006) and Jones (2011a), and, respectively, in Sections 4.6, 5.9 and 7.6.

## 2.5  Integration

The integration and application of interpreted foraminferal data in petroleum geology, mineral geology, engineering geology, environmental science and archaeology is discussed by Jones (1996), Jones (2006) and Jones (2011a), and, respectively, in Chapters 9, 10, 11, 12 and 13.

# 3

# Biology, morphology and classification

This chapter deals with biology, morphology and classification. It contains sections on biology; on morphology; and on classification. The section on biology includes sub-sections on the living organism and its soft-part morphology, on life strategies, and on life activities. The section on morphology includes sub-sections on descriptive terminology and morphological description, and on functional morphology. The section on classification includes sub-sections on Foraminifera in the scheme of living things, and on suprageneric classification of the Foraminifera.

The following is an abridged account of the biology, morphology and systematics, that is, the pure micropalaeontology, of the Foraminifera based on Jones (1996, 2006, 2011a) (see also the 'further reading' list at the end of the chapter).

## 3.1 Biology

The biology of living Foraminifera is comparatively poorly known, partly because the group has historically received more attention from geologists than biologists, and partly because of difficulties in reproducing living conditions in laboratory culture experiments (but see the further reading list at the end of the chapter).

### 3.1.1 The living organism, and its soft-part morphology

As intimated in the preface, Foraminifera are single-celled organisms similar to *Amoebae* or Amoebida, and testate *Amoebae* or Theacamoebida, but differing in the possession of granulo-reticulose rather than filose or lobose pseudopodia; and also by the possession of an agglutinated or secreted shell or test, bearing an opening or aperture through which they can communicate with their external milieu (see Section 3.2).

Foraminifera are single-celled organisms. The cells possess a single nucleus. They are composed of protoplasm, which is capable of being formed into pseudopodia alluded to above, or, literally, 'false feet', for the purposes of attachment, locomotion or feeding. Live protoplasm may be identified by means of stains such as Rose Bengal and CellTracker Green (see the further reading list at the end of the chapter).

### 3.1.2  Life strategies

Foraminifera have adopted a number of life strategies, including both benthic, or bottom-dwelling, and planktic, or free-floating (within the water column); and, among the benthics, both epifaunal, or surface-dwelling, and infaunal, or burrow-dwelling. Some, for example the larger Miliolida and Rotaliida, the so-called Larger Benthic Foraminifera (LBFs), have selected conservative generalist or 'K' strategies; others, for example, certain agglutinating and smaller calcareous benthics, have selected specialist opportunistic or 'r' strategies (Pianka, 1970; see also the further reading list at the end of the chapter). The 'K' selectors are characterised by slow development to reproductive maturity (at large size), sexual reproduction, small numbers of progeny, typically in more than one issue (iteroparity), long generation time, and a long life-span; the 'r' selectors by rapid development to reproductive maturity (at small size), asexual reproduction in some groups, large numbers of progeny, typically in one issue (semelparity), short generation time, and a short life-span.

### 3.1.3  Life activities

#### Feeding

Foraminifera are 'animal-like' protists or 'protozoans', constituting part of the zoobenthos or zooplankton, and as such are heterotrophs or consumers, as opposed to autotrophs or producers (see 'further reading' list at the end of the chapter). They typically eat Bacteria, Algae and organic detritus; and are in turn eaten by higher organisms. In cultures, cyanophyte and dinophyte Algae are generally rejected in favour of diatoms.

They practise a number of feeding strategies (Jones, 1984; Jones & Charnock, 1985; Jones, 1986, 2010; Murray *et al.*, 2011; and additional references cited therein; see also the further reading list at the end of the chapter). These include:

- *suspension-feeding*;
- *detritus-feeding or detritivory*, including carnivory, herbivory (browsing, grazing), and phytodetritivory;
- *symbiosis*, including chemosymbiosis, chloroplast symbiosis (kleptoplastidy), and photosymbiosis;

- *parasitism*;
- *denitrification of intracellular nitrate stores*;
- *direct uptake of amino acids or dissolved organic material (DOM).*

Incidentally, feeding strategy generally dictates, as least to some extent, life position and deployment of pseudopodia, and also test form (see also Section 3.2.2). Note, though, that ready deployment of (adhesive) pseudopodia is the only essential prerequisite for a carnivore, such that species exhibiting carnivorous feeding are of diverse morphology – witness, on the one hand, *Astrorhiza limicola* and on the other, *Pilulina argentea* (see the further reading list at the end of the chapter). Note also that species that take up DOM are morphologically indistinguishable from suspension-feeders, both being characterised by a comparatively large – cytoplasmic – surface area with respect to volume.

### *Growth*

Growth is either continuous, or attained by an increase in the size of a single chamber; or episodic, and attained by the addition of new chambers (see the further reading list at the end of the chapter).

### *Productivity*

Benthic foraminiferal productivity, or standing crop or stock, can be measured in the field or in the laboratory, and is expressed in terms of numbers of living individuals per unit area, typically 10 $cm^3$, or per unit area over time (see the further reading list at the end of the chapter). Planktic productivity or flux can also be measured.

### *Reproduction*

Reproduction typically takes place in alternate sexual and asexual stages, although under certain circumstances, for example of extreme environmental stress, it appears to take place entirely asexually (see the further reading list at the end of the chapter). Sexual reproduction provides a mechanism for mutation and evolution.

## 3.2 Morphology

As noted above, Foraminfera are characterised by the possession of an agglutinated or secreted shell or test, bearing an opening or aperture through which they can communicate with their external milieu. Incidentally, it is the aperture that gives the group its name, assigned by d'Orbigny in the early nineteenth century – Foramini-fera literally meaning 'aperture-bearing'.

The process of shell or test secretion, or biomineralisation or calcification, is presently poorly understood (but see the further reading list at the end of the chapter). Importantly, the process preserves an isotopic environmental signal that is of great value in palaeoenvironmental interpretation (see the further reading list at the end of the chapter; see also Sections 5.2, 5.7 and 5.8).

### 3.2.1 Descriptive terminology and morphological description

Foraminifera display great diversity in morphology. The simplest forms are undivided spheres, or tubes with apertures at one or both ends. More advanced forms are divided into chambers arranged in various ways. The final chambers communicate with the external world through apertures of various types. Internal chamber divisions of various types may be present, as may internal apertural structures. External ornamentation of various types may also be present.

In terms of morphological description, as intimated above, the wall structure or composition of the foraminiferal test can generally be categorised either as agglutinated or, archaically, arenaceous; or as secreted or calcareous (only a very few species are proteinaceous). The composition of calcareous test walls can be specified to be either calcitic or aragonitic, and the calcite as low- or high-Mg; the structure or texture, as microgranular, porcelaneous, granular or radial. The test itself can be described as either unilocular or single-chambered; or multilocular or many-chambered. The arrangement of the chambers can be described as (open or evolute, or closed or involute) planispiral, when in the form of a plane spiral; trochospiral, when in the form of a helical spiral; streptospiral, when irregular, or with regular alternations in the axis of coiling; 'milioline' or 'polymorphine', when wound in various planes about a longitudinal axis; triserial, when there are three per whorl; biserial, when there are two per whorl; or uniserial. The position of the aperture can be described as basal or interio-marginal, and either umbilical or peripheral, if at the base of the apertural face; or areal, if within the apertural face; or terminal.

Formal descriptions of new species include an explanation of the derivation of the chosen name, a synonymy, a diagnosis, and a summary of the observed stratigraphic and palaeoecological distribution of the species, as well as a physical description of its size and characters. The name itself is conventionally given in Linnean binomial form, and italicised, with the generic component first and capitalised, and the specific second and uncapitalised. The synonymy is essentially a listing of previous records of the new species under old names, and is important in that it helps to establish the new-species concept and its stratigraphic and palaeoecological range. The diagnosis is a succinct summary of how the new species differs from related species of the same genus, and is not the same as a physical description. The physical description should include comments on the amount of material available, and on any observed intra-specific variability. All of the specimens available to the author of the species constitute so-called syntypes. A single specimen from among the syntypes that well illustrates the characters of the species should be designated the holotype; the remainder, paratypes. All type material should be deposited in publicly accessible collections for the purposes of consultation.

Illustrations can be produced by any one or more of a range of techniques, depending on the desired effect, including drawing, *camera lucida* drawing, analogue and digital optical photography, digital image capture, video photography, X-ray photography, and electron photomicrography. Three-dimensional reconstruction can also be achieved.

It is worth emphasising at this point that accurate species identification is absolutely central to all aspects not only of pure but also of applied foraminiferology. Without it, we have no meaningful idea as to the palaeoecological distributions and stratigraphic ranges of species, and no meaningful foundation on which to base palaeoecological and stratigraphic interpretations (see also Chapters 5 and 7).

### 3.2.2 Functional morphology

The form or morphology of many organisms, including Foraminifera, is dictated by function; and, conversely, form can inform as to function (Thompson, 1917; see also Section 3.1).

#### Agglutinating Foraminifera

Four main functional morphological groups – or 'morphogroups' – of agglutinating Foraminifera have been recognised on the bases of empirical or experimental observation or of extrapolation, each characterised not only by a particular functional morphology but also by a particular life position and feeding strategy (Jones & Charnock, 1985; Jones, 1986; Kender *et al.*, in Kaminski & Coccioni, 2008; Kender *et al.*, 2008; Jones, 2010; Murray *et al.*, 2011; Fig. 3.1; see also the further reading list at the end of the chapter).

Morphogroup A is characterised by a tubular morphology, and by an epifaunal life position, and a suspension-feeding strategy. Morphogroup B is characterised by a globular to discidal morphology, and by an epifaunal life position, and a detritivorous – including phytodetritivorous – feeding strategy. Morphogroup C is characterised by an elongate morphology and a low surface-area to volume ratio, and by an infaunal life position, and a detritivorous feeding strategy. Morphogroup D is characterised by a conical morphology, and by an epiphytic life position, and a herbivorous feeding strategy.

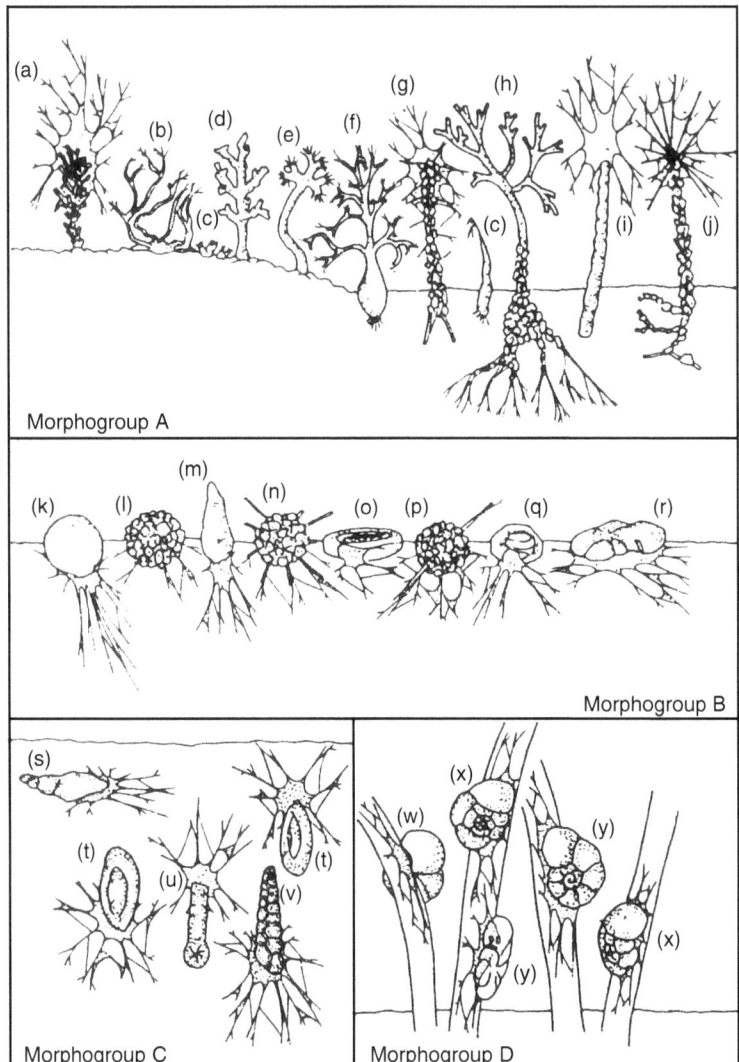

Fig. 3.1 **Morphogroups of agglutinating Foraminifera**. (a) *Halyphysema tumanowiczi*; (b) *Saccodendron* sp.; (c) *Pelosina* sp.; (d) *Dendrophrya erecta*; (e) *Halyphysema* sp.; (f) *Pelosina arborescens*; (g) *Jaculella obtusa*; (h) *Noto-dendrodes antarctikos*; (i) *Bathysiphon* sp.; (j) *Marsipella arenaria*; (k) *Saccammina alba*; (l) *Saccammina sphaerica*; (m) *Hippocrepina* sp.; (n) *Marsipella sphaerica anglica*; (o) *Ammodiscus* sp.; (p) *Psammosphaera parva*; (q) *Glomospira* sp.; (r) *Veleroninoides jeffreysi*; (s) *Reophax subfusiformis*; (t) *Miliammina fusca*; (u) *Ammobaculites exiguus*; (v) textulariid; (w) *Trochammina pacifica*; (x) *Trochammina inflata*; (y) *Entzia macrescens*.

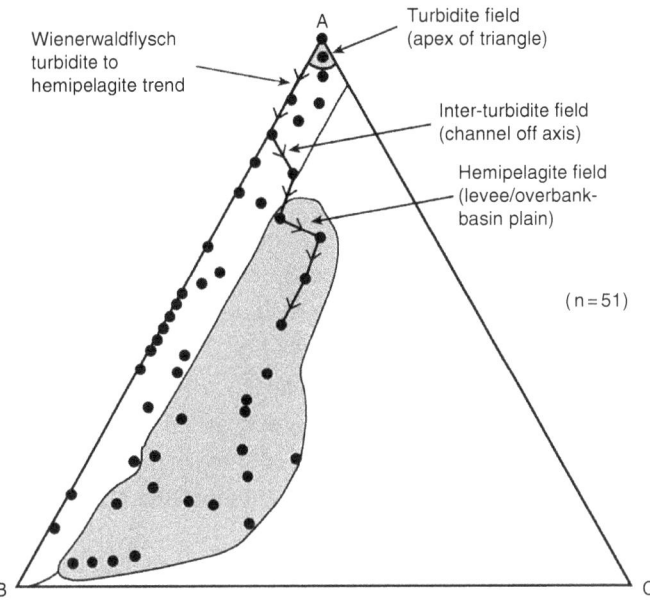

Fig. 3.2    **Triangular cross-plot of morphogroups of agglutinating Foraminifera in submarine fan sub-environments**. From Jones (2011).

Morphogroup A is characteristic of middle–lower bathyal environments, and
especially so of submarine fan sub-environments distinguished by significant amounts
of organic material in suspension. Morphogroup B occurs in all environments, but is
characteristic of deep marine, middle–lower bathyal and abyssal, environments distinguished by significant amounts of organic material on the sea floor. Morphogroup
C also occurs in all environments, but is characteristic of shallow to intermediate
marine, shelf–upper bathyal, environments distinguished by significant amounts of
organic material within the sediment on the sea floor. Morphogroup D is characteristic
of marginal marine environments, and especially so of delta–top sub-environments
distinguished by significant amounts of organic material in the form of vegetation.

Cross-plotting the proportions of selected morphogroups enables the discrimination of a range of depth zones (Jones & Charnock, 1985). It also enables the
discrimination of a range of sedimentary sub-environments in marginal marine,
peri-deltaic environments (Jones & Wonders, 1992; Jones, 2010; Murray *et al.*,
2011; see also the further reading list at the end of the chapter, and Sections 4.3.1
and 5.4.1); or in deep marine, submarine fan environments (Jones, in Jones &
Simmons, 1999; Jones *et al.*, in Powell & Riding, 2005; Jones, 2010; Murray *et al.*,
2011; Fig. 3.2.; see also Sections 4.3.3 and 5.4.3).

In the case of the Early Jurassic Dunlin delta of the subsurface of the North Sea,
and of analogous surface outcrops in Spitsbergen and Yorkshire, cross-plotting the

proportions of the equivalents of Jones and Charnock's morphogroups B, C and D enables the discrimination of vegetated delta-top and delta-front, and prodelta, sub-environments (see also Sections 9.3.3 and 9.5.1). Vegetated delta-top to delta-front sub-environments are characterised by significant incidences of epiphytic morphogroup D, proximal prodelta sub-environments by epifaunal morphogroup B, and distal prodelta sub-environments by infaunal morphogroups C.

In the case of the Palaeogene submarine fans of the subsurface of the North Sea, and of analogous surface outcrops in the Spanish Pyrenees, cross-plotting morphogroups A, B and C enables the discrimination of submarine fan channel, levee and overbank, or respectively equivalent turbiditic, inter-turbiditic and hemipelagic sub-environments (Fig. 3.2.; see also Sections 9.3.3, 9.5.4 and 9.7.3). Fan channel or turbiditic sub-environments are characterised by epifaunal suspension-feeding morphogroup A; levee or inter-turbiditic sub-environments by epifaunal morphogroup B; and overbank or hemipelagic sub-environments by small but significant incidences of infaunal morphogroup C.

The methodology is applicable throughout the Cenozoic and Mesozoic, and has even been experimented with in the Late Palaeozoic (see the further reading list at the end of the chapter).

### Smaller calcareous benthic Foraminifera

A number of authors have recognised morphogroups or 'morphotypes' of smaller calcareous benthics characterised not only by a particular functional morphology but also by a particular observed or inferred life position (see the further reading list at the end of the chapter). Epifaunal detritus-feeding smaller calcareous benthic morphotypes are characterised, like their agglutinated counterparts, by globular to flattened, biconvex or plano-convex morphologies; and infaunal ones, again like their agglutinated counterparts, by elongate morphologies (see above).

Note that the incidence of infaunal morphotypes appears to be more or less proportional to the organic content of the substrate, or inversely proportional to the oxygen content, and is a key input in the calculation of the 'Benthic Foraminiferal Oxygen Index (BFOI)' (see Section 4.5.1).

### Larger benthic Foraminifera (LBFs)

LBFs are characterised by comparatively large sizes, complex internal structures, and high internal surface-area to volume ratios, representing functional morphological adaptations to the accommodation of algal photosymbionts in shallow marine environments (see the further reading list at the end of the chapter).

### Planktic Foraminifera

As functional morphological adaptations to their life positions, shallow-water (within the water column) or epipelagic planktic Foraminifera, including the

Globigerinidae, are characterised by high surface-area to volume ratios, and deep-water (within the water column) or bathypelagic ones, including the Globorotalii-dae, by low surface-area to volume ratios.

## 3.3 Systematics

### 3.3.1 *Foraminifera in the scheme of living things*

Although there is no clear consensus as to their place in the scheme of life and their inter-relationships with other groups of organisms, Foraminifera are currently most commonly treated as a Class; and as such, incidentally, ought properly to be referred to as Foraminifer-ea rather than Foraminifer-a, although the latter is the much more widely used and accepted name for them (see the further reading list at the end of the chapter).

At higher hierarchical levels (phyla, kingdoms, domains, etc.), Foraminifera have been treated historically as belonging to the Rhizopoda, the Sarcodina or Sarcomastigophora, and the Protista or Protoctista ('animal-like' protists, or 'proto-zoans'), characterised by heterotrophism, or secondary consumption, rather than primary production; and in turn to the Eukaryota, characterised by the presence of nuclei. And more recently as belonging to the Retaria and Rhizaria, the SAR and AH/SAR, and the Bikonta; and in turn to the Eukaryota (see the further reading list at the end of the chapter). Recent molecular biological evidence indicates that their closest relatives, within the Retaria, are the Radiolaria.

### 3.3.2 *Suprageneric classification of the Foraminifera*

#### *Historical schemes*

As noted in Section 1.1, the earliest classification schemes were undertaken by the likes of d'Orbigny and Brady in the nineteenth century, and by Cushman in the early twentieth. Brady's uses essentially the same criteria as those that would be widely accepted today, that is, the nature of the wall, and of the chamber arrange-ment, and of the aperture. However, it also uses different weightings from those that would be widely accepted today, for example, placing undue weight on the significance on the perforation as opposed to the composition of the wall.

More modern, later twentieth-century schemes include those of Haynes (1981) (see also Haynes, 1990); Saidova (1981); Loeblich & Tappan (1987) (see also Loeblich & Tappan, 1964; Loeblich & Tappan, in Hedley & Adams, 1974; Loeblich & Tappan, in Takayanagi & Saito, 1992; Sen Gupta, in Sen Gupta, 1999 (modified version of Loeblich & Tappan, in Takayanagi & Saito, 1992); Lee, in Margulis *et al.* (1990); and Mikhalevich, in Tyszka *et al.* (2005). Haynes's and Loeblich & Tappan's schemes differ most significantly in the weighting that they apply to details of wall structure, which is essentially generic or lower in

Haynes's scheme, and familial or higher in Loeblich & Tappan's. At least in part as a consequence of this, Haynes's scheme recognises 99 families, and Loeblich & Tappan's 288. Mikhalevich's scheme is worth mentioning as it uses the innovative new criteria of 'polymerisation', that is, in this case the number not of nuclei or of genes inside nuclei but of chambers and apertures, and of secondary 'oligomerisation'. In other words, it manages to get around the problem encountered in other schemes as to the appropriate weighting to apply to the nature of the wall. For example, it classifies the genus *Miliammina*, which has an agglutinated wall but a 'milioline' chamber arrangement, not with the agglutinates but with the miliolines (in the 'Miliolata'), which recent molecular work has shown to be the proper thing to do (Habura *et al.*, 2006). (*Silicoloculina*, which has a siliceous wall but a milioline chamber arrangement, probably also belongs with the miliolines.) Perhaps more controversially, though, it also includes the agglutinated 'Ammodiscana' in the calcareous 'Spirillinata', the agglutinated 'Hormosinana' in the calcareous 'Nodosariata', and the agglutinated 'Textulariana' in the calcareous 'Rotaliata'.

## *Recent molecular biological work*

A significant amount of molecular biological work has been undertalen on living Foraminifera in the last twenty years (see, for example, Huber *et al.*, 1997; de Vargas *et al.*, 1999; Langer, 1999; Tsuchiya *et al.*, 2003; Darling *et al.*, 2004; Morard *et al.*, 2010; see also the further reading list at the end of the chapter).

   This work has had a significant impact on our understanding of the Foraminifera both at species and at higher levels. At species level, it has revealed in many cases the existence of more than one actual, biological species, or so-called 'cryptic' species, within what had formerly been interpreted as one morphological species (note, though, that at least in some cases the purportedly morphologically indistinguishable cryptic species are in fact, with care, morphologically distinguishable). At higher levels, recent molecular biological work has revealed in many cases that historical classification schemes are not entirely supportable, to say the least.

## *Current scheme*

Suprageneric classification within the Foraminifera is currently in something of state of flux, as historical schemes based essentially on the morphology of fossil forms (and phylogeny inferred from empirical observation of the fossil record), have not been entirely supported by recent molecular biological work on living ones (and phylogeny inferred from cladistics). Unfortunately, as yet, insufficient molecular biological work has been undertaken to allow for the construction of completely new classification, even of living species. It is therefore necessary, for pragmatic purposes, to adhere to an existing scheme, and one that is applicable to fossil and living species. The best of these, in my opinion, is that of Haynes (1981) and this is the one that I have adopted throughout this book (see key below). Note, though, that this scheme, good though it is, is in need of at least a little modification. Accordingly, I have moved the planktic Oberhauserellidae from the Robertinida to the Globigerinida, following BouDagher-Fadel *et al.* (1997). I could, and possibly should, also have moved the low trochospiral *Epistominella* from the Buliminida to the Rotaliida.

*Key*

1. Wall proteinaceous: Order Allogromiida (Recent).
    Superfamily Allogromiidea [Allogromiacea].
        Family Allogromiidae (Recent). Represenative genus: *Allogromia.*
2. Wall agglutinated; test unilocular or bilocular: Order Astrorhizida (Cambrian–Recent).
    Superfamily Ammodiscidea [Ammodiscacea].
        Unilocular: Family Saccamminidae (Ordovician–Recent). Represen-tative genera: *Ammopemphix, Colonammina* (Pl. 1.1), *Lagenam-mina* (Pl. 1.2), *Psammosphaera* (Pl. 1.3), *Saccammina* (Pl. 1.4), *Sorosphaera, Technitella, Tholosina, Thurammina* (Pl. 1.5).

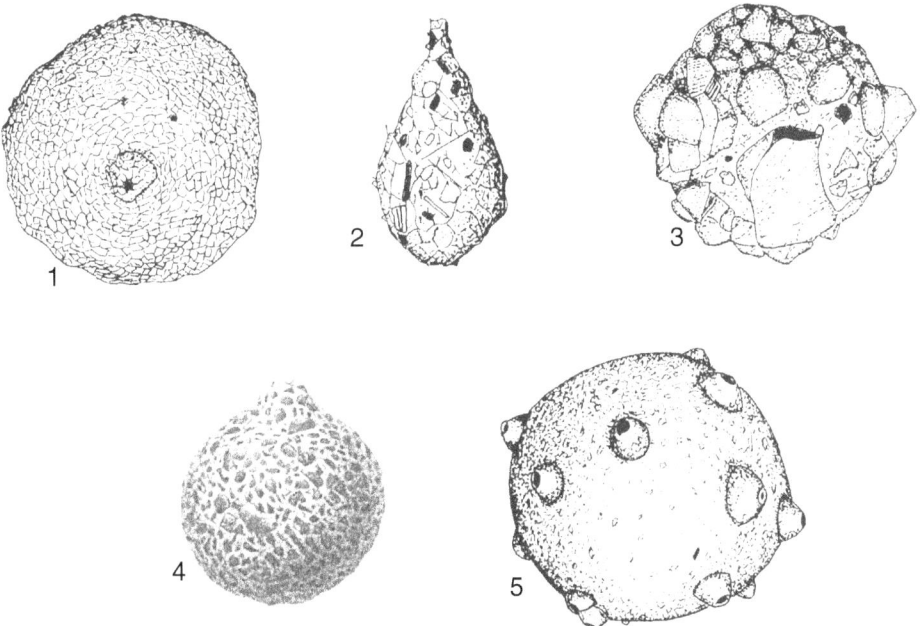

Pl. 1 **Saccamminidae**. Modified after Haynes (1981) and Jones (2006). 1.1 *Colonammina verruca*, ×162; 1.2 *Lagenammina ampullacea*, ×50; 1.3 *Psammo-sphaera fusca*, ×20; 1.4 *Saccammina sphaerica*, ×15; 1.5 *Thurammina papillata*, ×48.

Bilocular, globular first chamber, or proloculus, followed by tubular second chamber: Family Astrorhizidae (Cambrian–Recent). Representative genera: *Argillotuba, Astrammina, Astrorhiza* (Pl. 2.1), *Bathysiphon* (Pl. 2.2), *Botellina, Dendrophyra* (Pl. 2.3), *Halyphysema* (Pl. 2.4), *Hippocrepina* (Pl. 2.5), *Hyperammina* (Pl. 2.6), *Jaculella* (Pl. 2.7), *Kalamopsis, Micatuba, Oculosiphon, Platysolenites, Protobotellina, Psammosiphonella, Rhabdammina* (Pl. 2.8), *Rhizammina* (Pl. 2.9), *Saccorhiza* (Pl. 2.10).

Bilocular, globular proloculus followed by planispirally enrolled second chamber: Family Ammodiscidae (Silurian–Recent). Representative genera: *Ammodiscus* (Pl. 3.1), *Ammolagena* (Pl. 3.2), *Arenomeandrospira* (Pl. 3.3), *Glomospira* (Pl. 3.4), *Glomospirella* (Pl. 3.5), *Tolypammina* (Pl. 3.6), *Turritellella* (Pl. 3.7), *Usbekistania*.

3. Wall agglutinated; test multilocular; small to large, with or without complex interior: Order Lituolida (Carboniferous–Recent).

Essentially planispiral or biserial, in some cases uniserial: Superfamily Lituolidea [Lituolacea].

Planispiral or uniserial, rarely streptospiral; with or without alveolar wall: Family Lituolidae (Carboniferous–Recent). Representative genera: *Alveolophragmium, Ammobaculites* (Pl. 4.1), *Ammocycloloculina* (Pl. 4.2), *Ammomarginulina, Ammotium* (Pl. 4.3), *Bulbobaculites, Buzasina, Caudammina, Choffatella, Conglophragmium, Cribrostomoides, Cyclammina* (Pl. 4.4–4.6), *Cystammina, Dicyclina, Eratidus* (Pl. 4.7), *Evolutinella, Feurtillia, Flabellammina* (Pl. 4.8), *Flabellamminopsis, Haplophragmium, Haplophragmoides* (Pl. 4.9), *Hormosina* (Pl. 4.10), *Hormosinella, Lituola, Lituotuba* (Pl. 4.11), *Loftusia, Popovia, Pseudocyclammina, Recurvoidella, Recurvoides, Reophax* (Pl. 4.12), *Reticulophragmium, Reticulophragmoides, Sabellovoluta, Spiropsammina, Subreophax, Triplasia* (Pl. 4.13), *Trochamminoides* (Pl. 4.14), *Veleroninoides*.

Biserial to uniserial: Family Textulariidae (Cretaceous–Recent). Representative genera: *Bigenerina* (Pl. 5.1), *Parvigenerina, Plectinella, Spiroplectammina* (Pl. 5.4), *Spiroplectella* (Pl. 5.2), *Spiroplectinella* (Pl. 5.3), *Tawitawia, Textularia* (Pl. 5.5), *Vulvulina* (Pl. 5.6).

Essentially trochospiral: Superfamily Ataxophragmiidea [Ataxophragmiacea].

Low trochospiral: Family Trochamminidae (Carboniferous–Recent). Representative genera: *Adercotryma, Arenoparrella* (Pl. 6.1), *Dictyopsella, Entzia [Jadammina]* (Pl. 6.2), *Paratrochammina*

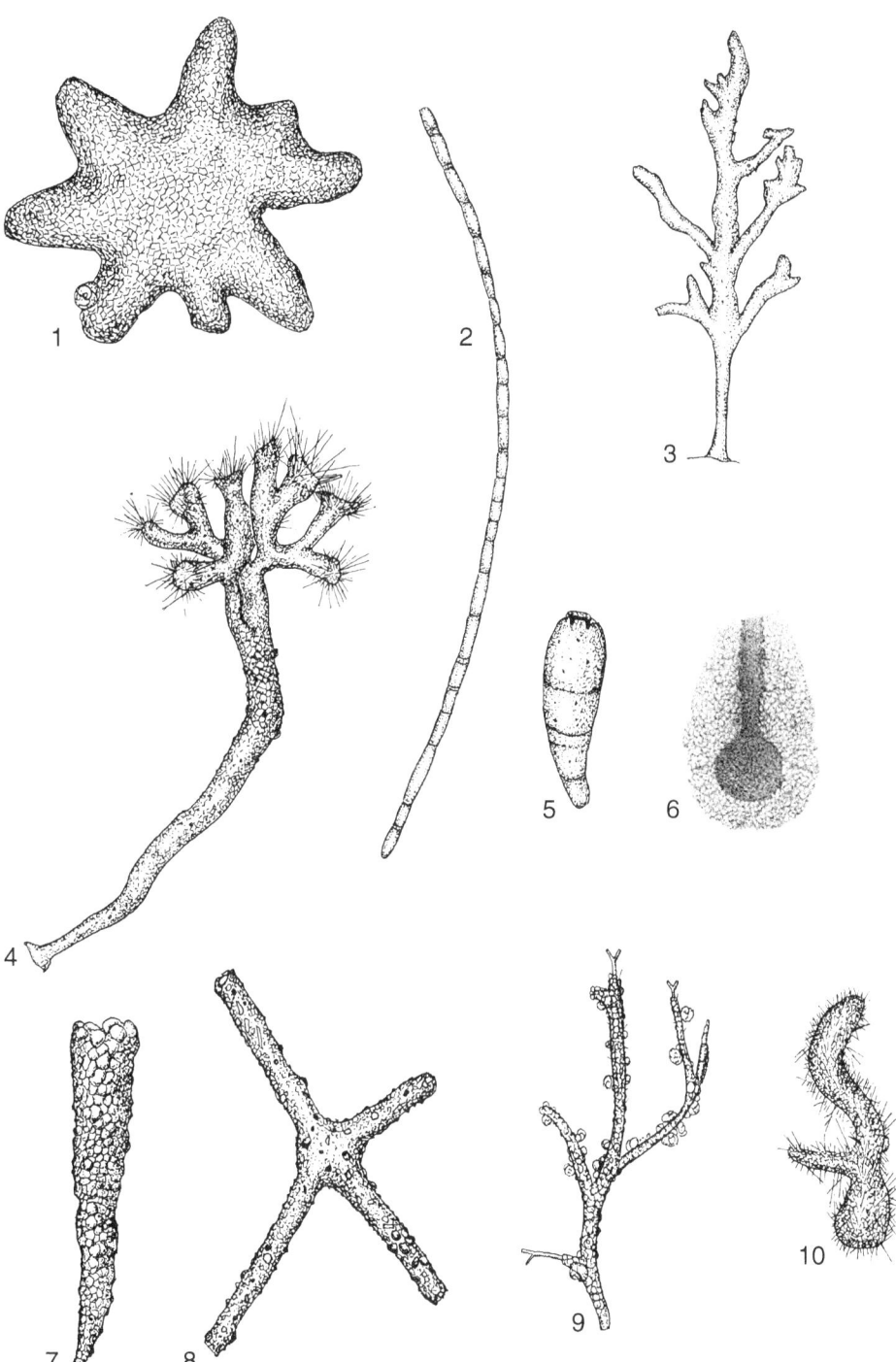

Pl. 2 **Astrorhizidae**. Modified after Haynes (1981) and Jones (2006). 2.1 *Astrorhiza arenaria*, ×8; 2.2 *Bathysiphon filiformis*, ×3; 2.3 *Dendrophyra erecta*, ×30; 2.4 *Halyphysema ramulosa*, ×20; 2.5 *Hippocrepina indivisa*, ×45; 2.6 *Hyperammina friabilis*, ×10; 2.7 *Jaculella acuta/obtusa*, ×15; 2.8 *Rhabdammina abyssorum*, ×10; 2.9 *Rhizammina algaeformis*, ×8; 2.10 *Saccorhiza ramosa*, ×10.

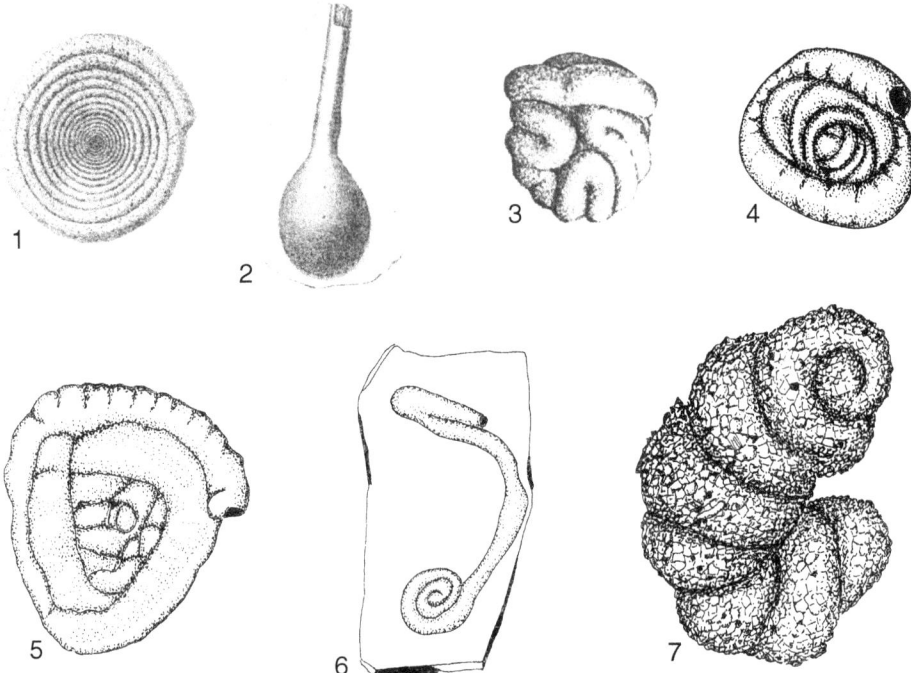

Pl. 3   **Ammodiscidae**. Modified after Haynes (1981), Jones & Wonders, in Hart *et al.* (2000) and Jones (2006). 3.1 *Ammodiscus anguillae*, ×15; 3.2 *Ammolagena clavata*, ×25; 3.3 *Arenomeandrospira glomerata*, ×116; 3.4 *Glomospira gordialis*, ×140; 3.5 *Glomospirella umbilicata*, ×68; 3.6 *Tolypammina vagans*, ×100; 3.7 *Turritellella spectabilis*, ×12.

(Pl. 6.3), *Tiptotrocha* (Pl. 6.4), *Tritaxis* (Pl. 6.5), *Trochammina* (Pl. 6.6–6.8).

High trochospiral to uniserial; aperture basal: Family Verneuilinidae (Triassic–Recent). Representative genera: *Cribrogoesella* (Pl. 7.1), *Dorothia* (Pl. 7.2), *Gravellina*, *Hensonia*, *Marssonella*, *Pseudotextulariella*, *Spiroplectinata* (Pl. 7.3), *Textulariella*, *Tritaxia* (Pl. 7.4), *Verneuilina* (Pl. 7.5–7.6), *Verneuilinoides*, *Verneuilinulla*.

High trochospiral to uniserial; aperture dentate: Family Valvulinidae (Triassic?–Recent). Representative genera: *Clavulina* (Pl. 8.1–8.2), *Clavulinoides*, *Cribrobulimina* (Pl. 8.3), *Dusenburyina*, *Martinottiella*, *Valvulammina* (Pl. 8.4–8.5), *Valvulina* (Pl. 8.6–8.7).

High trochospiral to uniserial; aperture areal to terminal: Family Eggerellidae (Cenozoic). Representative genera: *Eggerella* (Pl. 9.1), *Gaudryina* (Pl. 9.2), *Karreriella* (Pl. 9.3), *Karrerulina* (Pl. 9.4), *Multifidella* (Pl. 9.5), *Plectotrochammina* (Pl. 9.6), *Pseudoclavulina*.

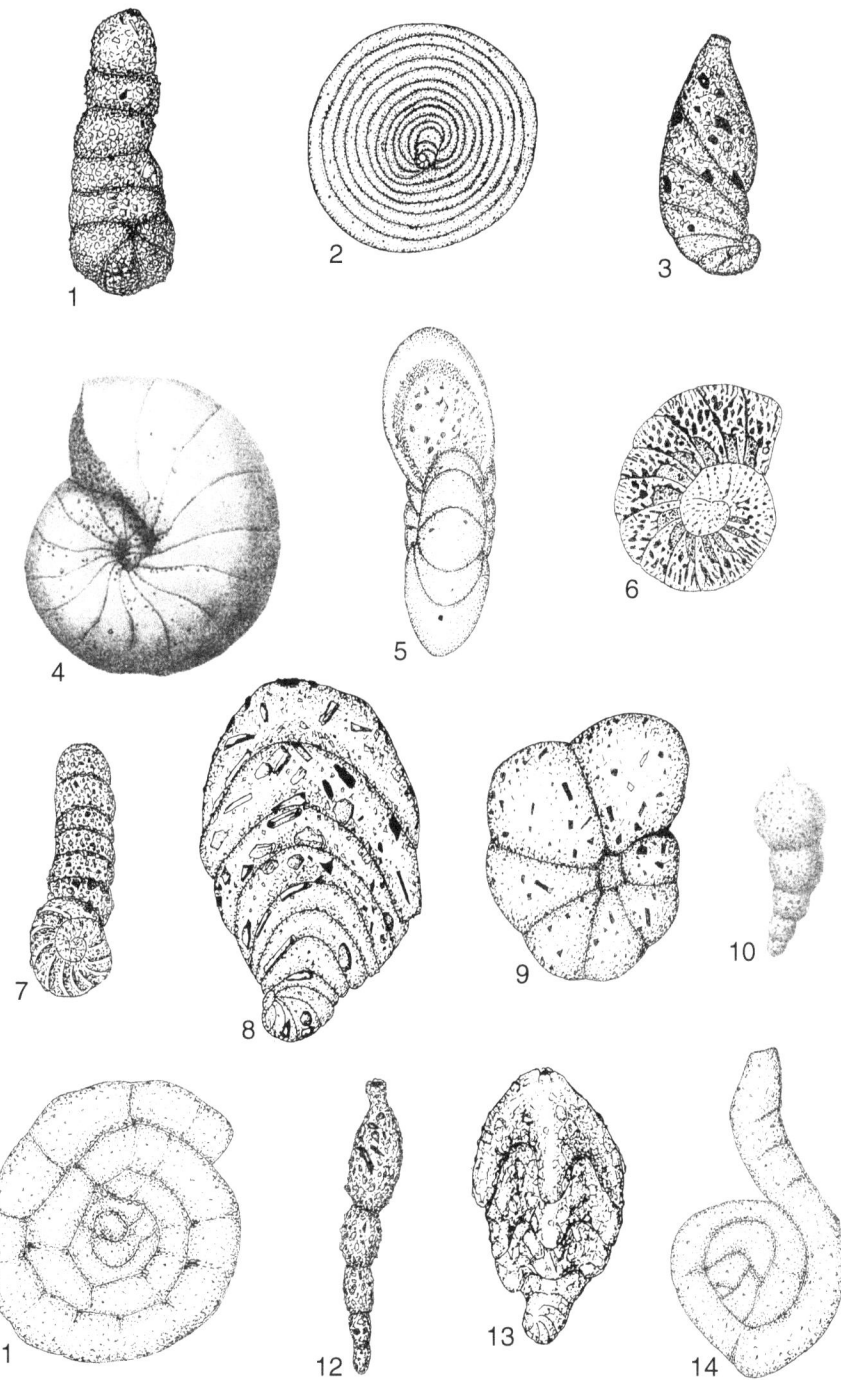

Pl. 4 **Lituolidae**. Modified after Haynes (1981) and Jones (2006). 4.1 *Ammo-baculites calcareus*, ×18; 4.2 *Ammocycloloculina erratica*, ×10; 4.3 *Ammotium cassis*, ×35; Figs. 4-6 *Cyclammina cancellata*, ×15; 4.7 *Eratidus foliaceus*, ×25; 4.8 *Flabellammina alexanderi*, ×32; 4.9 *Haplophragmoides canariense*, ×60; 4.10 *Hormosina globulifera*, ×15; 4.11 *Lituotuba lituiformis*, ×40; 4.12 *Reophax dentaliniformis*, ×40; 4.13 *Triplasia goodlandensis*, ×55; 4.14 *Trochamminoides proteus*, ×50.

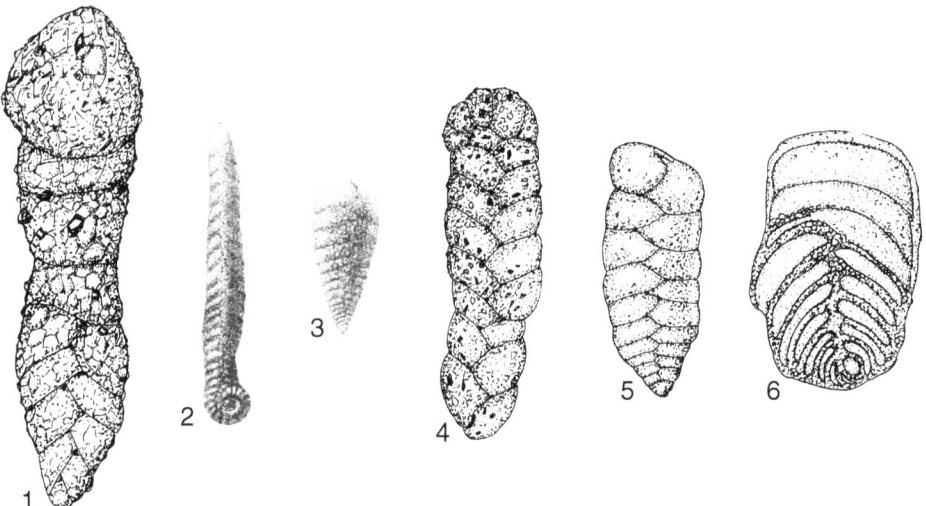

Pl. 5  **Textulariidae**. Modified after Haynes (1981) and Jones (2006). 5.1
*Bigenerina nodosaria*, ×70; 5.2 *Spiroplectella earlandi*, ×60; 5.3 *Spiroplectinella
wrightii*, ×35; 5.4 *Spiroplectammina biformis*, ×100; 5.5 *Textularia sagittula*,
×26; 5.6 *Vulvulina pennatula*, ×40.

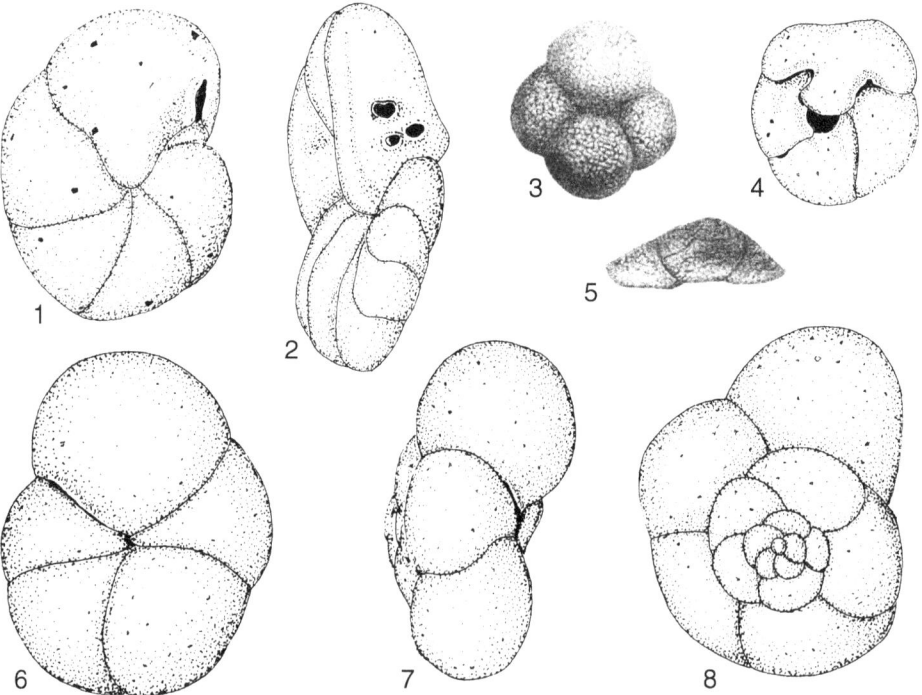

Pl. 6  **Trochamminidae**. Modified after Haynes (1981) and Jones (2006). 6.1
*Arenoparrella mexicana*, ×218; 6.2 *Entzia macrescens*, ×200; 6.3 *Paratrochammina challengeri*, ×25; 6.4 *Tiptotrocha comprimata*, ×60; 6.5 *Tritaxis challengeri*, ×50; 6.6–6.8 *Trochammina inflata*, ×100.

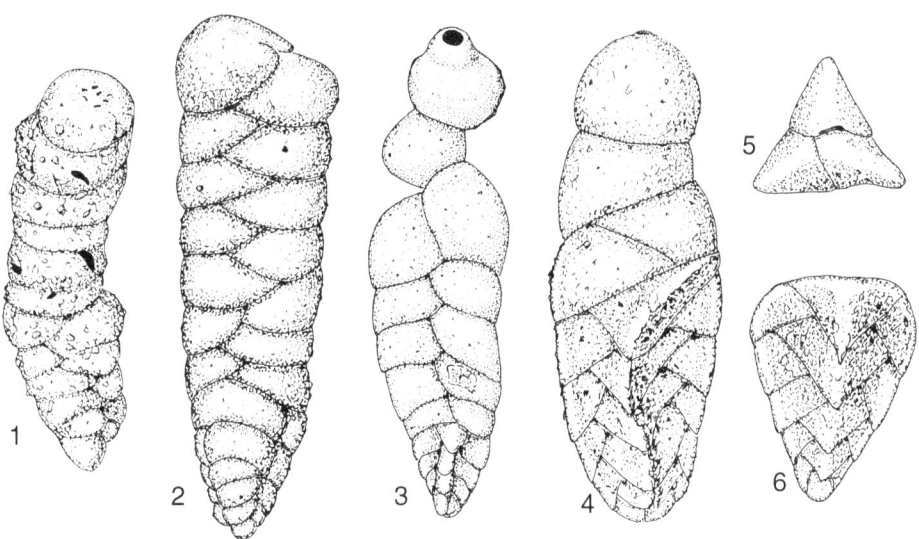

Pl. 7 **Verneuilinidae**. Modified after Haynes (1981). 7.1 *Cribrogoesella robusta*, ×17; 7.2 *Dorothia bradyana*, ×40; 7.3 *Spiroplectinata annectens*, ×74; 7.4 *Tritaxia capitosa*; ×62; 7.5–7.6 *Verneuilina tricarinata*, ×15.

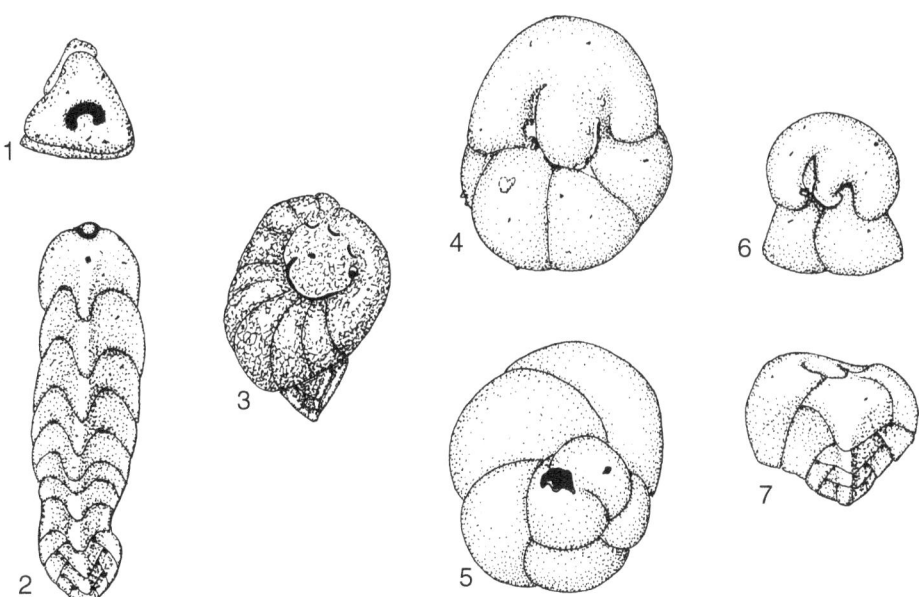

Pl. 8 **Valvulinidae**. Modified after Haynes (1981). 8.1-2 *Clavulina angularis*, ×48; 8.3 *Cribrobulimina mixta*, ×45; 8.4–8.5 *Valvulammina globularis*, ×50; 8.6–8.7 *Valvulina triangularis*, ×33.

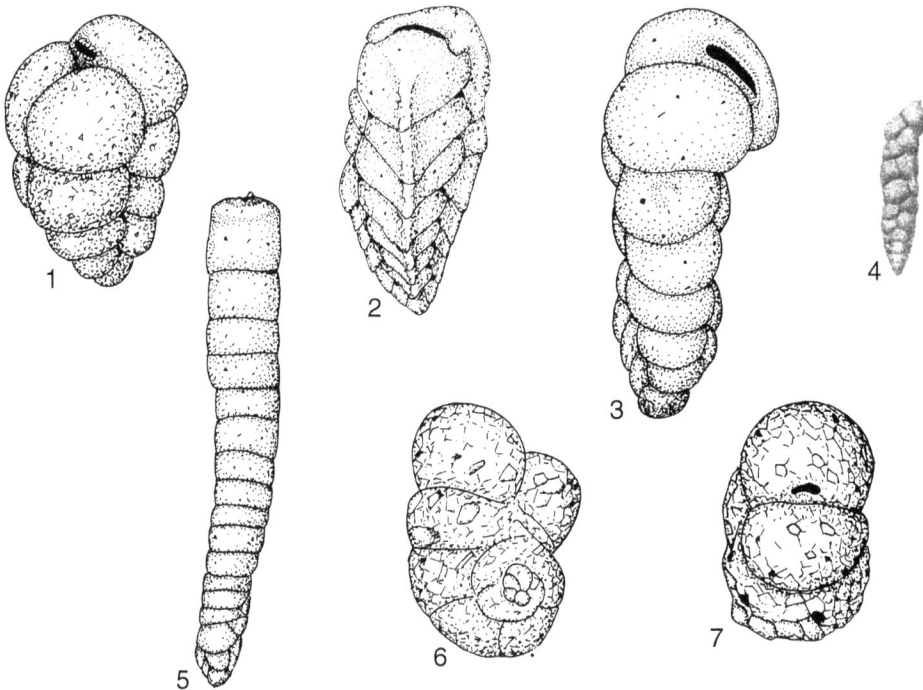

Pl. 9    **Eggerellidae**. Modified after Haynes (1981) and Jones (2006).9.1 *Eggerella bradyi*, ×25; 9.2 *Gaudryina atlantica*, ×14; 9.3 *Karreriella novangliae*, ×40; 9.4 *Karrerulina conversa*, ×40; 9.5 *Multifidella nodulosa*, ×20; 9.6–9.7 *Plectotrochammina subglobosa*, ×50.

High trochospiral; small to large, with or without complex interior; aperture 'bulimine': Family Ataxophragmiidae (Jurassic–Recent). Representative genera: *Arenobulimina* (Pl. 10.1), *Ataxophragmium, Eggerellina* (Pl. 10.2–10.3), *Eggerelloides*, *Kurnubia*, *Migros* (Pl. 10.4–10.5), *Pfenderina*, *Pseudobolivina* (Pl. 10.6), *Remesella*, *Tritaxilina*.

Essentially high trochospiral to uniserial; large, with complex interior; apertures multiple: Family Orbitolinidae (Jurassic–Palaeogene). Representative genera: *Coskinolina* (Pl. 11.1–11.2), *Coskinolinoides* (Pl. 11.3–11.4), *Dictyoconus*, *Kilianina*, *Lituonella* (Pl. 11.5–11.6), *Orbitolina* (Pl. 11.7–11.8), *Pseudolituonella*.

4. Wall secreted, calcareous, microgranular, in some cases secondarily radial; test unilocular or multilocular; chamber arrangement planispiral, streptospiral, trochospiral, biserial or uniserial; small to large, with or without complex interior: Fusulinida (Late Palaeozoic, Devonian–Permian).

Microgranular: Suborder Fusulinina.

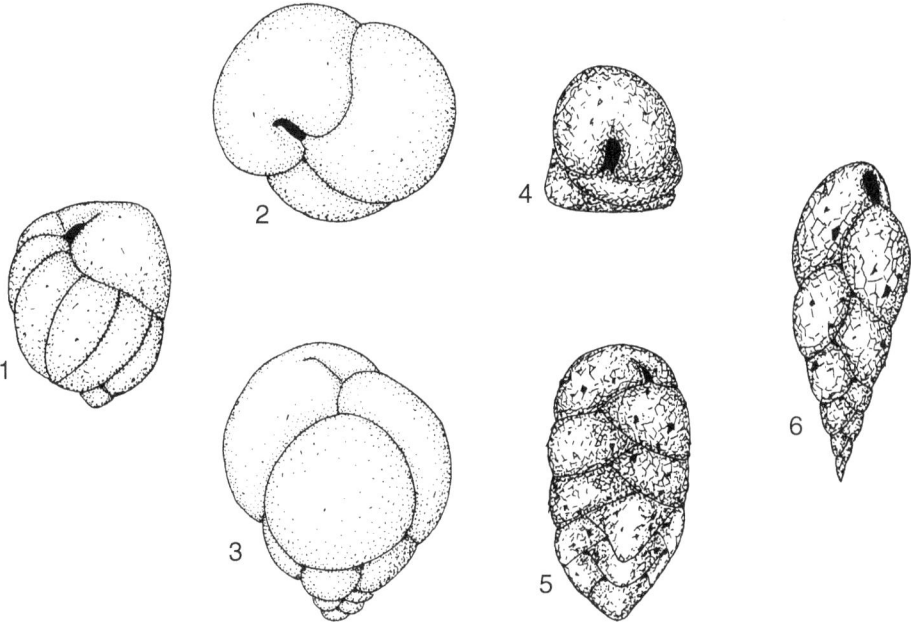

Pl. 10 **Ataxophragmiidae**. Modified after Haynes (1981). 10.1 *Arenobulimina presli*, ×100; 10.2–10.3 *Eggerellina brevis*, ×64; 10.4–10.5 *Migros midwayensis*, ×36; 10.6 *Pseudobolivina antarctica*, ×90.

Unilocular: Superfamily Parathuramminidea [Parathuramminacea].

　　Unilocular, chamber globular: Family Parathuramminidae (Late Palaeozoic). Representative genus: *Parathurammina*.

　　Unilocular, chamber planispirally enrolled: Family Tournayellidae (Late Palaeozoic). Representative genera: *Brunsiina, Pseudoglomospira, Septabrunsiina, Septatournayella, Tournayella* (Pl. 12.1).

　　Unilocular, chamber, tubular: Family Earlandiidae (Late Palaeozoic). Representative genera *Earlandia* (Pl. 12.2), *Paratikhinella*.

Multilocular; essentially streptospiral, planispiral or biserial, in some cases uniserial: Superfamily Endothyridea [Endothyracea].

　　Streptospiral or planispiral to uniserial: Family Endothyridae (Late Palaeozoic). Representative genera: *Bradyina* (Pl. 12.3–12.4), *Cribrospira, Endothyra* (Pl. 12.5–12.7), *Endothyranella, Endothyranopsis, Eoendothyranopsis, Glyphostomella, Haplophragmella* (Pl. 12.8), *Loeblichia* (Pl. 12.9), *Quasiendothyra*.

　　Biserial to uniserial; inflated: Family Palaeotextulariidae (Carboniferous–Permian). Representative genera: *Climacammina*

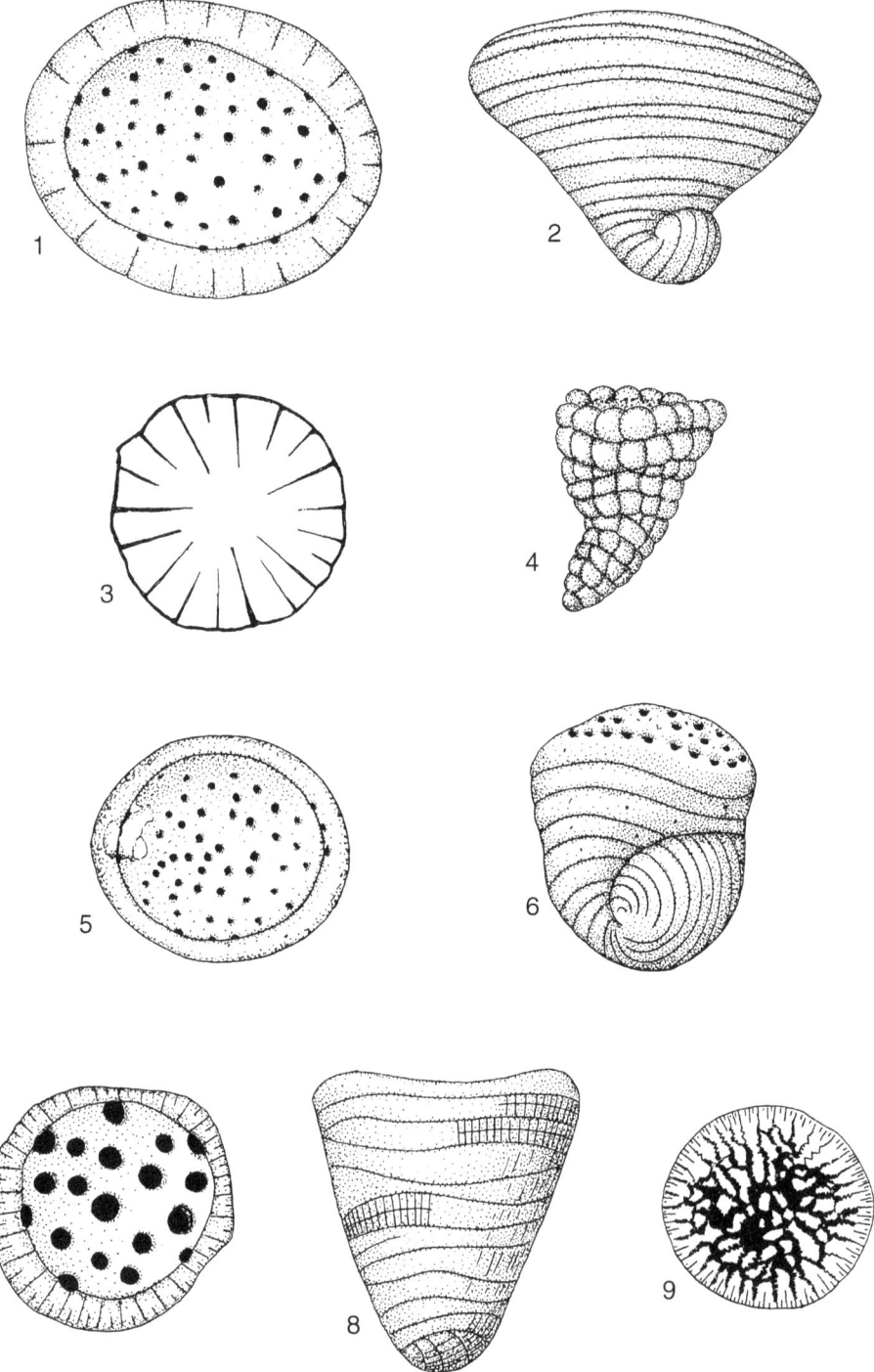

Pl. 11  **Orbitolinidae**. Modified after Haynes (1981). 11.1–11.2 *Coskinolina liburnica*, ×17; 11.3–11.4 *Coskinolinoides texanus*, ×100 and ×66; 11.5–11.6 *Lituonella roberti*, ×17; 11.7–11.9 *Orbitolina mosae*, ×66.

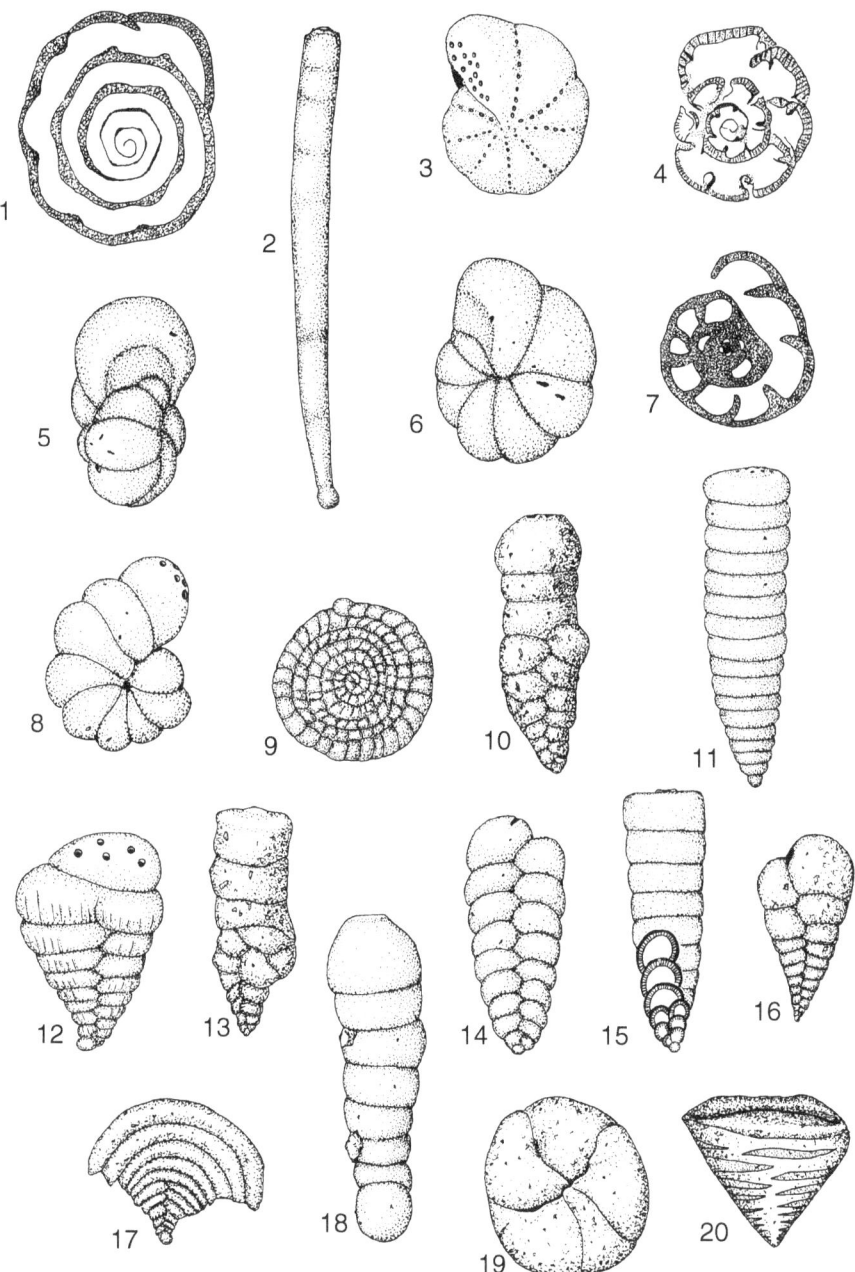

Pl. 12  **Tournayellidae (12.1), Earlandiidae (12.2), Endothyridae (12.3–9), Palaeotextulariidae (12.10–12.16), Semitextulariidae (12.17), Nodosinellidae (12.18) and Tetrataxidae (12.19–12.20)**. Modified after Haynes (1981). 12.1 *Tournayella spectabilis*, ×60; 12.2 *Earlandia perparva*, ×90; 12.3–12.4 *Bradyina rotula*, ×14; 12.5–7 *Endothyra bowmani*, ×80, ×80 and ×64; 12.8 *Haplophragmella panderi*, ×44; 12.9 *Loeblichia ammonoides*, ×64; 12.10 *Climacammina cylindrica*, ×26; 12.11 *Cribrogenerina sumatrana*; 12.12 *Cribrostomum textulariforme*, ×22; 12.13 *Deckerella clavata*, ×30; 12.14 *Deckerellina istiensis*, ×50; 12.15 *Palaeobigenerina geyeri*; ×14; 12.16 *Palaeotextularia grahamensis*, ×50; 12.17 *Semitextularia thomasi*, ×86; 12.18 *Nodosinella digitata*, ×48; 12.19–12.20 *Tetrataxis conica*, ×50.

(Pl. 12.10), *Cribrogenerina* (Pl. 12.11), *Cribrostomum* (Pl. 12.12), *Deckerella* (Pl. 12.13), *Deckerellina* (Pl. 12.14), *Palaeobigenerina* (Pl. 12.15), *Palaeospiroplectammina*, *Palaeotextularia* (Pl. 12.16).

   Biserial to uniserial; compressed: Family Semitextulariidae (Devonian). Representative genera: *Palaeopalmula*, *Semitextularia* (Pl. 12.17).

   Uniserial: Family Nodosinellidae (Late Palaeozoic). Representative genera: *Earlandinita*, *Frondilina*, *Nodosinella* (Pl. 12.18).

Multilocular; high trochospiral: Superfamily Tetrataxidea [Tetrataxacea].

   High trochospiral: Family Tetrataxidae (Carboniferous–Permian). Representative genera: *Polytaxis*, *Tetrataxis* (Pl. 12.19–12.20), *Valvulinella*.

Multilocular; essentially planispiral, in some cases streptospiral; small or large; with or without complex interior: Superfamily Fusulinidea [Fusulinacea].

Wall 'fusulinellid', with three layers (tectum, diaphonotheca and tectorium).

   Planispirally coiled to uncoiled; small, without complex interior: Family Ozawainellidae (Carboniferous–Permian). Representative genera: *Eostaffella*, *Millerella*, *Nankinella*, *Neostaffella*, *Ozawainella*, *Pseudostaffella*, *Rauserella* (Pl. 13.1), *Staffella* (Pl. 13.2), *Toriyamaia* (Pl. 13.3).

   Planispiral or streptospiral; large, with complex interior: Family Fusulinidae (Late Carboniferous–Permian). Representative genera: *Fusulina* (Pl. 13.4), *Fusulinella* (Pl. 13.5), *Palaeofusulina*, *Pseudofusulinella*, *Wedekindellina* (Pl. 13.6), *Yangchienia*.

Wall 'schwagerinid', with two layers (tectum and keriotheca); alveolar, and with chomata.

   Planispiral; large, with complex interior: Family Schwagerinidae (Late Carboniferous–Middle Permian). Representative genera: *Parafusulina*, *Pseudoschwagerina*, *Schwagerina* (Pl. 13.7), *Triticites* (Pl. 13.8).

Wall with one layer; alveolar, and with parachomata.

   Planispiral; large, with complex interior: spiral septula absent: Family Verbeekinidae (Middle–Late Permian). Representative genera: *Eoverbeekina* (Pl. 13.9), *Misellina* (Pl. 13.10–13.11), *Pseudodoliolina* (Pl. 13.12), *Verbeekina* (Pl. 13.13).

   Planispiral; large, with complex interior: spiral septula present: Family Neoschwagerinidae (Middle Permian). Representative genera: *Cancellina* (Pl. 14.1–14.2), *Neoschwagerina* (Pl. 14.3–14.4), *Sumatrina*, *Yabeina* (Pl. 14.5).

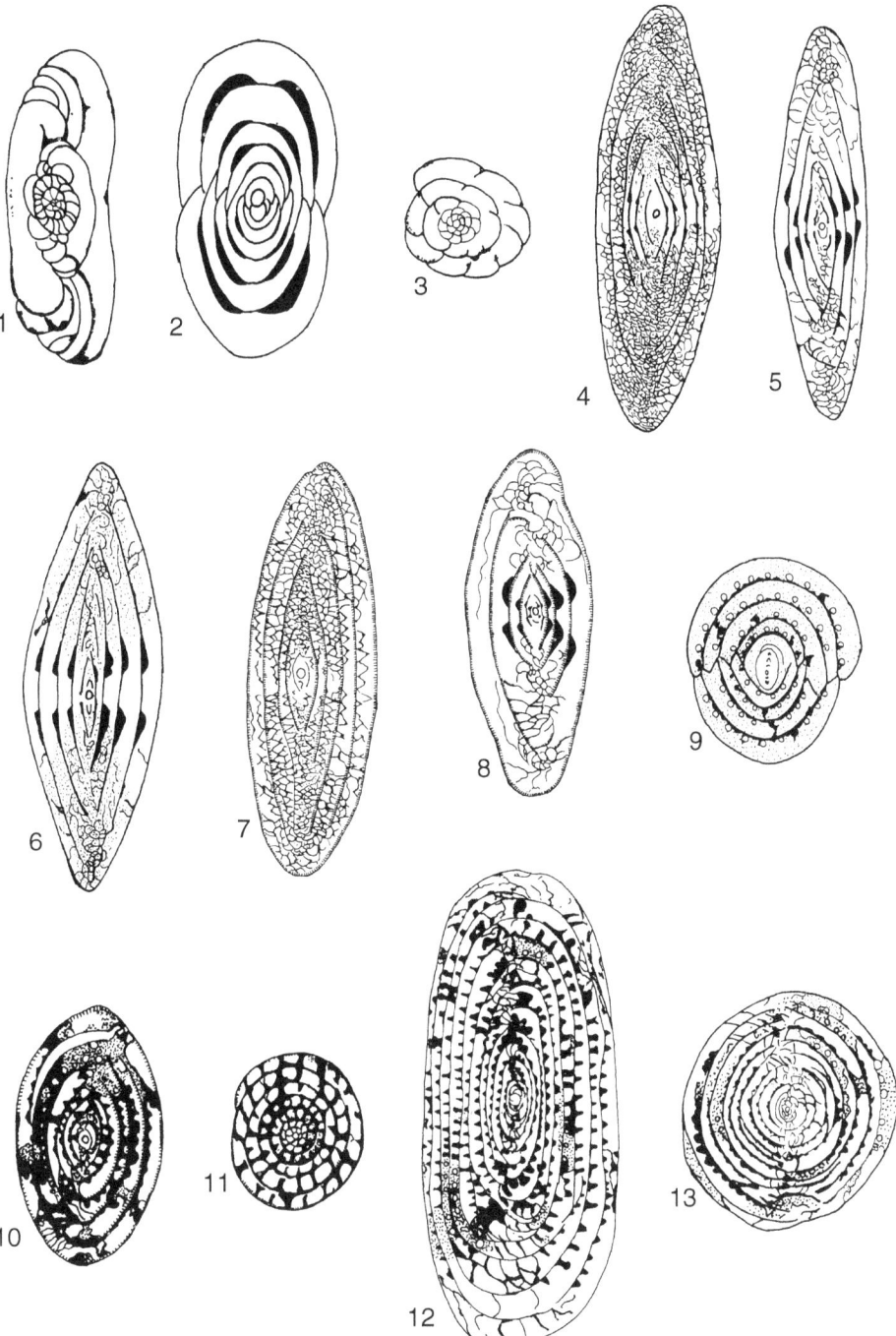

**Pl. 13** **Ozawainellidae (13.1–13.3), Fusulinidae (13.4–13.6), Schwagerinidae (13.7–13.8) and Verebeekinidae (13.9–13.13)**. Modified after Haynes (1981). 13.1 *Rauserella erratica*, ×25; 13.2 *Staffella expansa*, ×50; 13.3 *Toriyamaia laxiseptata*, ×25; 13.4 *Fusulina mysticensis*, ×10; 13.5 *Fusulinella juncea*, ×20; 13.6 *Wedekindellina matura*, ×30; 13.7 *Schwagerina huecaensis*, ×10; 13.8 *Triticites milleri*, ×20; 13.9 *Eoverbeekina intermedia*, ×12; 13.10–13.11 *Misellina ovalis*, ×15; 13.12 *Pseudodoliolina ozawai*, ×20; 13.13 *Verbeekina karinae*, ×12.

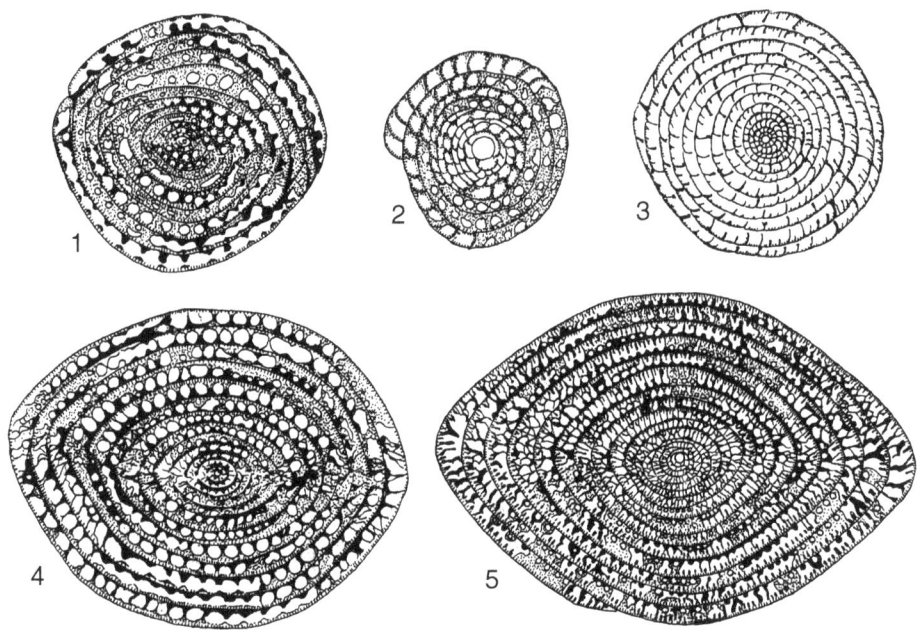

Pl. 14   **Neoschwagerinidae**. Modified after Haynes (1981). 14.1–14.2 *Cancellina primigera*, ×20; 14.3 *Neoschwagerina gifuensis*, ×15; 14.4 *Neoschwagerina haydeni*, ×20; 14.5 *Yabeina globosa*, ×10.

Secondarily radial: Suborder Archaediscina [including Lagenina or Lagenida of authors, for example Groves *et al.* (2003, 2004) and BouDagher-Fadel (2008)].

Unilocular: Superfamily Archaediscidea [Archaediscacea].

Unilocular, chamber planispirally or streptospirally enrolled: Family Archaediscidae (Carboniferous–Permian). Representative genera: *Archaediscus* (Pl. 15.1–15.2), *Permodiscus*.

Unilocular, chamber planispirally enrolled; conical: Family Lasiodiscidae (Carboniferous–Permian). Representative genera: *Howchinia* (Pl. 15.3), *Lasiodiscus*.

Multilocular, uniserial: Superfamily Colaniellidea [Colaniellacea].

Uniserial; compressed: Family Lunucamminidae (Late Palaeozoic). Representative genera: *Lunucammina* (Pl. 15.4), *Pachyphloia*.

Uniserial, with strongly overlapping chambers; inflated; interior partitioned: Family Colaniellidae (Late Palaeozoic). Representative genera: *Colaniella* (Pl. 15.5–15.6), *Multiseptida*.

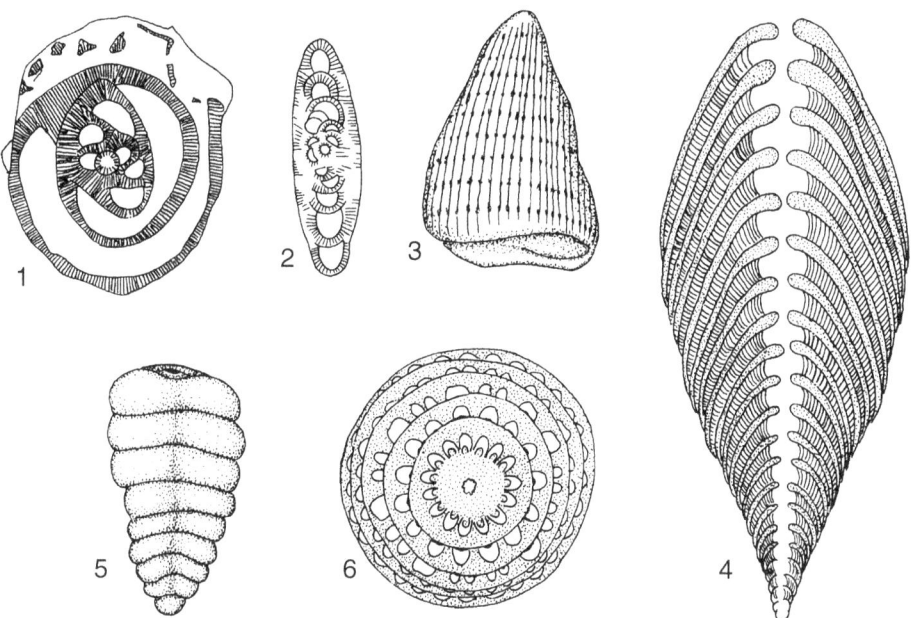

**Pl. 15 Archaediscidae (15.1–15.2), Lasiodiscidae (15.3), Lunucamminidae (15.4) and Colaniellidae (15.5–15.6)**. Modified after Haynes (1981). 15.1–15.2 *Archaediscus karreri*, ×64; 15.3 *Howchinia bradyana*, ×105; 15.4 *Lunucammina permiana*, ×105; 15.5–15.6 *Colaniella parva*, ×100.

5. Wall secreted, calcareous, porcelaneous (rarely, agglutinated); test bilocular or multilocular; chamber arrangement either essentially planispiral or 'milioline'; small to large, with or without complex interior: Order Miliolida (Carboniferous–Recent).

　Bilocular; or multilocular, with bilocular juvenarium followed by planispirally to uniserially arranged chambers, three or more per whorl in planispiral case: Superfamily Nubeculariidea [Nubeculariacea].

　　Bilocular, globular proloculus, followed by planispirally enrolled second chamber: Family Cornuspiridae (Carboniferous–Recent). Representative genera: *Agathammina* (Pl. 16.1–16.2), *Agathamminoides*, *Calcitornella* (Pl. 16.3), *Calcivertella*, *Cornuspira* (Pl. 16.4), *Cornuspirella*, *Cornuspiroides*, *Gordiospira* (Pl. 16.5), *Hemigordiopsis*, *Hemigordius* (Pl. 16.6–7), *Meandrospira*, *Vidalina*.

　　Multilocular, 'cornuspirid' juvenarium followed by planispiral, or irregularly planispiral, to uniserial chambers, three or more per whorl in planispiral case: Family Nubeculariidae (Jurassic–Recent).

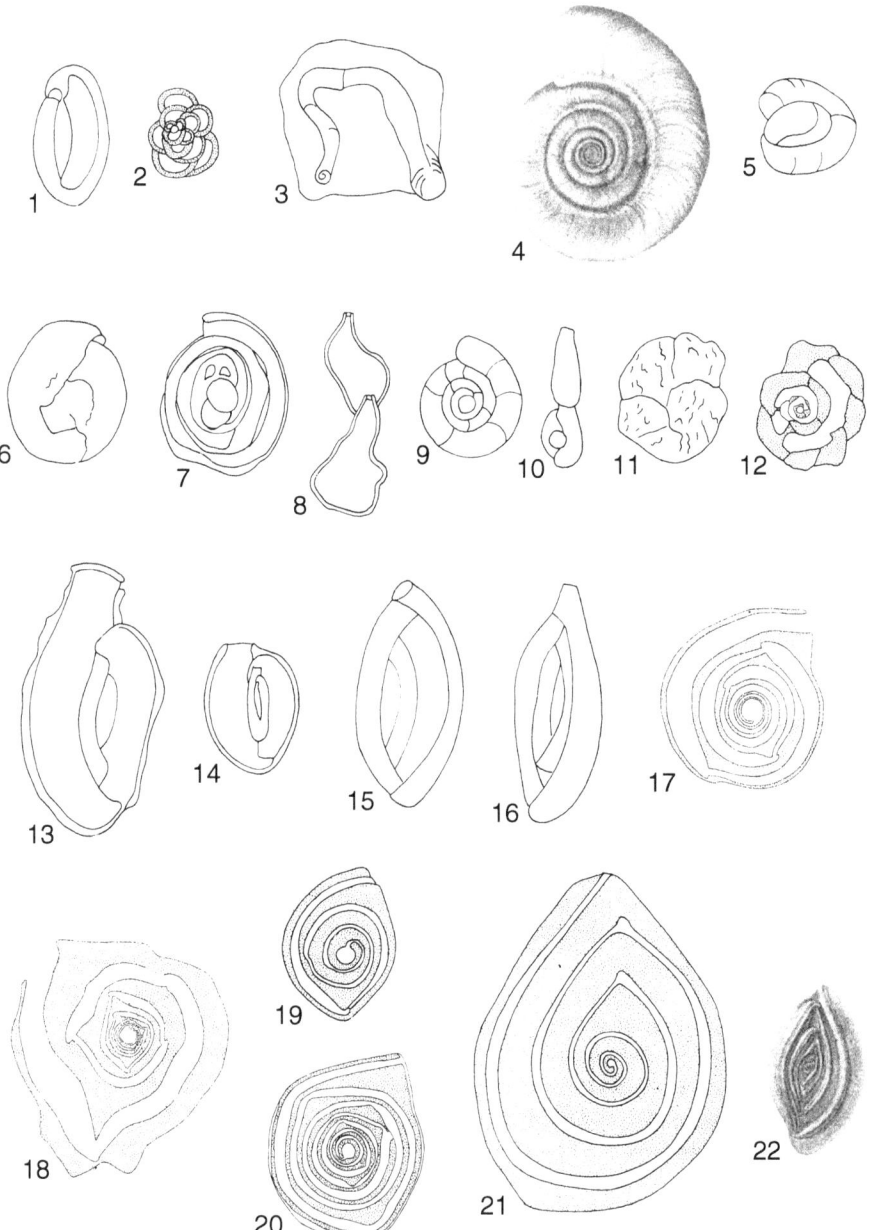

Pl. 16 **Cornuspiridae (16.1–16.7), Nubeculariidae (16.8–16.12), Palaeomiliolinidae (16.13–16.16) and Ophthalmidiidae (16.17–16.22).** Modified after Haynes (1981) and Jones (2006). 16.1–16.2 *Agathammina pusilla*, ×161; 16.3 *Calcitornella grahamensis*, ×25; 16.4 *Cornuspira foliacea*, ×15; 16.5 *Gordiospira fragilis*, ×102; 16.6–16.7 *Hemigordius schlumbergeri*, ×42; 16.8 *Calcituba polymorpha*, ×105; 16.9 *Fischerina rhodiensis*, ×53; 16.10 *Nodobacularia tibia*, ×70; 16.11–16.12 *Nubecularia lucifuga*, ×102; 16.13 *Edentostomina cultrata*, ×28; 16.14 *Edentostomina milletti*, ×28; 16.15 *Eosigmoilina explicata*, ×140; 16.16 *Palaeomiliolina occulta*, ×224; 16.17 *Eoophthalmidium tricki*, ×70; 16.18 *Hauerinella inconstans*, ×35; 16.19 *Ophthalmidium liasicum*, ×280; 16.20 *Ophthalmidium margatitiferum*, ×140; 16.21 *Praeophthalmidium orbiculare*; 16.22 *Spirophthalmidium acutimargo*, ×50.

Representative genera: *Calcituba* (Pl. 16.8), *Fischerina* (Pl. 16.9), *Nautiloculina*, *Nodobacularia* (Pl. 16.10), *Nodophthalmidium*, *Nubecularia* (Pl. 16.11–16.12), *Nubeculina*, *Nubeculinella*, *Planispirinella*, *Webbina*.

Multilocular, with bilocular juvenarium followed by planispirally to uniserially arranged chambers, no more than two per whorl in planispiral case: Superfamily Ophthalmidiidea [Ophthalmidiacea].

Multilocular, juvenarium followed by planispiral, or irregularly planispiral, chambers, two per whorl: Family Palaeomiliolinidae (Carboniferous–Recent). Representative genera: *Duoplanum*, *Edentostomina* (Pl. 16.13–16.14), *Eosigmoilina* (Pl. 16.15), *Karaburunia*, *Palaeomiliolina* (Pl. 16.16), *Psamminopelta*.

Multilocular, juvenarium followed by planispiral chambers, less than one per whorl in initial portion, two per whorl in final portion: Family Ophthalmidiidae (Triassic–Recent). Representative genera: *Discospirina*, *Eoophthalmidium* (Pl. 16.17), *Hauerinella* (Pl. 16.18), *Ophthalmidium* (Pl. 16.19–16.20), *Praeophthalmidium* (Pl. 16.21), *Spirophthalmidium* (Pl. 16.22).

Multilocular; planispirally coiled to uncoiled; interior partitioned: Superfamily Soritidea [Soritacea].

Evolute planispiral to uncoiled, or annular; commonly compressed, flabelliform; interior partially partitioned: Family Soritidae (Eocene–Recent). Representative genera: *Amphisorus* (Pl. 17.1–17.2), *Androsina*, *Archaias* (Pl. 17.3–17.5), *Cribrospirolina*, *Cyclorbiculina*, *Dendritina*, *Fusarchaias*, *Marginopora* (Pl. 17.6–17.7), *Orbitolites*, *Peneroplis* (Pl. 17.8), *Sorites* (Pl. 17.9–17.10), *Spirolina*.

Involute planispiral, or annular; interior partitioned: Family Meandropsinidae (Late Cretaceous). Representative genera: *Broeckina*, *Fallotia*, *Meandropsina*, *Praepeneroplis* (Pl. 17.11–12), *Vandenbroeckia*.

Involute planispiral to uniserial, or annular; interior partitioned and pillared: Family Praerhapydioninidae (Jurassic–Oligocene). Representative genera: *Haurania* (Pl. 17.13–17.15), *Praerhapydionina*, *Pseudorhipidionina* (Pl. 17.16).

Multilocular; milioline, or essentially planispiral; small or large; with or without complex interior; Superfamily Miliolidea [Miliolacea].

Milioline; interior with or without partitions; aperture with or without basal 'miliolinelline' flap: Family Miliolidae (Jurassic–Recent). Representative genera: *Adelosina*, *Affinetrina* (Pl. 18.1–18.2),

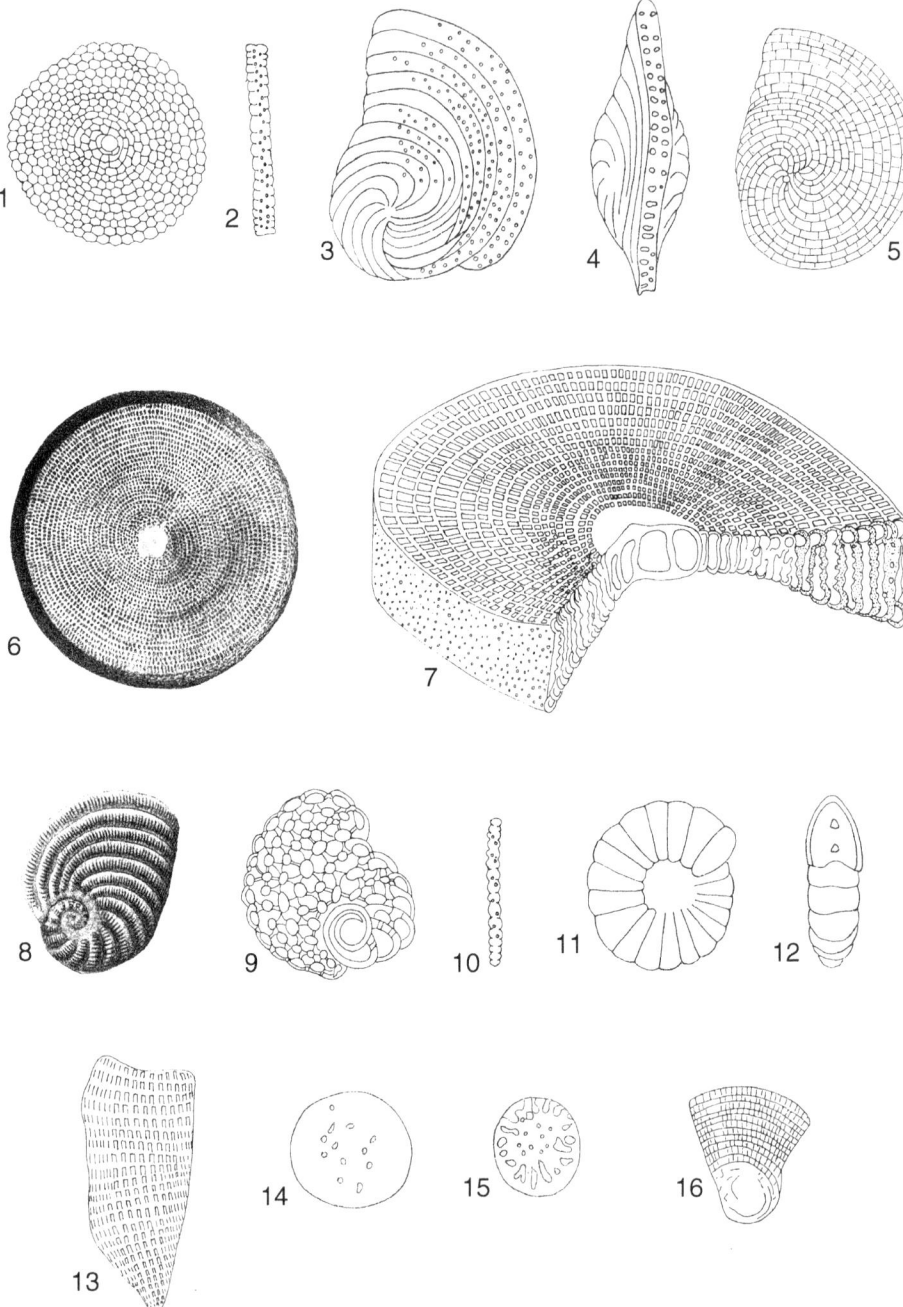

Pl. 17 **Soritidae (17.1–17.10), Meandropsinidae (17.11–17.12) and Praerha-pydioninidae (17.13–17.16)**. Modified after Haynes (1981) and Jones (2006). 17.1–17.2 *Amphisorus hemprichii*, ×11; 17.3–17.5 *Archaias angulatus*, ×21; 17.6–17.7 *Marginopora vertebralis*, ×20, ×63; 17.8 *Peneroplis planulatus*, ×50; 17.9–17.10 *Sorites marginalis*, ×35, ×21; 17.11–17.12 *Praepeneroplis senoniensis*, ×60; 17.13–17.15 *Haurania deserti*, ×56; 17.16 *Pseudorhipidionina macfadyeni*, ×17.

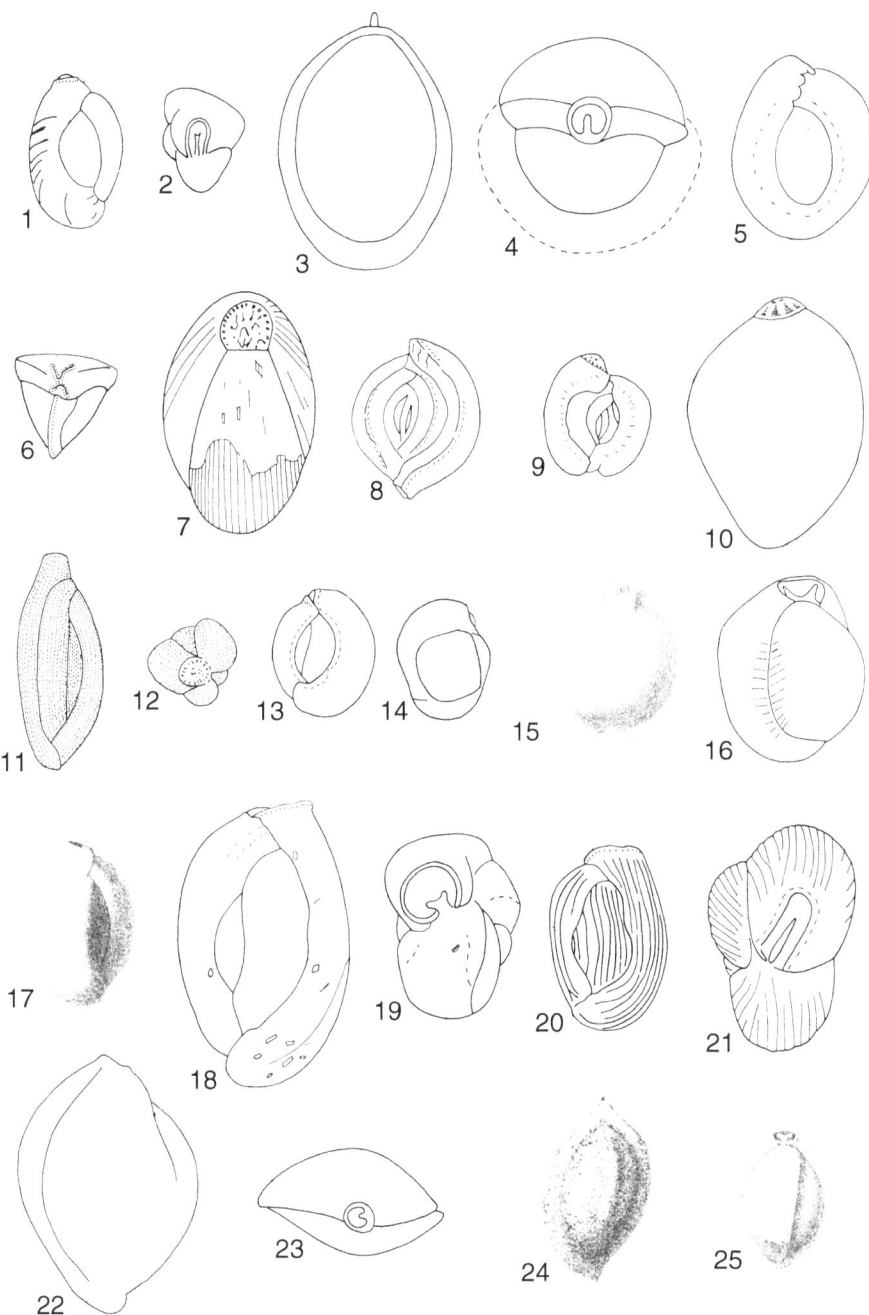

Pl. 18 **Miliolidae**. Modified after Haynes (1981) and Jones (2006). 18.1–18.2 *Affinetrina planciana*, ×105; 18.3–18.4 *Biloculina bulloides*; 18.5 *Crenetella mira*, ×210; 18.6 *Cruciloculina triangularis*, ×14; 18.7 *Fabularia ovata*, ×14; 18.8 *Flintia robusta*, ×70; 18.9 *Heterillina guespellensis*, ×18; 18.10 *Idalina antiqua*, ×11; 18.11-18.12 *Miliola saxorum*, ×70; 18.13 *Miliolinella subrotunda*, ×70; 18.14 *Nummoloculina contraria*, ×35; 18.15 *Pyrgo murrhina*, ×40; 18.16 *Pyrgo (Sinuloculina) williamsoni*, ×70; 18.17 *Quinqueloculina seminulum*, ×50; 18.18-18.19 *Quinqueloculina (Adelosina) aspera*, ×70; 18.20-18.21 *Quinqueloculina (Lachlanella) bicornis*, ×70, ×84; 18.22-18.23 *Sigmoilina sigmoidea*, ×63; 18.24 *Sigmoilopsis schlumbergeri*, ×50; 18.25 *Triloculina trigonula*, ×40.

*Ammomassilina, Austrotrillina, Biloculina* (Pl. 18.3–18.4), *Biloculinella, Crenetella* (Pl. 18.5), *Cribrolinoides, Cribropyrgo, Cruciloculina* (Pl. 18.6), *Cycloforina, Fabularia* (Pl. 18.7), *Flintia* (Pl.
18.8), *Hauerina, Heterillina* (Pl. 18.9), *Idalina* (Pl. 18.10), *Invol-
vohauerina, Lacazina, Lacazinella, Lachlanella, Miliammina,
Miliola* (Pl. 18.11–18.12), *Miliolinella* (Pl. 18.13), *Nevillina, Nummoloculina* (Pl. 18.14), *Periloculina, Planispirinoides, Podolia,
Pseudohauerina, Ptychomiliola, Pyrgo* (Pl. 18.15–18.16), *Pyrgoella, Quinqueloculina* (Pl. 18.17–18.21), *Riveroina, Schlumber-
gerina, Scutuloris, Sigmoilina* (Pl. 18.22–18.23), *Sigmoilopsis* (Pl.
18.24), *Sinuloculina, Spiroloculina, Spirosigmoilina, Tortonella,
Triloculina* (Pl. 18.25), *Varidentella.*

Essentially planispiral; large, with complex interior: Family Alveolinidae (Late Cretaceous–Recent). Representative genera: *'Alveolina'* (Pl. 19.1), *Alveolinella* (Pl. 19.2–19.3), *Borelis* (Pl. 19.4–
19.5), *Bullalveolina* (Pl. 19.6), *Cisalveolina, Flosculinella*
(Pl. 19.7), *Glomalveolina* (Pl. 19.8), *Multispirina, Ovalveolina*
(Pl. 19.9), *Praealveolina* (Pl. 19.10), *Pseudofabularia,
Simplalveolina, Streptalveolina, Subalveolina.*

Planispiral or streptospiral to uncoiled; commonly compressed, flabelliform; interior partitioned: Family Rhapydioninidae (Late Cretaceous). Representative genera: *Chubbina, Pseudedomia,
Rhapydionina [Rhipidionina].*

6. Wall secreted, calcareous, granular or radial; test unilocular to multilocular;
chamber arrangement typically planispiral, biserial or uniserial, and atypically
polymorphine; aperture typically terminal, and rounded, slit-like or radiate;
with or without associated internal tube: Order Nodosariida (Late Permian–
Recent).

Unilocular to multilocular; essentially planispiral or biserial, in some cases
uniserial: Superfamily Nodosariidea [Nodosariacea].

Unilocular to multilocular; planispiral to uniserial; aperture rounded
or radiate Family Nodosariidae (Permian–Recent). Representative
genera: *Amphicoryna* (Pl. 20.1), *Astacolus* (Pl. 20.2), *Chrysalogo-
nium, Citharina* (Pl. 20.3), *Cribrolenticulina, Cribrorobulina,
Dentalina* (Pl. 20.4), *Flabellinella* (Pl. 20.5), *Frondicularia* (Pl.
20.6), *Ichthyolaria, Lagena* (Pl. 20.7), *Lenticulina* (Pl. 20.8), *Marginulina* (Pl. 20.9), *Marginulinopsis, Neoflabellina, Nodosaria* (Pl.
20.10), *Palmula* (Pl. 20.11), *Pseudonodosaria, Robulus, Saracenaria* (Pl. 20.12), *Saracenella, Tristix, Vaginulina, Vaginulinopsis*
(Pl. 20.13).

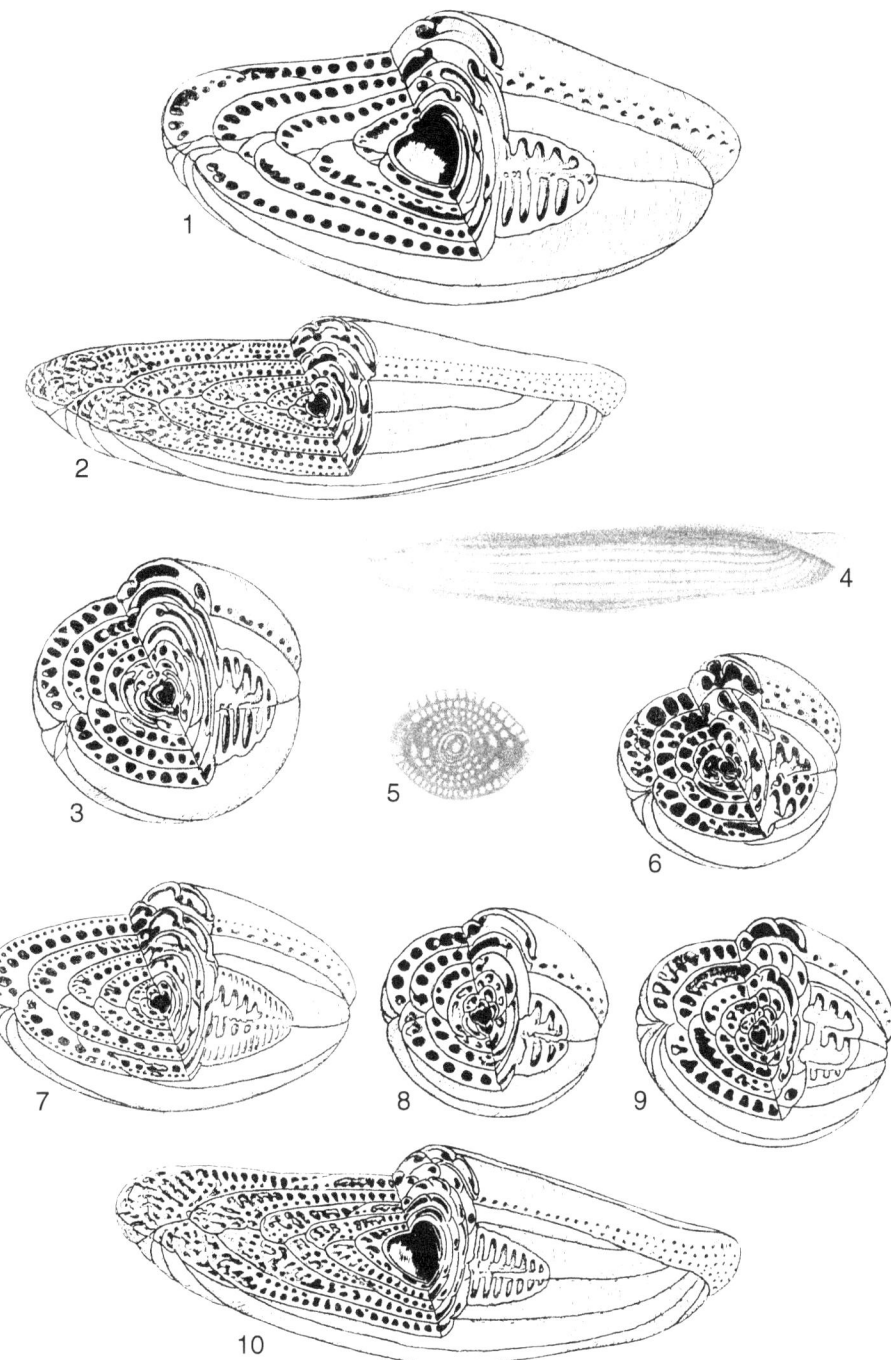

Pl. 19 **Alveolinidae**. Modified after Haynes (1981) and Jones (2006). 19.1 'Alveolina' schwageri, ×55; 19.2–19.3 Alveolinella quoyi, ×15; 19.4 Borelis melo curdica, ×85; 19.5 Borelis melo/bradyi, ×50; 19.6 Bullalveolina bulloides, ×80; 19.7 Flosculinella bontangensis, ×60; 19.8 Glomalveolina primaeva, ×55; 19.9 Ovalveolina ovum, ×50; 19.10 Praealveolina tenuis, ×45.

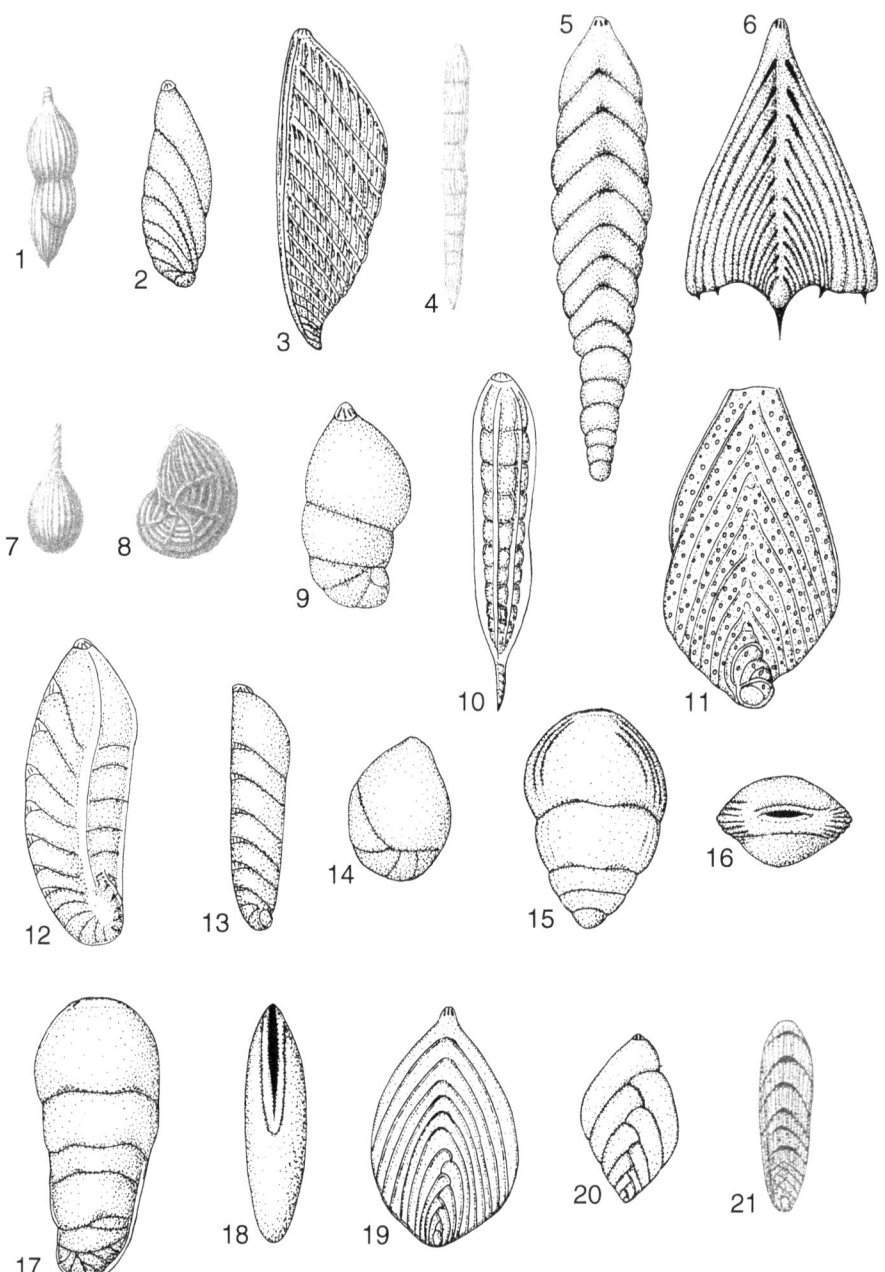

Pl. 20 **Nodosariidae (20.1–20.13), Lingulinidae (20.14–20.18) and Plecto-frondiculariidae (20.19–20.21)**. Modified after Haynes (1981) and Jones (2006). 20.1 *Amphicoryna scalaris*, ×50; 20.2 *Astacolus insolitus*, ×60; 20.3 *Citharina colliezi*, ×48; 20.4 *Dentalina flinti*, ×16; 20.5 *Flabellinella zitteliana*, ×25; 20.6 *Frondicularia sagittula*, ×15; 20.7 *Lagena sulcata*, ×60; 20.8 *Lenticulina anaglypta*, ×35; 20.9 *Marginulina obese*, ×30; 20.10 *Nodosaria lamnulifera*, ×15; 20.11 *Palmula rugosa*, ×50; 20.12 *Saracenaria italica*, ×20; 20.13 *Vaginulinopsis sublegumen*, ×30; 20.14 *Ellipsocristellaria sequana*, ×80; 20.15–20.16 *Lingulina seminuda*, ×12; 20.17 *Lingulinopsis carlofortensis*, ×20; 20.18 *Rimulina glabra*, ×50; 20.19 *Kyphopyxa christneri*, ×28; 20.20 *Lankesterina frondea*, ×80; 20.21 *Plectofrondicularia japonica*, ×48.

Multilocular; planispiral to uniserial; compressed; aperture slit-like: Family Lingulinidae (Permian–Recent). Representative genera: *Ellipsocristellaria* (Pl. 20.14), *Gonatosphaera*, *Lingulina* (Pl. 20.15–20.16), *Lingulinopsis* (Pl. 20.17), *Lingulonodosaria*, *Rimulina* (Pl. 20.18).

Multilocular; biserial to uniserial; compressed; aperture radiate: Family Plectofrondiculariidae (Cretaceous–Recent). Representative genera: *Amphimorphina*, *Dyofrondicularia*, *Kyphopyxa* (Pl. 20.19), *Lankesterina* (Pl. 20.20), *Plectofrondicularia* (Pl. 20.21).

Unilocular to multilocular; polymorphine: Superfamily Polymorphinidea [Polymorphinacea].

Multilocular; polymorphine; without internal tube: Family Polymorphinidae (Triassic–Recent). Representative genera: *Eoguttulina* (Pl. 21.1), *Glandulopleurostomella*, *Globulina* (Pl. 21.2), *Guttulina* (Pl. 21.3), *Polymorphina* (Pl. 21.4), *Pseudopolymorphina* (Pl. 21.5), *Pyrulina* (Pl. 21.6), *Pyrulinoides* (Pl. 21.7), *Sagoplecta* (Pl. 21.8), *Sigmoidella* (Pl. 21.9), *Sigmomorphina* (Pl. 21.10), *Spirofrondicularia*.

Unilocular to multilocular; polymorphine; with internal tube: Family Glandulinidae (Cretaceous–Recent). Representative genera: *Dainita* (Pl. 21.11), *Esosyrinx* (Pl. 21.12), *Fissurina* (Pl. 21.13), *Glandulina* (Pl. 21.14), *Globulotuba* (Pl. 21.15), *Laryngosigma* (Pl. 21.16–21.17), *Oolina* (Pl. 21.18), *Siphoglandulina*.

Form genera: *Bullopora* (Pl. 21.19), *Ramulina* (Pl. 21.20), *Sporadogenerina*, *Webbinella*.

7. Wall secreted, calcareous, granular or radial; test unilocular to multilocular; chamber arrangement typically high trochospiral, triserial, biserial or enrolled biserial, or uniserial; aperture basal to terminal; with or without associated internal tooth-plate: Order Buliminida (Jurassic–Recent).

Essentially triserially or high trochospirally coiled to uncoiled; aperture sub-terminal to terminal; with internal tooth-plate: Superfamily Buliminidea [Buliminacea].

Triserial, biserial or uniserial; inflated; aperture sub-terminal, loop-shaped or bulimine: Family Buliminidae (Jurassic–Recent). Representative genera: *Bulimina* (Pl. 22.1), *Cassidella*, *Delosina*, *Fursenkoina*, *Globobulimina*, *Neobulimina*, *Praebulimina*, *Praeglobobulimina* (Pl. 22.2), *Pyramidina*, *Rectobulimina*, *Stainforthia* (Pl. 22.3), *Virgulinella* (Pl. 22.4).

High trochospiral: Family Turrilinidae (Late Cretaceous–Recent). Representative genera: *Baggatella* (Pl. 22.5–22.6), *Buliminella*

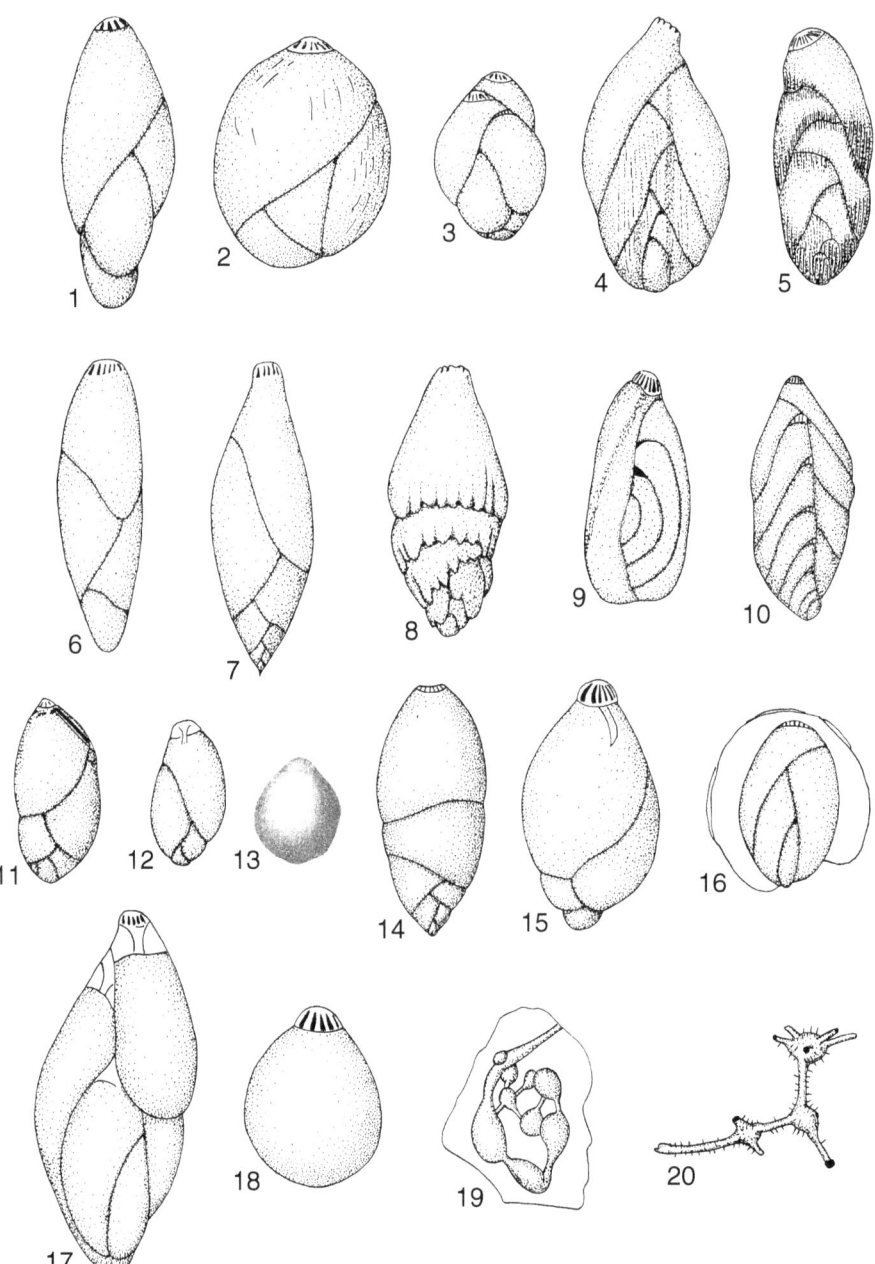

Pl. 21 **Polymorphinidae (21.1–21.10), Glandulinidae (21.11–21.18) and Form Genera (21.19–21.20)**. Modified after Haynes (1981) and Jones (2006). 21.1 *Eoguttulina anglica*, ×150; 21.2 *Globulina inaequalis*, ×100; 21.3 *Guttulina problema*, ×40; 21.4 *Polymorphina anceps*, ×30; 21.5 *Pseudopolymorphina ligua;*, ×18; 21.6 *Pyrulina cylindroides*, ×90; 21.7 *Pyrulinoides acuminatus*, ×180; 21.8 *Sagoplecta goniata*, ×95; 21.9 *Sigmoidella kagaensis*, ×50; 21.10 *Sigmomorphina pseudoregularis*, ×50; 21.11 *Dainita siberica*, ×47; 21.12 *Esosyrinx* sp., ×100; 21.13 *Fissurina laevigata*, ×70; 21.14 *Glandulina laevigata*, ×50; 21.15 *Globulotuba entosoleniformis*, ×150; 21.16 *Laryngosigma harrisi*, ×100; 21.17 *Laryngosigma lacteal*, ×100; 21.18 *Oolina laevigata*, ×170; 21.19 *Bullopora rostrata*, ×20; 21.20 *Ramulina globulifera*, ×35.

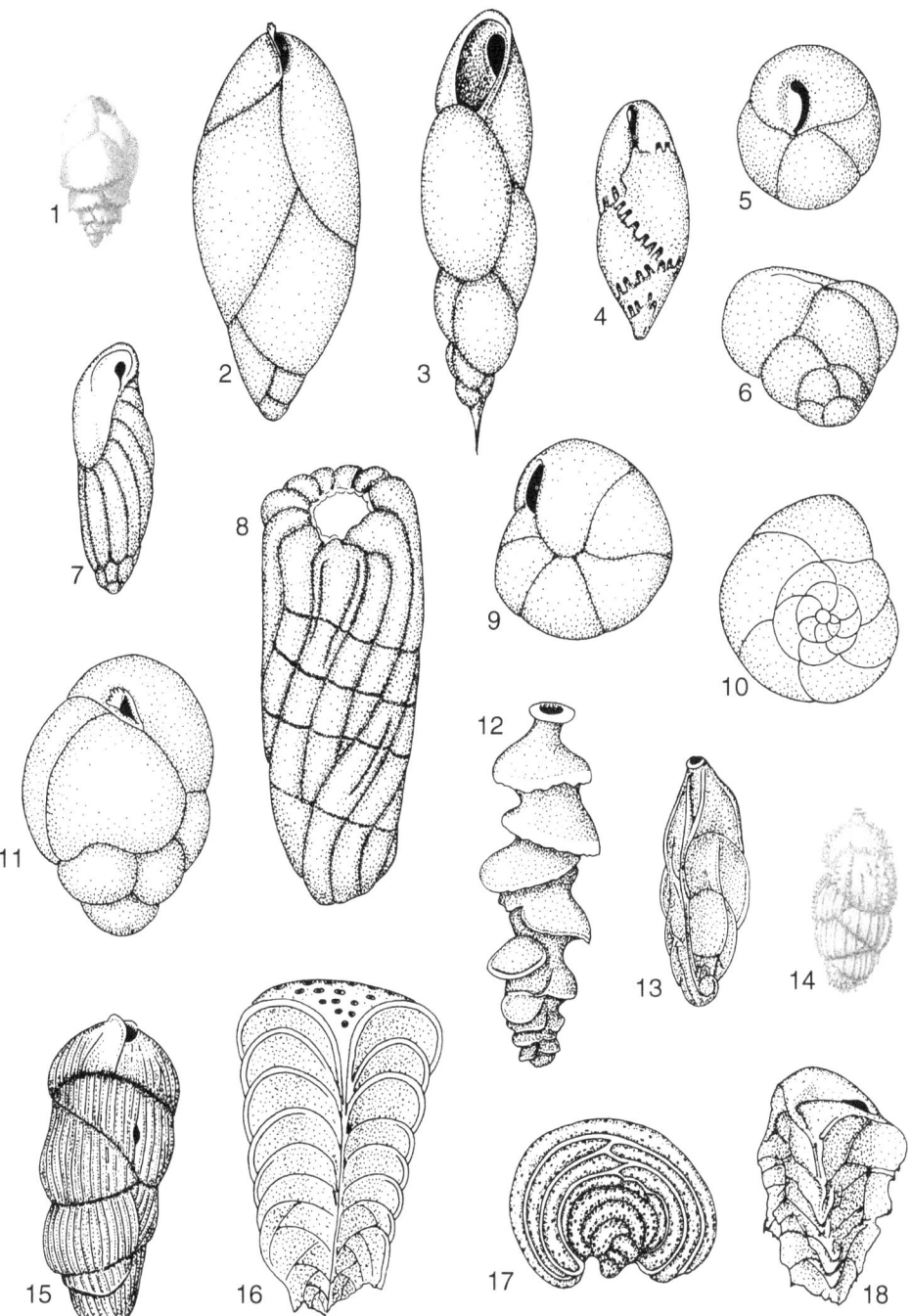

Pl. 22 **Buliminidae (22.1–22.4), Turrilinidae (22.5–22.11); Uvigerinidae (22.12–22.15) and Pavoninidae (22.16–22.18)**. Modified after Haynes (1981) and Jones (2006). 22.1 *Bulimina marginata*, ×80; 22.2 *Praeglobobulimina ovata*, ×120; 22.3 *Stainforthia concava*, ×200; 22.4 *Virgulinella pertusa*, ×64; 22.5–22.6 *Baggatella inconspicua*, ×300; 22.7 *Buliminella elegantissima*, ×200; 8 *Buliminoides williamsonianus*, ×450; 22.9–22.10 *Epistominella vitrea*, ×320; 22.11 *Turrilina brevispira*, ×312; 22.12 *Siphouvigerina fimbriata*, ×185; 22.13 *Trifarina angulosa*, ×83; 22.14 *Uvigerina bradyana*, ×40; 22.15 *Uvigerinella oveyi*, ×200; 22.16 *Chrysalidinella dimorpha*, ×140; 22.17 *Pavonina flabeliformis*, ×60; 22.18 *Reussella spinulosa*, ×100.

(Pl. 22.7), *Buliminellita*, *Buliminoides* (Pl. 22.8), *Caucasina*, *Epistominella* (Pl. 22.9–22.10), *Spirobuliminella*, *Turrilina* (Pl. 22.11).

Triserial, biserial or uniserial; inflated; aperture terminal, at end of neck: Family Uvigerinidae (Late Cretaceous–Recent). Representative genera: *Elhasella*, *Hopkinsina*, *Pseudouvigerina*, *Siphogenerina*, *Siphouvigerina* (Pl. 22.12), *Rectuvigerina*, *Trifarina* (Pl. 22.13), *Uvigerina* (Pl. 22.14), *Uvigerinella* (Pl. 22.15).

Triserial, biserial or uniserial; commonly either triangular in cross-section or compressed: Family Pavoninidae (Eocene–Recent). Representative genera: *Chrysalidinella* (Pl. 22.16), *Fijiella*, *Pavonina* (Pl. 22.17), *Reussella* (Pl. 22.18), *Tubulogenerina*.

Biserial to uniserial; aperture basal to terminal; with internal tooth-plate: Superfamily Bolivinitidea [Bolivinitacea].

Biserial to uniserial; compressed; aperture basal: Family Bolivinitidae (Jurassic–Recent). Representative genera: *Bolivina* (Pl. 23.1), *Bolivinita* (Pl. 23.2), *Bolivinoides* (Pl. 23.3), *Brizalina* (Pl. 23.4), *Rectobolivina* (Pl. 23.5), *Spirobolivina*.

Biserial to uniserial; inflated; aperture terminal, at end of neck: Family Eouvigerinidae (Late Cretaceous–Recent) Representative genera: *Eouvigerina* (Pl. 23.6), *Siphogenerinoides* (Pl. 23.7), *Siphonodosaria* (Pl. 23.8), *Stilostomella*.

Unilocular to multilocular; either enrolled biserial, or triserial to biserial or uniserial; aperture either basal or sub-terminal, in latter case, with hood; with or without internal tooth-plate: Superfamily Cassidulinidea [Cassidulinacea].

Multilocular; enrolled biserial, with or without internal tooth-plate (impossible to determine without dissection): Family Cassidulinidae *s.l.* (including Islandiellidae) (Cenozoic). Representative genera: *Cassidulina* (Pl. 23.9–23.10), *Cassidulinoides* (Pl. 23.11), *Ehrenbergina* (Pl. 23.12), *Favocassidulina*, *Globocassidulina*, *Islandiella* (Pl. 23.13–23.14), *Orthoplecta*, *Sphaeroidina* (Pl. 23.15).

Unilocular to multilocular; triserial to biserial or uniserial; aperture sub-terminal, with 'pleurostomelline' hood; without internal tooth-plate; in some cases with internal tube: Family Pleurostomellidae (Late Cretaceous–Recent). Representative genera: *Bandyella*, *Cribropleurostomella*, *Ellipsobulimina*, *Ellipsoglandulina* (Pl. 23.16–23.17), *Ellipsoidella*, *Ellipsolingulina* (Pl. 23.18), *Ellipsopolymorphina* (Pl. 23.19), *Nodosarella* (Pl. 23.20),

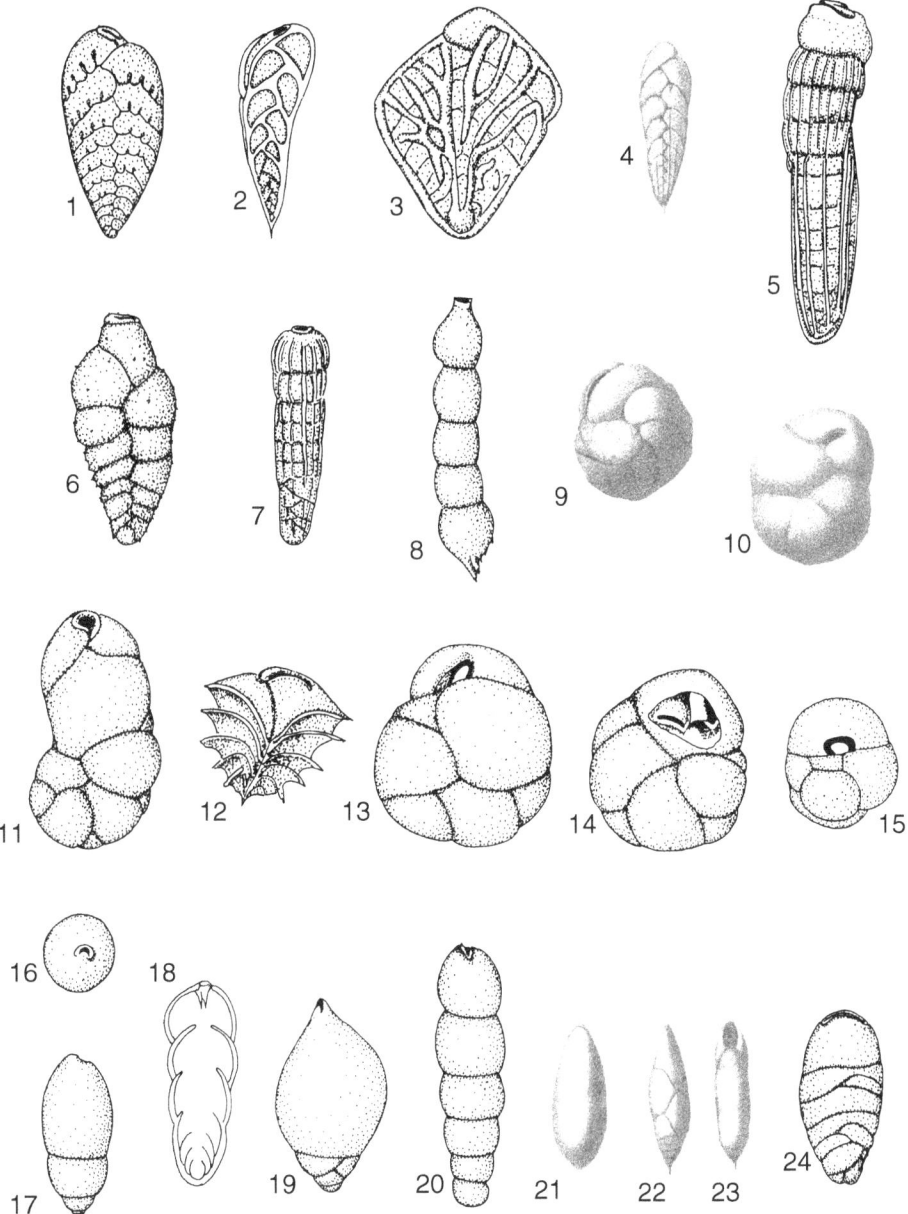

Pl. 23   **Bolivinitidae (23.1–23.5), Eouvigerinidae (23.6–23.8); Cassidulinidae *s.l.* (23.9–23.15) and Pleurostomellidae (23.16-2324)**. Modified after Haynes (1981) and Jones (2006). 23.1 *Bolivina robusta*, ×100; 23.2 *Bolivinita quadrilatera*, ×35; 23.3 *Bolivinoides draco*, ×100; 22.4 *Brizalina mexicana*, ×50; 23.5 *Rectobolivina raphana*, ×55; 23.6 *Eouvigerina zelandica*, ×180; 23.7 *Siphogenerinoides plummeri*, ×66; 23.8 *Siphonodosaria abyssorum*, ×22; 23.9 *Cassidulina teretis*, ×75; 23.10 *Cassidulina [Globocassidulina] subglobosa*, ×60; 23.11 *Cassidulinoides parkerianus*, ×150; 23.12 *Ehrenbergina pacifica*, ×60; 23.13-23.14 *Islandiella islandica*, ×50; 23.15 *Sphaeroidina bulloides*, ×44; 23.16–23.17 *Ellipsoglandulina laevigata*, ×44; 23.18 *Ellipsolingulina silvestrii*, ×150; 23.19 *Ellipsopolymorphina schlichti*, ×50; 23.20 *Nodosarella rotundata*, ×36; 23.21 *Parafissurina lateralis*, ×75; 23.22–23.23 *Pleurostomella acuminata*, ×70; 23.24 *Wheelerella magdalensis*, ×100.

*Parafissurina* (Pl. 23.21), *Pleurostomella* (Pl. 23.22–23.23), *Wheelerella* (Pl. 23.24).

8. Wall secreted, aragonitic: Order Robertinida (Middle Triassic–Recent).

Planispiral or trochospiral; aperture single or double, crenulate: Superfamily Duostominidea [Duostominacea].

Low trochospiral; aperture double: Family Duostominidae (Middle Triassic). Representative genera: *Diplotremina* (Pl. 24.1–24.2), *Duostomina* (Pl. 24.3–24.5).

Planispiral or low trochospiral; aperture single: Family Asymmetrinidae (Late Triassic). Representative genera: *Asymmetrina* (Pl. 24.6–24.7), *Plagiostomella* (Pl. 24.8).

Low trochospiral; aperture with internal tooth-plate: Superfamily Ceratobulimidea [Ceratobuliminacea].

Aperture basal, umbilical: Family Conorboididae (Late Jurassic–Cretaceous). Representative genera: *Conorboides* (Pl. 24.9–24.11), *Reinholdella* (Pl. 24.12–24.13).

Aperture basal, peripheral: Family Epistominidae (Jurassic–Recent). Representative genera: *Epistomina* (Pl. 24.14–24.16), *Epistominita*, *Epistominoides* (Pl. 24.17–24.18), *Hoeglundina* (Pl. 24.19), *Mironovella*.

Primary aperture basal, umblical; secondary aperture areal: Family Ceratobuliminidae (Late Cretaceous–Recent). Representative genera: *Ceratobulimina* (Pl. 24.20–24.21), *Ceratocancris* (Pl. 24.22), *Ceratolamarckina* (Pl. 24.23–24.25), *Lamarckina*.

Low or high trochospirally coiled, or uncoiled; with supplementary chamberlets: Superfamily Robertinidea [Robertinacea].

Low or high trochospirally coiled, or uncoiled; aperture basal: Family Robertinidae (Eocene–Recent). Representative genera: *Cerobertina*, *Geminospira* (Pl. 24.26), *Pseudobulimina*, *Robertina* (Pl. 24.27), *Robertinoides* (Pl. 24.28).

Planispiral or low trochospiral; primary aperture basal; secondary aperture areal: Family Alliatinidae (Neogene–Recent). Representative genera: *Alliatina* (Pl. 24.29–24.30), *Alliatinella* (Pl. 24.31), *Cushmanella*.

9. Wall secreted, calcareous, granular or radial; test unilocular or multilocular; chamber arrangement typically planspiral or low trochospiral; small to large, without or without complex interior: Order Rotaliida (Triassic–Recent).

Small, without complex interior: 'Smaller Rotaliida'.

Unilocular and planispiral; or multilocular, and planispiral to biserial: Superfamily Spirillinidea [Spirillinacea].

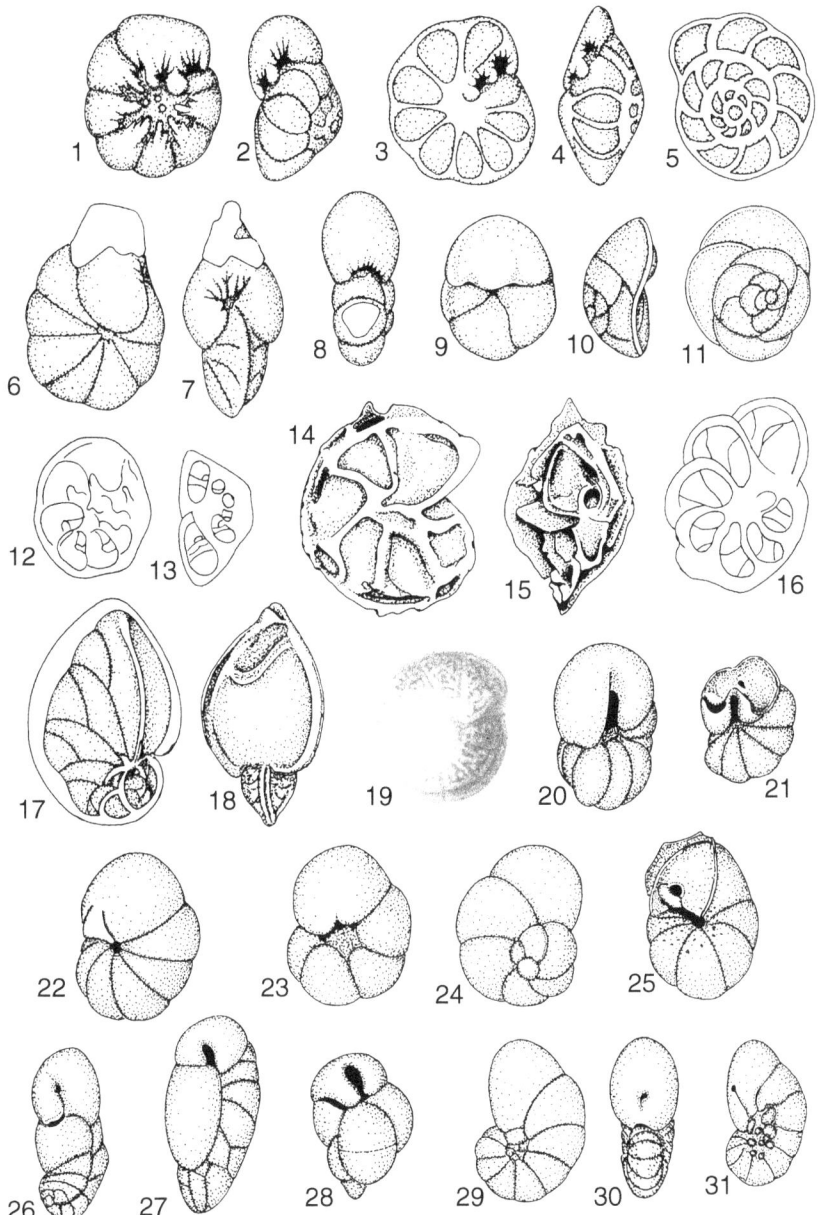

Pl. 24   **Duostominidae (24.1–24.5), Asymmetrinidae (24.6–24.8), Conorboididae (24.9–24.13), Epistominidae (24.14–24.19), Ceratobuliminidae (24.20–
24.25), Robertinidae (24.26–24.28) and Alliatinidae (24.29–24.31)**. Modified
after Haynes (1981) and Jones (2006). 24.1–24.2 *Diplotremina astrofimbriata*,
×125; 24.3–24.5 *Duostomina bicovexa*, ×125; 24.6–24.7 *Asymmetrina biomphalica*,
×80; 24.8 *Plagiostomella inflata*, ×125; 24.9–24.11 *Conorboides mitra*, ×87; 24.12–
24.13 *Reinholdella dreheri*, ×70; 24.14–24.16 *Epistomina spinulifera*, ×50, ×50,
×36; 24.17–24.18 *Epistominoides wilcoxensis*, ×93; 24.19 *Hoeglundina elegans*,
×35; 24.20–24.21 *Ceratobulimina contraria*, ×90; 24.22 *Ceratocancris clifdenensis*,
×70; 24.23–24.25 *Ceratolamarckina tuberculata*, ×115, ×115, ×100; 24.26 *Geminospira simaensis*, ×105; 24.27 *Robertina arctica*, ×75; 24.28 *Robertinoides normani*,
×50; 24.29–24.30 *Alliatina excentrica*, ×119; 24.31 *Alliatinella gedgravensis*, ×92.

Unilocular; planispiral: Family Spirillinidae (Triassic–Recent). Representative genera: *Alanwoodia*, *Spirillina* (Pl. 25.1), *Turrispirillina*.

Multilocular, planispiral to biserial, planispiral stage generally undivided: Family Patellinidae (Cretaceous–Recent). Representative genera: *Patellina* (Pl. 25.2–25.3), *Patellinoides*.

Multilocular; low to comparatively high trochospiral; typically bi-convex, although with evolute dorsal side slightly higher than involute ventral; aperture ventral; generally complex: Superfamily Discorbidea [Discorbacea].

Low trochospiral, aperture basal, umbilical; partially covered by umbilical chamber lobes: Family Discorbidae (Cretaceous–Recent). Representative genera: *Buccella* (Pl. 25.4), *Discorbinella* (Pl. 25.5), *Discorbis* (Pl. 25.6–25.7), *Discorbitura*, *Gavelinopsis*, *Laticarinina* (Pl. 25.8), *Neoconorbina*, *Rosalina* (Pl. 25.9).

Low trochospiral; aperture basal, umbilical; almost entirely covered by umbilical chamber lobes; lobes imperforate: Family Cancrisidae (Cretaceous–Recent). Representative genera: *Cancris* (Pl. 25.10–25.12), *Gavelinella*, *Gyroidinoides*, *Hansenisca* ['*Gyroidina*'] (Pl. 25.13), *Valvulineria* (Pl. 25.14).

Low trochospiral; aperture basal, umbilical; uncovered; umbilical areas of chambers imperforate: Family Bagginidae (Cretaceous–Recent). Representative genera: *Angulodiscorbis*, *Baggina* (Pl. 25.15–25.16), *Glabratella* (Pl. 25.17), *Heronallenia*.

Low to comparatively high trochospiral, with secondary chamberlets; apertures multiple, basal, umbilical, extending along sutures between chambers; uncovered: Family Cymbaloporidae (Late Cretaceous–Recent). Representative genera: *Cymbalopora* (Pl. 25.18), *Cymbaloporetta* (Pl. 25.19–25.21), *Fabianina*, *Halkyardia*.

Multilocular; essentially low trochospiral; typically bi-convex, although with evolute dorsal side slightly higher than involute ventral; aperture ventral, basal, areal or terminal; simple: Superfamily Asterigerinidea [Asterigerinacea].

Low trochospiral; aperture basal, umbilical, or areal: Family Eponididae (Cretaceous–Recent). Representative genera: *Cribrogloborotalia*, *Eponides* (Pl. 26.1–26.4), *Oridorsalis* (Pl. 26.5), *Poroeponides*, *Rectoeponides*.

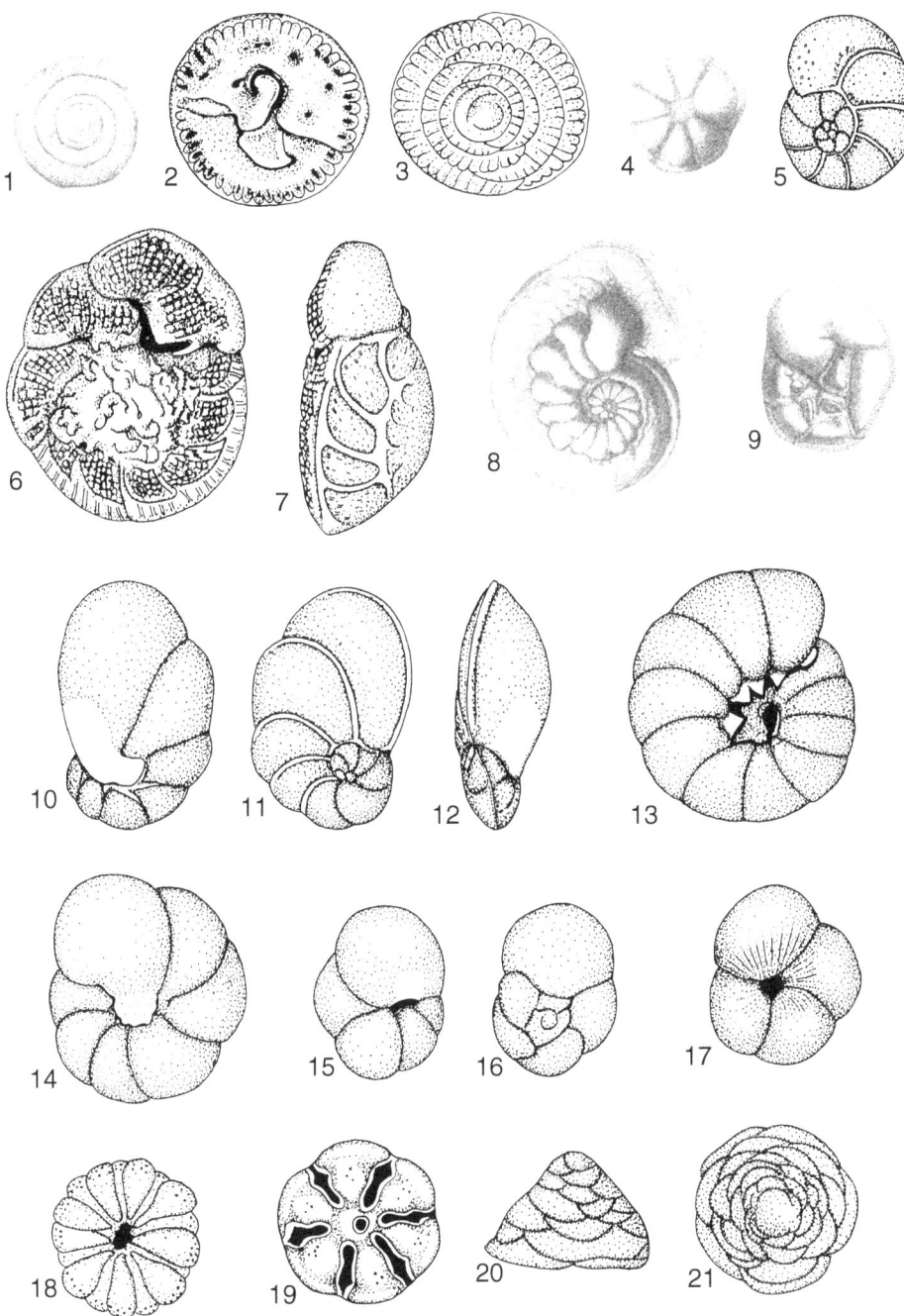

Pl. 25 **Spirillinidae (25.1), Patellinidae (25.2–25.3), Discorbidae (25.4–25.9), Cancrisidae (25.10–25.14), Bagginidae (25.15–25.17) and Cymbaloporidae (25.18–25.21)**. Modified after Haynes (1981) and Jones (2006). 25.1 *Spirillina vivipara*, ×100; 25.2–25.3 *Patellina corrugata*, ×100; 25.4 *Buccella tenerrima*, ×50; 25.5 *Discorbinella* sp., ×100; 25.6–25.7 *Discorbis wrighti*, ×160; 25.8 *Laticarinina pauperata*, ×35; 25.9 *Rosalina bradyi*, ×75; 25.10–25.12 *Cancris auriculus/oblongus*, ×66; 25.13 *Hansenisca soldanii/neosoldanii*, ×100; 25.14 *Valvulineria californica*, ×130; 25.15–25.16 *Baggina californica*, ×56; 25.17 *Glabratella crassa*, ×120; 25.18 *Cymbalopora bradyi*, ×50; 25.19–25.21 *Cymbaloporetta squammosa*, ×50.

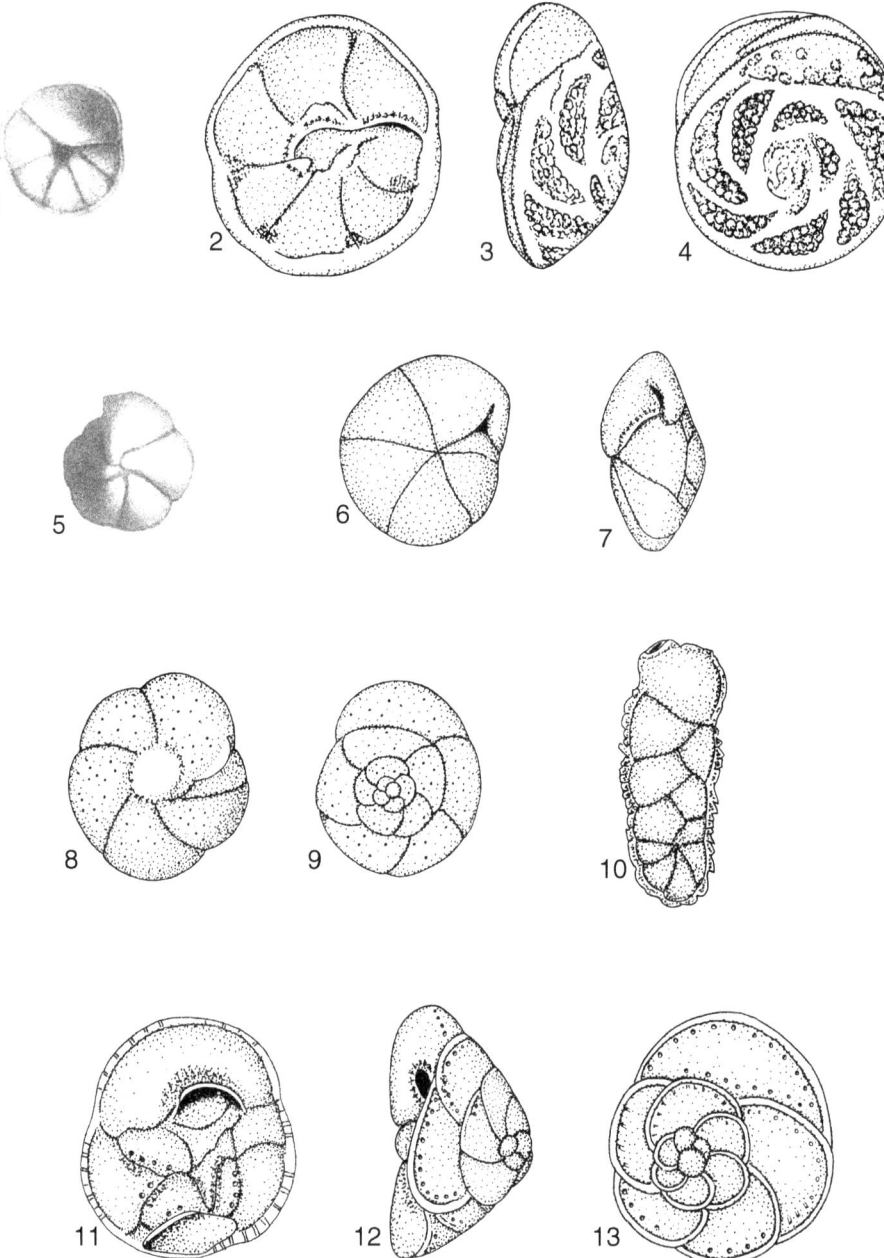

Pl. 26 **Eponididae (26.1–26.5), Alabaminidae (26.6–26.9), Siphoninidae (26.10) and Asterigerinidae (26.11–26.13)**. Modified after Haynes (1981) and Jones (2006). 26.1 *Eponides repandus*, ×35; 26.2–26.4 *Eponides repandus concameratus*, ×40; 26.5 *Oridorsalis umbonatus*, ×75; 26.6–26.7 *Alabamina obtusa*, ×100; 26.8–26.9 *Charltonina canterburyensis*, ×200; 26.10 *Siphonides biserialis*, ×200; 26.11–26.13 *Asterigerinata mamilla*, ×200.

Low trochospiral; aperture basal, umbilical, extending along fold in apertural face: Family Alabaminidae (Cretaceous–Recent). Representative genera: *Alabamina* (Pl. 26.6–26.7), *Conorotalites, Charltonina* (Pl. 26.8–26.9), *Globorotalites, Osangulariella.*

Low trochospiral to biserial or uniserial; aperture terminal: Family Siphoninidae (Late Cretaceous–Recent). Representative genera: *Pulsiphonina, Siphonides* (Pl. 26.10), *Siphonina, Siphoninella.*

Low trochospiral, with secondary chamberlets; aperture basal, umbilical: Family Asterigerinidae (Cretaceous–Recent). Representative genera: *Amphistegina, Asterigerina, Asterigerinata* (Pl. 26.11–26.13).

Multilocular; essentially low trochospirally or almost planispirally coiled, in some cases becoming irregularly uncoiled; typically plano-convex; with evolute dorsal side flat, and in some cases attached, and involute ventral side elevated; aperture medial to dorsal: Superfamily Orbitoididea [Orbitoidacea].

Low trochospirally or almost planispirally coiled, in some cases becoming irregularly uncoiled; aperture medial to dorsal; secondary apertures may or may not be present: Family Anomalinidae (Cretaceous–Recent). Representative genera: *Annulocibicides* (Pl. 27.1–27.2), *Anomalinella, Anomalinoides* (Pl. 27.3–27.5), *Cibicides* (Pl. 27.6–27.8), *Cibicidina, Cibicidoides* (Pl. 27.9), *Cyclocibicides, Cycloloculina, Dyocibicides* (Pl. 27.10), *Hanzawaia, Heterolepa* (Pl. 27.11–27.12), *Hyalinea* (Pl. 27.13), *Planulina, Rectocibicides* (Pl. 27.14), *Stensioina.*

Low trochospiral or annular; with supplementary chamberlets; apertures multiple; Family Planorbulinidae (Eocene–Recent). Representative genera: *Linderina* (Pl. 27.15), *Planorbulina* (Pl. 27.16), *Planorbulinella.*

Low trochospiral or annular, or, especially in the case of attached or encrusting forms, irregular; apertures replaced by coarse perforations: Family Acervulinidae (Eocene–Recent). Representative genera: *Acervulina* (Pl. 27.17), *Gypsina* (Pl. 27.18), *Planogypsina.*

Low trochospiral or, especially in the case of attached or encrusting forms, irregular; apertures multiple: Family Homotrematidae (Late Cretaceous–Recent). Representative genera:

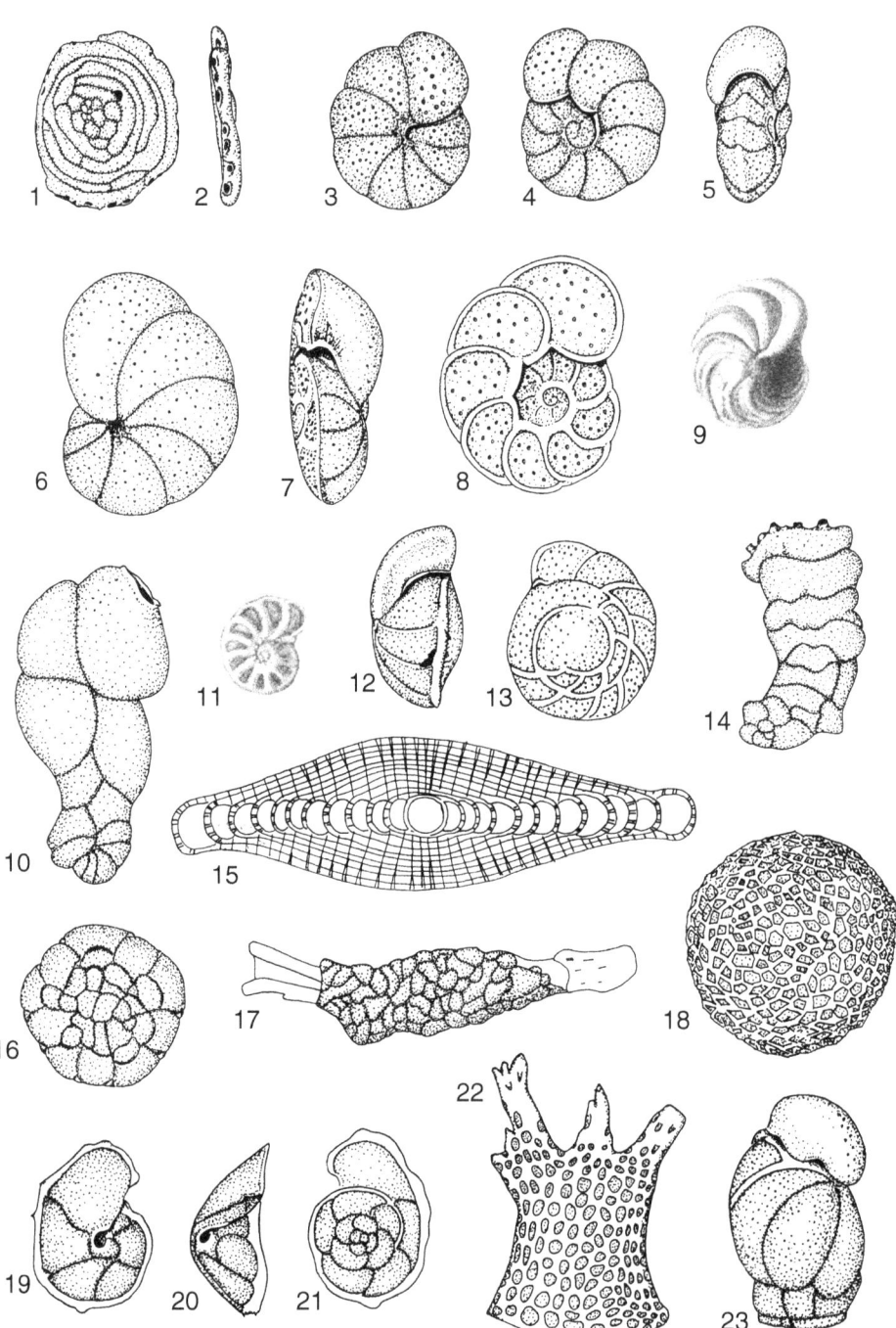

Pl. 27 **Anomalinidae (27.1–27.14), Planorbulinidae (27.15–27.16), Acervulinidae (27.17–27.18) and Homotrematidae (27.19–27.23)**. Modified after Haynes (1981) and Jones (2006). 27.1–27.2 *Annulocibicides projectus*, ×37; 27.3–27.5 *Anomalinoides pinguis*, ×70; 27.6–27.8 *Cibicides lobatulus*, ×160; 27.9 *Cibicidoides wuellerstorffi*, ×50; 27.10 *Dyocibicides biserialis*, ×50; 27.11–27.12 *Heterolepa dutemplei*, ×37; 27.13 *Hyalinea balthica*, ×50; 27.14 *Rectocibicides miocenicus*, ×50; 27.15 *Linderina brugesi*, ×150; 27.16 *Planorbulina distoma*, ×65; 27.17 *Acervulina inhaerens*, ×30; 27.18 *Gypsina vesicularis*, ×30; 27.19–27.21 *Carpenteria balaniformis*, ×20; 27.22 *Homotrema rubrum*, ×20; 27.23 *Rupertina stabilis*, ×40.

*Biarritzina*, *Carpenteria* (Pl. 27.19–27.21), *Homotrema* (Pl. 27.22), *Miniacina*, *Rupertina* (Pl. 27.23).

Multilocular; low trochospiral or involute planispiral; typically bi-convex; aperture medial to ventral; advanced forms with secondary apertures: Superfamily Nonionidea [Nonionacea].

Low trochospiral or involute planispiral; few chambers per whorl: Family Chilostomellidae (Cretaceous–Recent). Representative genera: *Allomorphina* (Pl. 28.1–28.2), *Allomorphinella*, *Chilostomella* (Pl. 28.3), *Chilostomelloides*, *Pullenia* (Pl. 28.4).

Involute planispiral or low trochospiral; many chambers per whorl: Family Nonionidae (Cretaceous–Recent). Representative genera: *Melonis* (Pl. 28.5), *Nonion* (Pl. 28.6–28.7), *Nonionella* (Pl. 28.8), *Nonionellina* (Pl. 28.9–28.10).

Involute planispiral or low trochospiral; many chambers per whorl; advanced forms with secondary apertures; and also with sub-sutural septal canals: Family Elphidiidae (Late Cretaceous?–Recent). Representative genera: *Astrononion, Cellanthus, Cribroelphidium, Cribrononion, Elphidiella* (Pl. 28.11), *Elphidium* (Pl. 28.12), *Faujasina* (Pl. 28.13–28.15), *Laffiteina, Ozawia, Parrellina, Polystomellina [Notorotalia]*.

Large, with complex interior, generally with and exceptionally without canal system: 'Larger Rotaliida' (Late Cretaceous–Recent).

Low trochospiral, planispiral or annular; with spiral and sub-sutural septal and umbilical canals, and in some cases radial canals: Superfamily Rotaliidea [Rotaliacea].

Low trochospiral; typically with septal fissures and umbilical plug: Family Rotaliidae (Late Cretaceous–Recent). Representative genera: *Ammonia* (Pl. 29.1–29.2), *Dictyoconoides* (Pl. 29.3), *Dictyokathina, Kathina, Lockhartia* (Pl. 29.4–29.5), *Pararotalia* (Pl. 29.6), *Pseudorotalia* (Pl. 29.7), *Rotalia* (Pl. 29.8–29.9), *Sakesaria*.

Low trochospiral or annular; typically conical: test large: Family Chapmaninidae (Middle Eocene–Miocene). Representative genera: *Chapmanina* (Pl. 29.10–29.11), *Ferayina* (Pl. 29.12).

Planispiral; typically with septal fissures and umbilical plug: Family Miscellaneidae (Late Cretaceous–Eocene).

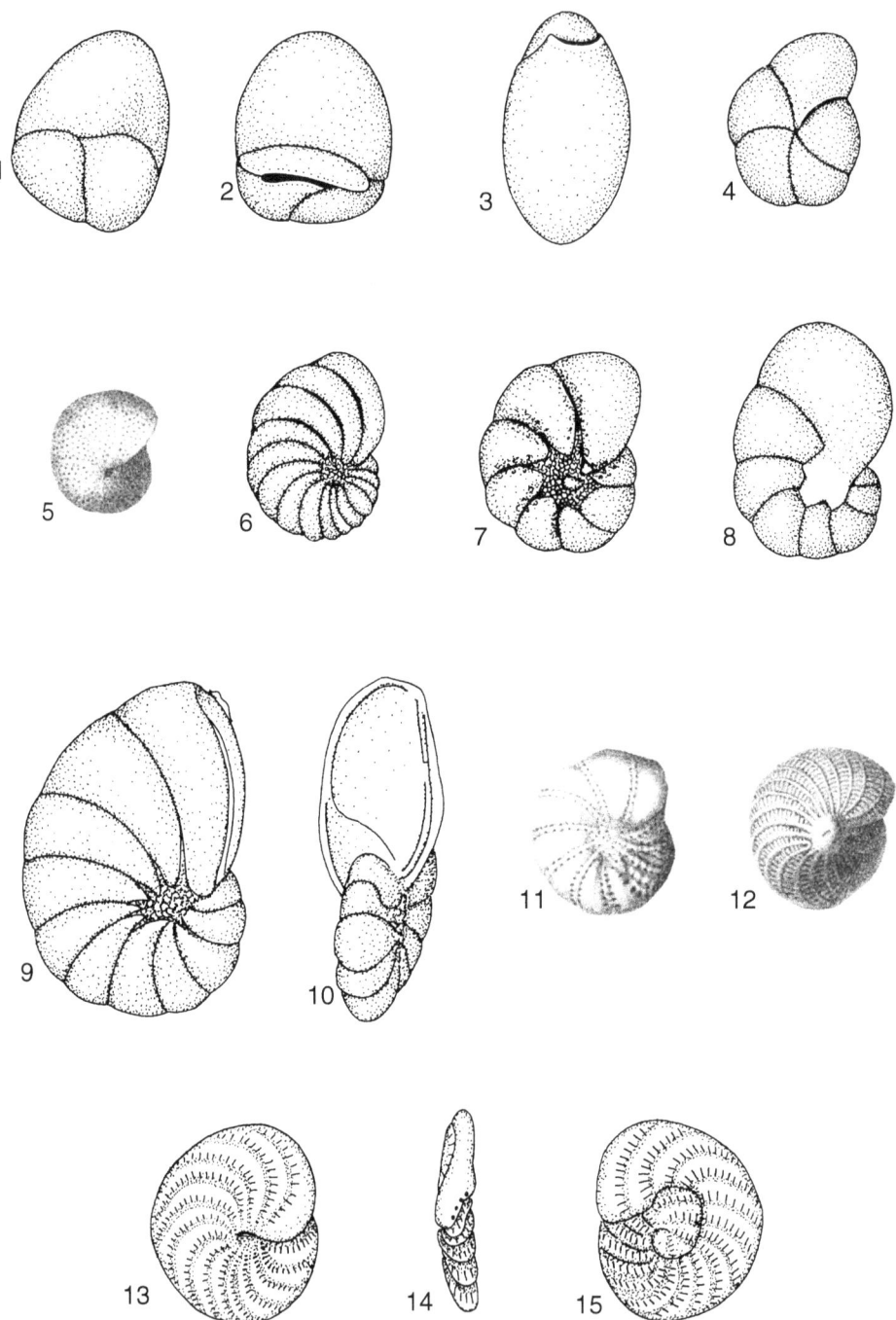

Pl. 28 **Chilostomellidae (28.1–28.4), Nonionidae (28.5–28.10) and Elphidii-dae (28.11–28.15)**. Modified after Haynes (1981) and Jones (2006). 28.1–28.2 *Allomorphina contraria*, ×78; 28.3 *Chilostomella oolina*, ×50; 28.4 *Pullenia quinqueloba*, ×66; 28.5 *Melonis pompilioides*, ×75; 28.6 *Nonion boueanus*, ×70; 28.7 *Nonion depressulus*, ×100; 28.8 *Nonionella digitata*, ×150; 28.9–28.10 *Nonionellina* sp., ×100; 28.11 *Elphidiella arctica*, ×30; 28.12 *Elphidium crispum*, ×40; 28.13–28.15 *Faujasina carinata*, ×50.

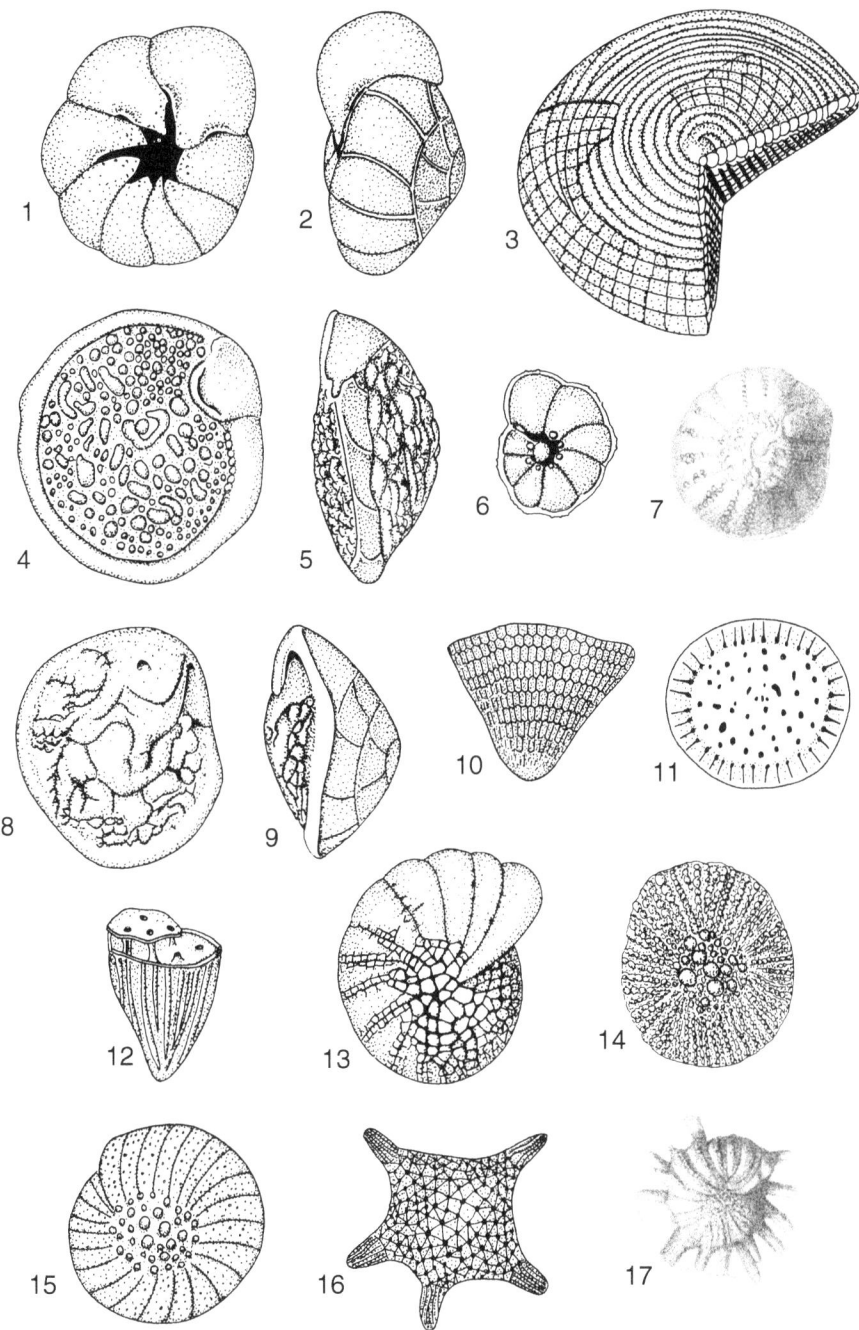

Pl. 29  **Rotaliidae (29.1–29.9), Chapmaninidae (29.10–29.12), Miscellaneidae (29.13–29.15) and Calcarinidae (29.16–29.17)**. Modified after Haynes (1981) and Jones (2006). 29.1–29.2 *Ammonia aberdoveyensis*, ×100; 28.3 *Dictyoconoides cooki*, ×10; 29.4–29.5 *Lockhartia haimei*, ×26; 29.6 *Pararotalia inermis*, ×128; 29.7 *Pseudorotalia schroeteriana*, ×25; 28.8–29.9 *Rotalia trochidiformis*, ×30; 29.10–29.11 *Chapmanina gassinensis*, ×35; 29.12 *Ferayina coralliformis*, ×105; 29.13 *Fissoelphidium operculiferum*, ×28; 29.14 *Miscellanea miscella*, ×20; 29.15 *Pseudosiderolites vidali*, ×8; 29.16 *Baculogypsina sphaerulata*, ×20; 29.17 *Calcarina spengleri*, ×30.

Representative genera: *Arnaudella, Biplanispira, Cuvillierina, Daviesina, Fissoelphidium* (Pl. 29.13), *Miscellanea* (Pl. 29.14), *Pseudosiderolites* (Pl. 29.15).

Low trochospiral to planispiral; with supplementary chamberlets; typically with spines in plane of coiling: Family Calcarinidae (Late Cretaceous–Recent). Representative genera: *Baculogypsina* (Pl. 29.16), *Baculogypsinoides, Calcarina* (Pl. 29.17), *Siderolites*.

Sub-annular, with eccentrically placed embryonic apparatus: Family Miogypsinidae (Late Oligocene–Middle Miocene). Representative genera: *Lepidosemicyclina, Miogypsina* (Pl. 30.1–30.2), *Miogypsinoides* (Pl. 30.3–4), *Miolepidocyclina*.

Annular: Family Lepidorbitoididae (Late Cretaceous). Representative genera: *Asterorbis, Helicorbitoides* (Pl. 30.5–30.6), *Lepidorbitoides* (Pl. 30.7–30.8), *Orbitocyclina*.

Planispiral, or annular; with sub-sutural canals; and also with marginal cords, and in some cases marginal spiral canals: Superfamily Nummulitidea [Nummulitacea].

Planispiral: Family Nummulitidae (Late Palaeocene–Recent). Representative genera: *Assilina* (Pl. 31.1–31.2), *Nummulites* (Pl. 31.3–31.4), *Operculina* (Pl. 31.5–31.7), *Operculinella* (Pl. 31.8), *Operculinoidea, Ranikothalia, Sulcoperculina* (Pl. 31.9).

Sub-annular, with eccentrically placed embryonic apparatus; with supplementary chamberlets: Family Cycloclypeidae (Eocene–Recent). Representative genera: *Cycloclypeus* (Pl. 31.10), *Heterostegina* (Pl. 31.11), *Katacycloclypeus, Radiocycloclypeus, Spiroclypeus* (Pl. 31.12–31.13).

Annular: Family Pseudorbitoididae (Late Cretaceous). Representative genera: *Pseudorbitoides* (Pl. 32.1–32.2), *Rhabdorbitoides, Sulcorbitoides, Vaughanina* (Pl. 32.3).

Annular; with supplementary chamberlets: Family Discocyclinidae (Late Palaeocene–Eocene). Representative genera: *Asterocyclina* (Pl. 33.1), *Asterophragmina, Discocyclina* (Pl. 33.2–33.3), *Pseudophragmina* (Pl. 33.4).

Without canal system.

Annular: Family Orbitoididae (Late Cretaceous). Representative genera: *Omphalocyclus* (Pl. 34.1–34.2), *Orbitoides* (Pl. 34.3–34.4), *Schlumbergeria, Simplorbites, Torreina*.

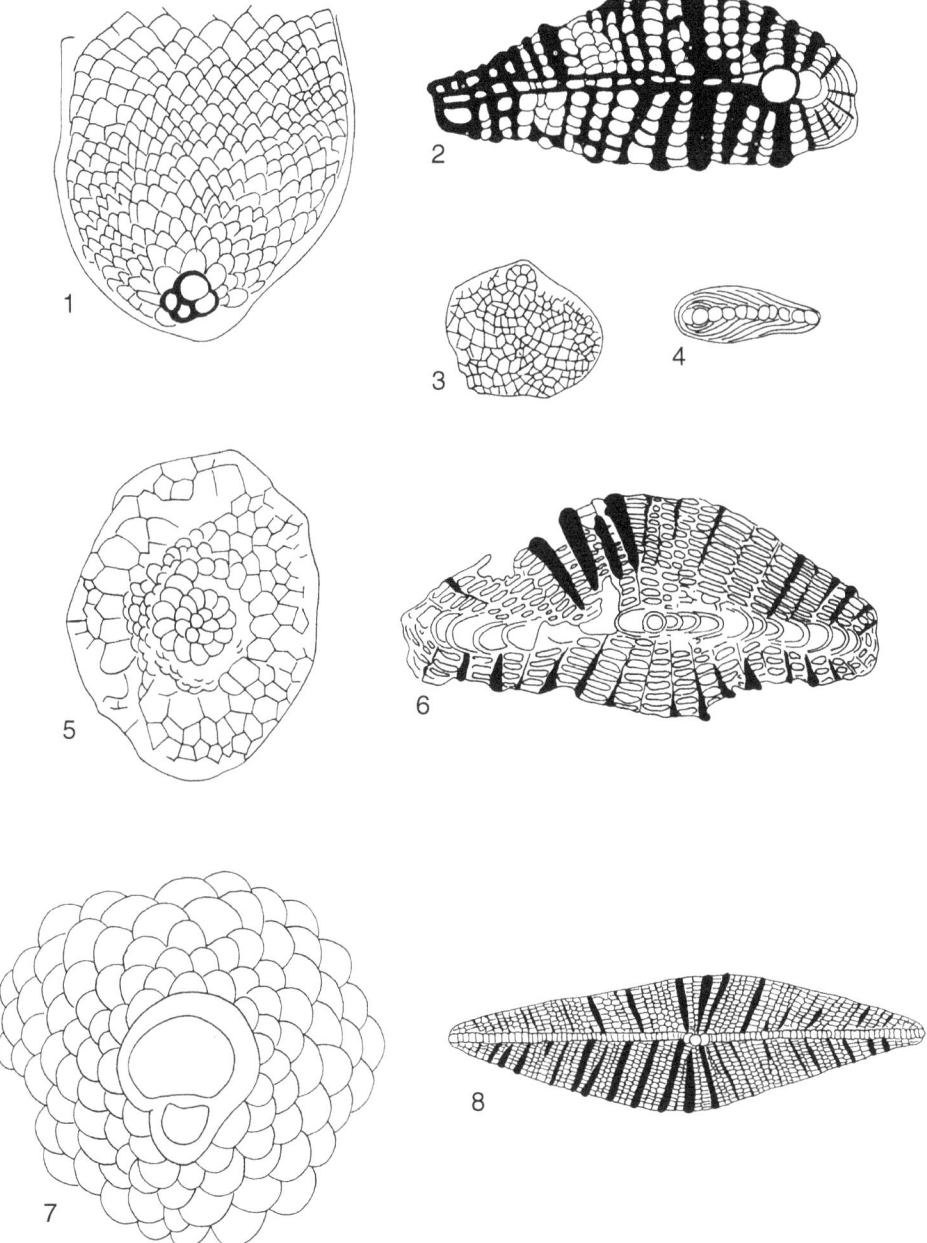

Pl. 30 **Miogypsinidae (30.1–30.4) and Lepidorbitoididae (30.5–30.8)**. Modified after Haynes (1981). 30.1 *Miogypsina antillea*, ×40; 30.2 *Miogypsina cushmani*, ×40; 30.3 *Miogypsinoides bantamensis*, ×15; 30.4 *Miogypsinoides lateralis*, ×15; 30.5–30.6 *Helicorbitoides longispiralis*, ×50, ×35; 30.7–30.8 *Lepidorbitoides socialis*, ×128, ×10.

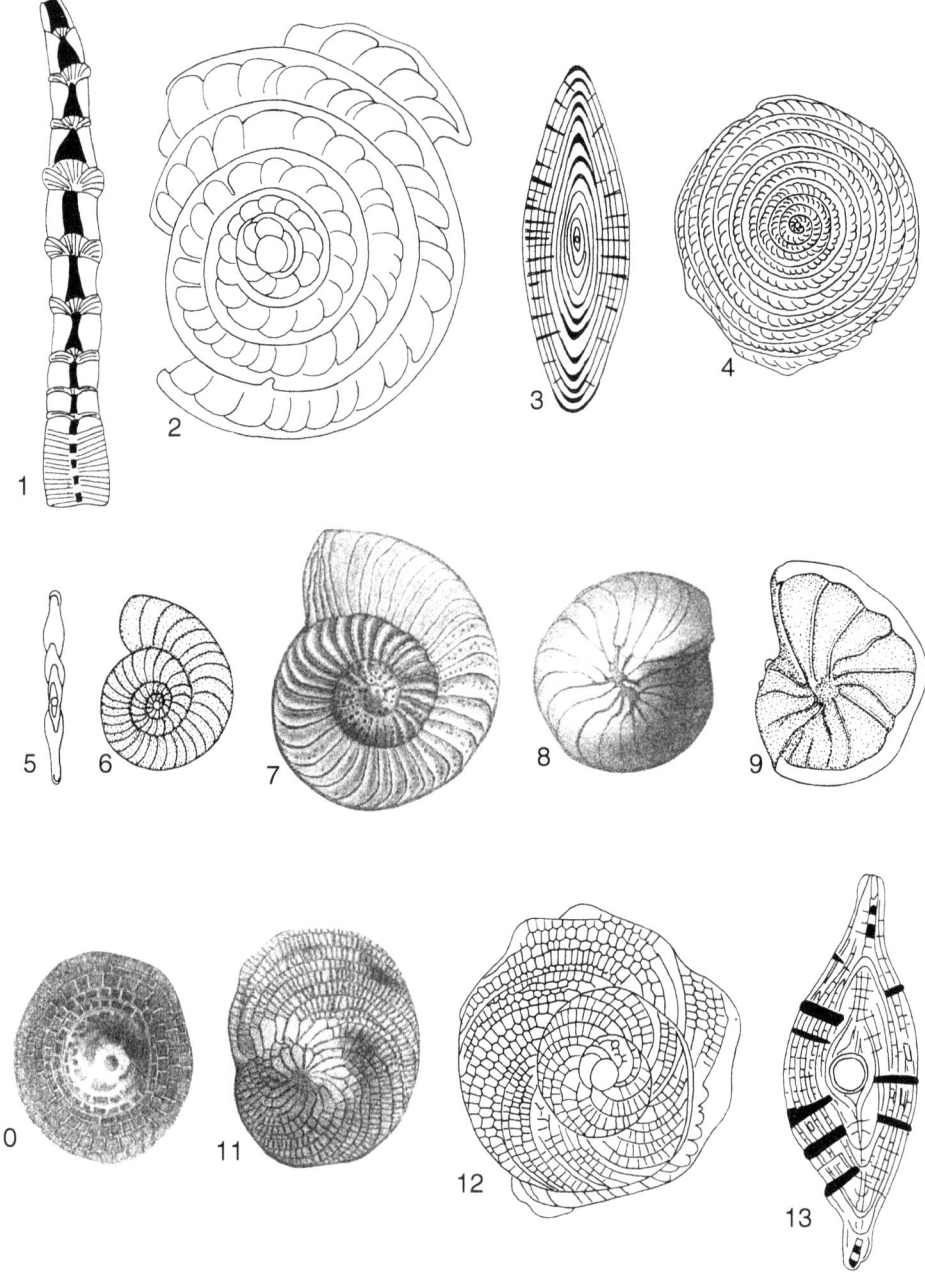

Pl. 31  **Nummulitidae (31.1–31.9) and Cycloclypeidae (31.10–31.13)**. Modified after Haynes (1981) and Jones (2006). 31.1–31.2 *Assilina spira*, ×7, ×13; 31.3–31.4 *Nummulites laevigatus*; ×5, ×4; 31.5–31.6 *Operculina ammonoides*, ×10; 31.7 *Operculina complanata/granulosa*, ×12; 31.8 *Operculinella cumingi*, ×18; 31.9 *Sulcoperculina dickersoni*, ×40; 31.10 *Cycloclypeus carpenteri*, ×12; 31.11 *Heterostegina depressa*, ×12; 31.12–31.13 *Spiroclypeus tidoeganensis*, ×16.

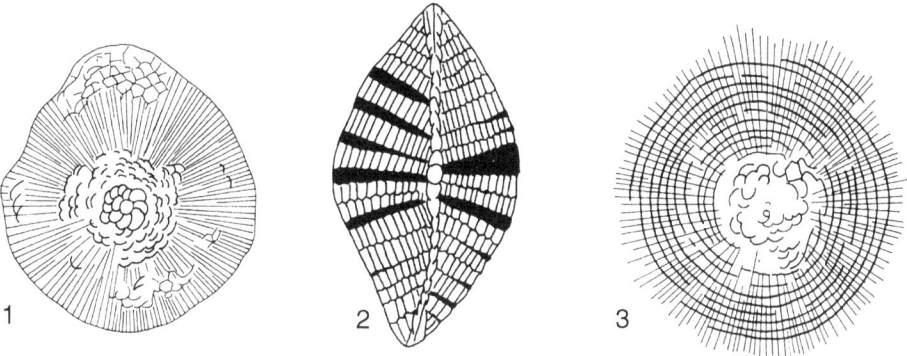

Pl. 32 **Pseudorbitoididae**. Modified after Haynes (1981). 32.1–32.2 *Pseudorbitoides israelskyi*, ×30, ×40; 32.3 *Vaughanina cubensis*, ×38.

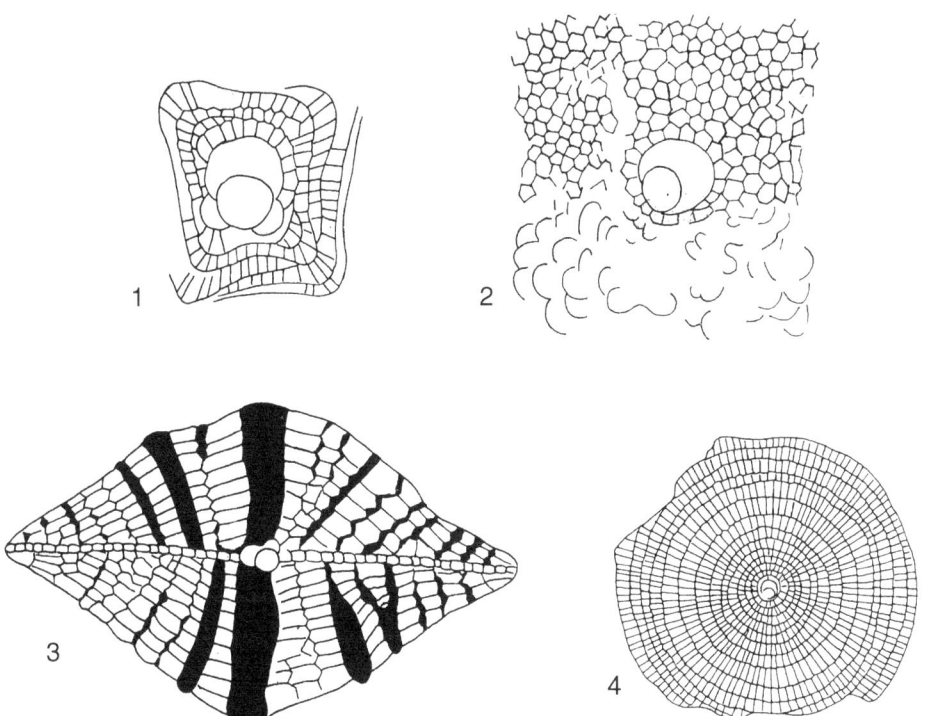

Pl. 33 **Discocyclinidae**. Modified after Haynes (1981). 33.1 *Asterocyclina asterisca*, ×85; 33.2–33.3 *Discocyclina douvillei*, ×52, ×43; 33.4 *Pseudophragmina flintensis*, ×40.

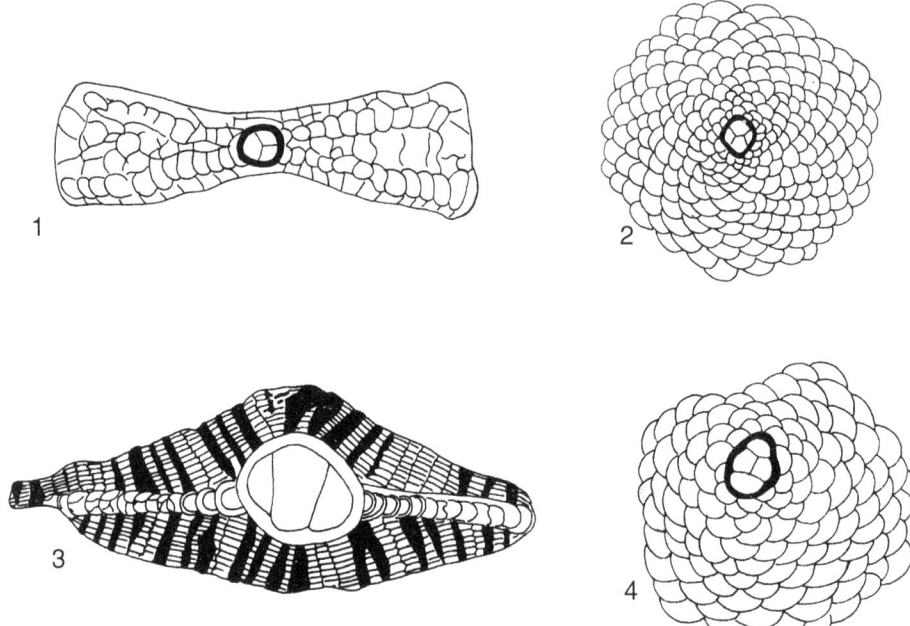

Pl. 34   **Orbitoididae**. Modified after Haynes (1981). 34.1–34.2 *Omphalocyclus macroporus*, ×20, ×13; 34.3 *Orbitoides apiculata*, ×22; 34.4 *Orbitoides media*, ×40.

Annular, with supplementary chamberlets: Family Lepidocyclinidae (Middle Eocene–Middle Miocene). Representative genera: *Eulepidina* (Pl. 35.4), *Helicolepidina*, *Helicostegina* (Pl. 35.1–35.2), *Lepidocyclina* (Pl. 35.3), *Nephrolepidina* (Pl. 35.5–35.6), *Pliolepidina* (Pl. 35.7), *Polylepidina* (Pl. 35.8).

10.  Habit planktic: Order Globigerinida (Middle Triassic–Recent).

Wall aragonitic: Family Oberhauserellidae (Middle Triassic–Jurassic). Representative genera: *Conoglobigerina* (Pl. 36.1–36.2), *Globuligerina* (Pl. 36.3), *Jurassorotalia*, *Oberhauserella* (Pl. 36.4–36.5), *Polskanella* (Pl. 36.6–36.7), *Praegubkinella*.

Wall generally calcitic, and exceptionally aragonitic; generally low trochospiral; typically compressed and keeled; aperture ventral; umbilical; partially or essentially completely covered by umbilical chamber lobes, or portici: Superfamily Hedbergellidea [Hedbergellacea].

Periphery typically weakly compressed, in some cases with weak single keel; aperture partially covered by portici: Family Hedbergellidae (Cretaceous, Hauterivian–Maastrichtian). Representative genera: *Clavihedbergella*, *Favusella*, *Hedbergella* (Pl. 37.1–37.2), *Praeglobotruncana* (Pl. 37.3–37.5), *Wondersella*.

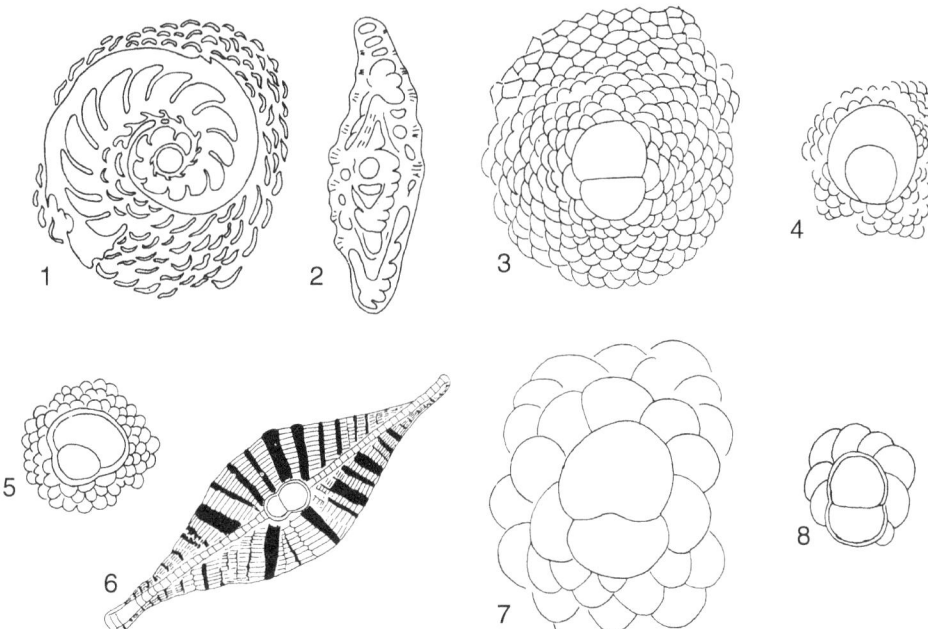

Pl. 35 **Lepidocyclinidae**. Modified after Haynes (1981). 35.1–35.2 *Helicoste-gina dimorpha*, ×43; 35.3 *Lepidocyclina mantelli*, ×40; 35.4 *Eulepidina formosa*, ×40; 35.5. *Nephrolepidina tournoueri*, ×40; 35. *Nephrolepidina verbeeki*, ×20; 35.7 *Pliolepidina proteiformis*, ×80; 35.8 *Polylepidina antillea*, ×65.

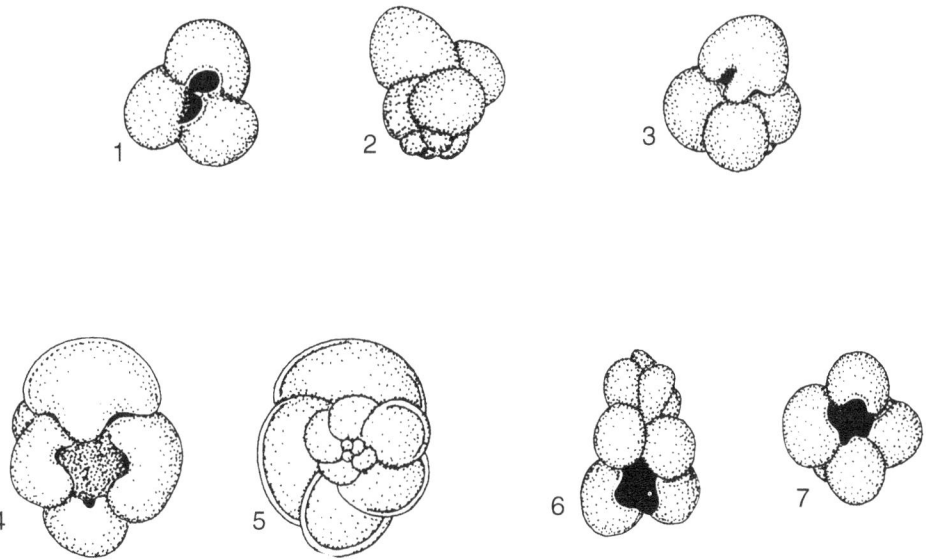

Pl. 36 **Oberhauserellidae**. Modified after Haynes (1981). 36.1–36.2 *Conoglobigerina dagestanica*, ×100; 36.3 *Globuligerina frequens*, ×100; 36.4–36.5 *Oberhauserella quadrilobata*, ×150; 36.6–36.7 *Polskanella altispira*, ×100.

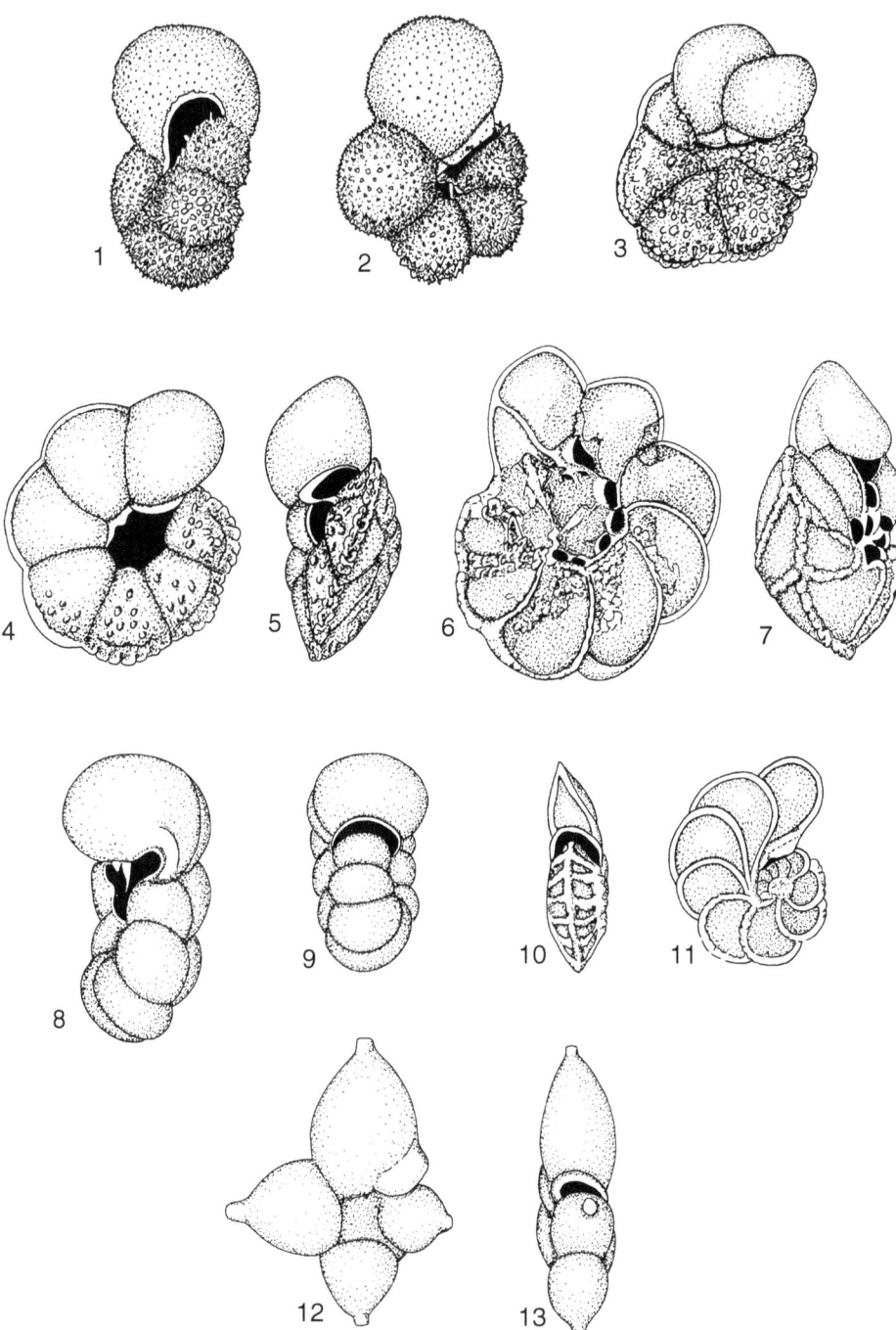

Pl. 37  **Hedbergellidae (37.1–37.5). Rotaliporidae (37.6–37.8), Planomalinidae (37.9–37.11) and Schakoinidae (37.12–37.13)**. Modified after Haynes (1981). 37.1–37.2 *Hedbergella delrioensis*, ×300; 37.3 *Praeglobotruncana delrioensis*, ×250; 37.4–37.5; *Praeglobotruncana stephani*, ×115; 37.6–37.7 *Rotalipora greenhornensis*, ×100; 37.8 *Ticinella roberti*, ×115; 37.9 *Globigerinelloides breggiensis*, ×100; 37.10–37.11 *Planomalina buxtorfi*, ×84; 37.12–37.13 *Schackoina cenomana*, ×200.

Periphery typically with single keel; primary aperture partially covered by plate formed by fusion of portici; with secondary marginal apertures: Family Rotaliporidae (Cretaceous, Albian–Turonian). Representative genera: *Anaticinella*, *Rotalipora* (Pl. 37.6–37.7), *Ticinella* (Pl. 37.8).

Periphery typically with double keel; primary aperture essentially completely covered by tegillum formed by complete fusion of portici; with secondary marginal and areal apertures: Family Globotruncanidae (Late Cretaceous, Turonian–Maastrichtian). Representative genera: *Abathomphalus*, *Archaeoglobigerina*, *Contusotruncana [Rosita]*, *Dicarinella*, *Globotruncana*, *Globotruncanella*, *Globotruncanita*, *Helvetoglobotruncana*, *Marginotruncana*, *Plummerita*, *Rugoglobigerina*, *Rugotruncana*, *Trinitella*.

Wall calcitic; low trochospiral, planispiral or biserial; typically compressed; aperture medial, partially covered with portici, or terminal; Superfamily Heterohelicidea [Heterohelicacea].

Planispiral, in some cases with low trochospiral initial portion; primary aperture medial, extending along sutures towards umbilicus, partially covered with portici; with secondary apertures: Family Planomalinidae (Cretaceous, Aptian?–Maastrichtian). Representative genera: *Biglobigerinella*, *Globigerinelloides* (Pl. 37.9), *Hastigerinoides*, *Planomalina* (Pl. 37.10–37.11).

Low trochospiral to planispiral; characteristically with radially elongate chambers bearing terminal tubulo-spines in plane of coiling: Family Schackoinidae (Cretaceous, Barremian–Cenomanian). Representative genera: *Leupoldina*, *Schackoina* (Pl. 37.12–37.13).

Planispiral to biserial, in some forms fan-shaped, and with or without supplementary chambers; aperture medial; with or without secondary apertures: Family Heterohelicidae (Cretaceous, Albian–Maastrichtian). Representative genera: *Bifarina*, *Gublerina* (Pl. 38.1), *Heterohelix* (Pl. 38.2), *Lunatriella*, *Planoglobulina*, *Pseudoguembelina* (Pl. 38.3), *Pseudotextularia*, *Ventilabrella* (Pl. 38.4).

Biserial; aperture terminal, at end of neck: Family Chiloguembelinidae (Palaeocene–Early Miocene). Representative genera: *Chiloguembelina* (Pl. 38.5–38.6), *Chiloguembelinella*.

Wall calcitic; essentially trochospiral; aperture ventral; umbilical: Superfamily Globigerinidea [Globigerinacea].

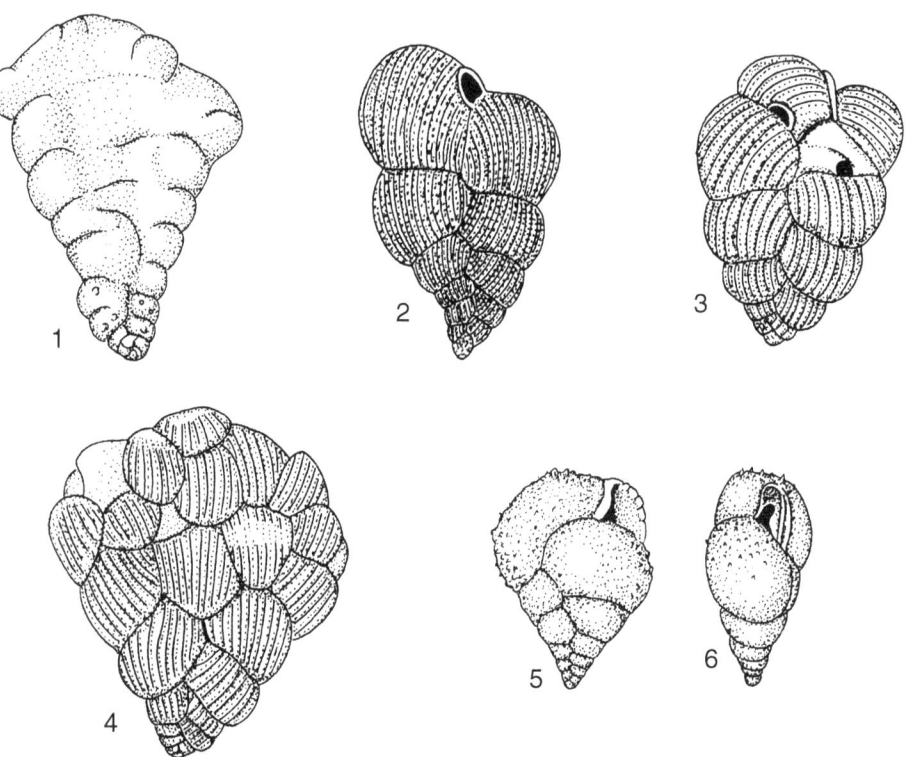

Pl. 38  **Heterohelicidae (38.1–38.4) and Chiloguembelinidae (38.5–38.6)**.
Modified after Haynes (1981). 38.1 *Gublerina ornatissima*, ×74; 38.2 *Heterohelix globulosa*, ×120; 38.3 *Pseudoguembelina excolata*, ×116; 38.4 *Ventilabrella eggeri*, ×100; 38.5–38.6 *Chiloguembelina midwayensis*, ×100.

High trochospiral: Family Guembelitriidae (Middle Jurassic–Oligocene). Representative genera: *Caucasella, Guembelitria* (Pl. 39.1), *Guembelitriella*.

Low trochospiral (rarely, streptospiral); typically inflated; and with strongly overlapping or embracing chambers; aperture ventral, umbilical; in some cases partially covered by bulla; with or without secondary sutural apertures on dorsal side: Family Globigerinidae (Palaeocene–Recent). Representative genera: *Beella, Biorbulina, Candeina, Catapsydrax* (Pl. 39.2), *Globigerapsis, Globigerina* (Pl. 40), *Globigerinatella, Globigerinatheka* (Pl. 39.3–39.4), *Globigerinita, Globigerinoides, Globigerinoita, Orbulina, Orbulinoides, Praeorbulina, Pulleniatina, Sphaeroidinella, Sphaeroidinellopsis, Subbotina, Turborotalita*.

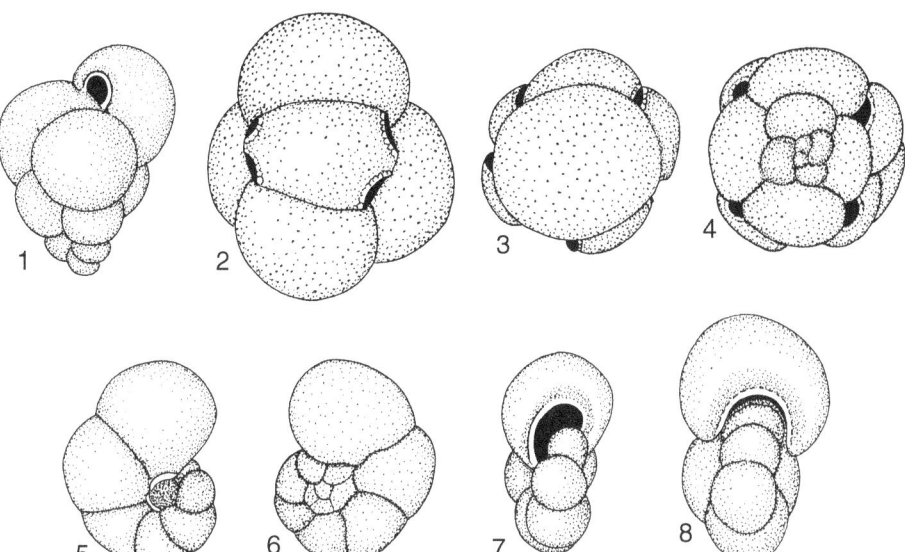

Pl. 39   **Guembelitriidae (39.1), Globigerinidae (39.2–39.4) and Hantkenini-dae (39.5–39.8)**. Modified after Haynes (1981). 39.1 *Guembelitria cretacea*, ×312; 39.2 *Catapsydrax dissimilis*, ×130; 39.3–39.4 *Globigerinatheka barri*, ×130; 39.5–39.8 *Globanomalina wilcoxensis*, ×150.

Low trochospiral; typically compressed and keeled; aperture extra-umbilical; with or without secondary sutural apertures on dorsal side: Family Globorotaliidae (Palaeocene–Recent). Representative genera: *Acarinina, Fohsella, Globoconella, Globoquadrina, Globorotalia, Menardella, Morozovella, Neogloboquadrina, Obandyella [Hirsutella], Paragloborotalia, Planorotalites, Truncorotalia, Truncorotaloides*.

Wall calcitic; essentially planispiral, in some cases with low trochospiral or streptospiral initial portion; aperture medial: Superfamily Hantkeninidea [Hantkeninacea].

Planispiral: Family Hantkeninidae (Late Palaeocene–Oligocene). Representative genera: *Aragonella, Clavigerinella, Cribrohantkenina, Globanomalina* (Pl. 39.5–39.8), *Hantkenina, Pseudohastigerina*.

Planispiral, with low trochospiral or streptospiral initial portion; characteristically with radially elongate chambers bearing terminal tubulo-spines in plane of coiling: Family Hastigerinidae (Miocene–Recent). Representative genera: *Bolliella, Globigerinella, Hastigerina, Hastigerinopsis*.

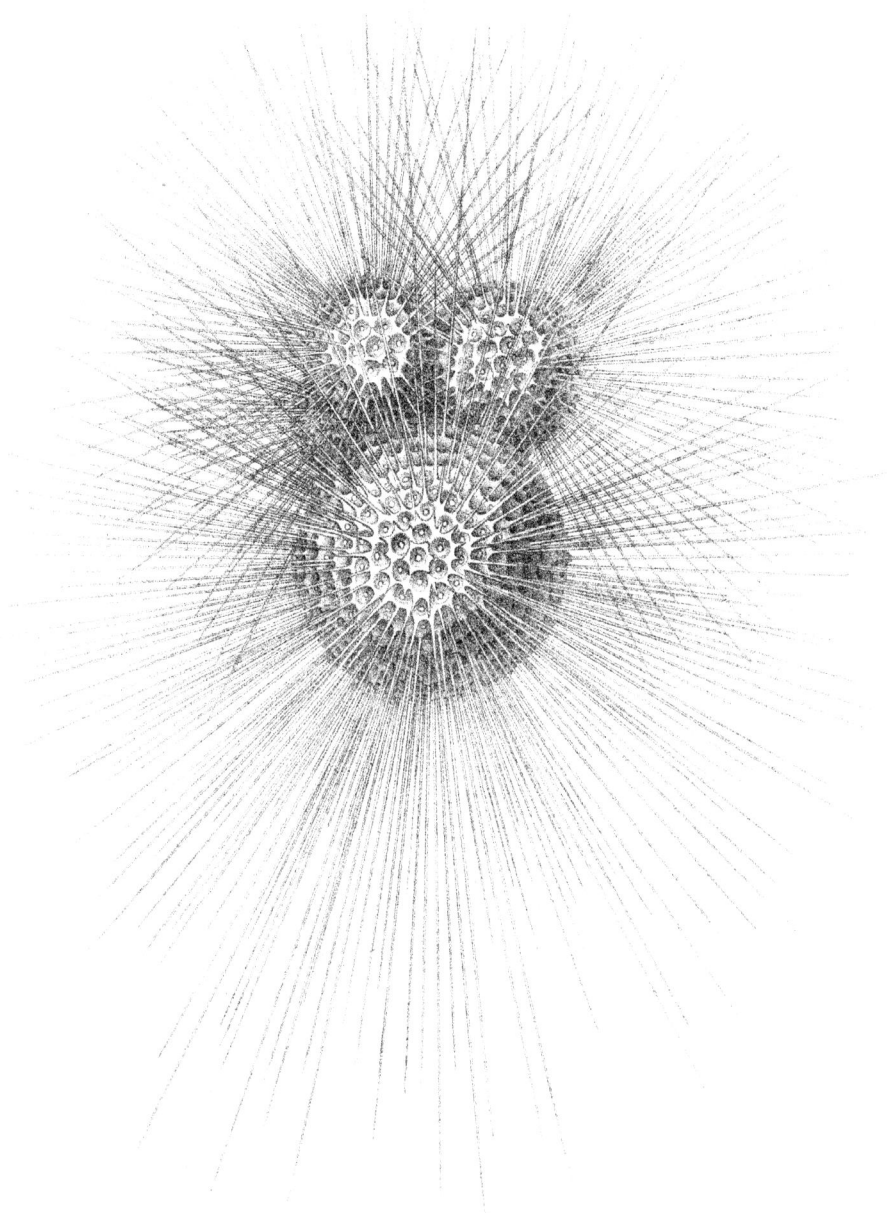

Pl. 40   **Globigerinidae**. *Globigerina bulloides*, x200. After Jones (2006).

## 3.4 **Further reading**
*General*

Boltovskoy & Wright, 1976; Bignot, 1985; BouDagher-Fadel, 2008, 2012; Brasier, 1980; Lipps, in Buzas & Sen Gupta, 1982; Giere, 2009; Goldstein & Bernhard, 1997; Haynes, 1981; Boersma, in Haq & Boersma, 1978; Boersma, in Haq & Boersma, 1998; Hedley, 1964; Hemleben *et al.*, 1989; Jorissen *et al.*, in Hillaire-Marcel & de Vernal, 2007; Kucera, in Hillaire-Marcel & de Vernal, 2007; Lee & Anderson, 1991; Lee & Hallock, 2000; Lee, in Levandowsky & Hutner, 1980; Lipps, 1993; Lipps & Goldstein, 2006; Loeblich & Tappan, 1964; Loeblich & Tappan, 1987; Murray, 1973, 1991, 2006; Be, in Ramsay, 1977; Goldstein, in Sen Gupta, 1999; Sen Gupta, in Sen Gupta, 1999.

*Biology*

Alve, 2010a; Alve & Bernhard, 1995; Arnold, 1954, 1966; Bernhard, 1988, 2000, 2003; Bernhard & Bowser, 1999; Bernhard *et al.*, 2006, 2012; Boltovskoy, 1963; Buchanan & Hedley, 1960; Kaminski & Wetzel, in Bubik & Kaminski, 2004; Lipps, in Buzas & Sen Gupta, 1982; Christiansen, 1971; Cushman, 1922; Debenay *et al.*, 2011; DeLaca, 1982; DeLaca *et al.*, 1980, 1981; Dettmering *et al.*, 1998; Doyle & Doyle, 1940; Faber *et al.*, 1985; Figueira *et al.*, 2012; Filipsson, 2008; Filipsson *et al.*, 2010; Fontanier *et al.*, 2008; Frankel, 1972, 1975b; Myers, in Galtsoff *et al.*, 1937; Goldstein, 1988; Goldstein & Corliss, 1994; Goldstein & Moodley, 1993; Gooday, 1983a, b, 1988, 1993, 1994; Gooday & Hughes, 2002; Gooday & Jorissen, 2012; Gooday & Lambshead, 1989; Gooday & Turley, 1990; Gooday *et al.*, 2010; Grell, 1954; Larkin *et al.*, in Hall *et al.*, 2010; Harney *et al.*, 1998; Ernst *et al.*, in Hart *et al.*, 2000; Hartman *et al.*, 2012; Haynes, 1965; Hedley, 1958; Arnold, in Hedley & Adams, 1974; Herguera & Berger, 1981; Hess *et al.*, 2010b; Honisch *et al.*, 2003; Jepps, 1942, 1953; Kitazato *et al.*, in Jorissen & Rohling, 2000; Kalogeropoulou *et al.*, 2010; Knight & Mantoura, 1985; Koho *et al.*, 2008, 2010; Larkin & Gooday, 2009; le Calvez, 1938, 1946, 1950; Lee, 2006; Anderson *et al.*, in Lee & Anderson, 1991; Lee & Anderson, in Lee & Anderson, 1991; Lee *et al.*, in Lee & Anderson, 1991; Lee & Muller, 1967; Lee & Pierce, 1963; Lee & Zucker, 1969; Lee *et al.*, 1961, 1963, 1966, 1969, 1991; Glover *et al.*, in Lesser, 2010; Leutenegger, 1977; Lipps, 1976; Lister, 1895, 1905, 1906; Matera & Lee, 1972; Mojtahid *et al.*, 2011; Moodley *et al.*, 1998; Murray, 2012; Murray & Bowser, 2000; Myers, 1935a, b, c, 1936, 1938, 1942a, b, 1943; Pillet *et al.*, 2010; Pina-Ochoa *et al.*, 2010; Reiss & Hottinger, 1984; Risgaard-Petersen *et al.*, 2006; Rottger *et al.*, 1990; Gooday, in Rowe & Pariente, 1992;

Levin & Gooday, in Rowe & Pariente, 1992; Sandon, 1932; Schaudinn, 1895, 1911; Schonfeld, 2012; Goldstein, in Sen Gupta, 1999; Hallock, in Sen Gupta, 1999; Smart & Gooday, 1997; Gooday, in Southward *et al.*, 2003; Spindler & Hemleben, 1980; Stouff *et al.*, 1999; Suhr *et al.*, 2003; Lipps, in Tevesz & McCall, 1983; Wefer *et al.*, 1982; Wilson & Horton, 2012; Winter, 1907; Beuck *et al.*, in Wisshak & Tapanila, 2008.

## Morphology

Angell, 1967; Diz *et al.*, 2012; Duguay, 1983; Hemleben *et al.*, in Leadbeater & Riding, 1986; ten Kuile, in Lee & Anderson, 1991; Berger, in Lipps *et al.*, 1979; Saraswati, 2010; Schumacher *et al.*, 2010; Hansen, in Sen Gupta, 1999; Rohling & Cooke, in Sen Gupta, 1999; Stanley, 2006; Vinogradov, 1953.

## Functional morphology

Bernstein *et al.*, 1978; Bromley & Nordmann, 1971; Bronnimann, 1978; Bronnimann & Maisonneuve, 1980; Bronnimann & Whittaker, 1983; Galeotti *et al.*, in Bubik & Kaminski, 2004; Buzas, 1974; Lipps, in Buzas & Sen Gupta, 1982; Sen Gupta, in Buzas & Sen Gupta, 1982; Preece *et al.*, in Cameron *et al.*, 1999; Cetean *et al.*, 2011; Christiansen, 1958, 1971; Coccioni & Galeotti, 1993; Collinson, 1980; Corliss, 1985, 1991; Corliss & Chen, 1988; Corliss & Fois, 1990; Dobson & Haynes, 1973; Ellison, 1972; Frankel, 1974, 1975a; Gallagher, 1998; Gooday & Haynes, 1983; Grun *et al.*, 1964, Nagy *et al.*, in Haas & Kaminski, 1997; Nagy *et al.*, in Hart *et al.*, 2000; Haward & Haynes, 1976; Hedley, 1958, 1962; Hedley & Wakefield, 1967; Heron-Allen & Earland, 1913; Hottinger, 2000; Hughes, 2010a; Nagy & Murray, in Kaminski & Filipescu, 2011; Kitazato *et al.*, in Jorissen & Rohling, 2000; Kazmierczak, 1973; Lister, in Lankester, 1903; Lipps, 1976; Matera & Lee, 1972; Maurer & Rettori, 2002; Menzies *et al.*, 1973; Muller, 1975; Nagy, 1992, 2005; Nagy & Seidenkrantz, 2003; Nagy *et al.*, 2001, 2010, 2011; Kitazato, in Oertli, 1984; Winkler, in Oertli, 1984, Olsson, 1976; Reolid *et al.*, 2008a; Rosoff & Corliss, 1992; Salami, 1976; Jorissen, in Sen Gupta, 1999; Severin, 1983; Lipps, in Tevesz & McCall, 1983; Lemanska, in Tyszka *et al.*, 2005; van den Akker *et al.*, 2000; van der Zwaan *et al.*, 1999; Walker & Bambach, 1974.

## Systematics

Adl *et al.*, 2005; Cavalier-Smith, 1993, 2003, 2004; Bhattacharya *et al.*, in Hedges & Kumar, 2009; Ishitani *et al.*, 2011; Lecointre & le Guyader, 2006; Lee *et al.*, in Lee *et al.*, 2002; Lipps *et al.*, 2011; Sen Gupta, in Sen Gupta, 1999.

*Recent molecular biological work*

Archibald *et al.*, 2003; Aurahs *et al.*, 2009, 2011; Berney & Pawlowski, 2003; Darling & Wade, 2008; Darling *et al.*, 1996, 1997, 1999, 2000, 2003, 2006, 2007; de Vargas & Pawlowski, 1998; de Vargas *et al.*, 1997, 2001, 2002; Ertan *et al.*, 2004; Flakowski *et al.*, 2005; Holzmann & Pawlowski, in Goldstein & Bernhard, 1997; Habura *et al.*, 2006, 2008; Hayward *et al.*, 2004a; Holzmann, 2000; Holzmann & Pawlowski, 2000; Holzmann *et al.*, 1998, 2001, 2003; Keeling, 2001; Kucera *et al.*, 2010; Kuroyanagi *et al.*, 2008; Pawlowski *et al.*, in Langer, 1995; Langer, 1999; Langer & Leppig, 2000; Langer *et al.*, 1993; Longet *et al.*, 2003; Darling *et al.*, in Moguilevsky & Whatley, 1996; Morard *et al.*, 2009; Pawlowski, 2000; Pawlowski & Holzmann, 2002, 2008; Pawlowski *et al.*, 1994a, b; 1997; 1999a, b; 2002a, b, c; 2003a; Quillevere *et al.*, 2010; Schweizer *et al.*, 2005, 2008, 2009, 2011; Seears, 2011; Stewart, 2000; Stewart *et al.*, 2001; Tsuchiya *et al.*, 2008; Wade *et al.*, 1996; Weiner *et al.*, 2010.

# 4

# Ecology

This chapter deals with ecology. It contains sections on (general comments on) ecological distribution, tolerances and preferences; on bathymetric distribution; on sedimentary environments; on (bio)geographic distribution; on oceanography; and on environmental interpretation technologies. The section on bathymetric distribution includes sub-sections on marginal marine or paralic environments; on shallow marine or neritic environments; and on deep marine, or bathyal and abyssal, environments. The section on sedimentary environments includes sub-sections on marginal marine, peri-deltaic; shallow marine, peri-reefal; and deep marine, submarine fan sub-environments. The section on oceanography includes a sub-section on the organic carbon cycle and oxygenation.

## 4.1 Ecology

Foraminiferal ecology is the study of Foraminifera in relation to their environment, including their distribution, tolerances and preferences; their abundance, that is, the number of specimens per sample or unit volume; and their diversity, that is, the number of species per sample (Jones, 1994, 1996, 2006, 2011a; and additional references cited therein; see also the further reading list at the end of the chapter).

It is ideally based on live populations, identified by staining, in order to establish live ranges. This is because total populations contain potentially allochthonous or transported dead specimens as well as autochthonous or *in situ* live specimens (see the further reading list at the end of the chapter).

*General comments on ecological distribution, tolerances and preferences*

Studies undertaken to date have shown that the principal controls on the ecological distributions of Foraminifera are: water salinity and bathymetry (see Section 4.2); sedimentary environment and substrate (see Section 4.3); biogeography and temperature (see Section 4.4); and oceanography (see Section 4.5). Other important abiotic or physico-chemical controls include, in the case of the calcareous benthic and planktic Foraminifera, calcium carbonate availability and pH; and in the case of those larger benthics and planktics which harbour photosymbionts, light availability, in turn controlled by turbidity and turbulence. Important biotic or biological controls include food availability, predation, and competitive interaction and exclusion.

As a unit, Foraminifera exhibit broad ecological tolerances. For example, in relation to water salinity and depth, they are distributed throughout all aqueous environments, fresh-water as well as marine (Jones, 1994, 1996, 2006, 2011a; see also below, and the further reading list at the end of the chapter). And in the marine realm, they are distributed throughout all marginal, shallow and deep environments (to depths in excess of 11 000 m in the Challenger Deep in the Mariana Trench; see the further reading list at the end of the chapter).

In contrast, as individual species or higher-level taxa, they often have narrow tolerances and preferences. Taxa with narrow depth, salinity and temperature preferences are termed stenobathyal, stenohaline and stenothermal, respectively; those with broad depth, salinity and temperature preferences, eurybathyal, euryhaline and eurythermal, respectively. Stenohaline taxa, with narrow, normal marine salinity preferences include all Nodosariida and Robertinida. Eurybathyal, stenothermal taxa, with broad depth but narrow temperature preferences, include those Cassidulinidae (Buliminida) that occur in shallow water in high latitudes and deep water in low latitudes.

### *Proteinaceous Foraminifera (Allogromiida)*

The ecology of proteinaceous Foraminifera, or Allogromiida, remains comparatively poorly known, although a number of recent studies have highlighted the hitherto substantially overlooked numerical importance of the group in fresh-water as well as marine environments, and even in some cases in damp, essentially terrestrial environments (see 'further reading' list at the end of the chapter).

### *Agglutinating Foraminifera (Astrorhizida and Lituolida)*

Modern smaller agglutinating Foraminifera, or Astrorhizida and Lituolida, occur in all marine environments, from marginal to deep, and some are tolerant of hyposalinity as well as normal marine salinity; and/or of hypoxia or dysoxia (Charnock & Jones, 1997; see also Section 4.5.1, and the further reading list at the end of the chapter).

They appear better able than their calcareous benthic counterparts to tolerate conditions of high fresh-water flux, and of high sediment and organic carbon flux, and associated lowered oxygen availability (also of lowered alkalinity, although this may be, at least in part, a preservational phenomenon). They are especially characteristic of marginal marine, peri-deltaic and of deep marine, submarine fan sedimentary environments.

Modern larger Lituolida (Larger Benthic Foraminifera or LBFs) are shallow to deep marine, and are tolerant of only normal marine salinity.

### *Smaller and larger calcareous benthic Foraminifera (Miliolida, Nodosariida, Buliminida, Robertinida and Rotaliida)*

Smaller Miliolida range from marginal to deep, and some are tolerant of hyposalininity or hypersalinity as well as normal marine salinity (see the further reading list at the end of the chapter). They are especially characteristic of platformal and peri-reefal carbonate environments, including back-reef lagoonal sub-environments.

Nodosariida range from shallow to deep, and are tolerant of only normal marine salinity. They are characteristic of deep marine environments, and some are tolerant of suboxic environments (see Section 4.5.1).

Buliminida also range from shallow to deep, and are generally tolerant of only normal marine salinity, although at least some are tolerant of hyposalininity or hypersalinity as well). They are characteristic of dysoxic environments, for example Oxygen Minimum Zones (OMZs) in the upper bathyal (see Section 4.5.1). Some species have been recorded living in entirely anoxic environments, for example in the OMZ off Chile.

Robertinida range from shallow to deep marine, and are tolerant of only normal marine salinity. They are characteristic of deep marine environments, and some are tolerant of suboxic environments (see Section 4.5.1).

Smaller Rotaliida range from marginal to deep, and some are tolerant of hyposalinity or hypersalinity as well as normal marine salinity, and some others, including, for example, *Hansenisca* ['*Gyroidina*'], *Oridorsalis* and *Chilostomella*, of suboxia or of hypoxia or dysoxia (see Section 4.5.1). They are characteristic of comparatively shallow marine, neritic to upper bathyal environments.

Larger Miliolida and Rotaliida (LBFs) are shallow marine, and some are tolerant of hypersalinity as well as normal marine salinity (see the further reading list at the end of the chapter). They are especially characteristic of platformal and peri-reefal carbonate environments, including back-reef lagoonal sub-environments. They all harbour photosynthetic algal symbionts of various types.

### *Planktic Foraminifera (Globigerinida)*

Planktic Foraminifera, or Globigerinida, live in the water column and sink to the sea floor when they die, and some are tolerant of hyposalininity or hypersalinity as well as normal marine salinity (see the further reading list at the end of the chapter). They occupy a range of depth environments within the water column. At least some harbour photosynthetic algal symbionts of various types.

## 4.2 Bathymetric distribution

The bathymetric distributions of living Foraminifera in the marine realm are comparatively well documented in the marine biological and biological oceano-graphic literature (see, for example, Jones, 1994; and additional references cited therein; Fig. 4.1; see also the further reading list at the end of the chapter). Unfortunately, though, the various sampling, processing, staining (for 'live' speci-mens), picking, identification and counting procedures used do not necessarily all conform to the same standards, and therefore the resulting distribution data-sets are not necessarily all directly comparable. Nonetheless, distribution patterns typically allow the differentiation of at least three bathymetric zones, namely, marginal, shallow and deep marine (see below). Supplementary bathymetric information is provided by bathymetrically related morphological trends (see the further reading list at the end of the chapter).

### *Proteinaceous Foraminifera (Allogromiida)*

Proteinaceous Foraminifera, or Allogromiida, occur in fresh-water and marine environments.

### *Agglutinating Foraminifera (Astrorhizida and Lituolida)*

As noted above, modern smaller agglutinating Foraminifera, or Astrorhizida and Lituolida, occur in all marine environments, from marginal to deep, although larger

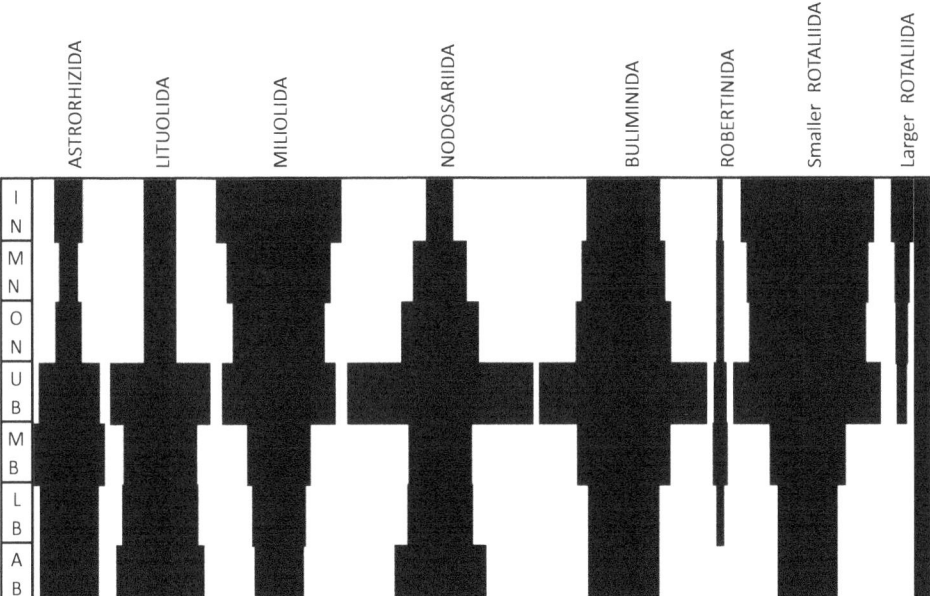

Fig. 4.1 **Specific-level diversity of Foraminifera in relation to bathymetry**. Based on data in Jones (1994). IN = inner neritic; MN = middle neritic; ON = outer neritic; UB = upper bathyal; MB = middle bathyal; LB = lower bathyal; AB = abyssal.

Lituolida (LBFs) are restricted to shallow to deep marine environments (Jones, 1994; but see also comments in Section 4.1). They attain their maximum diversity in, and are arguably most characteristic of, deep marine, bathyal and abyssal environments. Indeed, one family, the Verneuilinidae, is restricted to deep marine environments. However, some other families, most notably the Astrorhizidae, occur in some numbers in shallow marine, including inner neritic, environments, and indeed the Ataxophragmiidae are entirely restricted to shallow marine environments. Moreover, the Trochamminidae occur in some numbers in marginal marine environments.

### Smaller and larger calcareous benthic Foraminifera
### (Miliolida, Nodosariida, Buliminida, Robertinida and Rotaliida)

Smaller Miliolida range from marginal to deep marine (Jones, 1994). Their diversity is approximately inversely proportional to depth. They attain their maximum diversity in, and are characteristic of, shallow marine, inner neritic environments. However, some families, such as the Ophthalmidiidae, are actually restricted to deep marine environments.

Nodosariida range from shallow to deep marine (Jones, 1994). Their diversity is approximately proportional to depth in shallow marine, neritic environments. They attain their maximum diversity in, and are characteristic of, deep marine, upper bathyal environments. Indeed, some families, such as the Lingulinidae and Plectofrondiculariidae, are entirely restricted to deep marine, bathyal to abyssal environments. However, other families, most notably the Polymorphinidae, occur in some numbers in shallow marine, neritic environments.

Buliminida also range from shallow to deep marine (Jones, 1994). Their diversity is approximately proportional to depth in shallow marine, neritic environments, and inversely so in deep marine, bathyal to abyssal ones. They attain their maximum diversity in, and are characteristic of, deep marine, upper bathyal environments. Indeed, the Pleurostomellidae are entirely restricted to deep marine, bathyal to abyssal environments. However, other families, most notably the Turrilinidae, Pavoninidae and Millettiidae, are restricted to the shelf to upper slope.

Robertinida range from shallow to deep marine, neritic to bathyal, although not abyssal environments (below the Aragonite Compensation Depth of approximately 2000 m) (Jones, 1994). Their diversity is approximately proportional to depth in shallower marine environments, and inversely so in deeper ones. They attain their maximum diversity in, and are characteristic of, deep marine, middle bathyal environments.

Smaller Rotaliida range from marginal to deep marine (Jones, 1994). Their diversity is approximately inversely proportional to depth. They attain their maximum diversity in, and are characteristic of, comparatively shallow marine, neritic to upper bathyal environments. Indeed, some families, such as the Discorbidae, Glabratellidae, Eponididae, Epistomariidae, Planorbulinidae, Acervulinidae, Nonionidae and Elphidiidae are essentially entirely restricted to comparatively shallow marine environments. However, other families, such as the Alabaminidae, Anomalinidae and Chilostomellidae, are more diverse in, and characteristic of, deeper marine, bathyal to abyssal environments. Moreover, some genera of the essentially shallow marine Discorbidae (*Gavelinopsis*, *Laticarinina*), Eponididae (*Ioanella*, *Oridorsalis*) and Nonionidae (*Melonis*), are actually more characteristic of deep than of shallow marine environments.

Larger Miliolida and Rotaliida (LBFs) are restricted to shallow marine environments (Jones, 1994). They attain their maximum diversity in, and are characteristic of, inner neritic environments. They all harbour photosymbionts of various types that restrict them to the photic zone, that is, no deeper than 130 m in clear water, and shallower still in turbid or turbulent water (light penetration, determined by water clarity, turbidity and turbulence, is actually the primary control on their distribution rather than depth as such) (see the further reading list at the end of the chapter). Larger Miliolida, which harbour chlorophyte, rhodophyte, pyrrhophyte or bacilliarophyte (green or red algal or diatom) symbionts, are characteristic of the euphotic and sub-euphotic zones; larger Rotaliida, which harbour bacillariophyte (diatom) symbionts, of the sub-euphotic zone. (Conical or globular morphotypes are also characteristic of the euphotic zone, and discoidal or flattened ones of the sub-euphotic zone.) Interestingly, the tests of the larger Rotaliida are more enriched in photosynthetic carbon-12 (or impoverished in carbon-13) than those of the larger Miliolida, arguably indicating a greater photosynthetic capacity (another possibility is that the larger Rotaliida use carbon-12-enriched photosynthetic carbon from their symbionts in test construction, whereas the larger Miliolida use carbon from sea water). This is consistent with the observation that the larger Rotaliida are characteristic of oligotrophic, and the larger Miliolida of mesotrophic to eutrophic, environments, such that the larger Rotaliida are more, and the larger Miliolida less, reliant on their symbionts as a food source. In other words, the symbiosis may be obligate in the larger Rotaliida and facultative in the larger Miliolida.

### *Planktic Foraminifera (Globigerinida)*

Planktic Foraminifera, or Globigerinida, live in the water column and sink to the sea floor when they die (Jones, 1994; see also the further reading list at the end of the chapter). They occupy a range of depth environments within the water column, including shallow, or epipelagic; intermediate, or mesopelagic; and deep, or bathypelagic. At least some harbour photosymbionts of various types that restrict them to the photic zone. Most Globigerinidae and Hastigerinidae, which harbour dinophyte symbionts, are characteristic of the euphotic zone (epipelagic and mesopelagic environments); some Globigerinidae (*Candeina, Pulleniatina*) and Hastigerinidae, and all Globorotaliidae, which harbour chrysophyte symbionts, of the sub-euphotic zone (bathypelagic environment).

### 4.2.1 Marginal marine or paralic environments

Modern marginal marine or paralic environments are characterised by Lituolida, Miliolida and Rotaliida, and to a lesser extent by Nodosariida and Buliminida (Jones, 1994, 1996, 2006, 2011a; Jones & Whittaker, in Whittaker & Hart, 2010; see also the further reading list at the end of the chapter). Lituolida include Lituolidae (*Ammoastuta, Ammobaculites, Ammomarginulina, Ammotium, Haplophragmoides, Reophax*), Textulariidae (*Textularia*), Trochamminidae (*Arenoparrella, Entzia [Jadammina], Tiptotrocha, Trochammina, Trochamminita*) and Eggerellidae (*Eggerelloides, Pseudoclavulina*); Miliolida, Cornuspiridae (*Cornuspira*) and Miliolidae (*Miliammina, Miliolinella, Quinqueloculina*); Rotaliida, Discorbidae (*Discorbis, Discorinopsis, Glabratella*), Anomalinidae (*Cibicides*), Nonionidae (*Nonion*), Elphidiidae (*Cribroelphidium, Elphidium*) and Rotaliidae (*Ammonia, Asterorotalia, Aubignyna, Pseudoeponides*); Nodosariida, Glandulinidae (*Fissurina*); and Buliminida, Buliminidae (*Stainforthia*) and Bolivinitidae (*Bolivina*). One species, *Cribroelphidium gunteri*, has been recorded in salinities locally as low as 1–2 parts per thouasand (ppt) in Lake Winnipegosis in Manitoba in Canada, into which it may have been introduced by migratory birds in the Holocene (see the further reading list at the end of the chapter).

Foraminiferal abundance is typically at least locally high in marginal marine environments, although diversity is typically comparatively low, and dominance correspondingly comparatively high (see, for instance, Murray, 1973; and additional references cited therein).

Various marginal marine sub-environments can be distinguished, including, generally in relation to depth and tidal range, supratidal, intertidal and subtidal, including paludal ((salt-)marsh and, in low latitudes, mangrove swamp), estuarine and lagoonal; and specifically in relation to sedimentary environment, peri-deltaic (see further reading list at the end of the chapter). Marginal marine sub-environments are also differentiable on the basis of the shell chemistry of Foraminifera.

### Salt-marsh and mangrove swamp sub-environments

Salt-marsh and mangrove swamp sub-environments are characterised by low to high abundance, and low to moderate diversity, assemblages of Foraminifera (generally low abundance and diversity of dominantly agglutinating Foraminifera at the high ends, and high abundance, although still only low to moderate diversity, of agglutinating and calcareous benthic Foraminifera at the low ends) (see the further reading list at the end of the chapter).

Salt-marshes and mangrove swamps are extremely sensitive indicators of sea-level, and are important proxies for – empirically- or transfer-function-based – sea-level reconstructions, especially for the Quaternary (see Sections 11.2.1 and 13.1.2).

### *Estuarine sub-environments*

Estuarine sub-environments are also characterised by low to high abundance, and low to moderate diversity, assemblages of Foraminifera (generally low abundance and diversity of dominantly agglutinating Foraminifera at the inner ends, and high abundance, although still only low to moderate diversity, of agglutinating and calcareous benthic Foraminifera at the outer ends) (see the further reading list at the end of the chapter).

### *Lagoonal sub-environments*

Lagoonal sub-environments are similarly characterised by low to high abundance, and low to moderate diversity, assemblages of Foraminifera; hypersaline ones especially by Miliolida, often apparently associated with sea-grasses, as in the Caribbean, Gulf of Mexico, Mid-East Gulf and Indo-Pacific (see the further reading list at the end of the chapter).

## 4.2.2 Shallow marine or neritic environments

Modern shallow marine or neritic (shelfal) environments are characterised by Miliolida and Rotaliida, and to a lesser extent by Astrorhizida, Lituolida, Nodosariida, Buliminida and Robertinida (Jones, 1994, 1996, 2006, 2011a; Jones & Whittaker, in Whittaker & Hart, 2010; see also the further reading list at the end of the chapter).

Foraminiferal abundance is typically at least locally high in shallow marine environments, and diversity is typically also high, and dominance correspondingly low. Note, though, that diversity is lower in hyposaline shallow marine environments, that is, those characterised by less than normal marine salinity, such as, for example, the Baltic Sea.

Various shallow marine sub-environments can be distinguished, including, generally in relation to depth, inner and outer neritic, and specifically in relation to sedimentary environment, peri-reefal.

### *Inner neritic sub-environments*

Modern inner neritic sub-environments are characterised especially by smaller and larger Miliolida and Rotaliida.

Modern outer neritic sub-environments are characterised especially by smaller Miliolida and Rotaliida, and to a lesser extent by Astrorhizida, Lituolida, Nodosariida, Buliminida and Robertinida.

### 4.2.3  Deep marine, or bathyal and abyssal, environments

Modern deep marine or bathyal and abyssal (slope and abyssal plain) environments are characterised by Astrorhizida, Lituolida, Nodosariida, Buliminida and Robertinida, and to a lesser extent by smaller Miliolida and Rotaliida (Jones, 1994, 1996, 2006, 2011a; see also the further reading list at the end of the chapter). Astrorhizida and Lituolida attain their maximum diversity in, and are most characteristic of, deep marine, and especially submarine fan and associated environments, with the Verneuilinidae entirely restricted to deep marine environments. Nodosariida attain their maximum diversity in, and are most characteristic of, deep marine environments, with the Lingulinidae and Plectofrondiculariidae entirely restricted to deep marine environments. Buliminida attain their maximum diversity in, and are most characteristic of, deep marine, especially OMZ environments on the upper slope, with the Pleurostomellidae entirely restricted to deep marine environments. Robertinida attain their maximum diversity in, and are most characteristic of, deep marine, bathyal environments, although they are not found in abyssal environments (below the Aragonite Compensation Depth of approximately 2000 m). (Smaller) Miliolida attain their maximum diversity in, and are most characteristic of, shallow marine, inner neritic environments, although the Ophthalmidiidae are actually restricted to deep marine environments. Smaller Rotaliida attain their maximum diversity in, and are most characteristic of shallow marine environments, although the Alabaminidae, Anomalinidae and Chilostomellidae are more characteristic of deep marine environments, as are some genera of the essentially shallow marine Discorbidae (*Gavelinopsis*, *Laticarinina*), Eponididae (*Ioanella*, *Oridorsalis*) and Nonionidae (*Melonis*).

Foraminiferal abundance is typically at least locally high in deep marine environments, and diversity is typically also high, and dominance correspondingly low.

Various deep marine sub-environments can be distinguished, including, generally in relation to depth, upper, middle and lower bathyal, abyssal, cold (hydrocarbon) seep, hydrothermal vent and 'nekton fall' (see below); and, specifically in relation to sedimentary environment, submarine fan (see 4.3.3).

Modern upper bathyal sub-environments are characterised especially by Astrorhizida, Lituolida, Nodosariida, Buliminida and Robertinida, and to a lesser extent by smaller Miliolida and Rotaliida; Oxygen Minimum Zones (OMZ) especially by Buliminida (see also below).

### Middle bathyal sub-environments

Modern middle bathyal sub-environments are characterised especially by Astrorhizida, Lituolida, Nodosariida, Buliminida and Robertinida, and to a lesser extent by smaller Miliolida and Rotaliida.

### Lower bathyal sub-environments

Modern lower bathyal sub-environments are characterised especially by Astrorhizida, Lituolida, Nodosariida and Buliminida, and to a lesser extent by smaller Miliolida and Rotaliida, and by Robertinida. Robertinida are not found below the Aragonite Compensation Depth of approximately 2000 m.

### Abyssal sub-environments

Modern abyssal sub-environments are characterised especially by Astrorhizida and Lituolida, and to a lesser extent by smaller Miliolida and Rotaliida, Nodosariida and Buliminida. Smaller Miliolida and Rotaliida, Nodosariida and Buliminida are not found below the Calcite Compensation Depth (CCD) of approximately 4000 m.

### Cold (hydrocarbon) seeps

Modern cold seeps (and apparently associated coral reefs), which are essentially restricted to deep marine environments, are characterised by opportunistic Foraminifera either directly or indirectly exploiting the available geochemical energy source (Jones, 1992; Jones, in Jenkins, 1993; Jones, 1996, 2006, 2011a; see also the further reading list at the end of the chapter).

Although the results of the various studies of the Foraminifera associated with seeps are not all directly comparable, as the studies did not involve directly comparable analytical and interpretive procedures, the following observations and comparisons can be offered:

- Samples from seep and control sites are distinct when compared with one another using specific-level similarity indices.
- In the North Sea and Gulf of Mexico, samples from seeps are distinct in terms of lower abundance and diversity of Foraminifera, higher dominance, and lower equitability (that is, evenness of species distribution). However, in the Blake Ridge area of the western North Atlantic, and in the Adriatic in the Mediterranean, samples from seeps proved distinct in terms of higher abundance of Foraminifera, one possible explanation being that the bacterial mat associated with the seeps here provides an additional food source. Moreover, in the Porcupine basin in the North Atlantic, samples from seeps proved distinct in terms of lower abundance, but higher diversity, of Foraminifera, one possible explanation for the higher diversity being that the hard substrate provided by the carbonates associated with the seeps provides niches for a wider variety of

species. Note in this context that attached species are significantly both more abundant and more diverse in the seep samples. Note also that species attached to vestimentiferan tube-worms well above the sediment-water interface, possibly in order to avoid the toxic conditions there, have recently been recorded at seep sites in the Gulf of Mexico.

- In the North Sea, Gulf of Mexico, North Atlantic and North Pacific, samples from seeps are distinct in terms of proportionately higher abundances of epifaunal Foraminifera, and lower abundances of infaunal Foraminifera.
- In some studies, the epifaunal Rotaliida *Cibicidoides pachyderma*, *Elphidium ex gr. clavatum* and *Hyalinea balthica* appear positively correlated with seepage at more than one site, as, incidentally, do the epifaunal agglutinating Foraminifer *Arenomeandrospira glomerata*; and the infaunal Buliminida *Bulimina marginata* and *Trifarina angulosa* and the infaunal Rotaliida *Chilostomella oolina* appeared negatively correlated with seepage at more than one site.
- However, in other studies, the infaunal Buliminida *Cassidulina laevigata carinata s.l.* and *Rutherfordoides cornuta* appear positively correlated with seepage (or at least able to tolerate the toxicity, hypoxia or dysoxia, and either hypo- or hyper- salinity associated with it).
- Moreover, in other studies on California and Oregon in the North Pacific, and in the Adriatic in the Mediterrranean, the infaunal Buliminida *Bolivina subargentea* and *Uvigerina peregrina* appear on the basis of carbon isotopic evidence to be actually metabolising seeping methane, either directly or indirectly through the incidental ingestion of the bacterium *Beggiatoa*.

### *Hydrothermal vents and 'nekton falls'*

Like cold seeps (see above), hydrothermal vents and nekton falls support benthic meio- and macro- biotas that ultimately depend for their survival on chemo-autotrophic or chemosynthetic Bacteria rather than photosynthetic plants or Algae, and thus on a geochemical rather than a biochemical energy source, and are essentially only found where geochemical energy sources are available on the sea floor. In recent years there have been several studies on benthic Foraminifera associated with modern hydrothermal vents and nekton falls (Jones, 2006, 2011a; see also the further reading list at the end of the chapter).

## 4.3 Sedimentary environments

### *4.3.1 Marginal marine peri-deltaic sub-environments*

Peri-deltaic sub-environments are characterised by low to high abundance, and low to moderate diversity, assemblages of Foraminifera (generally low abundance and diversity of dominantly agglutinating Foraminifera on the delta top, and high abundance, although still only low to moderate diversity, of agglutinating and calcareous benthic Foraminifera on the delta front and prodelta) (Jones, 1996, 2006, 2011a; see also the further reading list at the end of the chapter).

In the case of the modern Ebro delta, onshore and offshore areas are distinguishable on the bases of the benthic Foraminifera that they contain, and also of triangular cross-plots of wall structure groups of the same, namely, agglutinating (Textulariina of Loeblich & Tappan, 1964 = Astrorhizida and Lituolida of Haynes, 1981); porcelaneous (Miliolina = Miliolida); and hyaline (Rotaliina = Nodosariida, Buliminida, Robertinida and Rotaliida).

In the case of the modern Mississippi delta, channels, levees and interdistributary bays on the delta top are distinguishable on the bases of the agglutinating Foraminifera that they contain, and of triangular cross-plots of 'morphogroups' of the same.

### 4.3.2 Shallow marine, peri-reefal sub-environments

Peri-reefal sub-environments are characterised by smaller and, especially, larger Miliolida and Rotaliida (Jones, 1996, 2006, 2011a; see also the further reading list at the end of the chapter). Miliolida include Soritidae (*Amphisorus*, *Archaias*, *Marginopora*, *Peneroplis*, *Sorites*) and Alveolinidae (*Alveolinella*, *Borelis*); Rotaliida, Elphidiidae (*Elphidium*), Rotaliidae (*Asterorotalia*, *Pararotalia*, *Pseudorotalia*, *Rotalinoides*), Calcarinidae (*Baculogypsina*, *Calcarina*), Nummulitidae (*Operculina*, *Operculinella*) and Cycloclypeiidae (*Heterostegina*). The larger Miliolida and Rotaliida (LBFs) all harbour photosymbionts of various types that restrict them to the photic zone. Larger Miliolida, which harbour green or red algal or diatom symbionts, are characteristic of the euphotic and sub-euphotic zones, and of back-reef, including lagoonal, sub-environments; larger Rotaliida, which harbour diatom symbionts, of the sub-euphotic zone, and of reef and fore-reef sub-environments.

Wilson's standard facies belts are distinguishable on the bases of the Foraminifera that they contain, and of triangular cross-plots of morphogroups of the same.

### 4.3.3 Deep marine, submarine fan sub-environments

Modern submarine fan sub-environments are characterised by autochthonous deep marine benthic Foraminifera, especially by Deep-Water Agglutinating Foraminifera or DWAFs; by allochthonous, but contemporaneous, shallow marine benthic and planktic Foraminifera; and, occasionally, by allochthonous, and non-contemporaneous, Foraminifera (Jones, 1996, 2006, 2011a; see also the further reading list list at the end of the chapter).

Fan channel, levee and overbank, and respectively corresponding turbiditic, inter-turbiditic and hemipelagic sub-environments, have been described from a wide range of modern fan complexes. Fan channel or turbiditic sub-environments are characterised by interpreted allochthonous Foraminifera only. Levee and over-bank, or inter-turbiditic and hemipelagic, sub-environments are characterised by increasing incidences of autochthonous Foraminifera, especially DWAFs. They are also, as for example in the El Huco system, where the organic carbon flux is high and oxygenation correspondingly low, locally and exceptionally characterised by calcareous benthics such as the opportunistic infaunal detritus-feeding and dysoxia-tolerant Buliminida *Cassidulina* and *Globobulimina*.

Debris-flow deposits or debrites in the Nile deep-sea fan are characterised by autochthonous and allochthonous Foraminifera (Ducassou *et al.*, 2013). The allochthonous ones are derived from water depths up to 2000 m shallower, and 200 km away.

## 4.4 (Bio)geographic distribution

Many groups of organisms, including Foraminifera, have restricted or endemic biogeographic distributions, which render them of use in the characterisation of palaeobiogeographic realms, or of provinces within these realms (Jones, 1994; Jones, in Agusti *et al.*, 1999; Jones, 2006; Whidden *et al.*, 2009; Jones, 2011a; Whidden & Jones, 2012; and additional references cited therein; see also the further reading list at the end of the chapter).

As a whole, LBFs are restricted to the low-latitude Indo-Pacific, Mediterranean or American/Caribbean provinces, which last-named, incidentally, was essentially contiguous with the Indo-Pacific prior to the closure of the Strait of Panama in the Pliocene. Their overall geographic distribution appears to be limited by sea-surface temperature, with a minimum temperature of 18 °C apparently required for reproduction. Note in this context also that their diversity is at least approximately proportional to temperature, with a high proportion (94%) of high diversity sites, a moderate proportion (65%) of moderate diversity sites, and a low proportion (47%) of low diversity sites in waters warmer than 25 °C; and that it can therefore be used as a proxy for temperature, and indeed palaeotemperature (see also Section 5.7). Note also, though, that LBF diversity is sensitive not only to temperature but also other factors such as facies, evolution, dispersal/migration, duration of studied interval and/or size of studied area, sampling artefact and taxonomic artefact.

And while agglutinating Foraminifera, smaller calcareous benthics and plank-tics as a whole are cosmopolitan, many individual forms have restricted, provincial distributions (see the further reading list at the end of the chapter). Nine biogeo-graphic provinces can be identified on the basis of planktic Foraminifera, namely, from north to south, Arctic, Subarctic, northern-hemisphere transitional, northern-hemisphere sub-tropical, tropical, southern-hemisphere sub-tropical, southern-hemisphere transitional, sub-Antarctic, and Antarctic. Polar and sub-polar assemblages are characterised by low diversity, transitional assemblages by inter-mediate, and sub-tropical and tropical assemblages by high diversity. Present diversity gradients thus parallel temperature gradients, and are correspondingly steep.

## 4.5 Oceanography

Oceanography is the science of the seas, and of their physics, chemistry and biology. The science may be said to have come of age with the publications of the *Reports of the Scientific Results of the Voyage of H.M.S.* Challenger, including the report on the Foraminifera published by Brady in 1884 (see Jones, 1994, and also the further reading list at the end of the chapter). Much of the more recent oceanographic work undertaken on the Foraminifera has been directed toward the understanding of biogeochemistry, in particular the organic carbon cycle (see below). Some studies, though, have focussed on water masses and upwelling, or on the carbonate cycle (see, for example, Jones & Pudsey, 1994; see also the further reading list at the end of the chapter). Jones & Pudsey (1994) found foraminiferal assemblages associated with 'Fresh Shelf Water' on the Antarctic Peninsula shelf to be characterised by agglutinating and calcareous benthic Foraminifera, including Hormosinidae, *Quinqueloculina* and *Trifarina angulosa*; ones associated with 'Circum-Polar Deep Water' (CPDW) by agglutinating and calcareous benthic Foraminifera, including *Pyrgo murrhina*, *Uvigerina* and *Cibicides ex gr. lobatulus*, and ones associated with 'Antarctic Bottom Water' (AABW) by agglutinating Foraminifera only, including Astrorhizidae, *Cribrostomoides subglobosus* and *Martinottiella communis*.

### *4.5.1 Organic carbon cycle and oxygenation*

Organic carbon or OC is produced by primary biological productivity in the terrestrial and marine realms (see the further reading list at the end of the chapter). Note in this context that there is also a significant input of OC of terrigenous origin into the marine realm, associated not only with wind- and river-borne sediments, but also with turbidites, and with hyperpycnal turbidites generated by earthquakes and landslides on steep-sided oceanic islands, and by storms. In the terrestrial realm, the production of OC is controlled by the productivity of land plants, and is typically highest in places or at times of optimal climate. In the marine realm, the production of OC is controlled by the productivity of marine Algae, and is highest in places or at times of optimal climate, or of enhanced nutrient supply associated with riverine input, or with upwelling. Present-day upwelling is indicated by phytoplankton blooms on satellite images. There are various models for upwelling. *Zonal upwelling* is driven by currents, for example along the 'equatorial divergence' in low latitudes or 'polar divergence' in high latitudes. *Meridional upwelling* is driven by winds, for example in the trade-wind and monsoon belts in low latitudes. The north-easterly trades drive north-westerly directed surface currents in the northern hemisphere, and the south-easterly trades south-westerly directed surface currents in the southern hemisphere, as a result of a combination of Coriolis and Ekman (frictional drag) effects. The westerly directed component of the surface currents in turn drives easterly directed upwelling of deep currents along the western margins of landmasses in both hemispheres.

Organic carbon is preserved, and organic-rich sediments produced, when there is an excess of production over consumption (see the further reading list at the end of the chapter). Consumption and preservation of OC is essentially controlled by the availability of oxygen, and consumption is lowest, and preservation highest, in places of dysoxia or anoxia, such as the Black Sea or Indian Ocean. In the case of the Indian Ocean, consumption is particularly low, and preservation particularly high, in the particularly well developed OMZ there.

Modern organic-rich/oxygen-poor environments and sediments, including those of OMZs, are characterised especially by Buliminida, together with certain agglutinating Foraminifera and Rotaliida (see the further reading list at the end of the chapter; see also Sections 5.6.1 and 9.1.1).

Oxygen-poor environments have been sub-divided in the micropalaeontological literature into anoxic sub-environments, indicated by dissolved oxygen concentrations of 0–0.1 ml/l; dysoxic or hypoxic sub-environments, indicated by dissolved oxygen concentrations of 0.1–0.3 ml/l; and suboxic sub-environments, indicated by dissolved oxygen concentrations of 0.3–1.5 ml/l (see, for example, Kaiho, 1991, 1994, 1999; see also the further reading list at the end of the chapter). Benthic Foraminifera (most of them, incidentally, infaunal) are found in anoxic, dysoxic and suboxic sub-environments. Only certain Buliminida and smaller Rotaliida are found in anoxic environments. Certain agglutinating Foraminifera, Buliminida and smaller Rotaliida are found in dysoxic environments. Certain agglutinating Foraminifera, Nodosariida, Buliminida, Robertinida and smaller Rotaliida are found in suboxic environments. (All groups of Foraminifera are found in oxygen-rich or oxic environments, apart from those certain Buliminida specifically adapted to anoxic environments.) Certain agglutinating Foraminifera are characteristic of dysoxic biofacies associated with riverine input, 'fresh-water overhang', salinity stratification and bottom-water stagnation in marginal marine, peri-deltaic and associated, environments. Certain Buliminida are characteristic of dysoxic biofacies associated with OMZs in deep marine, especially upper-slope, environments.

Indications of the degree of oxygenation of the environment are provided by the Benthic Foraminiferal Oxygen Index or BFOI (Kaiho, 1991, 1994, 1999). The

BFOI is given by the formula BFOI $= \{[S/(S + D] - 1\} \times SO$, where S is suboxic, and D, dysoxic. S includes the Nodosariida *Dentalina*, *Lenticulina* and *Nodosaria*; the Buliminida *Bulimina* and *Stilostomella*; the Robertinida *Hoeglundina*; and the smaller Rotaliida *Hansenisca* ['*Gyroidina*'] and *Oridorsalis*. D includes the Buliminida *Fursenkoina*, *Globobulimina*, *Rutherfordia*, *Bolivina*, *Suggrunda* and *Cassidulina*; and the Rotaliida *Chilostomella*. Indications of the degree of oxygenation are also provided, at least in the case of the Mississippi, by the *Ammonia/Elphidium* or A/E Index (Sen Gupta *et al.*, 1996; Platon & Sen Gupta, in Rabalais & Turner, 2001), and by the *Pseudononion–Epistominella–Buliminella* or PEB Index (Osterman *et al.*, 2001; Osterman, 2003; Osterman *et al.*, 2005, 2009; Osterman *et al.*, 2009). Further indications are provided by patterns of pore density in selected calcareous benthics (Kuhnt *et al.*, 2010).

Anthropogenically meditated eutrophication and anoxia is discussed in Section 12.2.1.

## 4.6 Environmental interpretation technologies

A range of quantitative techniques has been applied in the field of environmental interpretation, including abundance, diversity, dominance and equitability; artificial neural networks (ANNs) and self organising maps (SOMs); eigenvector and multivariate techniques, including cluster analysis; planktic:benthic (P:B) ratios; and similarity indices (see Jones, 1996, 2006, 2011a; and additional references cited therein; see also below, and the further reading list at the end of the chapter).

*Abundance, diversity, dominance and equitability*

Abundance refers to the number of specimens in a sample, preferably normalised to sample size; and standing stock to the number of specimens in a unit volume. The abundance of individual species is expressed either in absolute terms, or in relative terms, that is, as a percentage of the total abundance. Diversity refers to the number of species. It is measured and expressed as simple, Fisher, Shannon–Wiener, Simpson or SHE diversity. In its simplest form, dominance refers to the relative abundance of the dominant species. Equitability refers to the evenness of distribution of species in a sample.

Abundance, diversity, dominance and equitability are useful guides to environment (see the further reading list at the end of the chapter). Abundance and diversity are typically highest in deep water, although depth is only one of a number of variables controlling them. Note also that abundance and diversity are not always co-variant. High abundance associated with low diversity often indicates some form of environmental stress, such as temperature, salinity, oxicity or toxicity, as do high dominance and low equitability. Fisher diversity has been used

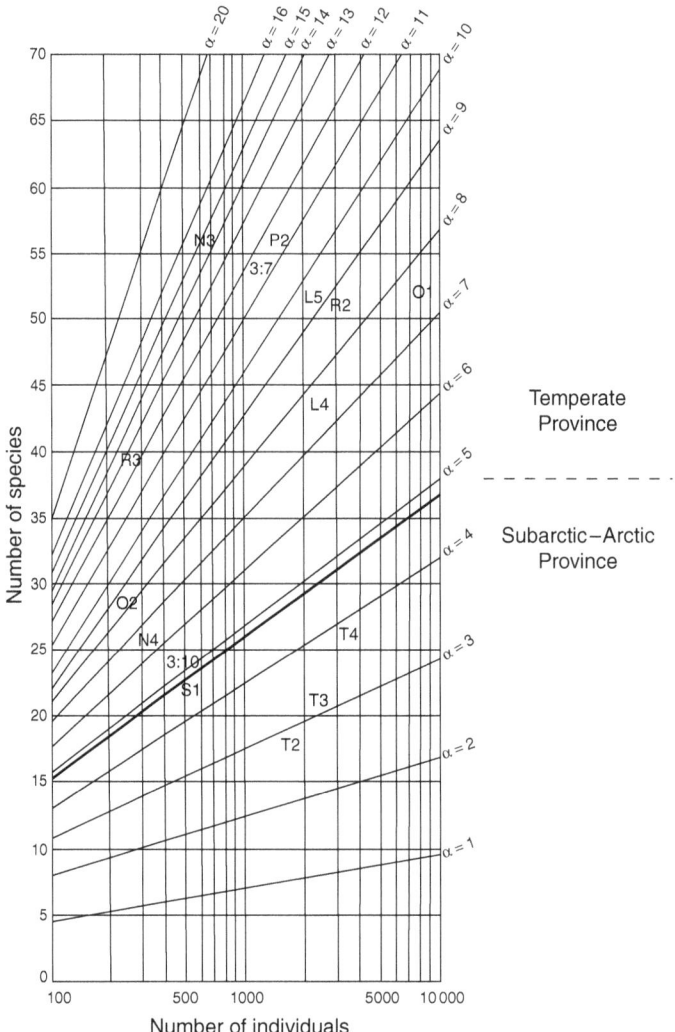

Fig. 4.2 **Example of Fisher (α) diversity plot**. Here used to distinguish bio-geographic provinces in the north-east Atlantic and Arctic. T2, T3, T4 etc. are studied sample. Sites with Fisher diversity > 5 (bold line) distinguish the Temperate Province; those with diversity < 5, the Subarctic–Arctic Province (see Jones & Whittaker, in Whittaker & Hart, 2010; see also Section 13.1.1 and Fig. 13.1). From Jones (1983).

to distinguish biofacies, bathymetric zones and biogeographic provinces (see, for example, Jones, 1983; Fig. 4.2.).

*Artificial neural networks (ANNs) and self organising maps (SOMs)*

These networks and maps have been used in studies on benthic foraminiferal distributions (see the further reading list at the end of the chapter).

### *Eigenvector and multivariate techniques*

Eigenvector and multivariate techniques include cluster analysis, correspondence analysis, factor analysis and polytopic vector analysis. Various of the techniques have been used for biofacies identification (see the further reading list at the end of the chapter). For example, hierarchical R- and Q-mode cluster analysis has been used to identify four biofacies associated with particular sedimentary environments on the modern Po delta.

### *Planktic:benthic (P:B) ratios*

The planktic:benthic (P:B) ratio tends to increase with increasing depth, or distance from shoreline, and has been used as a measure of the same (see the further reading list at the end of the chapter). Note, though, that the trend toward a higher ratio in deeper water is non-linear, such that values are not necessarily unique to any particular bathymetry or bathymetric zone. For example, the ratio is anomalously low in those parts of oceans outboard of major rivers, where planktic foraminiferal productivity is low, because salinity is low (on account of the effect of the 'freshwater plume'). It is also anomalously low in OMZs on the upper parts of continental slopes, where benthic foraminiferal productivity is high, because meiofaunal predation is low. And around the lysocline and Calcite Compensation Depth (CCD) in the deep sea, where planktic foraminiferal preservation is lower than benthic foraminiferal preservation (because planktics have a higher susceptibility to dissolution than benthics, on account of their higher surface area with respect to volume).

### *Similarity indices*

Similarity indices have been used in biofacies, bathymetric zone or biogeographic province identification (see, for example, Jones, 1983).

## 4.7 Further reading

### *General*

Akimoto *et al.*, 2001; Bandy & Arnal, 1960; Berger, 1969; Boltovskoy & Wright, 1976; BouDagher-Fadel, 2008, 2012; Buzas, 1968; Buzas & Sen Gupta, 1982; Buzas *et al.*, 1993; Frerichs, 1970; Be & Tolderlund, in Funnell & Riedel, 1971; Gooday *et al.*, 2010a; Gooday *et al.*, 2008b; Larkin *et al.*, in Hall *et al.*, 2010; Hartman *et al.*, 2012; Hayward, 1986, 1990; Myers & Cole, in Hedgpeth, 1957; Hemleben *et al.*, 1989; Jorissen *et al.*, in Hillaire-Marcel & de Vernal, 2007; Kucera, in Hillaire-Marcel & de Vernal, 2007; Kalogeropoulou *et al.*, 2010; Glover *et al.*, in Lesser, 2010; Linke & Lutze, 1993; Lipps *et al.*, 1979; Murray, 1973, 1991, 2000, 2001, 2006; Murray & Alve, 2011; Myers, 1942c; Phleger,

1960; Be, in Ramsay, 1977; Schonfeld, 2012; Sen Gupta, 1977, 1999; Todo *et al.*, 2005; Wilson & Horton, 2012.

### *Ecological distribution – proteinaceous Foraminifera*

Cedhagen *et al.*, 2002; da Silva *et al.*, 2006; Gooday, 2002b; Gooday & Aranda da Silva, 2009; Gooday & Bowser, 2005; Gooday *et al.*, 2000b, 2001a, 2004a, 2004b, 2005, 2006, 2008a, 2009a, 2010b, 2011, Gooday & Pawlowski, 2004; Holzmann & Pawlowski, 2002; Larkin & Gooday, 2004; Meisterfeld *et al.*, 2001; Pawlowski *et al.*, 1999a, b, 2002b, c, 2010; Rothe *et al.*, 2011; Sabbatini *et al.*, 2002, 2007, 2010; Sergeeva *et al.*, 2005, 2010.

### *Agglutinating Foraminifera*

Nagy *et al.*, 2010.

### *Calcareous benthic Foraminifera*

Beavington-Penney & Racey, 2004; BouDagher-Fadel, 2008; Brasier & Donahue, 1985; Brasier & Green, 1993; Burollet *et al.*, 1986; Debenay, 1988; Mateu-Vicens *et al.*, in Demchuk & Gary, 2009; Hallock, 1988; Hallock & Glenn, 1986; Hallock & Peebles, 1993; Hallock Muller, 1976; Hohenegger, 2005; Levy *et al.*, 1988; Poag & Tresslar, 1981; Reiss & Hottinger, 1984; Saraswati, 2010; Hallock, in Sen Gupta, 1999; Sen Gupta, in Sen Gupta, 1999; Truffleman *et al.*, 1991.

### *Planktic Foraminifera*

Kucera, in Hillaire-Marcel & de Vernal, 2007; BouDagher-Fadel, 2012.

### *Bathymetric distribution – general*

Boltovskoy *et al.*, 1991; Hemleben *et al.*, 1989; Reiss & Hottinger, 1984; Hallock, in Sen Gupta, 1999; van Marle, 1988, 1991; van Morkhoven *et al.*, 1986.

### *Marginal marine environments (including peri-deltaic sub-environments)*

Adegoke *et al.*, 1976; Barbosa *et al.*, 2005; Billman *et al.*, 1980; Boltovskoy, 1984; Boudreau *et al.*, 2001; Brasier, 1975; Bronnimann & Whittaker, 1993; Brönnimann *et al.*, 1992; Callard *et al.*, 2011; Cann *et al.*, 2000; Carbonel & Moyes, 1987; Culver, 1990; Debenay & Guiral, 2006; Debenay *et al.*, 1987, 1996, 2001, 2004; Buchan & Lewis, in Demchuk & Gary, 2009; Duchemin *et al.*, 2005; Duleba & Debenay, 2003; Edwards *et al.*, 2004; Fatela *et al.*, 2009; Ghosh *et al.*, 2009; Haynes & Dobson, 1969; Hiltermann & Haman, 1986; Hippensteel *et al.*, 2000; Horton & Edwards, 2006; Horton *et al.*, 1999, 2005, 2007; Hughes, in Kaminski & Coccioni, 2008; Mohamed *et al.*, in Kaminski & Filipescu, 2011; Kemp *et al.*, 2009; Langer & Lipps, 2006; Lee *et al.*, 1969; Leorri *et al.*, 2008;

Debenay *et al.*, in Martin, 2000; Matera & Lee, 1972; Muller, 1975; Horton & Edwards, in Olson & Leckie, 2003; Martin *et al.*, in Olson & Leckie, 2003; Massey *et al.*, 2006; Patterson *et al.*, 1997, 2000; Perry *et al.*, 2008; Phleger, 1955, 1963; Petters, in Oti & Postma, 1995; Rossi & Horton, 2009; Rossi & Vaiani, 2008; Scott *et al.*, 1991; Semensatto *et al.*, 2009; Sen Gupta, in Sen Gupta, 1999; Skinner, 1966; Southall *et al.*, 2006; van Gorsel, 1988; Haman, in Verdenius *et al.*, 1983; Wilson *et al.*, 2008; Woodroffe *et al.*, 2005; Zaninetti, 1979.

*Shallow marine environments (including peri-reefal sub-environments)*

Beavington-Penney & Racey, 2004; BouDagher-Fadel, 2008; Brasier & Donahue, 1985; Brasier & Green, 1993; Burollet *et al.*, 1986; Debenay, 1988; Mateu-Vicens *et al.*, in Demchuk & Gary, 2009; Hallock, 1988; Hallock & Glenn, 1986; Hallock & Peebles, 1993; Hallock Muller, 1976; Hohenegger, 2005; Levy *et al.*, 1988; Poag & Tresslar, 1981; Reiss & Hottinger, 1984; Saraswati, 2010; Sen Gupta, 1977; Hallock, in Sen Gupta, 1999; Sen Gupta, in Sen Gupta, 1999; Truffleman *et al.*, 1991.

*Deep marine environments (including submarine fan sub-environments)*

Bhaumik & Sen Gupta, 2005; Bouma *et al.*, in Bouma *et al.*, 1985; Kohl *et al.*, in Bouma *et al.*, 1985; Fillon, in Demchuk & Gary, 2009; Duros *et al.*, 2012; Gieskes *et al.*, 2011; Sen Gupta *et al.*, in Goldstein & Bernhard, 1997; Hill *et al.*, 2003; Ingram *et al.*, 2010; Kiel, 2010; Koho, 2008; Koho *et al.*, 2007; Lobegeier & Sen Gupta, 2008; Lundquist *et al.*, 1997; Margreth *et al.*, 2009, 2011; Martin *et al.*, 2010; McGann *et al.*, 2010; Morigi *et al.*, 2010, 2012; Panieri & Camerlanghi,  2010; Panieri & Sen Gupta, 2008; Qing Li *et al.*, 2010; Rathburn *et al.*, 2000, 2010; Roberts *et al.*, 2009; Ruggeberg *et al.*, 2007; Schonfeld *et al.*, 2011; Sen Gupta & Aharon, 1994; Sen Gupta *et al.*, 2007; Stewart *et al.*, 2011; Torres *et al.*, 2010; van Dover, 2000; Vella, 1964; Vilela & Maslin, 1997; Beuck *et al.*, in Wisshak & Tapanila, 2008; Wollenburg & Tiedemann, 2010; Xiang Rong *et al.*, 2010; Yanko & Flexer, 1992.

*(Bio)geographic distribution*

Belasky, 1996; BouDagher-Fadel, 2008, 2012; Gooday & Jorissen, 2012; Langer & Hottinger, 2000; Murray, 2013; Be, in Ramsay, 1977; Renema, 2002; Renema, in Renema, 2007; Stehli *et al.*, in Schopf, 1972; Arnold & Parker, in Sen Gupta, 1999; Culver & Buzas, in Sen Gupta, 1999.

*Oceanography*

Abu-Zied *et al.*, 2008; Altenbach & Struck, 2001; Altenbach *et al.*, 1999, 2012; Alve, 1995, 2010a; Alve & Bernhard, 1995; Anderson, 1975; Armynot de Chatelet *et al.*, 2009; Ashckenazi-Polivoda *et al.*, 2010; Gooday, in Austin & James, 2008; Bernhard, 1986, 1989, 1993, 1996; Bernhard & Alve, 1996; Bernhard & Bowser, 1999; Bernhard *et al.*, 1997, 2003; Bruckner & Mackensen, 2008; Brunner *et al.*, 2006; Burdige, 2006; Clift *et al.*, 2002; Corliss & Chen, 1988; Corliss & Fois, 1990; den Dulk *et al.*, 1998, 2000; Denne & Sen Gupta, 1991, 1993; Erbacher & Nelskamp, 2006; Emerson & Hedges, 2008; Ernst *et al.*, 2005; Fontanier *et al.*, 2005; Galeron *et al.*, 2001; Goineau *et al.*, 2011; Bernhard *et al.*, in Goldstein & Bernhard, 1997; Gooday, 2002a, 2003; Gooday & Hughes, 2002; Gooday *et al.*, 2000a, 2001a, 2009b, 2009c; Harris, 2005; Ernst *et al.*, in Hart *et al.*, 2000; Heinz *et al.*, 2001; Jorissen *et al.*, in Hillaire-Marcel & de Vernal, 2007; Kucera, in Hillaire-Marcel & de Vernal, 2007; Husum & Hald, 2012; Jannink *et al.*, 1998; de Rijk *et al.*, in Jorissen & Rohling, 2000; Kitazato *et al.*, in Jorissen & Rohling, 2000; Mackensen *et al.*, in Jorissen & Rohling, 2000; Venec-Peyre & Caulet, in Jorissen & Rohling, 2000; Wollenberg & Kuhnt, in Jorissen & Rohling, 2000; Kaminski *et al.*, in Kaminski *et al.*, 1995; Koho *et al.*, 2008, 2010; Komiya *et al.*, 2008; Jorissen *et al.*, in Langer, 1995; Leiter & Altenbach, 2010; Naidu & Malmgren, in Langer, 1995; Langer *et al.*, 1997; Levin *et al.*, 2009; Licari *et al.*, 2003; Loubere *et al.*, in Loubere, 1995; Loubere, 1997; Mackensen *et al.*, 1990; Milliman & Farnsworth, 2011; Mojtahid *et al.*, 2009, 2010; Moodley & Hess, 1992; Morigi *et al.*, 2001; Nigam *et al.*, 2007; Nomaki *et al.*, 2005, 2011; Perez-Cruz & Machain-Castillo, 1990; Phleger & Soutar, 1973; Platon *et al.*, 2005; Poag, 1985; Quinterno & Gardner, 1987; Rathburn *et al.*, 2001; Resig, 1992; Risgaard-Petersen *et al.*, 2006; Rosoff & Corliss, 1992; Salgueiro *et al.*, 2008; Schmiedl *et al.*, 2000; Schonfeld, 2001; Schulz & Zabel, 2000; Schumacher *et al.*, 2007; Bernhard & Sen Gupta, in Sen Gupta, 1999; Jorissen, in Sen Gupta, 1999; Loubere & Fariduddin, in Sen Gupta, 1999; Sen Gupta & Machain-Castillo, 1993; Sen Gupta *et al.*, 1981; Smart & Gooday, 1997, 2006; Curry *et al.*, in Summerhayes *et al.*, 1992; Hermelin, in Summerhayes *et al.*, 1992; Steens *et al.*, in Summerhayes *et al.*, 1992; Szarek *et al.*, 2007; van der Zwaan *et al.*, in Tyson & Pearson, 1991; Ufkes *et al.*, 1998; Williams & Follows, 2011; Woulds *et al.*, 2007; Yamasaki *et al.*, 2008; Zhang *et al.*, 2010.

*Environmental interpretation technologies*

Berger & Diester-Haass, 1988; Borcic *et al.*, 2010; Grimsdale & van Morkhoven, 1955; Hayward & Triggs, 1994; Murray, 1976; Rossi & Vaiani, 2008; Parker & Arnold, in Sen Gupta, 1999; Speijer *et al.*, 2010a; van der Zwaan *et al.*, 1990; Stehli, in van Hinte, 1966.

# 5

# Palaeobiology, palaeoecology (or palaeoenvironmental interpretation)

This chapter deals with palaeobiology, palaeoecology (or palaeoenvironmental interpretation). It contains sections on Foraminifera in palaeoenvironmental interpretation; on palaeobathymetry; on sedimentary environments; on palaeo (bio)geography; on palaeoceanography; on biogeochemical proxies in palaeoenvironmental interpretation; on palaeoclimatology; and on palaeoenvironmental interpretation and associated visualisation technologies. The section on palaeobathymetry includes sub-sections on marginal marine or paralic environments; on shallow marine or neritic environments; and on deep marine, or bathyal and abyssal, environments. The section on sedimentary environments includes sub-sections on marginal marine, peri-deltaic; shallow marine, peri-reefal; and deep marine, submarine fan sub-environments. The section on palaeoceanography includes a sub-section on the organic carbon cycle and oxygenation.

## 5.1 Palaeoenvironmental interpretation

Palaeoecology or palaeoenvironmental interpretation involves the use of fossils in the ordering of the rock record in space.

### *Facies fossils*

Palaeoenvironmentally useful so-called facies fossils share two common characteristics:

firstly, restricted ecological distributions (for example, to a particular bathymetric zone or biogeographic province);

and secondly, low evolutionary turnover rates, and hence long stratigraphic ranges.

The most useful, for practical purposes, are also abundant, well preserved, and easy to identify.

Many of them are benthic.

## 5.2 Foraminifera in palaeoenvironmental interpretation

Foraminiferal palaeoecology, or palaeoenvironmental interpretation, involves the use of Foraminifera in the ordering of the rock record in space (Jones, 1996, 2006, 2011a; see also the further reading list at the end of the chapter). In the case of extant, typically Pleistogene and Neogene, and atypically Palaeogene and Mesozoic, taxa, it relies on Charles Lyell's fundamental 'Principle of Uniformitarianism', in other words on modern analogy (see Chapter 4). In the case of extinct, Palaeozoic, taxa, palaeoenvironmental interpretation relies on the functional morphological principle (see Section 3.2.2), or on associated fossils and sedimentary facies. Ancient Foraminifera are interpreted on these bases as having occupied a range of more or less marine environments, and as having adopted a variety of life strategies, life positions and feeding strategies (see Section 4.1 and also below).

It is important to note that palaeoenvironmental interpretation using Foraminifera can be impaired by natural, taphonomic, factors or biases, such as *post-mortem* transportation of species; non-contemporaneous reworking; and fossil-preservation bias, or the loss of species to the rock record in the fossilisation process, which can select against disintegration- or dissolution-prone species, and for resistant species (see the further reading list at the end of the chapter). (Many Foraminifera disintegrate almost immediately after death and have low preservation potential, including the entire Order Allogromiida, and also the agglutinated 'komoki': see the further reading list at the end of the chapter.)

Palaeoenvironmental interpretation can also be impaired by artificial factors or biases, such as sampling or sample-preparation bias, or the loss of species in the preparation process (see Section 2.2); sample-contamination bias; and species-identification bias, or subjectivity and inaccuracy in the identification of species, in turn leading to interpretation bias (see Sections 2.3–2.4 and 3.2.1).

### *Proteinaceous Foraminifera (Allogromiida)*

There are no ancient proteinaceous Foraminifera, or Allogromiida, known to the author.

### *Agglutinating Foraminifera (Astrorhizida and Lituolida)*

Ancient smaller agglutinating Foraminifera, or Astrorhizida and Lituolida, are interpreted on the basis of analogy with their modern counterparts as having been marine, marginal to deep, and in some cases as having been tolerant of hyposalinity as well as normal marine salinity, and/or of hypoxia or dysoxia; and are further interpreted as especially characteristic of marginal marine, peri-deltaic sedimentary environments or of deep marine, submarine fan sedimentary environments (Charnock & Jones, in Hemleben *et al.*, 1990; Jones & Wonders, 1992; Jones *et al.*, in Jenkins, 1993; Jones, 1996; Jones, in Ali *et al.*, 1998; Jones *et al.*, in Jones & Simmons, 1999; Jones *et al.*, in Powell & Riding, 2005; Jones, 2006, 2011a; and additional references cited therein; see also Section 4.1, and the further reading list at the end of the chapter). Note that ancient so-called 'flysch-type-' or *Rhabdammina-*' faunas, widely recorded in Cretaceous and Cenozoic flysch-type deposits from a variety of settings around the world, and composed almost exclusively of unilocular and uniserial multilocular agglutinating Foraminifera, have no modern counterparts, and are difficult to interpret palaeoenvironmentally. However, they have conventionally been interpreted, essentially on the basis of analogy with the agglutinating components of modern mixed foraminiferal faunas, as of deep-water aspect, hence their alternative name of Deep-Water Agglutinating Foraminifera or DWAF faunas; and as associated with submarine fan, deposits, preferentially deposited during (tectono)eustatically mediated sea-level low-stands (see also Section 5.4.3). The absence of a calcareous benthic foraminiferal and foraminiferal component has been variously attributed either to primary factors preventing their presence in the first place, such as competitive exclusion or an inability to tolerate environmental stress; or to secondary factors preventing their preservation, such as early diagenetic carbonate dissolution, either below the Calcite Compensation Depth (CCD), typically in depths in excess of 4 km, or below what might be termed the 'effective CCD', essentially at almost any depth. In the case of the effective CCD, it is caused by acids produced by early diagenetic reactions involving the decomposition of organic matter in restricted and likely dysoxic environments. The restriction and likely dysoxia in the environment is in turn caused either by stagnation associated with basin configuration, or by salinity stratification associated with excess fresh-water run-off from a major river system (which, incidentally, would also have the effect of eliminating the normal marine phytoplankton that, as – seasonal – phytodetritus, is known to constitute an important food source for at least some calcareous benthic foraminiferal species).

Ancient larger Lituolida (Larger Benthic Foraminifera or LBFs) are interpreted as shallow to deep marine, and as having been tolerant of only normal marine salinity (see Section 4.1).

### *Smaller and larger calcareous benthic Foraminifera (Fusulinida, Miliolida, Nodosariida, Buliminida, Robertinida and Rotaliida)*

Ancient smaller Fusulinida are interpreted on the bases of analogy with their closest living relatives, and of associated fossils and sedimentary facies, as marine, marginal to deep; and are further interpreted as especially characteristic of shallow marine, back-reef lagoonal carbonate sub-environments (Jones, 1996, 2006, 2011a; and additional references cited therein; see also Section 4.1, and the further reading list at the end of the chapter).

Ancient smaller Miliolida are interpreted on the basis of analogy with their modern counterparts as marine, marginal to deep, and in some cases as having been tolerant of hyposalinity or hypersalinity as well as normal marine salinity; and are further interpreted as characteristic of shallow marine, back-reef lagoonal carbonate sub-environments (see Section 4.1).

Ancient Nodosariida are interpreted on the basis of analogy with their modern counterparts as marine, shallow to deep, and as having been tolerant only of normal marine salinity; and are further interpreted as characteristic of deep marine environments (see Section 4.1). At least some, including, for example, *Dentalina*, *Lenticulina* and *Nodosaria*, are further interpreted as having been tolerant of suboxic environments.

Ancient Buliminida are also interpreted on the basis of analogy with their modern counterparts as marine, shallow to deep, and as having been tolerant only of normal marine salinity; and are further interpreted as especially characteristic of dysoxic environments, most notably Oxygen Minimum Zones (OMZs) in the upper bathyal (see Section 4.1). *Siphogenerinoides* has been hypothesised to have opportunistically fed on the phosphatised faeces of clupeoid fish in the dysoxic environments of the Late Cretaceous, Santonian–Maastrichtian of the upper Magdalena basin of Colombia.

Ancient Robertinida are interpreted on the basis of analogy with their modern counterparts as marine, shallow to deep, and as having been tolerant only of normal marine salinity; and are further interpreted as characteristic of deep marine environments (see Section 4.1). At least some, including, for example, *Hoeglundina*, are further interpreted as having been tolerant of suboxic environments.

Ancient smaller Rotaliida are interpreted on the basis of analogy with their modern counterparts as marine, marginal to deep, and in some cases as having been tolerant of hyposalinity or hypersalinity as well as normal marine salinity, or of hypoxia or dysoxia; and are further interpreted as characteristic of comparatively shallow marine environments (see Section 4.1).

Ancient larger Fusulinida, Miliolida and Rotaliida (LBFs) are interpreted on the bases of analogy with their closest living relatives, and of associated fossils and sedimentary facies, as restricted to shallow marine environments, and in some cases as having been tolerant of hypersalinity as well as normal marine salinity; and are further interpreted as especially characteristic of shallow marine, platformal and peri-reefal carbonate environments (Fusulinida and Miliolida of back-reef lagoonal sub-environments; Rotaliida of reef and fore-reef sub-environments) (see Section 4.1).

*Planktic Foraminifera (Globigerinida)*

Ancient planktic Foraminifera, or Globigerinida, are interpreted on the basis of analogy with their modern counterparts as having lived in the water column and sunk to the sea floor when they died, and in some cases, for example that of the Cretaceous *Hedbergella*, as having been tolerant of hyposalinity as well as normal marine salinity; and are further interpreted as having occupied a range of depth environments within the water column (Jones, 1996, 2006, 2011a; and additional references cited therein; see also Section 4.1, and the further reading list at the end of the chapter).

## 5.3 Palaeobathymetry

Ancient Foraminifera are interpreted on the bases of analogy with their modern counterparts, or of associated fossils and sedimentary facies, as having occupied at least three palaeobathymetric zones, namely, marginal, shallow and deep marine (Jones, 1996, 2006, 2011a; and additional references cited therein; see also Section 4.2, the further reading list at the end of the chapter, and Sections 5.4. and 9.3). Note in this context that proxy modern bathymetric distribution data cannot be meaningfully used in palaeobathymetric interpretation in those cases in which distributions are determined by factors other than depth. For example, within the area of influence of the Mississippi delta in the Gulf of Mexico, where distributions are determined by the so-called 'delta effect' rather than depth, and where depth limits can be either elevated or depressed depending on the local sedimentary regime. Or elsewhere in the North Atlantic, where distributions are determined by temperature rather than depth, and where depth limits can be either elevated or depressed depending on the local temperature.

*Proteinaceous Foraminifera (Allogromiida)*

There are no ancient proteinaceous Foraminifera, or Allogromiida, known to the author.

*Agglutinating Foraminifera (Astrorhizida and Lituolida)*

Ancient – smaller and larger – agglutinating Foraminifera are interpreted on the basis of analogy with their modern counterparts as most characteristic of deep marine, bathyal and abyssal environments (Jones, 1996, 2006, 2011a; see also Section 4.2, and the further reading list at the end of the chapter). Indeed, some families, for example the Lituotubidae and Verneuilinidae, are interpreted as having been restricted to deep marine environments. Note, though, that the Astrorhizidae probably also occurred in shallow marine environments, and that the Ataxophragmiidae was probably restricted to shallow marine environments.

Note also that the Trochamminidae probably also occurred in some numbers in marginal marine environments.

### Smaller and larger calcareous benthic Foraminifera (Fusulinida, Miliolida, Nodosariida, Buliminida, Robertinida and Rotaliida)

As noted above, ancient smaller Fusulinida are interpreted on the bases of analogy with their closest living relatives, and of associated fossils and sedimentary facies as marine, marginal to deep (Jones, 1996, 2006, 2011a; see also Section 4.2, and the further reading list at the end of the chapter).

Ancient smaller Miliolida are interpreted as characteristic of shallow marine, inner neritic environments. Note, though, that at least some representatives of the Ophthalmidiidae could actually have been restricted to deep marine environments.

Ancient Nodosariida are interpreted on the bases of analogy with their modern counterparts and of associated fossils as characteristic of deep marine, upper bathyal environments, for example those in the Permian of south China. Indeed, the Lingulinidae and Plectofrondiculariidae, are interpreted as having been restricted to deep marine environments. Note, though, that the Polymorphinidae, probably occurred in shallow marine environments. Note also that at least some ancient Nodosariida have been interpreted, essentially on the basis of associated fossils, as shallow marine, for example those in the Jurassic of north-west Europe.

Ancient Buliminida are interpreted on the bases of analogy with their modern counterparts as characteristic of deep marine environments. Indeed, the Pleurostomellidae are interpreted as having been restricted to deep marine environments. Note, though, that the Turrilinidae, Pavoninidae and Millettiidae probably also occurred in shallow marine environments.

Ancient Robertinida are interpreted on the basis of analogy with their modern counterparts as characteristic of deep marine, middle bathyal environments. Note, though, that at least some have been interpreted, essentially on the basis of associated fossils, as shallow marine, for example those in the Jurassic of north-west Europe.

Ancient smaller Rotaliida range are interpreted on the basis of analogy with their modern counterparts as characteristic of comparatively shallow marine environments. Indeed, the Discorbidae, Glabratellidae, Eponididae, Epistomariidae, Planorbulinidae, Acervulinidae, Nonionidae and Elphidiidae are interpreted as having been restricted to comparatively shallow marine environments. Note, though, that the Alabaminidae, Anomalinidae and Chilostomellidae probably also occurred in deeper marine environments. In addition, some genera of the essentially shallow marine Discorbidae (*Gavelinopsis*, *Laticarinina*), Eponididae (*Ioanella*, *Oridorsalis*) and Nonionidae (*Melonis*), probably also occurred in deeper marine environments.

Ancient larger Fusulinida, Miliolida and Rotaliida (LBFs) are interpreted on the bases of analogy with their modern counterparts, and of associated fossils and sedimentary facies, as characteristic of shallow marine environments (Jones, 1996, 2006, 2011a; see also Section 4.2, and the further reading list at the end of the chapter). They are also interpreted as having harboured photosynthetic algal symbionts that restricted them to the photic zone, that is, no deeper than 130 m in clear water, and shallower still in turbid or turbulent water (light penetration, determined by water clarity, turbidity and turbulence, is actually the primary control on their distribution rather than depth as such). The larger Fusulinida and Miliolida are interpreted as having harboured mainly green or red algal symbionts that restricted them to the euphotic zone; and the larger Rotaliida, diatom symbionts that restricted them to the sub-euphotic zone. Incidentally, the tests of some ancient larger Rotaliida are more enriched in photosynthetic carbon-12 (or impoverished in carbon-13) than others, arguably indicating a greater photosynthetic capacity. Coincidentally or otherwise, these species evolved and became extinct later than the others, arguably indicating that symbiosis played an important role in the evolution and extinction of the larger Rotaliida.

### Planktic Foraminifera (Globigerinida)

Ancient planktic Foraminifera, or Globigerinida, are interpreted on the basis of analogy with their modern counterparts as having occupied a range of depth environments within the water column (Jones, 1996, 2006, 2011a; see also Section 4.2, and the further reading list at the end of the chapter). At least some are also interpreted as having harboured algal photosymbionts that restricted them to the photic zone. For example, the Globigerinidae are interpreted as having harboured symbionts that restricted them to the epipelagic to mesopelagic euphotic zone, and the Globorotaliidae, ones that restricted them to the bathypelagic sub-euphotic zone, in the Cenozoic. (Note, though, that isotopic evidence suggests that Globigerinidae were bathypelagic, and the Globorotaliidae epipelagic, in the Palaeogene, assuming isotopic equilibrium between their tests and the ambient environment.) And the unkeeled Hedbergellidae and associated taxa are interpreted as having harboured symbionts that restricted them to the epipelagic to mesopelagic euphotic zone, and the keeled Globotruncanidae as associated taxa, ones that restricted them to the bathypelagic sub-euphotic zone, in the Cretaceous. Note, incidentally, that the former, epipelagic to mesopelagic, taxa, including *Rugoglobigerina*, *Pseudoguembelina*, *Planoglobulina* and *Racemiguembelina*, are conspicuous by their absence in expanded OMZs.

### 5.3.1 Marginal marine or paralic environments

Ancient marginal marine or paralic environments are characterised by taxonomically or morphologically similar foraminiferal faunas to modern ones, although in certain cases lacking some then not yet extant taxa, or possessing some now extinct ones (Jones, 1996, 2006, 2011a; see also Section 4.2.1, the further reading list at the end of the chapter, and Section 5.4.1). They are also characterised by at least locally high foraminiferal abundance, comparatively low diversity and correspondingly comparatively high dominance.

Foraminiferal faunas from salt-marsh and estuarine environments of the Holocene of Virginia and North Carolina are characterised by *Entzia [Jadammina] macrescens, Miliammina fusca* and *M. petila* (high marsh), *Trochammina inflata* and *Tiptotrocha comprimata* (low marsh), and *Elphidium excavatum* (estuarine). Note that salt marshes and mangrove swamps are extremely sensitive indicators of sea level, and are important proxies for – empirically- or transfer-function-based – sea-level reconstructions, especially for the Quaternary (see also Sections 11.2.1. and 13.1.2).

Faunas from the Late Cretaceous of the 'Western Interior Seaway' of Arizona, Colorado and Utah in the United States are characterised by *Ammobaculites, Haplophragmoides, Textularia, Trochammina, Verneuilinoides* and *Miliammina*.

### 5.3.2 Shallow marine or neritic environments

Ancient shallow marine environments are characterised by taxonomically or morphologically similar foraminiferal faunas to modern ones, although in certain cases lacking some then not yet extant taxa or possessing some now extinct ones (Jones, 1996, 2006, 2011a; see also Section 4.2.2, the further reading list at the end of the chapter, and Section 5.4.2). They are also characterised by at least locally high foraminiferal abundance and diversity, and correspondingly low dominance.

### 5.3.3 Deep marine, or bathyal and abyssal, environments

Ancient deep marine environments are characterised by taxonomically or morphologically similar foraminiferal faunas to modern ones, although in certain cases lacking some then not yet extant taxa, or possessing some now extinct ones (Jones, 1996, 2006, 2011a; see also Section 4.2.3, the further reading list at the end of the chapter, and Section 5.4.3). They are also characterised by at least locally high foraminiferal abundance and diversity, and correspondingly low dominance.

Upper bathyal sub-environments are characterised by Nodosariida, and dysoxic environments, including OMZs in upper bathyal sub-environments, by Buliminida;

and deeper bathyal sub-environments, although not abyssal ones, by Robertinida (see also Section 5.3).

Deep marine, submarine fan sub-environments are characterised by DWAFs (see Section 5.4.3).

*Cold (hydrocarbon) seeps, hydrothermal vents and 'nekton falls'*

Ancient cold seeps (and apparently associated coral reefs), hydrothermal vents and nekton falls are characterised by taxonomically or morphologically similar foraminiferal faunas to modern ones. Analyses of the tests of Foraminifera associated with cold seep deposits in the Pliocene of the Cascadia accretionary margin in Washington in the Pacific indicate that at least *Globobulimina pacifica* (infaunal Buliminida) and *Cibicidoides mckannai* and *Nonion basispinatum* (epifaunal Rotaliida) appear to have been metabolising methane.

## 5.4 Sedimentary environments

### 5.4.1 *Marginal marine, peri-deltaic sub-environments*

Ancient peri-deltaic sub-environments are characterised by taxonomically or morphologically similar foraminiferal faunas to modern ones (Jones & Wonders, 1992; Jones *et al.*, in Jenkins, 1993; Jones, 1996; Jones, in Ali *et al.*, 1998; Jones *et al.*, in Jones & Simmons, 1999; Jones, 2006, 2011a; and additional references cited therein; see also Section 4.3.1, and the further reading list at the end of the chapter).

In the case of the ancient, Miocene–Pliocene, and of the modern, Orinoco delta, fluvially dominated delta-top sub-environments are characterised by the testate amoeban *Centropyxis*; tidally dominated delta-top sub-environments by *Miliammina* and *Trochammina* foraminiferal assemblages; delta-front sub-environments, by *Buliminella* assemblages; proximal prodelta sub-environments by *Eggerella* assemblages; and distal prodelta sub-environments by *Glomospira* and *Alveovalvulina/Cyclammina* assemblages (see also Sections 9.3.5 and 9.5.2).

In the case of the Early Jurassic Dunlin delta of the subsurface of the North Sea, and of analogous surface outcrops in Spitsbergen and Yorkshire, interpreted vegetated delta-top to delta-front sub-environments are characterised by significant incidences of epiphytic morphotypes of agglutinating Foraminifera; proximal prodelta assemblages by epifaunal morphotypes; and distal prodelta sub-environments by infaunal morphotypes (see also Sections 9.3.3 and 9.5.1). The different sub-environments fall in different fields on a triangular cross-plot of epiphytic, epifaunal and infaunal morphotypes. The agglutinating Foraminifera characterising the delta-top and delta-front sub-environments include *Ammodiscus* and *Trochammina*, which are interpreted on the basis of analogy with modern counterparts in Drammensfjord in Norway and in Aso-kai lagoon in Japan as tolerant of hyposalinity and of hypoxia or dysoxia.

### 5.4.2 Shallow marine, peri-reefal sub-environments

Ancient peri-reefal sub-environments are characterised by taxonomically or morphologically similar foraminiferal faunas to modern ones (Jones, 1996, 2006, 2011a; and additional references cited therein; see also Section 4.3.2, the further reading list at the end of the chapter, and Sections 9.3.1 and 9.3.2). Note in this context, though, that the representativeness of modern ('ice-house') peri-reefal environments in relation to ancient ('greenhouse') ones is questionable because of the influence of Plio-Pleistocene glaciation and deglaciation, and associated high-frequency sea-level fluctuations. In particular, the rise of continentality through the glacial period appears to have brought about an increased run-off and nutrient enhancement of surface waters, and hence a flourishing of mesotrophic algae at the expense of oligotrophic corals.

Ancient back-reef lagoonal sub-environments are characterised by larger Fusulinida and Miliolida; and reef and fore-reef sub-environments by larger Rotaliida (see also Sections 9.3.1 and 9.3.2). The possibility that certain ancient LBFs might have harboured algal symbionts is discussed by Cowen, in Tevesz & McCall (1983) (see the further reading list at the end of the chapter).

### Nummulites and nummulite banks

Certain larger Rotaliida (Nummulitidae) were sufficiently abundant as to be rock-forming in the Palaeogene of the circum-Mediterranean (Jones, 1996, 2006, 2011a; and additional references cited therein; see also the further reading list at the end of the chapter). Importantly, the so-called banks that they formed are locally of commercial significance as petroleum reservoirs (see also Section 9.1.1).

Nummulites or 'coin-stones' constitute an extant sub-group of larger Rotaliida ranging in age from Late Palaeocene to Recent (the genus *Nummulites* ranges in age from Late Palaeocene to Early Oligocene, global standard calcareous nannoplankton Zones NP8–NP24). Modern representatives host diatom symbionts, and thus in terms of bathymetry characterise sub-euphotic, commonly reefal and fore-reefal carbonate environments. In terms of biogeography, they characterise low latitudes. Ancient nummulites exhibited essentially Pan-Tethyan distributions. The most northerly occurrences were on Rockall Bank in the Late Palaeocene ('*Nummulites*' *rockallensis*), and in the Hampshire basin in the Eocene (*Nummulites planulatus, N. laevigatus, N. variolarius, N. prestwichianus* and *N. rectus*). The best known of the approximately 300 nominal species, *Nummulites gizehensis*, occurred in sufficient abundance in the Middle Eocene of Egypt as to have formed the building stone of which the Great Pyramids of Gizeh were later built.

As intimated above, nummulites were locally sufficiently abundant, essentially throughout the Early–Middle Eocene of the circum-Mediterranean, as to have formed accumulations called 'banks'. Nummulite banks are constituted of variable proportions of unbroken nummulites and other microfossils, broken and fragmented nummulites ('nummulithoclastic debris'), lime mud, and cement. They tend to be dominated either by comparatively large individuals of a single species, generally 'B' forms generated by sexual reproduction, or by small individuals, generally 'A' forms generated by asexual reproduction. Carbonate sedimentologists have historically tended to interpret the size sorting as evidence of hydrodynamic sorting, the B-form-dominated accumulations as winnowed and essentially autochthonous, and the A-form-dominated accumulations as re-deposited and allochthonous. Note in this context that the hydrodynamic behaviour of nummulites, as demonstrated by flume-tank experiments, is more closely comparable to that of silt than of comparably sized sand particles, on account of their high internal porosities (a function of their functional morphology), and low pick-up velocities. (Note, though, that it is also possible to interpret A-form-dominated accumulations as representing unmodified death assemblages of populations reproducing exclusively asexually under conditions of extreme environmental stress, and as reflecting reproductive strategy rather than taphonomy.)

### 5.4.3 Deep marine, submarine fan sub-environments

Ancient submarine fan sub-environments are characterised by taxonomically or morphologically similar foraminiferal faunas to modern ones (Charnock & Jones, in Hemleben *et al.*, 1990; Jones, 1996; Jones, in Ali *et al.*, 1998; Jones, in Jones & Simmons, 1999; Jones, 2001, 2003a, b; Jones *et al.*, 2003; Jones *et al.*, in Powell & Riding, 2005; Jones, 2006; Kender *et al.*, in Kaminski & Coccioni, 2008; Kender *et al.*, 2008; Jones, 2011a; and additional references cited therein; see also Section 4.3.3, and the further reading list at the end of the chapter).

Fan channel, levee and overbank, and respectively corresponding turbiditic, inter-turbiditic and hemipelagic sub-environments, have been described from a wide range of ancient fan complexes.

In the Palaeogene of the subsurface of the North Sea, and of analogous surface outcrops in the Spanish Pyrenees, fan channel and turbiditic sub-environments are characterised by low-diversity assemblages dominated by epifaunal suspension-feeding morphotypes of DWAFs; levee and inter-turbiditic environments by intermediate-diversity assemblages dominated by epifaunal detritus-feeding morphotypes; and overbank and hemipelagic sub-environments by high-diversity assemblages containing significant proportions of infaunal detritus-feeding morphotypes (Charnock & Jones, in Hemleben *et al.*, 1990; Jones, 1996; Jones, in Jones & Simmons, 1999; Jones *et al.*, in Powell & Riding, 2005; Jones, 2006, 2011a; and additional references cited therein; see also the further reading list at the end of the chapter, and Sections 9.3.3, 9.5.4 and 9.7.3). The different sub-environments fall in different fields on a triangular cross-plot of epifaunal suspension-feeding, epifaunal detritus-feeding, and detritus-feeding morphotypes (respectively, 'morphogroups' A, B and C of Jones & Charnock, 1985: see Fig. 3.2).

In the Neogene–Pleistogene of the Gulf of Mexico, turbiditic sub-environments are distinguished either by allochthonous foraminiferal assemblages or by autochthonous ones dominated by DWAFs (*Ammodiscoides, Ammodiscus, Ammolagena, Bathysiphon, Cribrostomoides, Cyclammina, Cystammina, Dorothia, Glomospira, Haplophragmoides, Hormosina, Hyperammina, Martinottiella, Paratrochammina, Recurvoides, Reophax, Rhizammina, Saccammina, Spiroplectammina, Spirosigmoilina, Trochammina* and *Veleroninoides*); hemipelagic sub-environments by DWAFs and calcareous benthics (the Nodosariida *Lenticulina* and *Planularia*, and the Rotaliida *Cibicides, Cibicidoides, Eponides, Laticarinina, Melonis, Oridorsalis, Osangularia* and *Planulina*); and confined, dysoxic and sapropelic, hemipelagic, sub-environments, also by DWAFs and calcareous benthics (the Nodosariida *Fissurina, Globulina, Guttulina* and *Pseudoglandulina*, the Buliminida *Bolivina, Bulimina, Cassidulina, Ehrenbergina, Fursenkoina, Globobulimina, Globocassidulina, Praeglobobulimina, Rectuvigerina, Siphogenerina, Sphaeroidina, Transversigerina* and *Uvigerina*, the Robertinida *Ceratobulimina*, and the Rotaliida *Chilostomella, Gyroidinoides, Hansenisca* ['*Gyroidina*'] and *Siphonina*) (see the further reading list at the end of the chapter; see also Section 9.3.6). Pelagic sub-environments on the adjoining basin-plain are characterised by planktics, as well as by benthics (including the Rotaliida *Cibicidoides, Eponides, Nuttallides* and *Oridorsalis*).

Elsewhere, turbiditic sub-environments are characterised by only allochthonous, and locally hydrodynamically sorted, calcareous benthic and planktic Foraminifera.

## 5.5  Palaeo(bio)geography

Many groups of fossils, including Foraminifera, had restricted or endemic palaeo-(bio)geographic distributions, which renders them of use in the characterisation of palaeobiogeographic realms, such as the Palaeo-Tethyan, Tethyan, Austral and Boreal realms, or of provinces within these realms (Jones, 1996; Jones, in Agusti *et al.*, 1999; Jones, 2006; Jones *et al.*, 2006; Jones, 2011a; and additional references cited therein; see also the further reading list at the end of the chapter). This in turn renders them of use in the constraint of plate-tectonic and terrane

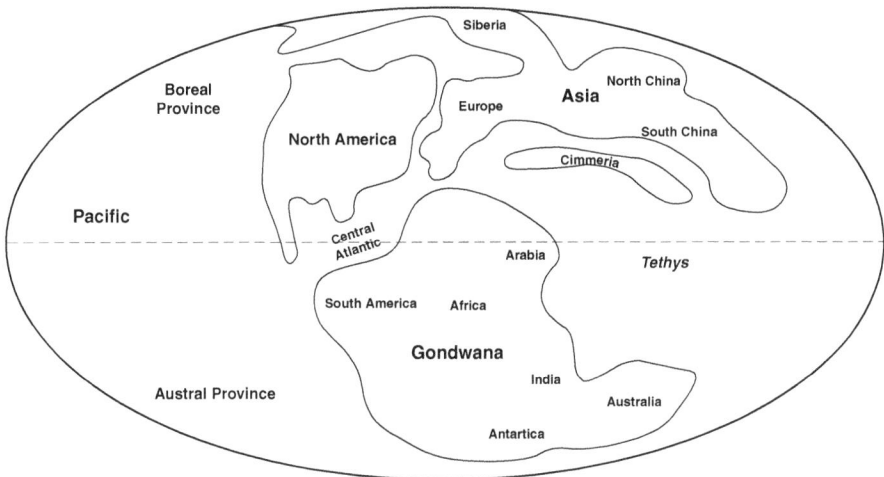

Fig. 5.1 **Jurassic palaeo(bio)geography**. From Jones (2006).

reconstructions, importantly, providing at least some measure of longitude as well as latitude (not provided by palaeomagnetism).

In the Late Palaeozoic, the larger benthic Fusulinida as a whole exhibited a Palaeo-Tethyan distribution (Jones, 1996, 2006, 2011a; see also the further reading list at the end of the chapter).

In the Mesozoic, the larger benthics as a whole exhibited a Tethyan distribution, and many individual smaller benthics or planktics either Austral, Boreal or Tethyan distributions, or, for example, provincial Arctic, Boreal-Atlantic and Boreal-Pacific distributions within the Boreal realm (the Arctic being characterised by agglutinating Ammodiscidae, Lituolidae, Trochamminidae and Ataxophragmiidae; the Boreal-Atlantic by calcareous benthic Epistominidae; and the Boreal-Pacific by endemic Rzehakinidae) (Jones, 1996, 2006, 2011a; Figs. 5.1–5.2; see also the further reading list at the end of the chapter). Note, though, that there was probably at least local or intermittent communication between the realms, as indicated, for instance, by the occurrence of the Early Cretaceous essentially Tethyan planktic *Lilliputianella globulifera* [*Hedbergella maslakovae*] in the Austral realm.

Through the Jurassic–Cretaceous, the Tethyan realm was variously characterised in the Early Jurassic, Pliensbachian by the larger benthic *Orbitopsella*; in the Middle Jurassic, Bathonian by *Paracoskinolina (occitana)*, *Spiraloconulus*, *Satorina* and *Pfenderina (salernitana)*; in the Late Jurassic, Oxfordian–Kimmeridgian, by *Alveosepta*, *Kurnubia*, *Parurgonia*, *Labyrinthina*, *Mangashtia* and *Everticyclammina*; in the latest Jurassic–earliest Cretaceous, Tithonian–Valanginian by *Anchispirocyclina*; in the Early Cretaceous, Hauterivian–early Barremian by *Choffatella*

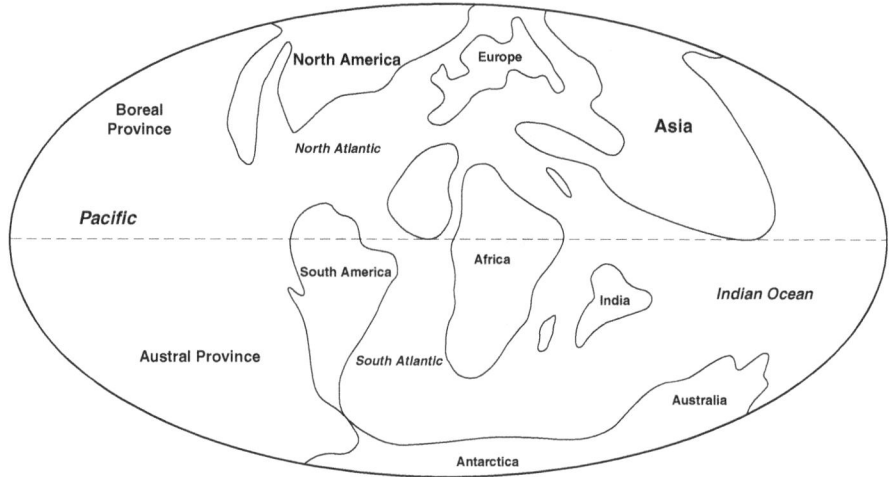

Fig. 5.2    **Cretaceous palaeo(bio)geography.** From Jones (2006).

*(pyrenaica), Pseudocyclammina, Pfenderina* and *Valdanchella*; in the late Barremian–early Aptian by *Valserina* and *Orbitolinopsis*; in the late Barremian–early Aptian by *Valserina, Palaeodictyoconus, Choffatella (decipiens), Palorbitolina, Orbitolinopsis* and *Palaeodictyoconus*; in the late Aptian–middle Albian by *Orbitolina (Mesorbitolina), Coskinolinoides, Nummoloculina, Neoiraqia, Neorbitolinopsis, Pseudochoffatella, Archaealveolina, Barkerina, Simplorbitolina, Coskinolinella, Paracoskinolina (tunesiana)* and *Archaealveolina*; in the Early–Late Cretaceous, late Albian–early Cenomanian by *Neorbitolinopsis, Orbitolina, Neoiraqia* and *Valdanchella*; in the Late Cretaceous, middle–late Cenomanian, by *Ovalveolina, Praealveolina* and *Cisalveolina*; and in the Santonian by *Lacazina, Lamarmorella* and *Murgella* (see the further reading list at the end of the chapter). (Incidentally, the Early Jurassic *Orbitopsella* has been interpreted, on the bases of analogy with living soritids, functional morphology, and associated fossils and sedimentary facies, as having accommodated algal photosymbionts, and as having constituted part of a shallow marine ?oligotrophic community associated with lithiotid bivalve 'reefs'.)

In the Cenozoic, the larger benthics as a whole again exhibited a Tethyan distribution, although many individual forms exhibited provincial Indo-Pacific, Mediterranean or American/Caribbean distributions within the realm, with the commonality between the Indo-Pacific and Mediterranean provinces a key indicator of the evolution and eventual extinction of the Tethyan realm over the Oligocene–Miocene (Jones, in Agusti *et al.*, 1999; Jones *et al.*, 2006; Fig. 5.3).

Many individual smaller benthics or planktics exhibited either Austral, Boreal or Tethyan distributions. In the later Cenozoic, many planktics exhibited

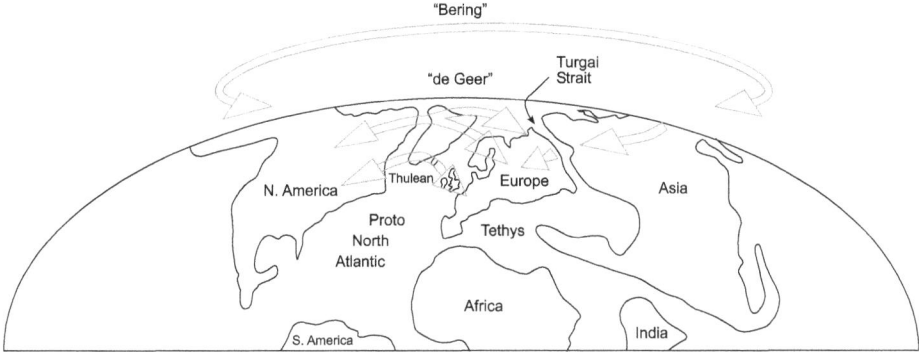

Fig. 5.3 **Palaeogene palaeo(bio)geography of the northern hemisphere**. Showing land bridges, inferred in part from foraminiferal data. From Jones (2006).

polar, sub-polar, temperate, sub-tropical and tropical distributions in both northern and southern hemispheres (see the further reading list at the end of the chapter).

## 5.6 Palaeoceanography

Foraminifera have proved of considerable use in palaeoceanography, not least through the Deep Sea Drilling Project (DSDP) and succeeding Ocean Drilling Program (ODP) (see, for example, Charnock & Jones, 1997; Kender *et al.*, 2007a, b, 2009; Jones, 2011a; and additional references cited therein; see also the further reading list at the end of the chapter). Much of the palaeoceanographic work undertaken on the group to date has been directed toward the understanding of biogeochemistry, in particular the organic carbon cycle (see Section 5.6.1), or palaeoclimatology (see Section 5.7). Some studies, though, have focussed on water masses and upwelling, or on the carbonate cycle (see, for example, Kender *et al.*, 2007a, b, 2009). Past upwelling can be indicated by micropalaeontological evidence (Kender *et al.*, 2007a, b, 2009).

### 5.6.1 Organic carbon cycle and oxygenation

Ancient, at least Cenozoic and Cretaceous, organic-rich/oxygen-poor environments and rocks, including those of OMZs and 'Oceanic Anoxic Events' (OAEs), are characterised by taxonomically similar benthic foraminiferal faunas to Recent ones, and especially by Buliminida, together with certain agglutinating Foraminifera and Rotaliida; and also by low Benthic Foraminiferal Oxygen Index or BFOI values (Charnock & Jones, 1997; Kender *et al.*, 2007a, b, 2009; Jones, 2011a; see also Section 4.5.1, the further reading list at the end of the chapter, and Section 9.1.1). Cretaceous OMZs and OAEs are further characterised by planktic

Foraminifera with radially elongate chambers, and by certain globotruncanids, including *Rugoglobigerina*, and heterohelicids, including *Pseudoguembelina*, *Planoglobulina* and *Racemiguembelina* (see the further reading list at the end of the chapter).

Organic-rich, oxygen-poor environments and rocks, including potential petroleum source-rocks, in the Late Cretaceous of the north-western African margin are characterised by *Praebulimina, Bolivina, Gabonita, Pyramidina, Eouvigerina, Loxostomum, Afrobolivina, Orthokarstenia, Neobulimina, Buliminella* (all Buliminida), and by *Cibicides beaumontianus* and *Gavelinella dakotaensis* (Rotaliida) (see the further reading list at the end of the chapter).

Organic-rich, oxygen-poor environments and rocks in the Middle Miocene of Angola in West Africa are characterised by anomalously low foraminiferal abundance and diversity and BFOI values, interpreted as indicating enhanced upwelling and/or an expansion of the OMZ at this time, perhaps in response to global warming in turn associated with Columbia River flood basalt volcanism (Kender *et al.*, 2007a, b, 2009; see also the further reading list at the end of the chapter).

## 5.7  Palaeoclimatology

Ancient Foraminifera are inferred to have had exacting ecological requirements and tolerances, and to have responded rapidly to changing environmental, and especially climatic, conditions; and, moreover, their tests preserve stable isotope records of changing temperature; such that they are extremely useful in palaeoclimate and palaeotemperature interpretation, in palaeoclimatology, at least in the Cenozoic and Mesozoic (Jones, 1996; Jones, in Agusti *et al.*, 1999; Jones, 2006; Whidden *et al.*, 2009; Jones, 2011a; Whidden & Jones, 2012; and additional references cited therein; see also the further reading list at the end of the chapter). Ratios of coiling directions in planktic Foraminifera are also of use in palaeotemperature interpretation.

LBF diversity has been used as a proxy for temperature in the greenhouse world of the early to mid Cenozoic of the Tethyan realm, although it is important to note that it is sensitive not only to temperature but also other factors such as facies, evolution, dispersal/migration, duration of studied interval and/or size of studied area, sampling artefact and taxonomic artefact (Jones, in Agusti *et al.*, 1999; Whidden *et al.*, 2009; Whidden & Jones, 2012; see also the further reading list at the end of the chapter). Significantly, the zones of highest diversity (and highest turnover) in the early Cenozoic correspond to the times of highest palaeotemperatures inferred from isotope data: the Mid-Palaeocene Biotic Event (MPBE) at approximately 58.2 Ma; the Palaeocene–Eocene Thermal Maximum (PETM), at approximately 55.5 Ma; the Early Eocene Climatic Optimum (EECO), at approximately 52–50 Ma; and the Middle Eocene Climatic Optimum (MECO), at approximately 40.1 Ma. The latitudinal migration of planktic biogeographic belts (polar, sub-polar, transitional, sub-tropical, tropical), has been used as a proxy for temperature change in the ice-house world of the late Cenozoic (see the further reading list at the end of the chapter). Temperature and associated diversity gradients in the ice-house world are and were steep, while in contrast those of the greenhouse world were significantly less so.

As noted above, the tests of ancient Foraminifera preserve stable isotope records of changing temperature or ice volume, the heavier isotope of oxygen, oxygen-18 or $^{18}$O, for example, occurring in higher proportions in tests during 'ice ages' or glaciations, because of differential evaporation into the atmosphere and eventual sequestration into glaciers of the lighter isotope, $^{16}$O. Temperature curves derived from the isotopic analysis of foraminiferal tests have now been constructed for the entire Cenozoic (which, incidentally, in the case of that curve derived from oxygen isotopic analysis, also serves as the basis for a workable Marine Isotope Stage or MIS, or Oxygen Isotope Stage or OIS, climatostratigraphy, at least for the Quaternary ice age: see Section 13.1.1). These show long-term climatic maxima in the Palaeocene–Middle Eocene, Early–early Middle Miocene, and Early Pliocene; and minima in the Late Eocene–Oligocene, late Middle to Late Miocene, and Late Pliocene–Pleistocene. They also show medium- to short-term climatic minima in the glacials and stadials; and maxima in the interglacials and interstadials, of the Pleistocene (and Holocene) (and ultra-short-term minima in the Dansgaard–Oeschger and Heinrich events of the Pleistocene). According to power spectrum analyses, the variance in the medium- to short-term data is attributable to 100 ka (50%), 40 ka (25%) and 20 ka (10%) Milankovitch cycles, of which the 40 ka is associated with variation in the obliquity of the Earth's axis, and the 20 ka with variation in the orientation or precession in the Earth's axis.

## 5.8 Biogeochemical proxies in palaeoenvironmental interpretation

Foraminiferal biogeochemical proxies for, among other things, palaeo-salinity and palaeo-productivity, as distinct from palaeotemperature, may be said to be in various stages of development (see the further reading list at the end of the chapter).

Proxies for palaeo-salinity, including Sr isotope ratios, F/Ca ratios, and paired $^{18}$O values and Mg/Ca ratios, have been described by Reinhardt *et al.* (1998a, b), Flecker & Ellam (1999), Lea, in Sen Gupta (1999), Reinhardt *et al.* (2003) and Quillmann *et al.* (2012). Quillmann *et al.* (2012) used paired $^{18}$O values and Mg/Ca ratios from *Cibicides lobatulus* to infer cooling and freshening on the north-western Iceland shelf at 8.2 ka, associated with the catastrophic failure of the ice dams to proglacial lakes Agassiz and Ojibway during the final deglaciation of the Laurentide ice sheet in Canada.

Proxies for palaeo-productivity or nutrient level, including Ba and Cd values, and transition-metal (Fe, Co, Ni, Cu and Ag) concentrations, have been described by Lea, in Sen Gupta (1999). Barium behaves like a nutrient, while the behaviour of cadmium in the ocean closely mimics that of phosphate in the water column.

## 5.9  Palaeoenvironmental interpretation and associated visualisation technologies

A range of quantitative techniques has been applied in the field of palaeoenvironmental interpretation in academia and in industry, including abundance, diversity, dominance and equitability; eigenvector and multivariate techniques, including cluster analysis, correspondence analysis, factor analysis and polytopic vector analysis; fuzzy C means (FCM); fuzzy logic; IPS; planktic:benthic (P:B) ratios; and transfer functions (see Jones, 1996, 2006, 2011a; and additional references cited therein; see also below, and the further reading list at the end of the chapter). A range of visualisaton tools has also been applied, including IPS overlain on work-station seismic.

*Abundance, diversity, dominance and equitability*

As noted in Section 4.6, abundance refers to the number of specimens in a sample, preferably normalised to sample size. The abundance of individual species is expressed either in absolute terms, or in relative terms, that is, as a percentage of the total abundance. Diversity refers to the number of species. It is measured and expressed as simple, Fisher, Shannon–Wiener, Simpson or SHE diversity (SHE analysis for Biozone Identification (SHEBI) is discussed in Section 7.6). In its simplest form, dominance refers to the relative abundance of the dominant species. Equitability refers to the evenness of distribution of species in a sample.

Abundance, diversity, dominance and equitability are all useful guides to environment, and in particular to bathymetry, and also to sequence stratigraphy and systems tract (ST) (Jones, 1996, 2006, 2011a; see also the further reading list at the end of the chapter, and Section 8.2). Abundance and diversity are typically highest in deep water and/or in maximum flooding surfaces (MFSs), although depth is only one of a number of variables controlling abundance and diversity. Note also that abundance and diversity are not always co-variant. High abundance associated with low diversity often indicates some form of environmental stress, such as temperature, salinity, oxicity or toxicity, as do high dominance and low equitability.

In the oil industry, SHE diversity has been used in the palaeoenvironmental interpretation of the Cenozoic of Trinidad, including the reservoir in the Dolphin field in the Columbus basin (see the further reading list at the end of the chapter; and also Section 9.3.5).

### *Eigenvector and multivariate techniques*

Eigenvector and multivariate techniques include cluster analysis, correspondence analysis, factor analysis and polytopic vector analysis. Various of the techniques have been used in the oil industry for biofacies identification, for example in the Cenozoic of the Gulf of Mexico (Jones, 1996, 2006, 2011a; see also the further reading list at the end of the chapter, and Section 9.3.6). Correspondence analysis has also been used for biozone identification (see Section 7.6).

### *Fuzzy C means (FCM)*

Fuzzy C means (FCM) is a clustering technique used in the interpretation of mixed assemblages. It has been applied in the oil industry in the interpretation of (palynological) palaeoenvironmental data from the Jurassic reservoir of Hawkins field in the North Sea (Jones, 2011a; see also the further reading list at the end of the chapter, and Section 9.3.3).

### *Fuzzy logic*

Fuzzy logic has been used in the oil industry in the palaeoenvironmental interpretation of the foraminiferal assemblages of the Palaeogene reservoir in Fleming field in the North Sea, building on previous qualitative observations on agglutinating foraminiferal morphogroup distributions (Jones, 2011a; see also the further reading list at the end of the chapter, and Section 9.3.3). It has enabled the identification of intra-reservoir mudstones constituting potential barriers and baffles to fluid flow, which has implications for integrated reservoir characterisation and for reservoir simulation. Even more importantly, it has also enabled the refinement of the simulation model, so as to better match production history, and the revision of the production strategy and well-planning programme.

### *IPS*

IPS is a presently proprietary automated system for palaeoenvironmental interpretation based on bathymetric distribution data in look-up tables. The output is in the form of a palaeobathymetric curve capable of being imported into the work-station environment for integration with well log and seismic data. Trends in the curve can be used to identify facies and/or stratigraphic discontinuities, and to constrain sequence stratigraphic interpretation and characterise systems tracts (STs) (see also

Section 7.6). Running cosine theta and Otsuka plots, which are essentially displays of correlation coefficients or similarity indices between samples, can also be used to identify facies and/or stratigraphic discontinuities.

IPS has been used in the oil industry in the palaeoenvironmental interpretation of the Cretaceous–Cenozoic of the Indus basin in Pakistan, in the Cenozoic reservoir of the Dolphin field in the Columbus basin in Trinidad, and in the Cenozoic of the Gulf of Mexico (Jones, 2006, 2011a; see also the further reading list at the end of the chapter, and Sections 9.3.2, 9.3.5 and 9.3.6).

### Planktic:benthic (P:B) ratios

The planktic:benthic (P:B) ratio tends to increase with increasing depth, or distance from shoreline, and has been used as a measure of the same (Jones, 1996, 2006, 2011a; see also Section 4.6, and the further reading list at the end of the chapter). It has been used in the oil industry in the palaeoenvironmental interpretation of the Cenozoic of the Gulf of Mexico (see the further reading list at the end of the chapter; see also Section 9.3.6).

### Transfer functions

Transfer functions based on the depth distributions of Foraminifera have been used in sea-level reconstructions, for example in the Quaternary and Tertiary (Jones, 2006, 2011a; see also the further reading list at the end of the chapter, and Section 13.1.2).

## 5.10 Further reading

### General

Bandy & Arnal, 1960; Belanger, 2011; Berkeley *et al.*, 2009; BouDagher-Fadel, 2008, 2012; Buzas & Sen Gupta, 1982; Chamney, 1976; Frerichs, 1970; Hayward, 1986, 1990; Myers & Cole, in Hedgpeth, 1957; Jorissen *et al.*, in Hillaire-Marcel & de Vernal, 2007; Kucera, in Hillaire-Marcel & de Vernal, 2007; Berger, in Lipps *et al.*, 1979; Lipps *et al.*, 1979; Lowemark *et al.*, 2008; Murray, 1973, 1982, 1991, 2006; Murray & Alve, 2011; Murray *et al.*, 2011; Nguyen *et al.*, 2009; Pawlowski *et al.*, 2003b; Pipperr, 2011; Pipperr & Reichenbacher, 2010; Reolid *et al.*, 2010, 2012a; Martin, in Sen Gupta, 1999; Sliter *et al.*, 1975; Tendal, 1979; Tendal & Hessler, 1977; Toth *et al.*, 2010; Wang & Murray, 1983.

### Agglutinating Foraminifera

Nagy *et al.*, in Hart *et al.*, 2000; Nagy & Murray, in Kaminski & Filipescu, 2011; Nagy, 1985a, b, 1992, 2005; Nagy & Johansen, 1991; Nagy & Seidenkrantz, 2003; Nagy *et al.*, 1984, 2001, 2010, 2011.

*Calcareous benthic and planktic Foraminifera*

Beavington-Penney & Racey, 2004; BouDagher-Fadel, 2008, 2012; Corfield, in Hart, 1987; d'Hondt & Zachos, 1998; Houston & Huber, 1998; Keller, in Kennett, 1985; Koutsoukos *et al.*, 1989; Leckie, 1987; Martinez, 2003; Rohling *et al.*, 2004; Ruckheim *et al.*, 2006.

*Palaeobathymetry*

Abramovich *et al.*, 2003; Ando *et al.*, 2010; Bandy, 1970; McNeil, in Dixon *et al.*, 1985; Gallagher, 1998; Gu Songzhu *et al.*, 2007; Haig, 2003, 2005, 2012; Funnell, in Hallam, 1967; Hart, 1980a, b; Haywick & Henderson, 1991; Olsson & Usmani, in Ishizaki & Saito, 1992; Johnson, 1976; Kaminski & Gradstein, 2005; Podobina, in Kaminski *et al.*, 1995; Kotthoff *et al.*, 2010; Koutsoukos, 1985b; Koutsoukos & Merrick, 1986; Koutsoukos *et al.*, 1990, 1991; Kovacs, 2005; Carrillo *et al.*, in Langer, 1995; Leckie, 1987; Moullade, in Oertli, 1984; Nielsen *et al.*, 2008; Katz *et al.*, in Olson & Leckie, 2003; Thompson & Abbott, in Olson & Leckie, 2003; Olsson & Nyong, 1984; Pflum & Frerichs, 1976; Pujos-Lamy, 1984; Kaminski *et al.*, in Rogl & Gradstein, 1988; Olsson & Wise, in Ross & Haman, 1987; Olsson *et al.*, in Ross & Haman, 1987; Saint-Marc, 1992; Scheibnerova, 1978; Ingle, in Sliter, 1980; Sliter & Baker, 1972; Smith *et al.*, 2010; van Hinsbergen *et al.*, 2005; van Morkhoven *et al.*, 1986, Hart & Bailey, in Wiedmann, 1979; Olsson, in Wilgus *et al.*, 1988.

*Marginal marine environments (including peri-deltaic sub-environments)*

Kohl *et al.*, in Anderson & Fillon, 2004; Callard *et al.*, 2011; Fiorini *et al.*, 2010; Nagy *et al.*, in Hart *et al.*, 2000; Johnson *et al.*, 2005; Nagy & Murray, in Kaminski & Filipescu, 2011; Nagy *et al.*, in Kaminski *et al.*, 1995; Nagy, 1985a, b, 1992, 2005; Nagy & Johansen, 1991; Nagy & Seidenkrantz, 2003; Nagy *et al.*, 1984, 2001, 2010, 2011; Tibert *et al.*, in Olson & Leckie, 2003; Robinson & McBride, 2006; Govindan & Bhandari, in Rogl & Gradstein, 1988; Rossi & Vaiani, 2008; Batjes, in Saunders, 1968; Tibert & Leckie, 2004.

*Shallow marine environments (including peri-reefal sub-environments)*

Abdulsamad & Barbieri, 1999; Aigner, 1983; Bailey *et al.*, 1989; Beavington-Penney, 2004; Beavington-Penney & Racey, 2004; Beavington-Penney *et al.*, 2005, 2006; Brasier, 1975; Briguglio & Hohenegger, 2009, 2011; Buxton & Pedley, 1989; Carozzi *et al.*, 1976; Chaproniere, 1975, 1980; Dawson, 1993; Dawson & Racey, 1993; Dawson *et al.*, 1993; Decrouez & Lanterno, 1979; Eva, 1976, 1980; Gargouri, 1988; Geel, 2000; Ghose, 1977; Grosheny & Tronchetti, 1988; Gusic *et al.*, 1988; Hallock & Glenn, 1985, 1986; Hallock & Pomar, 2008;

Hamaoui, 1976; Moody, in Hart, 1987; Pautal, in Hart, 1987; Jorry *et al.*, 2003; Loucks *et al.*, in MacGregor *et al.*, 1998; Mateu-Vicens *et al.*, 2011; Nazri Ramli & Ho Kiam Fui, 1984; Benjamini, in Neale & Brasier, 1981; Berthou, in Oertli, 1984; Billotte, in Oertli, 1984; Luterbacher, in Oertli, 1984; Pelissie *et al.*, in Oertli, 1984; Serra-Kiel & Regaunt, in Oertli, 1984; Tronchetti, in Oertli, 1984; Papazzoni, 2010; Racey, 2001; Racey *et al.*, 2001; Reuter *et al.*, 2010; Enos, in Roehl & Choquette, 1985; Romero *et al.*, 2002; Saint-Marc, 1982; Mzoughi & Kohler, in Salem & Oun, 2003; Sartorio & Venturini, 1988; Scheibner *et al.*, 2003; Scott, 1990; Singh, 1984; Cowen, in Tevesz & McCall, 1983; Trifonova & Vaptzarova, 1988; van Gorsel, 1988; Vennin *et al.*, 2003.

*Deep marine environments (including submarine fan sub-environments and analogues)*

Allison *et al.*, 2008; Alve, 1999; Kohl *et al.*, in Anderson & Fillon, 2004; Anschutz *et al.*, 2002; Kohl *et al.*, in Bouma *et al.*, 1985; Brouwer, 1965; Brunner & Ledbetter, 1987; Brunner & Normark, 1985; Butt, 1981; Fillon, in Demchuk & Gary, 2009; Drinia *et al.*, 2007; Duros *et al.*, 2012; Ellison & Peck, 1983; Ernst *et al.*, 2002; Fontanier *et al.*, 2005, 2008; Gradstein & Berggren, 1981; Grun *et al.*, 1964; Hayward *et al.*, 2011; Hess & Kuhnt, 1996; Hess *et al.*, 2001, 2005; Hilterman, 1968; Ingram *et al.*, 2010; Thompson, in Ishizaki & Saito, 1992; Schmiedl *et al.*, in Jorissen & Rohling, 2000; Kaminski, 1985; Kaminski & Gradstein, 1987; Kaminski & Schroder, 1987; Ksiazkiewicz, 1975; Kitazato, in Langer, 1995; Levin *et al.*, 1991; Linke *et al.*, in Loubere, 1995; Natland, 1963; Winkler, in Oertli, 1984; Ortiz *et al.*, 2011; Panieri *et al.*, 2009; Price *et al.*, 2008; Rios-Netto, *et al.*, 2010; Rogerson *et al.*, 2006; Kaminski *et al.*, in Rogl & Gradstein, 1988; Schafer, 1982; Tyszka *et al.*, 2010; Vella, 1964; Gradstein *et al.*, in Verdenius *et al.*, 1983.

*Palaeo(bio)geography*

Adams, 1976, 1980; Adams *et al.*, 1990; Adams, in Adams & Ager, 1967; Gobbett, in Adams & Ager, 1967; Anglada & Radrianasolo, 1985; Bassoulet *et al.*, 1985; BouDagher-Fadel, 2008, 2012; Rogl & Steininger, in Brenchley, 1984; Caus *et al.*, 2009; Hart *et al.*, in Crame & Owen, 2002; Dalby *et al.*, 2009; Goldbeck & Langer, in Demchuk & Gary, 2009; Longoria & Monreal, in Demchuk & Gary, 2009; Steininger & Rogl, in Dixon & Robertson, 1984; Drooger, 1979; Gradstein *et al.*, 1999; Haig, 1979; Adams, in Hallam, 1973; Dilley, in Hallam, 1973; Gobbett, in Hallam, 1973; Ross, in Hallam, 1973; van Gorsel, in Hedley & Adams, 1978; Kucera, in Hillaire-Marcel & de Vernal, 2007; Hudson *et al.*, 2009; Adams, in Ikebe & Tsuchi, 1984; Kaminski & Gradstein, 1987; Kureshy, 1980; Pignatti, in Matteucci *et al.*, 1994; Adams *et al.*, in Meulenkamp, 1983; Dilley, in Middlemiss *et al.*, 1971; Moullade *et al.*, 1985;

Thompson & Abbott, in Olson & Leckie, 2003; Renema, in Renema, 2007; Renema *et al.*, 2008; Rogl, 1998; Ross, 1967; Scheibnerova, 1970; Steininger *et al.*, 1985; Bignot & Pomerol, in Whittaker & Hart, 2010; Adams *et al.*, in Whybrow & Hill, 1999.

### *Palaeoceanography*

Brown *et al.*, 2011; Holbourn *et al.*, in Cameron *et al.*, 1999; Preece *et al.*, in Cameron *et al.*, 1999; Cannariato *et al.*, 1999; Coccioni *et al.*, 2006; Coxall *et al.*, 2007; d'Hondt & Zachos, 1998; Elkhazri *et al.*, 2013; Erbacher *et al.*, 1999; de Romero *et al.*, 2003; Berger, in Emiliani, 1981; Vincent & Berger, in Emiliani, 1981; Friedrich, 2010; Friedrich *et al.*, 2003, 2011; Gebhardt *et al.*, 2004; Gradstein *et al.*, 1999; Harris, 2005; Jorissen *et al.*, in Hillaire-Marcel & de Vernal, 2007; Kucera, in Hillaire-Marcel & de Vernal, 2007; Holbourn *et al.*, 2001; Huang *et al.*, 2003; Crespo de Cabrera *et al.*, in Huber *et al.*, 1999; Huber *et al.*, in Huber *et al.*, 1999; Kuhnt & Wiedman, in Huc, 1995; Jarvis *et al.*, 1988; Jenkyns, 1980; Jorissen, 1999; Kaiho, 1991; Katz & Thunell, 1984; Komiya *et al.*, 2008; Koutsoukos *et al.*, 1990; de Romero & Galea-Alvarez, in Langer, 1995; Leckie *et al.*, 2002; Lob & Mutterlose, 2012; Loubere, 1995; Luning *et al.*, 2004; Macellari & de Vries, 1987; Mann & Stein, 1997; Martinez, 2003; Mullineaux & Lohmann, 1981; Nagy *et al.*, 2009, 2010; Requejo *et al.*, 1994; Reolid *et al.*, 2008b; Reolid *et al.*, 2012b; Rodrigues, 1993; Lea, in Sen Gupta, 1999; Skelton, 2003; Smart & Ramsay, 1995; Tribovillard *et al.*, 1991; Tyszka, 1994; Vismara-Schilling & Coulbourn, 1991; Widmark, 1995; Widmark & Speijer, 1997; Stefanelli & Kapotondi, 2008; Wilson, 2008a; Zapata *et al.*, 2003.

### *Palaeoclimatology*

Adams *et al.*, 1990; Bicchi *et al.*, 2003; Boersma & Premoli Silva, 1983, 1991; Olsson, in Buzas & Sen Gupta, 1982; CLIMAP, 1976; 1981; Cronblad & Malmgren, 1981; Chapman, in Culver & Rawson, 2000; de Man & van Simaeys, 2004; Emiliani, 1955; Epstein *et al.*, 1953; Filipsson *et al.*, 2010; Fisher *et al.*, 2003; Hallock *et al.*, 1991; Hay & Floegel, 2012; Jorissen *et al.*, in Hillaire-Marcel & de Vernal, 2007; Kucera, in Hillaire-Marcel & de Vernal, 2007; Husum *et al.*, 2010; Imbrie *et al.*, 1973; Feldman *et al.*, in Longoria & Gamper, 1998; Martin *et al.*, 1997; McCrea, 1950; Thompson & Abbott, in Olson & Leckie, 2003; Ruddiman & Sarnthein, 1986; Sancetta *et al.*, 1972; Stehli *et al.*, in Schopf, 1972; Chumakov, in Semikhatov & Chumakov, 2004; Lea, in Sen Gupta, 1999; Rohling & Cooke, in Sen Gupta, 1999; Shackleton & Opdyke, 1973; Urey, 1947; Hart, in Williams *et al.*, 2007; Kucera & Schonfeld, in Williams *et al.*, 2007; Sellwood & Valdes, in Williams *et al.*, 2007; Sluijs *et al.*, in Williams *et al.*, 2007.

*Biogeochemical proxies in palaeoenvironmental interpretation*

Brown *et al.*, 2011; Cronblad & Malmgren, 1981; Grocke & Wortman, 2008; Katz *et al.*, 2010; Martin *et al.*, 1997; Rohling & Cooke, in Sen Gupta, 1999.

*Palaeoenvironmental interpretation technologies*

Baldi & Hohenegger, 2008; Borcic *et al.*, 2010; Fluegeman, in Demchuk & Gary, 2009; Gary *et al.*, in Demchuk & Gary, 2009; Edwards *et al.*, 2004; Gebhardt, 2006; Hohenegger, 2005; Hohenegger *et al.*, 2008; Horton *et al.*, 1999; Leorri *et al.*, 2008; Lesslar, 1987; Massey *et al.*, 2006; Rossi & Horton, 2009; Salgueiro *et al.*, 2008; Fang, in Olson & Leckie, 2003; Olson *et al.*, in Olson & Leckie, 2003; Wakefield, in Olson & Leckie, 2003; Perez-Asensio *et al.*, 2012; Punyasena *et al.*, 2012; Crux *et al.*, in Ratcliffe & Zaitlin, 2010; Jones *et al.*, in Repetski, 1996; Robinson & Kohl, 1978; Sejrup *et al.*, 2004; Parker & Arnold, in Sen Gupta, 1999; Lagoe *et al.*, in Takayanagi & Saito, 1992; Wakefield & Monteil, 2002; Wakefield *et al.*, 2001; Wilson, 2008b, 2010, 2011, Wilson & Costelloe, 2011a; Wilson *et al.*, 2012; Zellers & Gary, 2007; Zivkovic & Glumac, 2007.

# 6

# Palaeobiological or evolutionary history of the Foraminifera

This chapter deals with the palaeobiological or evolutionary history of the Foraminifera.

## 6.1 Evolutionary history of the Foraminifera

The following is an abridged account of the evolutionary history of the Foraminifera, which constitute part of the Mesozoic–Cenozoic or modern evolutionary biota of Sepkoski (1981), based on Jones (2006, 2011a), and in turn on data in Haynes (1981) and Loeblich & Tappan (1964, 1987) (see also the further reading list at the end of the chapter).

The fossil record of the Foraminifera is sufficiently good, and sufficiently well documented, to allow detailed observations to be made on the evolution and control of at least the familial-level diversity of the group through time (Fig. 6.1). Familial-level diversity data from a database based on Haynes's scheme has been plotted against time, enabling observations on trends through time. Diversity data from Loeblich & Tappan's scheme has also been plotted for comparative purposes (not shown here). 'Taxonomic bias' is evident only in the cases of the interpreted apparent rather than real end-Carboniferous disappearance and Holocene appearance events in the Loeblich & Tappan data, and other than in these cases can be effectively eliminated.

Although molecular biology indicates that Foraminifera evolved, from a cercozoan ancestor, in the Neoproterozoic, around 800 Ma, there is incontrovertible morphological evidence for their existence only from the Early Cambrian, 'Nemakit–Daldynian' to 'Tommotian' onwards (see below).

They diversified in the Palaeozoic, slowly in the Cambrian–Silurian, steadily in the Devonian, and then rapidly, especially in the case of the Fusulinida, in the Carboniferous–Permian, before suffering significant losses in the end-Permian mass extinction, one of the 'big five' of Raup & Sepkoski (1982), when 64% of all families were wiped out, all of them Fusulinida.

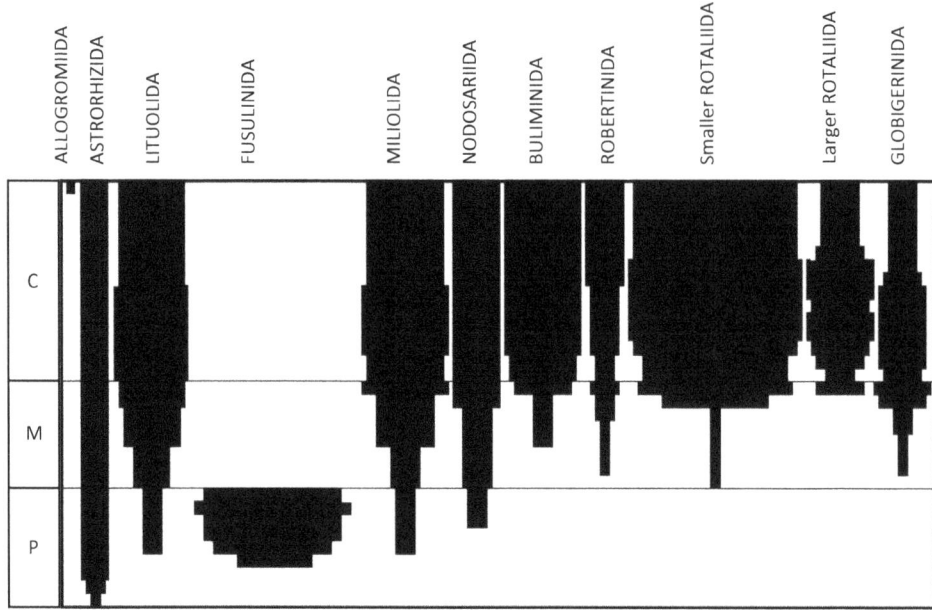

Fig. 6.1    **Familial-level diversity of Foraminifera through time**. Based on data in Haynes (1981). P = (Late) Palaeozoic; M = Mesozoic; C = Cenozoic.

Foraminifera then recovered and diversified again through the Mesozoic, slowly in the Triassic, which is best described a 'recovery interval', steadily in the Jurassic, and then rapidly in the Cretaceous, especially in the case of the larger Rotaliida and Globigerinida, before again suffering significant losses in the end-Cretaceous mass extinction, another of the big five, when 19% of all families were wiped out, all of them either larger Rotaliida, which lost 43% of all their families, or Globigerinida, which lost 86% of theirs.

Foraminifera then recovered and diversified yet again in the Cenozoic, slowly in the Palaeocene, which is best described a recovery interval, and then rapidly in the Eocene, according to data based on Haynes's scheme attaining their maximum diversity in the Eocene, and declining over the interval from the Oligocene to the present, during which 13% of all families have become extinct, all of them specialist larger Rotaliida and Globigerinida.

### *Agglutinating Foraminifera (Astrorhizida and Lituolida)*

As intimated above, the agglutinating Foraminifera appear on morphological evidence to have evolved in the Cambrian, and diversified in the Palaeozoic, and, especially, Mesozoic and Cenozoic, attaining their maximum diversity in the Palaeogene (see the further reading list at the end of the chapter). The oldest known is *Platysolenites* from the Early Cambrian, Nemakit–Daldynian to Tommotian of Newfoundland, Great Britain, Scandinavia, Estonia and Russia. Taphonomic, teleological and ultrastructural studies suggest that ancient *Platysolenites* had a similar grade of organisation and mode of life to modern *Bathysiphon* (an epifaunal detritus feeder), although morphological evidence, specifically the presence of an agglutinated proloculus, indicates that it was probably not directly related.

### Smaller and larger calcareous benthic Foraminifera (Fusulinida, Miliolida, Nodosariida, Buliminida, Robertinida and Rotaliida)

The Fusulinida evolved in the Devonian, and diversified rapidly in the Carboniferous to Early Permian, attaining their maximum diversity in the Middle Permian, before dying out during the end-Permian mass extinction (see the further reading list at the end of the chapter). The rapid diversification of the interpreted specialist K-strategist shallow marine larger benthic Fusulinida in the Carboniferous may have been on account of the occupation of niches vacated by stromatoporoids and tabulate corals during the Late Devonian mass extinction. It could be argued that the disproportionate loss that they suffered in the end-Permian mass extinction may actually have been on account of their specialisation in and dependence on vulnerable shallow marine habitats (so-called 'selectivity by trait').

The Miliolida evolved in the Carboniferous, and diversified in the Mesozoic, attaining their maximum diversity during the Late Cretaceous, before suffering some losses in the end-Cretaceous mass extinction, and recovering in the Palaeogene, when they attained their maximum diversity.

The Nodosariida evolved in the Permian, and diversified in the Triassic and Jurassic, before attaining their maximum diversity in the Cretaceous. They moved from shallow to deep environments in the Cretaceous, perhaps in response to competition from the diversifying Rotaliida.

The Buliminida evolved in the Jurassic, and diversified rapidly in the Late Cretaceous, although they actually attained their maximum diversity in the Palaeogene. Their turnover in the Late Cretaceous may have been driven by the intermittent Oceanic Anoxic Events or OAEs taking place at this time. Interestingly, a significant number of specific-level extinctions took place among the deep marine Buliminida in the Pleistocene.

The Robertinida evolved in the Middle Triassic, and diversified in the Jurassic to Cretaceous, before suffering some losses in the end-Cretaceous mass extinction, and recovering in the Palaeogene, attaining their maximum diversity in the Neogene. Like the Nodosariida, they moved from shallow to deep environments in the Cretaceous.

The smaller Rotaliida evolved in the Triassic, and diversified in the Jurassic and, especially, Cretaceous, before suffering some losses in the end-Cretaceous mass extinction, and recovering in the Palaeogene, when they attained their maximum diversity.

The larger Rotaliida evolved in the Late Cretaceous, and diversified rapidly, before suffering significant losses in the end-Cretaceous mass extinction, and recovering in the Palaeogene, when they attained their maximum diversity. The evolution of the interpreted specialist K-strategist shallow marine larger benthic Rotaliida may have been on account of the occupation of niches newly created by the sea-level rise of the time. It could be argued that the disproportionate loss that they suffered in the end-Cretaceous mass extinction may actually have been on account of their specialisation in and dependence on vulnerable shallow marine habitats (see comments under Fusulinida above).

### Planktic Foraminifera (Globigerinida)

The Globigerinida evolved in the Middle Triassic (?) or Middle Jurassic, and diversified rapidly in the Cretaceous, when they attained their maximum diversity, before suffering significant losses, and indeed almost becoming extinct in the end-Cretaceous mass extinction, and recovering in the Palaeogene (see the further reading list at the end of the chapter). The evolution of the Globigerinida in the Cretaceous may have been on account of the occupation of niches newly created by the sea-level rise of the time; evolutionary turnover, by associated intermittent OAEs. It could be argued that the disproportionate loss that they suffered in the end-Cretaceous mass extinction may actually have been on account of their specialisation in and dependence on vulnerable habitats (see comments under Fusulinida and larger Rotaliida above).

### 6.2 Further reading

*General*

Hart & Williams, in Benton, 1993; Bosak *et al.*, 2012; Brasier, in Bosence & Allison, 1995; McKinney, in Briggs & Crowther, 2001; Kaminski, in Bubik & Kaminski, 2004; Brown, 1995; Erwin, 1998; Gaucher & Germs, in Gaucher *et al.*, 2010; Gaucher & Sprechman, 1999; Goldstein & Bernhard, 1997; Jablonski & Raup, 1995; Kiessling, 2006; Murray, in Jenkins & Murray, 1989; Lipps & Goldstein, 2006; Parente *et al.*, 2008; Pawlowski & Gooday, 2009; Pawlowski *et al.*, 2003b; Prokoph *et al.*, 2004; Jablonski, in Rothschild & Lister, 2003; Sen Gupta, in Sen Gupta, 1999; Tappan & Loeblich, 1988; von Koenigswald *et al.*, 1963; Wray *et al.*, 1995.

*Agglutinating Foraminifera*

BouDagher-Fadel, 2008; Haynes, 1981; Holcova, 2003, 2004, 2010; Kaminski *et al.*, 2010; McIlroy *et al.*, 2001; Scott *et al.*, 2003.

## Calcareous benthic Foraminifera

Bond & Wignall, 2009; BouDagher-Fadel, 2008; Groves & Altiner, 2005; Groves & Lee, 2008; Groves & Wang Yue, 2009; Hayward, 2001, 2002; Hayward & Kawagata, 2005; Hayward *et al.*, 2010; Kawagata *et al.*, 2005; Krainer & Vachard, 2011; Marquez, 2005; Schonfeld, in Moguilevsky & Whatley, 1996; Ruban, 2011; Vachard *et al.*, 2010; Weinholz & Lutze, 1989.

## Planktic Foraminifera

Banner & Lowry, 1985; BouDagher-Fadel, 2012; BouDagher-Fadel *et al.*, 1997; Caron & Homewood, 1983; Coxall *et al.*, 2007; Hart *et al.*, in Crame & Owen, 2002; Darling *et al.*, 1997; di Bari, 1999; Hart, 1980b; Hart *et al.*, 2003, 2012; Haynes, 1981; Hillebrandt & Urlichs, 2008; Korchagin *et al.*, 2003; Leckie *et al.*, 2002; Schmidt *et al.*, 2004; Hart & Ball, in Summerhayes & Shackleton, 1986; Wernli, 1988.

# 7

# Biostratigraphy

This chapter deals with biostratigraphy. It contains sections on Foraminifera in biostratigraphy; on the Palaeozoic, Mesozoic and Cenozoic; on biostratigraphic and associated visualisation technologies; and on stratigraphic time-scales.

## 7.1 Biostratigraphy

Biostratigraphy involves the use of fossils in the ordering of the rock record in time (Jones, 1996, 2006, 2011a; see also McGowran, 2005). It relies on William Smith's fundamental 'Law of Strata Identified by Organised Fossils', which posits that particular ages of rock, regardless of facies, can be characterised at any given locality, and correlated from one locality to another, by means of index or marker fossils (see below). Like palaeoenvironmental interpretation, it can be impaired by natural, taphonomic, factors or biases; and also by artificial factors or biases, such as species identification bias, or subjectivity and inaccuracy in the identification of species, in turn leading to interpretation bias (see, for example, Ginsburg, 1997; see also Sections 2.3, 2.4 and 3.2.1).

Biostratigraphy has widespread applications in academia, in field geology, and in marine geology. It is and has long been critical to field geological mapping, for example that undertaken by the British Geological Survey in the British Isles (Wilkinson, 2011), and, in collaboration with local surveys, overseas, as in the Borbon, Manabi and Progreso basins in Ecuador (Whittaker, 1988). It makes extensive use both of macrofossils, which can be identified in the field, and also of microfossils such as Foraminifera, which are generally identified in the laboratory (note in this context, though, that certain microfossils, such as Larger Benthic Foraminifera (LBFs) are sufficiently large as to be identifiable as least to generic level in hand specimen in the field, for example, larger Fusulinida in the Permian of New Mexico and west Texas in the south-western United States; larger agglutinating Foraminifera in the Cretaceous of the Middle East; and larger

Miliolida and Rotaliida in the Palaeogene of the Middle East, north Africa, northern South America and the Caribbean, and the Pyrenees, and in the Neogene of the Far East). In the field and/or in the laboratory, it assists not only in the characterisation of age and in correlation but also in the construction of an integrated tectono-sequence stratigraphic framework, and in the constraint of tectonic evolution, for example in the constraint of the collision-related uplift in the Miocene–Pliocene of Timor, by means of detailed palaeobathymetric interpretation (Haig, 2012).

Biostratigraphy also has widespread applications in industry, in petroleum geology (see Chapter 9), in minerals geology (see Chapter 10), in engineering geology (see Chapter 11), and in archaeology (see Chapter 13).

### *Biostratigraphic index or marker fossils*

Biostratigraphically useful so-called index or marker fossils share two common characteristics:

- firstly, high evolutionary turnover rates, and hence short stratigraphic ranges;
- and secondly, unrestricted ecological distributions (for example, across a range of bathymetric zones or biogeographic provinces.

The most useful, for practical purposes, are also abundant, well preserved, and easy to identify.

Many of them are planktic.

### *Biostratigraphic zonation or biozonation*

The fundamental unit in biostratigraphy is the biozone (Fig. 7.1). The main types of biozone are defined on:

- evolutionary or First Appearance Datums or FADs, also known as First Occurrences or FOs, or, in the oil industry, as Last Down-hole Occurrences or LDOs, or as 'bases', of marker species (see above);
- extinction or Last Appearance Datums or LADs, also known as Last Occurrences or LOs, or, in the oil industry, First Down-hole Occurrences or FDOs, or 'tops';
- total, partial or concurrent (overlapping) ranges; and abundances or acmes.

In the oil industry, biozones are defined on FDOs or tops, or on first consistent or common down-hole occurrences or FCDOs, because bases can appear low in ditch-cuttings samples due to down-hole contamination or 'caving' (see Chapter 9). Note, though, that tops can appear high due to reworking. Also, tops can appear low in the event that the rock record is either incomplete, or contains unfavourable facies

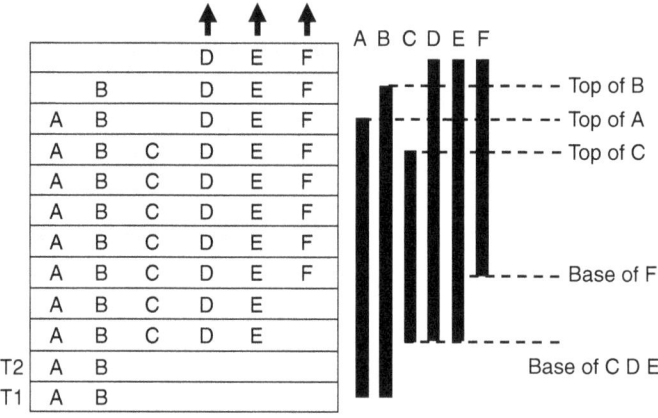

Fig. 7.1 **The basis of biostratigraphic zonation**. T1, T2 etc. are successive time-slices represented by rock units. A–F are fossils occurring in the various time-slices. Their overall ranges, between their first and last appearances, or bases and tops, are indicated on the right. Biostratigraphic zones or biozones can be defined, for example, between the bases of C, D and E, and the base of F, etc. From Jones (2006).

(in which latter case they are referred to as 'ecologically depressed'). Integration of bio- and seismic-sequence- stratigraphy enables the discrimination of apparent from real tops (see Section 8.2 and Chapter 9).

## *Biostratigraphic correlation*

Biostratigraphic correlation is superior to lithostratigraphic correlation in that it allows the identification of rock units that are correlatable in time and space, or are isochronous; and the discrimination of those that are not, or are diachronous.

   Correlation is critically important in the construction of accurate and meaningful time-slice Gross Depositional Environment (GDE) and Common Risk Segment (CRS) maps in play fairway evaluation, in petroleum exploration (see 9.1.2., 9.2. and 9.3). It is also important in integrated reservoir description or characterisation, and in reservoir modelling or simulation, in reservoir exploitation (see Sections 9.1.3, 9.4 and 9.5).

## 7.2 Foraminifera in biostratigraphy

Stratigraphic range data are available in a number of public-domain databases, including CHRONOS (www.chronos.org), the Ocean Drilling Stratigraphic

Network (www.odsn.de), PANGAEA (www.pangaea.de), and Plankrange (http://palaeo.gly.bris.ac.uk/Data/plankrange.html).

### *Proteinaceous Foraminifera (Allogromiida)*

There are no ancient proteinaceous Foraminifera, or Allogromiida, known to the author.

### *Agglutinating Foraminifera (Astrorhizida and Lituolida)*

The smaller agglutinating Foraminifera are regionally useful in biostratigraphy, in the deep waters of the Mesozoic–Cenozoic of various parts of the high-latitude Austral and Boreal realms, and of the low-latitude Tethyan realm; while the larger representatives are inter-regionally useful, in the shallow waters of the Mesozoic–Cenozoic of the Tethyan realm.

### *Smaller and larger calcareous benthic Foraminifera (Fusulinida, Miliolida, Nodosariida, Buliminida, Robertinida and Rotaliida)*

The smaller Fusulinida are only extremely locally useful in biostratigraphy; while the larger representatives are inter-regionally useful, in the shallow waters of the Late Palaeozoic of the Palaeo-Tethyan realm.

The smaller Miliolida are only extremely locally useful in biostratigraphy; while the larger representatives are inter-regionally useful in the shallow waters of the Mesozoic–Cenozoic of the Tethyan realm.

The Nodosariida are locally useful in biostratigraphy, for example in the shallow waters of the Mesozoic of north-west Europe and formerly contiguous North America, in the Boreal realm.

The Buliminida are locally useful in biostratigraphy, for example in the dysoxic waters of the Mesozoic–Cenozoic of various parts of the Austral, Boreal and Tethyan realms.

The Robertinida are only locally useful in biostratigraphy, for example in the shallow waters of the Mesozoic of north-west Europe and formerly contiguous North America, in the Boreal realm.

The smaller Rotaliida are regionally useful in biostratigraphy, in the shallow to deep waters of the Mesozoic–Cenozoic of various parts of the Austral, Boreal and Tethyan realms; while the larger representatives are inter-regionally useful, in the shallow waters of the Mesozoic–Cenozoic of the Tethyan realm.

### *Planktic Foraminifera (Globigerinida)*

The Globigerinida are essentially globally useful in biostratigraphy in the Cretaceous–Cenozoic (see also below). The Oberhauserellidae are locally useful in the Triassic–Jurassic.

*Summary*

The larger agglutinating Foraminifera, Fusulinida, Miliolida and Rotaliida are inter-regionally useful in stratigraphy, in the Late Palaeozoic, Mesozoic and Cenozoic (see, respectively, Sections 7.3, 7.4 and 7.5).

The Globigerinida are essentially globally useful, in the Mesozoic and Cenozoic (see Sections 7.4 and 7.5). They exhibit short stratigraphic ranges, and, because of their free-floating habit, wide ecological distributions and facies independence, making them ideal marker fossils.

## 7.3 Palaeozoic

Biozonation schemes based on the Fusulinida have been established for the Late Palaeozoic that have inter-regional, Palaeo-Tethyan applicability (see the further reading list at the end of the chapter; see also Sections 9.3.1 and 9.3.2).

In my working experience, they have proved of particular use in the following areas:

• the Middle East (see Section 9.3.1) and north Africa; and the Indian subcontinent (see Section 9.3.2).

## 7.4 Mesozoic

Biozonation schemes based on the larger agglutinating Foraminfera, Miliolida and Rotaliida, have been established for the parts of the Mesozoic that have inter-regional, Tethyan applicability (see the further reading list at the end of the chapter; see also Sections 9.3.1. and 9.3.2). Various of the schemes have been calibrated against the absolute chronostratigraphic or geochronological time-scale by means of planktic foraminiferal or calcareous nannoplankton biostratigraphy.

Biozonations based on the Globigerinida have been established that have essentially global applicability (Fig. 7.2; see also the further reading list at the end of the chapter).

In my working experience, they have proved of particular use in the following areas:

• the Middle East (see Section 9.3.1) and north Africa;
• the Indian subcontinent (see Section 9.3.2);
• the North Sea (see Section 9.3.3);
• the South Atlantic salt basins (see Section 9.3.4);
• the eastern Venezuelan basin (see Section 9.3.5); and the Gulf of Mexico (see 9.3.6).

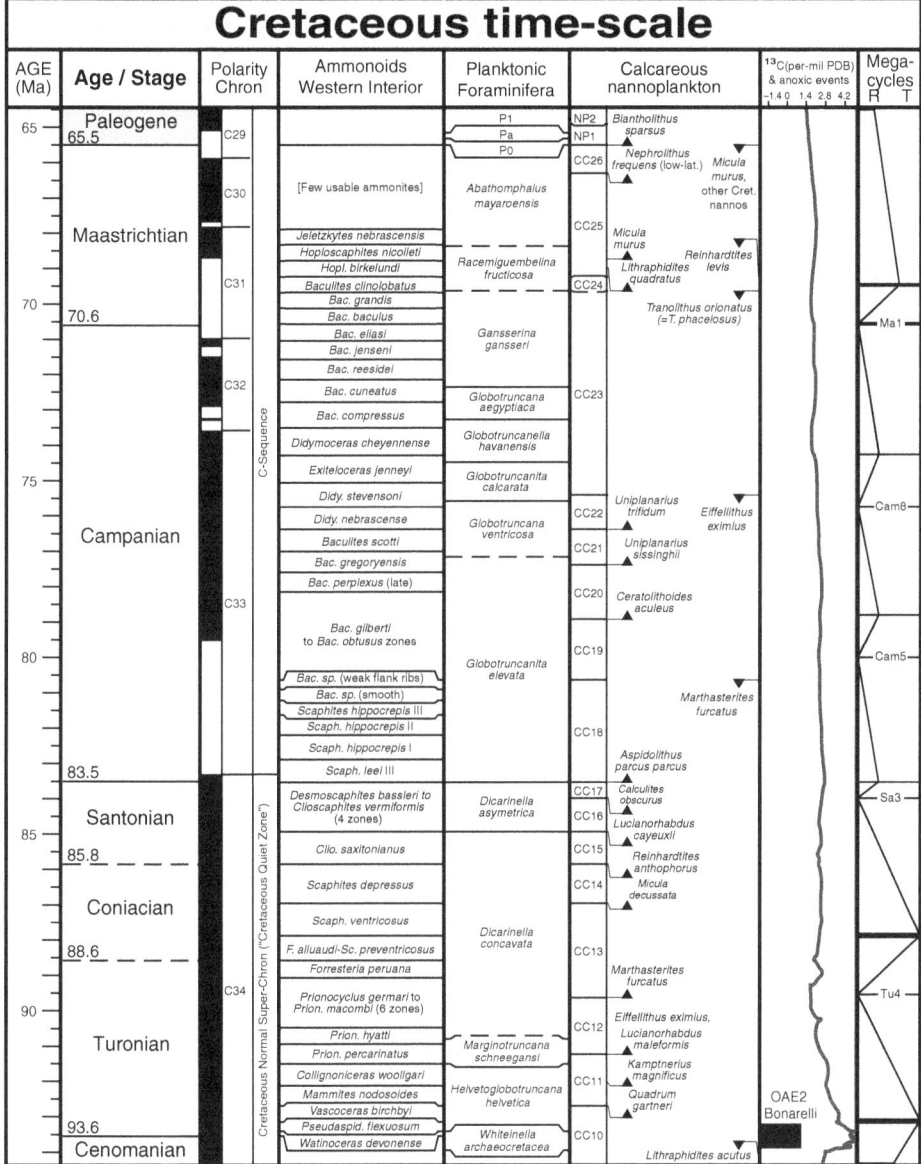

Fig. 7.2 **Cretaceous time-scale**. Reproduced with permission from Ogg *et al.* (2008).

## 7.5 Cenozoic

Biozonation schemes based on the larger agglutinating Foraminifera, Miliolida and Rotaliida, including biometrically or morphometrically based ones, have been established for the Cenozoic that have inter-regional, Tethyan applicability

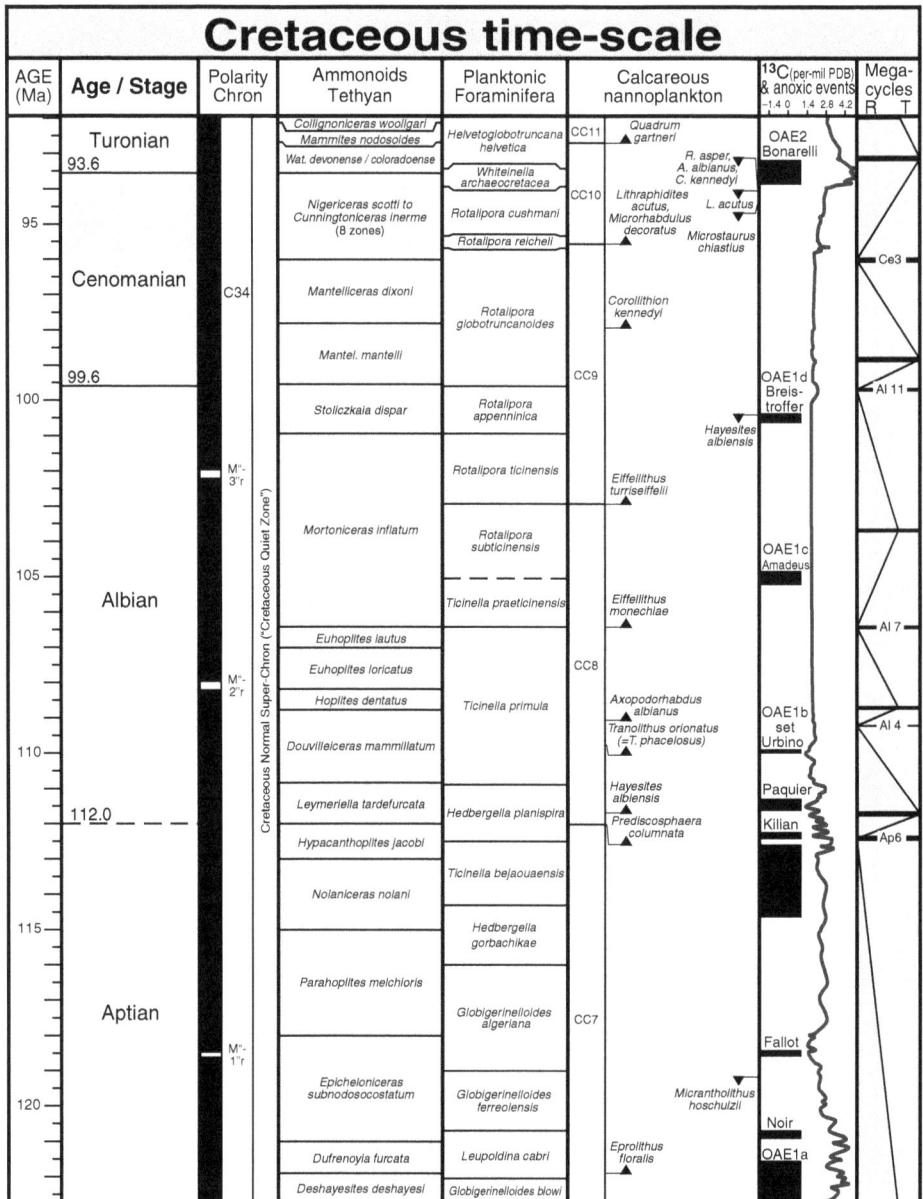

# Cretaceous time-scale

| AGE (Ma) | Age / Stage | Polarity Chron | Ammonoids Tethyan | Planktonic Foraminifera | Calcareous nannoplankton | ¹³C(per-mil PDB) & anoxic events −1.4 0  1.4 2.8 4.2 | Mega-cycles R   T |
|---|---|---|---|---|---|---|---|
| | Turonian | | Collignoniceras woollgari | Helvetoglobotruncana helvetica | CC11 | OAE2 Bonarelli | |
| 93.6 | | | Mammites nodosoides | | Quadrum gartneri | | |
| | | | Wat. devonense / coloradoense | Whiteinella archaeocretacea | | R. asper, A. albianus, C. kennedyi | |
| 95 | | | Nigericeras scotti to Cunningtoniceras inerme (8 zones) | Rotalipora cushmani | CC10 | Lithraphidites acutus, Microrhabdulus decoratus L. acutus Microstaurus chiastius | |
| | Cenomanian | C34 | | Rotalipora reicheli | | | Ce3 |
| | | | Mantelliceras dixoni | Rotalipora globotruncanoides | | Corollithion kennedyi | |
| | | | Mantel. mantelli | | | | |
| 99.6 | | | | | CC9 | OAE1d Breis-troffer | |
| 100 | | | Stoliczkaia dispar | Rotalipora appenninica | | Hayesites albiensis | Al 11 |
| | | M″-3″r | | Rotalipora ticinensis | | Eiffellithus turriseiffelii | |
| 105 | Albian | | Mortoniceras inflatum | Rotalipora subticinensis | | OAE1c Amadeus | Al 7 |
| | | | | Ticinella praeticinensis | | Eiffellithus monechiae | |
| | | | Euhoplites lautus | | | | |
| | | | Euhoplites loricatus | | CC8 | Axopodorhabdus albianus | |
| | | M″-2″r | Hoplites dentatus | Ticinella primula | | Tranolithus orionatus (=T. phacelosus) | OAE1b set Urbino | Al 4 |
| 110 | | | Douvilleiceras mammillatum | | | Hayesites albiensis | Paquier |
| | | | Leymeriella tardefurcata | Hedbergella planispira | | Prediscosphaera columnata | Kilian |
| 112.0 | | | Hypacanthoplites jacobi | | | | Ap6 |
| | | | Nolaniceras nolani | Ticinella bejaouaensis | | | |
| 115 | | | | Hedbergella gorbachikae | | | |
| | | | Parahoplites melchioris | Globigerinelloides algeriana | CC7 | Fallot | |
| | Aptian | | | | | | |
| | | M″-1″r | Epicheloniceras subnodosocostatum | Globigerinelloides ferreolensis | | Micrantholithus hoschulzii | |
| 120 | | | | | | Noir | |
| | | | Dufrenoyia furcata | Leupoldina cabri | | Eprolithus floralis | OAE1a |
| | | | Deshayesites deshayesi | Globigerinelloides blowi | | | |

(vertical label in Polarity Chron column) Cretaceous Normal Super-Chron ("Cretaceous Quiet Zone")

Fig. 7.2 (*cont.*)

(Fig. 7.3; see also the further reading list at the end of the chapter, and Sections 9.3.1 and 9.3.2). Various of the schemes have been calibrated against the absolute chronostratigraphic or geochronological time-scale by means of planktic foraminiferal biostratigraphy, magnetostratigraphy or strontium isotope stratigraphy.

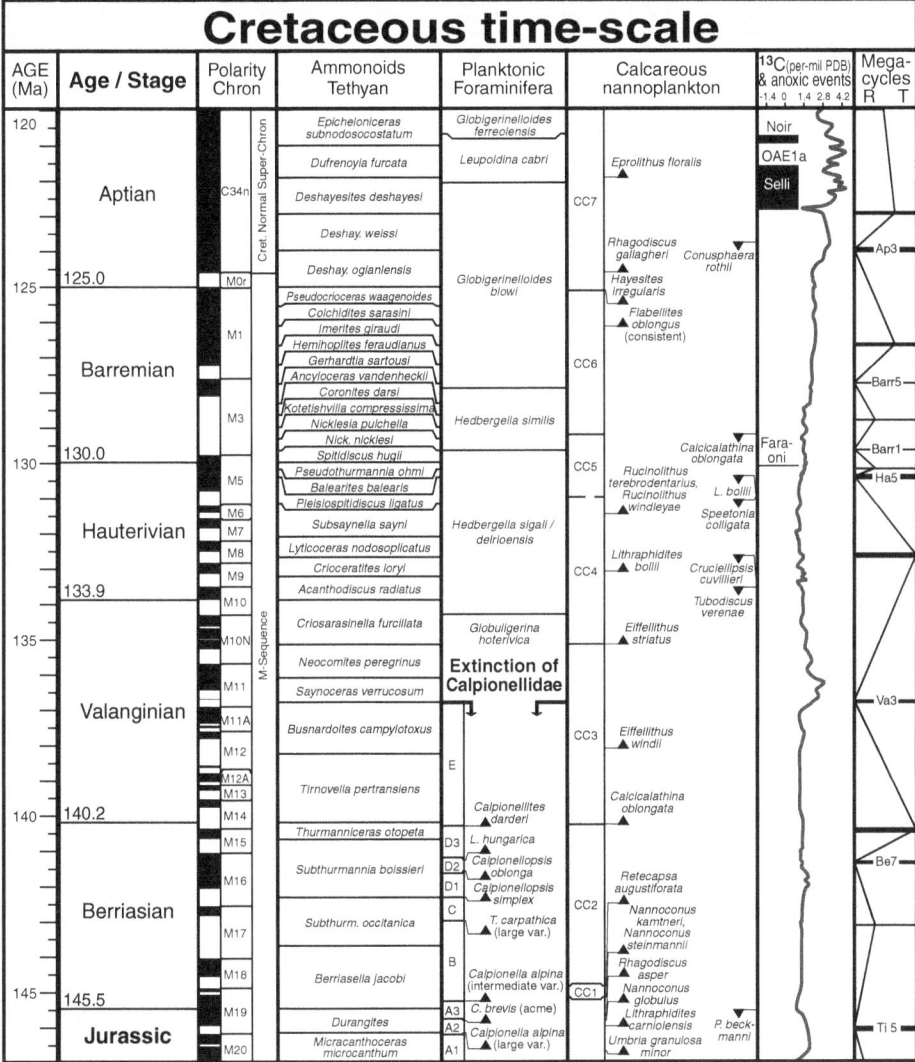

Fig. 7.2 (cont.)

Biozonations based on the Globigerinida have been established that have essentially global applicability (Figs. 7.4 and 7.5; see also the further reading list at the end of the chapter).

The global standard scheme has a mean zonal duration or resolution, in part a function of evolutionary turnover rate, of the order of 1 Ma. It has been calibrated against the geochronological time-scale by means of so-called 'astronomical tuning'.

| | Stage | Zonation | | Larger Foraminifera | |
|---|---|---|---|---|---|
| | Aquitanian | SBZ 24 | | *Miogypsina* gr. *gunteri–tani* | |
| **Oligocene** | Chattian | SBZ 23 | | *Miogypsinoides, Lepidocyclinids Nummulites bouillci* | |
| | Rupelian | SBZ 22 | b | *Lepidocyclinids, Nummulites vascus* | *Cycloclypeus* |
| | | | a | *N. fichteli, N. bouillei* | *Bullalveolina* |
| | | ? | | | |
| | | SBZ 21 | | *Nummulites vascus, N. fichteli* | |
| **Eocene** | Priabonian | SBZ 20 | | *Nummulites retiatus, Heterostegina gracilis* | |
| | | SBZ 19 | | *Nummulites fabianii, N. garnieri, Discocyclina pratti minor* | |
| | Bartonian | SBZ 18 | | *Nummulites biedai, N. cyrenaicus* | |
| | | SBZ 17 | | *Alveolina elongata, A. fragilis, A. fusiformis, Discocylina pulcra baconica Nummulites perforatus, N. brogniarti, N. biarritzensis* | |
| | Lutetian | SBZ 16 | | *Nummulites herbi, N. aturicus, Assilina gigantea, Discocylina pulcra balatonica* | |
| | | SBZ 15 | | *Alveolina prorrecta, Nummulities millecaput, N. travertensis,* | |
| | | SBZ 14 | | *Alveolina munieri, Nummulities beneharnensis, N. boussaci, Assilina spira spira* | |
| | | SBZ 13 | | *Alveolina stipes, Nummulities laevigatus, N. uranensis* | |
| | Cuisian | SBZ 12 | | *Alveolina violae, Nummulities manfredi, N. campesinus, N. caupennensis, Assilina major, A. cuvillieri* | |
| | | SBZ 11 | | *Alveolina cremae, A. dainelli, Nummulites praelaevigatus, N. nitidus, N. archiaci, Assilina laxispira* | |
| | | SBZ 10 | | *Alveolina schwageri, A. indicatrix, Nummulites burdigalensis burdigalensis, N. planulatus, Assilina placentula, Discocylina archiaci archiaci* | |
| **Paleocene** | Ilerdian | SBZ 9 | | *Alveolina trempina, Nummulites involutus, Assilina adrianensis* | |
| | | SBZ 8 | | *Alveolina corbarica, Nummulites exilis, N. atacicus, Assilina leymeriei* | |
| | | SBZ 7 | | *Alveolina moussoulensis, Nummulites praecursor, N. carcasonensis* | |
| | | SBZ 6 | | *Alveolina ellipsoidalis, A. pasticillata, Nummulites minervensis* | |
| | | SBZ 5 | | *Orbitolites gracilis, Alveolina vredenburgi, Nummulites gamardensis* | |
| | Thanetian | SBZ 4 | | *Glomalveolina levis, Nummulites catari, Assilina yvettae* | |
| | | SBZ 3 | | *Glomalveolina primaeva, Fallotella alavensis, Miscellanea yvettae* | |
| | Selandian | SBZ 2 | | *Miscellanea globularis, Ornatononion minutus, Paralockhartia eos, Lockhartia akbari* | |
| | Danian | SBZ 1 | | *Bangiana hanseni, Laffiteina bibensis* | |

Fig. 7.3 **Biostratigraphic zonation of the Palaeogene of Tethys by means of Larger Benthic Foraminifera (LBFs)**. Reproduced with permission from Gradstein *et al.* (2004).

In my working experience, they have proved of particular use in the following areas:

the Middle East (see Section 9.3.1) and north Africa;

the Indian subcontinent (see Section 9.3.2);

the North Sea (see Section 9.3.3);

the South Atlantic salt basins (see Section 9.3.4);

the eastern Venezuelan basin (see Section 9.3.5); and the Gulf of Mexico (see Section 9.3.6).

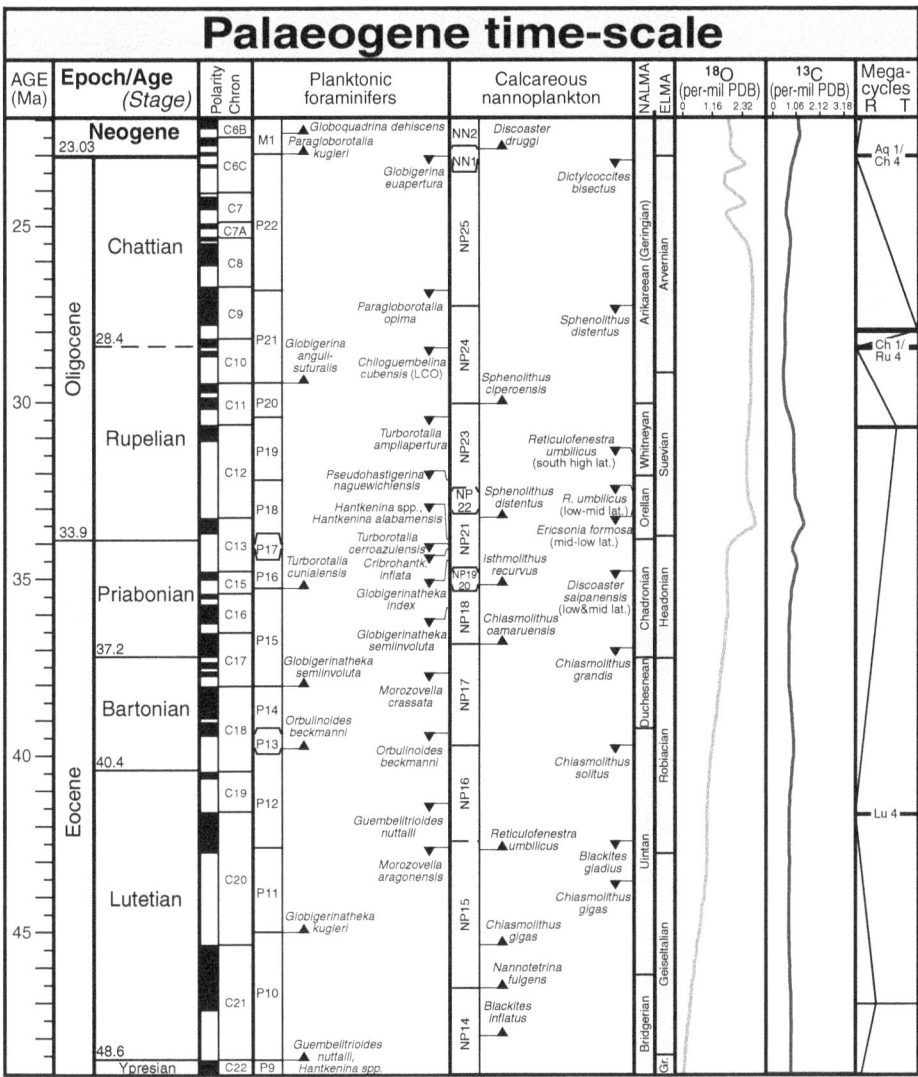

Fig. 7.4 **Palaeogene time-scale**. Reproduced with permission from Ogg *et al.* (2008).

## 7.6 Biostratigraphic and associated visualisation technologies

A range of quantitative techniques has been applied in the field of biostratigraphy in academia and in industry, including constrained optimisation (CONOP), correspondence analysis (CA), graphic correlation (GC), IPS, probabilistic correlation, ranking and scaling (RASC), and SHE analysis for biozone identification (SHEBI) (Jones, 1996, 2006; Kender *et al.*, in Kaminski & Coccioni, 2008; Kender *et al.*,

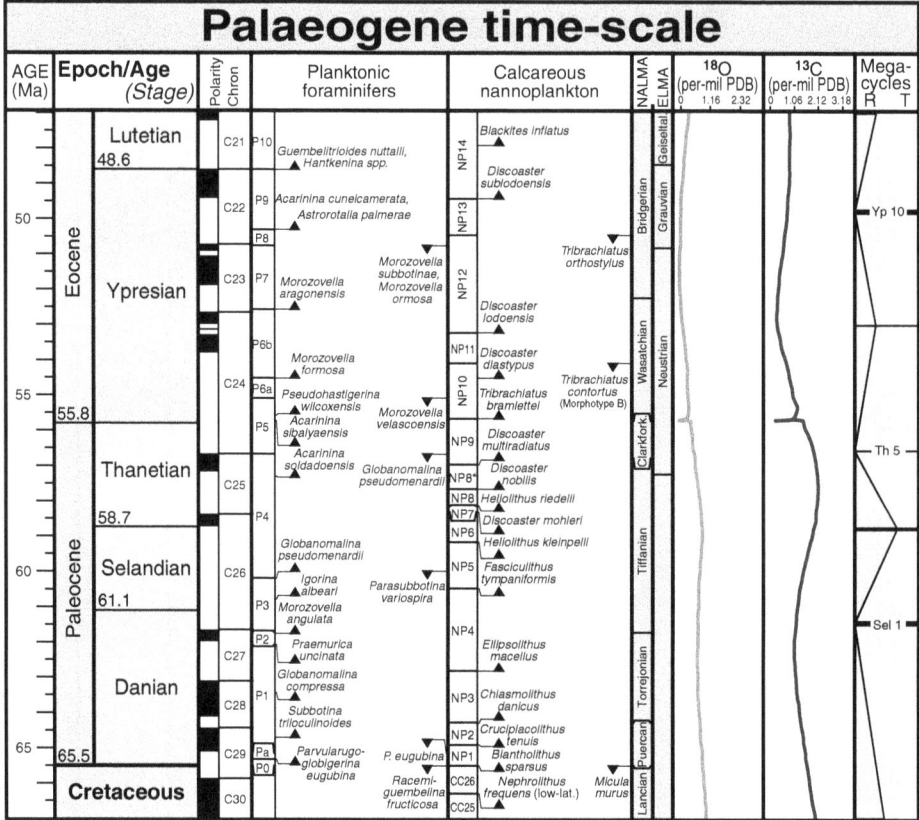

Fig. 7.4 (*cont.*)

2008; Jones, 2011a; and additional references cited therein; see also the further reading list at the end of the chapter, and Section 9.3). Each of these techniques is discussed in turn below, apart from IPS, which is discussed in Section 5.9, and probabilistic correlation.

A range of visualisaton tools has also been applied, including fossil age group plots, overlain on work-station seismic data; GC, overlain on work-station seismic; and stratigraphy to seismic (StS)™ (see the further reading list at the end of the chapter; see also Sections 9.3.3, 9.3.4, 9.3.5 and 9.3.6).

## Constrained optimisation (CONOP)

This provides a deterministic sequence of biostratigraphic events in sections, using a methodology similar to that of automated GC, but with the best-fit LOC found with a 'simulated annealing heuristic search' (Jones, 2006, 2011a; see also the further reading list at the end of the chapter, and Section 9.3.6).

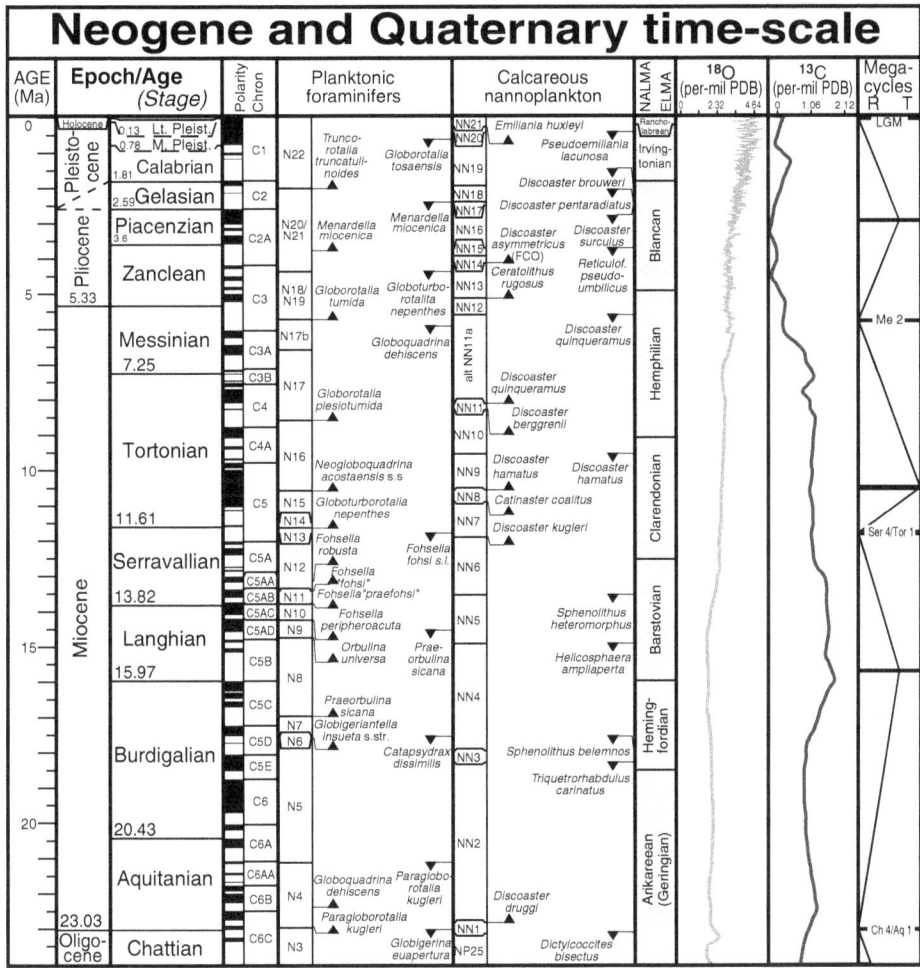

Fig. 7.5 **Neogene and Quaternary time-scale**. Reproduced with permission from Ogg *et al.* (2008).

### *Correspondence analysis (CA)*

This is a form of cluster analysis that can be used for biozone identification (Jones, 2006, 2011a; see also the further reading list at the end of the chapter). Examples of applications of CA in the oil industry have been published by Kender *et al.*, in Kaminski & Coccioni (2008) and Kender *et al.* (2008) (Fig. 7.6; see also the further reading list at the end of the chapter, and Section 9.3.4).

### *Graphic correlation (GC)*

This involves correlation of an outcrop or well section against a global composite standard section by treating the two as co-ordinate axes and plotting the fossil

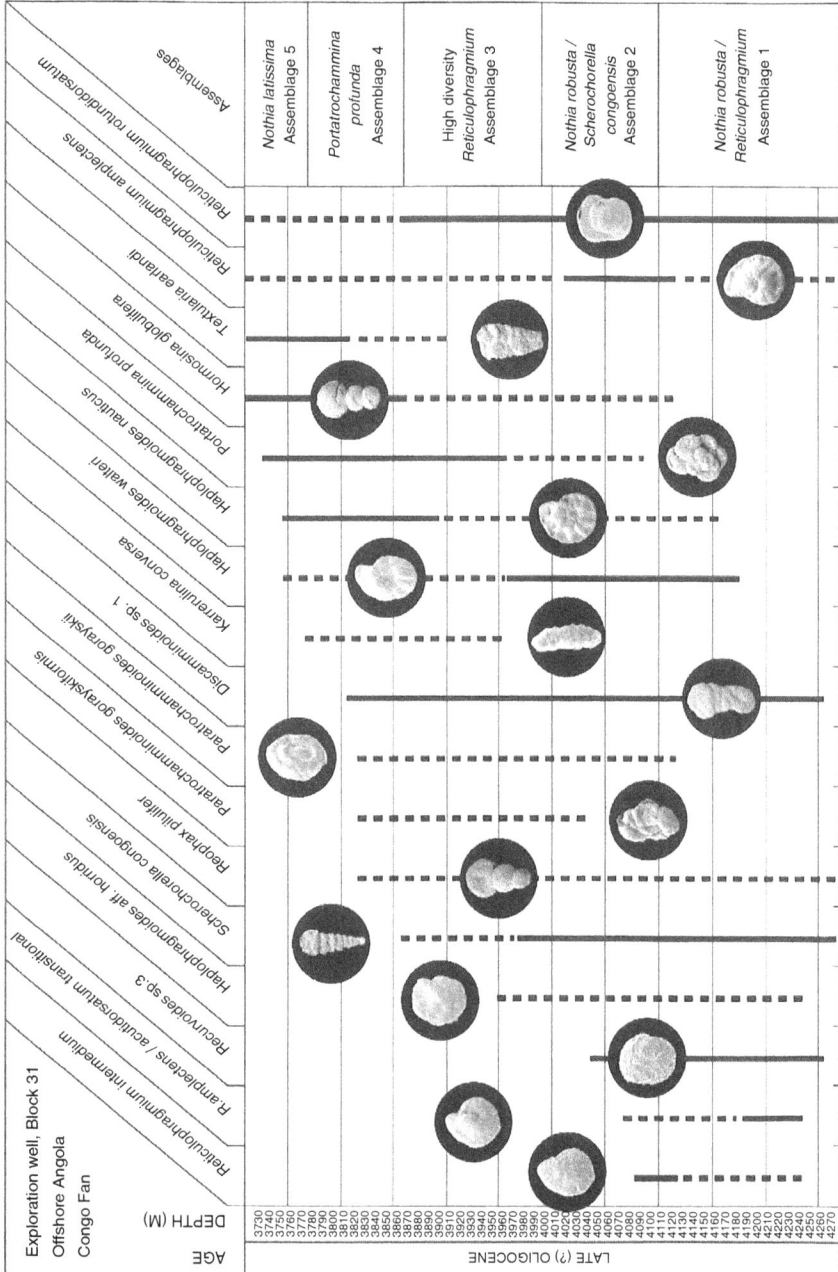

Fig. 7.6 **Stratigraphic zonation of the Oligocene of Angola by means of correspondence analysis (CA)**. Reproduced with permission from Kender *et al.*, in Kaminski & Coccioni (2008).

events common to both as a series of points (Jones, 1996, 2006, 2011a; see also the further reading list at the end of the chapter). The 'line of correlation' (LOC), fitted between the points so as to honour palaeontological and other geological data, provides information as to sediment accumulation rates and stratigraphic breaks that is of use in sequence stratigraphy and in petroleum systems analysis. Rates are indicated by, and indeed can be calculated from, the gradient of the LOC; breaks, and/or consensed sections, from plateaux. Projecting 'composite standard units' (CSUs) through the LOC into the section provides a precise and accurate indication of which units are represented and which unrepresented in the section, and of the degree of expansion or condensation. The GC software package StrataPlot is available publicly, and can be downloaded from www.strataplot.egi.utah.edu.

In the oil industry, GC has become fully automated (and, as intimated above, capable of being imported into the work-station environment for integration with seismic and well log data) (see Fig. 8.1). Alternative LOCs are fitted using path-finding or slotting algorithms (see the further reading list at the end of the chapter; see also Sections 9.3.1, 9.3.3 and 9.3.6). The 'best fit' is selected by appropriate weighting of defining events. Confidence limits can be assigned.

### *Ranking and scaling (RASC)*

Ranking (RA) provides a probabilistic sequence of biostratigraphic events in sections, determined by manual scoring or computer-driven matrix permutation; scaling (SC), a measure of the inter-event distance determined by cross-over frequency (partly a function of sample spacing) (Jones, 1996, 2006, 2011a; see also the further reading list at the end of the chapter, and Sections 9.3.3 and 9.3.6). The RASC software can be downloaded from www.rasc.uio.no.

### *SHE analysis for biozone identification (SHEBI)*

This is a form of diversity analysis that can be used for biozone identification (see the further reading list at the end of the chapter; see also Section 9.3.5).

## 7.7 Stratigraphic time-scales

The principal published absolute chronostratigraphic or geochronological time-scales incorporating biostratigraphic data and covering substantial periods of time are those of Harland *et al.* (1982), Haq *et al.* (1987), Harland *et al.* (1990), Berggren *et al.* (1995), de Graciansky *et al.* (1998), Gradstein *et al.* (2004), Ogg *et al.* (2008) and Gradstein *et al.* (2012) (Figs. 7.3–7.6.).

The Haq *et al.*, de Graciansky *et al.* and Gradstein *et al.* time-scales are widely used, and indeed have become unofficial standards, in the oil industry, undoubtedly because they also incorporate sequence stratigraphic data (see below).

Further up-to-date stratigraphic information, including information on Global Stratotype Sections and Points (GSSPs), is available through the International Commission on Stratigraphy or ICS website (www.stratigraphy.org).

## 7.8 Further reading

### *Palaeozoic*

Gobbett, in Adams & Ager, 1967; BouDagher-Fadel, 2008; Brenckle, 2005; Brönnimann *et al.*, 1978; Forbes, 1960; Kanmera *et al.*, 1976; Davydov, 1988a, b; Dawson, 1993; Burns & Nestell, in Demchuk & Gary, 2009; Ross & Ross, in Demchuk & Gary, 2009; Davydov *et al.*, in Gradstein *et al.*, 2004, 2012 Wardlaw *et al.*, in Gradstein *et al.*, 2004; Groves & Brenckle, 1997; Izart *et al.*, 1998; Jin Yugan *et al.*, 1997; Douglass, in Kauffman & Hazel, 1977; Martinez-Dias *et al.*, 1983, 1985, 1996; Lys, in Oertli, 1984; Davydov *et al.*, in Olson & Leckie, 2003; Ross, 1967; Ross & Ross, in Ross & Haman, 1987, 1995; Kobayishi, in Ross *et al.*, 1997; Makhlina, in Strogen *et al.*, 1996; Rukina, in Strogen *et al.*, 1996; Termier *et al.*, 1978; Toriyama, 1984; Toriyama *et al.*, 1975; Vachard *et al.*, 1993.

### *Mesozoic*

Banner & Desai, 1988; Banner *et al.*, 1993; Bralower *et al.*, in Berggren *et al.*, 1995; Gradstein *et al.*, in Berggren *et al.*, 1995; Bolli, 1969; Caron, in Bolli *et al.*, 1985; BouDagher-Fadel, 2008, 2012; BouDagher-Fadel *et al.*, 1997; Bassoullet, in Cariou & Hantzpergue, 1997; Caus *et al.*, 1988; Ogg *et al.*, in Gradstein *et al.*, 2004; Gradstein *et al.*, 2012; van Gorsel, in Hedley & Adams, 1978; Longoria, 1984; Longoria & Gamper, 1975; Moullade *et al.*, 1985; Nederbragt, 1990; Postuma, 1971; Premoli-Silva & Bolli, 1973; Reichel, 1937; Robaszynski & Caron, 1979; Robaszynski *et al.*, 1984; Thierry, in Roure, 1994; Schroder, 1975; Schroder & Neumann, 1985; Septfontaine, 1988; Septfontaine *et al.*, 1991; Sliter, 1989; van Hinte, 1972; Wonders, 1980.

### *Cenozoic*

Adams, 1970; Berggren, 1969; Berggren & Miller, 1988; Berggren *et al.*, in Berggren *et al.*, 1995; Billman *et al.*, 1980; Blow, 1979; Bolli, 1969; Bolli & Saunders, in Bolli *et al.*, 1985; Iaccarino, in Bolli *et al.*, 1985; Jenkins, in Bolli *et al.*, 1985; Rogl, in Bolli *et al.*, 1985; Toumarkine & Luterbacher, in Bolli *et al.*, 1985; BouDagher-Fadel, 2002, 2008, 2012; BouDagher-Fadel & Banner, 1999; BouDagher-Fadel & Price, 2010; Blow, in Brönnimann & Renz, 1969; Cahuzac & Poignant, 1997; Chaproniere, 1980, 1981; Cifelli & Scott, 1986; Drobne, 1988; Drooger, 1979; Fermont, 1982; Luterbacher *et al.*, in Gradstein *et al.*, 2004; Gradstein *et al.*, 2012; Lourens *et al.*, in Gradstein *et al.*, 2004; Haak & Postuma,

1975; Hottinger, 1962, 1977; Hottinger & Drobne, 1988; Adams, in Ikebe & Tsuchi, 1984; Isik & Hakyemez, 2011; Kennett & Srinivasan, 1983; Less, 1987; Less & Kovacs, 1995; Morley Davies, 1971; Olsson *et al.*, 1999; Payros *et al.*, 2007; Pearson, 1993; Pearson *et al.*, 2006; Postuma, 1971; Premoli-Silva & Bolli, 1973; Raju, 1974; Reichel, 1937; Rodriguez-Pinto *et al.*, 2012; Saraswati, 2012. Schaub, 1951, 1981; Scheibner & Speijer, 2009; Scott *et al.*, 1990; Serra-Kiel *et al.*, 1998; White, in Simmons, 1994; Stainforth *et al.*, 1975; van Vessem, 1978; Wade *et al.*, 2011; White, 1992; Pujalte *et al.*, in Wing *et al.*, 2003; Zakrevskaya *et al.*, 2009.

*Biostratigraphic technologies*

Agterberg & Gradstein, 1988, 1999; Cooper *et al.*, 2001; Crux *et al.*, 2005; Gradstein & Agterberg, in Cubitt & Reyment, 1982; Edwards, 1984; Gradstein & Kaminski, 1989; Gradstein *et al.*, 1992, 1994, 2008; Gradstein *et al.*, in Gradstein *et al.*, 2004; Groves & Brenckle, 1997; Harper & Crowley, 1985; Sadler & Cooper, in Harries, 2003; Highton *et al.*, 1997; Miller, in Kauffmann & Hazel, 1977; Neal, in Knox *et al.*, 1996; Carney & Pierce, in Mann & Lane, 1995; Kemple *et al.*, in Mann & Lane, 1995; Martin & Fletcher, in Mann & Lane, 1995; Neal *et al.*, in Mann & Lane, 1995; Neal *et al.*, 1994; Wakefield *et al.*, in Nikravesh *et al.*, 2003; Olson *et al.*, in Olson & Leckie, 2003; Wakefield, in Olson & Leckie, 2003; Petters, 1982; Platon & Sikora, 2005; Gary *et al.*, in Powell & Riding, 2005; Punyasena *et al.*, 2012; Crux *et al.*, in Ratcliffe & Zaitlin, 2010; Watson *et al.*, in Ratcliffe & Zaitlin, 2010; Scott, in Robertson *et al.*, 1990; Gradstein *et al.*, in Rogl & Gradstein, 1988; Shaw, 1964; Simmons, in Simmons, 1994; Thomas *et al.*, 1988; Wakefield & Monteil, 2002; Wakefield *et al.*, 2001; Watson & Nairn, 2005; Wescott & Boucher, 2000; Wescott *et al.*, 1998; Williamson, 1987; Wilson, 2008b, 2011; Wilson & Costelloe, 2011a, b; Wilson *et al.*, 2010, 2012.

# 8

## Sequence stratigraphy

This chapter deals with sequence stratigraphy. It contains a section on Foraminifera in sequence stratigraphy.

### 8.1 Sequence stratigraphy

Sequence stratigraphy attempts to subdivide the rock record into genetically related – commonly unconformity-bounded – rock units or sequences (Jones *et al.*, in Jenkins, 1993; Jones, 1996, 2006, 2011a; and additional references cited therein; see also the further reading list at the end of the chapter). The unconformities that form the basis of sequence stratigraphy are generated by base-level fall, in turn driven by glaciation or glacio-eustasy, and/or by structuration or tectonism (see also the comments on 'General and clastic sequence stratigraphy' below).

#### *Definitions*

*Sequences and sequence boundaries* A Vailian *sequence* is defined as a 'stratigraphic unit composed of a relatively conformable succession of genetically related strata bounded at its top and base by unconformities or their correlative conformities'. Vailian *sequence boundaries* (SBs) (bounding unconformities) are readily recognisable on seismic stratigraphic criteria (erosional truncation below, and transgressive onlap above). They are also recognisable on palaeontological, sedimentological and petrophysical criteria (see also Section 8.2).

A Gallowayan genetic stratigraphic *sequence* is defined as 'a package of sediments recording a significant episode of basin-margin outbuilding and basin-filling'. Gallowayan *SBs* (maximum flooding surfaces (MFSs): see below) are also readily recognisable on seismic stratigraphic criteria (landward shifts in facies below, and downlap, and basinward shifts in facies, above), and on palaeontological, sedimentological and petrophysical criteria (see also Section 8.2).

The spatio-temporal distribution of sequences is dictated by *accommodation space*, which is defined as 'the space ... available for potential sediment accumulation'. Accommodation space is in turn dictated by the complex interplay of a number of dynamically variable processes, including (glacio-)eustasy, tectonism, sediment compaction and sediment supply. *Eustasy*, that is, rising or falling absolute sea level, can be either creative and destructive of accommodation space. *Tectonism*, that is – independent – subsidence and falling, or uplift and rising, absolute land level, can also be either creative and destructive of accommodation space. *Sediment compaction* is always creative. *Sediment supply* is always destructive. At any given point in a basin, if the rate of sediment supply exceeds that of the creation of accommodation space by the processes described above, there will be a basinward shift in bathymetrically ordered facies belts. This process is termed *progradation*. Conversely, if the rate of sediment supply exceeds that of the creation of accommodation space, there will be a landward shift in facies belts. This process is termed *retrogradation*. If the rates of sediment supply and the creation of accommodation space are in equilibrium, facies belts will remain in place, and build vertically. This process is termed *aggradation*.

*Systems tracts*    Sequences can be internally subdivided into so-called *systems tracts (STs)*. An ST is defined as 'a linkage of contemporaneous depositional systems' or bathymetrically ordered facies belts, for example, continental–paralic–shelfal–bathyal–abyssal. The principal types are the *low-stand systems tract (LST)*, the *shelf-margin systems tract* or *shelf-margin wedge (SMST or SMW)*, the *transgressive systems tract (TST)* and the *high-stand systems tract (HST)*. The LST can be further subdivided into a lower, *low-stand fan (LSF)* and an upper, *low-stand wedge* or *low-stand prograding complex (LSW or LPC)*. (Note, though, that marine geologists interpret LSFs and LSWs as attached rather than detached systems: Nelson & Damuth, 2003.) The LST and TST are separated by the *transgressive surface (TS)*, or essentially equivalent *maximum regressive surface (MRS)*. The TST and HST are separated by the *maximum flooding – or transgressive – surface (MFS)*. *Condensed sections (CSs)* commonly occur at the TS and MFS.

The LST is defined as 'the lowermost systems tract in a depositional sequence ... if it lies directly on a Type 1 Sequence Boundary', and the related SMST or SMW as 'the lowermost systems tract in a depositional sequence associated with a Type 2 Sequence Boundary'; the TST as 'the middle systems tract of both Type 1 and Type 2 sequences'; and the HST as 'the upper systems tract in either a Type 1 or a Type 2 sequence'. Note that a Type 1 SB is generated by a sea-level fall to a position basinward, and a Type 2 SB by a sea-level fall to a position landward, of the shelf-edge or 'offlap break' of the preceding sequence.

The TS is defined as 'the first significant marine flooding surface ... within a sequence', and the MFS as 'the surface corresponding to the time of maximum flooding'. CSs are defined as 'thin marine stratigraphic units consisting of pelagic to hemipelagic sediments characterised by very low sedimentation rates ... most areally extensive at the time of maximum regional transgression'. They commonly contain 'abundant and diverse planktic and benthic microfossil assemblages', which enable easy identification, and also, importantly, calibration against global standard biostratigraphic zonation schemes (see also Section 8.2).

Gamma log trends can be useful in the characterisation of STs and CSs. LSTs and SMSTs are typically characterised by low gamma values; TSTs by increasing values; and HSTs by decreasing values. CSs, that is, TSs and MFSs, are typically characterised by high gamma values. Note, though, that CSs are atypically characterised by low gamma values, where represented by limestones rather than shales (Jones *et al.*, in Jenkins, 1993).

### General and clastic sequence stratigraphy

Vail, Haq and others of the so-called 'Exxon school' have generated a conceptual model of the development of sequences in relation to cycles of sea-level change. This model, or modifications of it, can be used to predict the spatio-temporal distribution of lithofacies, even in tectonically active areas such as the North Sea rift province (see Section 9.3.3), the Columbus basin growth-fault province, offshore Trinidad (see Section 9.3.5), or the Gulf of Mexico growth-fault province (see Section 9.3.6).

Vail, Haq and others have also generated curves showing variations in global coastal onlap and sea level through time, at least for the Carboniferous–Recent, that enhance the predictive utility of the model (Ross & Haman, 1987, presenting the biostratigraphic age control). (Independent sea-level curves have also been constructed for the Cambrian, Ordovician, Silurian and Devonian.)

They have argued that at least the 'third-order' observed cyclicity is, essentially, (glacio-) eustatically – rather than tectonically – mediated, and that the (glacio-) eustatic signal, while it may be strengthened or weakened by the tectonic signal, is always stronger than it.

This argument has been criticised by a number of authors, on the grounds that it fails to explain the discrepancy between the frequencies of the modelled third-order cycles (of the order of a million years) and the observed incontrovertibly glacio-eustatically mediated cycles (of the order of tens to hundreds of thousands of years) of the 'ice-house' world of the Pleistogene; that it also fails to explain the observed cyclicity in the 'greenhouse' world of the Mesozoic (when obviously there would have been no glaciation or glacio-eustasy); and that it takes insufficient account of local tectono- as against regional glacio-eustasy.

This criticism has in turn been countered by arguments that 'fourth- to sixth-order' cycles are in fact resolvable, although only in areas of high sedimentation rate, such as the Plio-Pleistocene of the Gulf of Mexico, the Niger delta, or the south Caspian; and that third-order cycles are mediated not only by (glacio-) eustasy but also by tectonism, in the form of – episodic – local and regional stress release.

### *Carbonate sequence stratigraphy*

Sarg, and Handford & Loucks, have generated conceptual models of the development of carbonate sequences (see also Jones, 1993). Handford & Loucks's models are comprehensive, and take into account criticisms that earlier ones relied too much on analogy with models for clastic sequences, which, in view of the more complex controls on carbonate sequences, is probably only partly appropriate. In particular, they take into account observations that modern and ancient carbonate systems, unlike clastic systems, deposit comparatively insignificant volumes of sediment during LSTs, when they are subject to chemical as against physical erosion, and much more significant volumes during HSTs.

### *Mixed sequence stratigraphy*

Jacquin *et al.*, Garcia-Mondejar & Fernandez-Mendiola, Holland and McLaughlin *et al.* have generated models of the development of mixed clastic–carbonate sequences. McLaughlin *et al.* argued that the primary control, at least in the Ordovician of Kentucky and Ohio, is (glacio-) eustatic, with only a secondary tectonic component associated with Taconian tectonism.

Tucker has generated a model for mixed carbonate-evaporite sequences.

### *Seismic facies analysis*

As noted above, seismic reflection terminations form the basis of sequence stratigraphic analysis, while reflection patterns form the basis of seismic facies analysis (Jones, 2006, 2011a). To simplify, reflection geometry (i.e., whether clinoform etc.) provides an indication of depositional process; amplitude, an indication of the product of the process, that is, lithology (and also, porosity and pore fluid content). Time-slices through three-dimensional seismic volumes allow the identification of individual sedimentary – and reservoir – bodies such as marginal marine deltas, shallow marine reefs and deep marine, submarine fans. Other seismic attributes such as coherency and spectral decomposition can also be used in palaeoenvironmental interpretation.

### *Chronostratigraphic diagrams*

Chronostratigraphic diagrams describe and predict the distributions of sequences of rocks, and especially of source-, reservoir- and cap-rocks, in time, conventionally displayed on the vertical axis, and in space, conventionally displayed on the horizontal axis (Jones, 2011a). Distributions in time are constrained in part by biostratigraphy; and distributions in space in part by palaeobiology.

## 8.2  Foraminifera in sequence stratigraphy

Foraminifera and other microfossils have been used in sequence stratigraphy principally in order to characterise master surfaces such as SBs and MFSs, and STs; and thus to define meaningful time-slices for gross depositional environment (GDE) and common risk segment (CRS) mapping as part of the play fairway evaluation process (Jones *et al.*, in Jenkins, 1993; Jones, 1996; Simmons *et al.*, in Hart *et al.*, 2000; Jones *et al.*, in Bubik & Kaminski, 2004; Jones, 2006; Lehmann *et al.*, 2006a–c; Jones, 2011a; see also the further reading list at the end of the chapter).

It is important to integrate micropalaeontological data into seismic stratigraphic and facies interpretation, not least in order to identify apparent or 'ecologically depressed', as distinct from real, extinctions or tops, and thus to prevent mis-correlations. Nowadays this integration is typically undertaken in the work-station environment (Fig. 8.1; see also Section 7.6).

### *Sequence boundaries*

Vailian SBs in clastic sequences are typically characterised by the abrupt juxtaposition of younger, more proximal fossil assemblages, sometimes containing reworked specimens, on older, more distal ones ('faunal discontinuity events'). Note, though, that the contrast in assemblages above and below the SB varies along depositional dip, typically being more pronounced in proximal, up-dip settings, where a biostratigraphically resolvable unconformity may also be in evidence, and less so in distal, down-dip settings, where the relationship may be closer to that of a correlative conformity. Note also that in the most proximal, up-dip settings, the assemblage above the SB is actually more distal than that below, as it is not directly associated with the SB but rather with the succeeding TS.

Gallowayan SBs, or MFSs, are typically characterised by CSs, and by abundant and diverse fossil assemblages. Note, though, that high fossil abundance and diversity is not in itself diagnostic of MFSs (it is also characteristic of debris flows, for example). MFSs are diagnosed by fossil assemblages indicating maximum flooding, or bathymetry; and, at least in down-dip settings, the maximum abundance of plankton, enabling calibration against global standard biozonation schemes established in the open ocean.

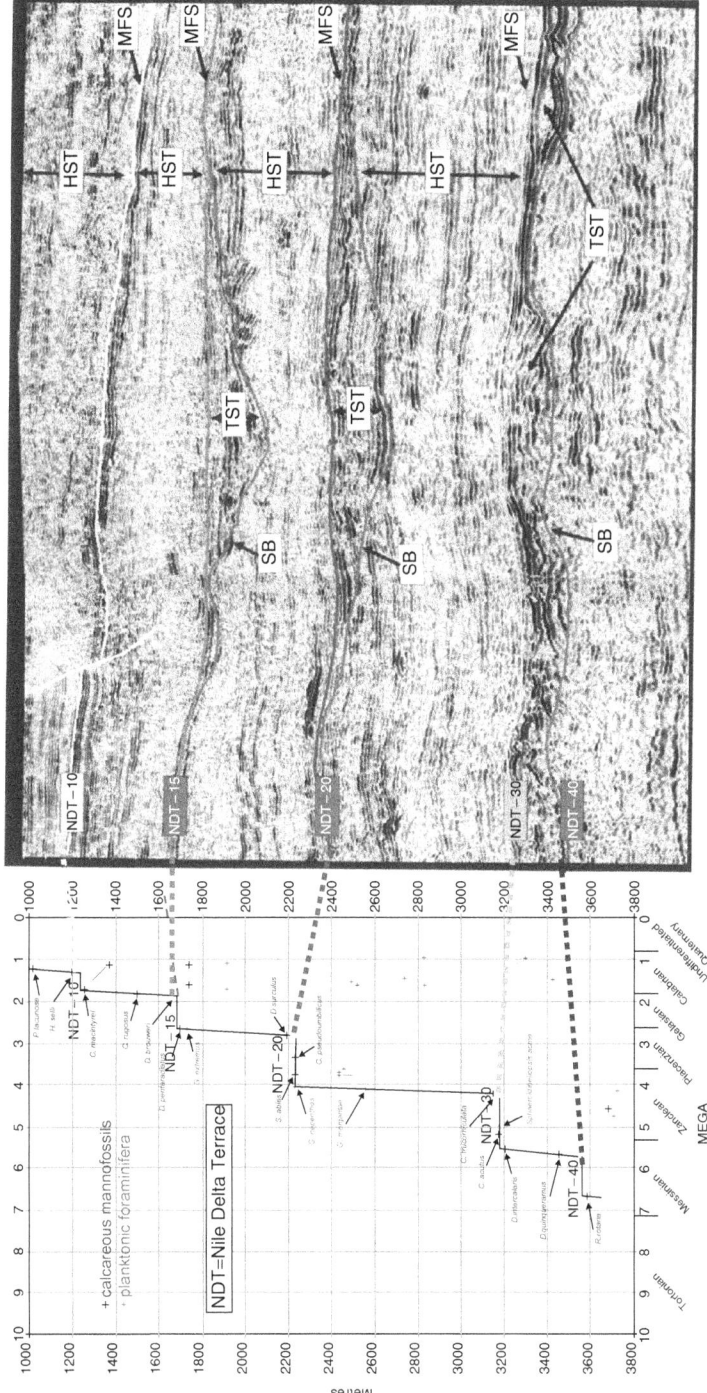

Fig. 8.1 **Integated biostratigraphic and sequence stratigraphic interpretation of the Plio-Pleistocene of offshore Egypt**. Modified after Wescott *et al.* (1998). SB = sequence boundary; TST = transgressive systems tract; MFS = maximum flooding surface; HST = high-stand systems tract. NDT = Nile Delta Terrace (identified on basis of graphic correlation).

## Systems tracts

Only present in former basinal settings, LSTs are represented by erosional hiatuses in former marginal settings. They are typically characterised by eroded and reworked shallow-water Foraminifera, and by *in situ* Deep-Water Agglutinating Foraminifera or DWAFs (see, for example, McNeil, in Dixon *et al.*, 1985; van Gorsel, 1988; Olson & Thompson, in Koutsoukos, 2005; Raju, in Raju *et al.*, 2005). The deep-water component is typically of shallower aspect than that of the uderlying sequence, although the difference is not necessarily always resolvable (because bathymetric resolution is comparatively poor – of the order of a few hundred metres – in deep water).

Typically, TSTs are characterised by deepening-upward bathymetric trends (retrogradation). In the clastic and carbonate sequences of the Neogene of south-east Asia and the Indian subcontinent, they are characterised by 'clear water' Foraminifera, including Larger Benthic Foraminifera (LBFs) (van Gorsel, 1988; Raju, in Raju *et al.*, 2005). In the shallow-water carbonate sequences of the Early Cretaceous of Tethys, they are characterised by deepening-upward trends and by accompanying depth-related trends towards flattened lenticular morphologies in orbitolinid LBFs (Simmons *et al.*, in Hart *et al.*, 2000; Jones *et al.*, in Bubik & Kaminski, 2004).

Typically, HSTs are characterised by shoaling-upward bathymetric trends (progradation). In the clastic and carbonate sequences of the Neogene of south-east Asia and the Indian subcontinent, they are characterised by 'turbid water' Foramini-fera, including *Uvigerina* and *Bolivina* (van Gorsel, 1988; Raju, in Raju *et al.*, 2005). In the shallow-water carbonate sequences of the Early Cretaceous of Tethys, they are characterised by shoaling-upward trends and by accompanying depth-related trends towards globular morphologies in orbitolinid LBFs (Simmons *et al.*, in Hart *et al.*, 2000; Jones *et al.*, in Bubik & Kaminski, 2004). Also in the shallow-water carbonate sequences of the Jurassic Surmeh formation of the Zagros mountains of south-west Iran, HSTs are characterised by shoaling-upward cycles (Lehmann *et al.*, 2006a–c). Here, cycles in the lower part of the formation, typically tens of metres thick, are marked at the base by flooding basinal facies representing palaeobathymetries of at least several tens of metres, and at the top by shoal facies representing palaeobathymetries of the order of ten metres. Cycles in the upper part, typically of the order of ten metres thick, are marked at the base by flooding fore-shoal or shoal facies representing palaeobathymetries of the order of ten metres, and at the top by back-shoal lagoonal or sub-tidal facies representing palaeobathymetries close to sea-level. Diagenesis appears to have involved early cementation of – subaerially exposed – grainstone shoal facies at the tops of cycles within the upper part of the formation, and later dolomitisation of underlying fore-shoal facies. Importantly, the dolomitisation process was enhancive of porosity and reservoir quality.

## 8.3 Further reading

### *General*

Embry, in Armentrout & Rosen, 2002; Catuneanu, 2006; Catuneanu *et al*., 2009, 2011; Cloetingh, 1986; Coe, 2003; Crews *et al*., 2000; Vail *et al*., in Einsele, 1991; Embry, 2009; Emery & Myers, 1996; Galloway, 1989a; Garcia-Mondejar & Fernandez-Mendiola, 1993; Simmons, in Gradstein *et al*., 2012; Haq *et al*., 1987; Holland, 1993; Hubbard, 1988; Jacquin *et al*., 1991; Handford & Loucks, in Loucks & Sarg, 1993; McLaughlin *et al*., 2004; Miall, 1986, 1990, 1992, 2010; Brown & Fisher, in Payton, 1977; Mitchum *et al*., in Payton, 1977; Vail *et al*., in Payton *et al*., 1977; Bertram, in Roberts & Bally, 2012; Vail *et al*., in Schlee, 1984; Sloss, 1963; Embry, in Steel *et al*., 1995; Mitchum *et al*., in Weimer & Posamentier, 1993; Sharland *et al*., 2001; Tucker, 1991; Wheeler, 1958; Haq *et al*., in Wilgus *et al*., 1988; Jervey, in Wilgus *et al*., 1988; Loutit *et al*., in Wilgus *et al*., 1988; Sarg, in Wilgus *et al*., 1988; van Wagoner *et al*., in Wilgus *et al*., 1988.

### *Foraminifera in sequence stratigraphy*

Brown *et al*., 1995; Crux *et al*., 2005; Cubaynes *et al*., 1990; West *et al*., in Dean & Arthur, 1998; Burns & Nestell, in Demchuk & Gary, 2009; Grafe, 1999; Hart, 2000; Simmons *et al*., in Hart *et al*., 2000; Moreno-Vasquez, in Langer, 1995; Ross & Ross, in Langer, 1995; Zellers, in Langer, 1995; Davydov *et al*., in Olson & Leckie, 2003; Ross & Ross, in Olson & Leckie, 2003; Wakefield, in Olson & Leckie, 2003; Rey *et al*., in Posamentier *et al*., 1993; Crux *et al*., in Ratcliffe & Zaitlin, 2010; Watson *et al*., in Ratcliffe & Zaitlin, 2010; Robaszynski *et al*., 1990; Sharland *et al*., 2001; Simmons *et al*., 1992; Armentrout, in Weimer & Link, 1991; Wakefield & Monteil, 2002; Watson & Nairn, 2005; Wescott *et al*., 1998.

# 9

# Applications and case studies in petroleum geology

This chapter deals with applications and case studies in petroleum geology. It contains sections on the principles and practice of petroleum geology; on applications and on case studies of applications in petroleum exploration; on applications and on case studies of applications in reservoir exploitation; on applications and on case studies of applications in well-site operations; on unconventional petroleum geology; and on micropalaeontology and Health, Safety and Environmental (HSE) issues in the petroleum industry. The section on the principles and practice of petroleum geology includes sub-sections on play components; on petroleum exploration; on reservoir exploitation; and on well-site operations. The section on applications in petroleum exploration includes sub-sections on chronostratigraphy and palaeoenvironmental interpretation; and on operational biostratigraphy. The section on case studies of applications in petroleum exploration includes case studies from the Middle East; the Indian subcontinent; the central and northern North Sea; the South Atlantic salt basins; the eastern Venezuelan basin; and the Gulf of Mexico. The section on applications in reservoir exploitation includes a sub-section on integrated reservoir characterisation. The section on case studies of applications in reservoir exploitation includes case studies from the marginal to shallow marine, peri-deltaic clastic reservoirs in the Gullfaks, Snorre and Statfjord fields in the Norwegian sector, and in the Ninian and Thistle fields in the UK sector, in the North Sea; the marginal to shallow marine, peri-deltaic clastic reservoir in the Pedernales field in Venezuela in northern South America; the shallow marine, peri-reefal carbonate reservoirs in the Al Huwaisah, Dhulaima, Lekhwair and Yibal fields in Oman, and in Shaybah field in Saudi Arabia, in the Middle East; and the deep marine, submarine fan reservoir in Forties field in the UK sector in the North Sea. The section on applications in well-site operations includes a sub-section on 'biosteering'. The section on case studies of applications in well-site operations includes case histories from the shallow marine carbonate reservoir in the Sajaa field in the United Arab Emirates in the Middle East; the deep marine carbonate reservoir in the Valhall field in the Norwegian sector in the North Sea; and the deep marine clastic reservoir in the Andrew field in the UK sector in the North Sea. The section on unconventional petroleum geology includes a sub-section on shale gas. The section on micropalaeontology and HSE issues in the petroleum industry includes sub-sections on Environmental Impact Assessment (EIA); on site investigation; on pressure prediction; on well-site operations; and on environmental monitoring.

## 9.1  Principles and practice of petroleum geology

### *9.1.1  Play components*

There are essentially three components to a successful play, namely, a working petroleum source-rock and system, to provide the petroleum charge; a reservoir, to store, and ultimately to flow, same; and a cap-rock (seal) and trap, to prevent it from escaping (Jones, 1996, 2006, 2011a; see also the further reading list at the end of the chapter). Petroleum source-rocks and systems (also, palaeontological inputs into petroleum systems analysis), reservoir-rocks (also, nummulite reservoirs), cap-rocks (also, micropalaeontological characterisation of cap-rocks), and traps are discussed in more detail in turn below. Unconventional petroleum geology, in which the source-rock is also the reservoir, is discussed in Section 9.8. Site investigation in petroleum geology is discussed in Sections 9.9. and 11.1.4; EIA and environmental monitoring, in Sections 9.9, 12.1.1 and 12.2.1.

### *Petroleum source-rocks*

Petroleum source-rocks are those from which petroleum is derived. They typically consist of abundant organic material or kerogen, that can be of either terrestrial land-plant or humic, or marine algal or sapropelic, origin. This is transformed, or matured, into oil and gas by heating on burial. The product will depend on the nature of the source-rock, and on the transformation ratio, or degree of maturation, in turn determined by burial history and by basin type. Typically, oil is generated from so-called 'Type I' or 'Type II' source-rocks, whose organic material is of marine algal origin; and from 'Type III' source-rocks, whose organic material is of terrestrial land-plant origin. Note, though, that gas can be generated from Type I or Type II source-rocks subjected to a high degree of maturation; and that oil can be generated from Type III source-rocks such as coals. Maturation – and migration – can be modelled using various publicly available and proprietary one-, two- and three-dimensional basin or petroleum systems modelling software packages.

*Stratigraphic controls on source-rock development*   There are both intrinsic and extrinsic stratigraphic controls on source-rock development. There is intrinsic control within the sequence stratigraphic framework, with preferential development during transgressive systems tracts or TSTs (see the further reading list at the end of the chapter). There is also extrinsic control outwith this sequence stratigraphic framework, although still within an 'Earth systems' framework, with preferential development during the 30% of Phanerozoic time represented by the Silurian, Late Devonian–Early Carboniferous, Late Carboniferous–Early Permian, Middle–Late Jurassic, 'Middle' Cretaceous and Oligo-Miocene, accounting for a disproportionate 90%+ of the world's discovered oil (see the further reading list at the end of the chapter).

### *Petroleum systems*

A petroleum system has been variously defined as 'a dynamic petroleum generating and concentrating system functioning in geological space and time', or 'a pod of active source rock and the resulting oil and gas accumulations'. The petroleum-system concept is 'a means of formalising the relationship between the geologic elements in time and space that are required for the development of a commercial petroleum accumulation' (see the further reading list at the end of the chapter). Petroleum systems analysis is concerned with the presence of these elements, that is, source-, reservoir- and cap-rock and trap, and with their effectiveness (see also comments on 'play fairway analysis or evaluation' below), and may also be said to be concerned with how much petroleum a basin has generated, where it has migrated, and where it is trapped, and in what quantity and phase. The relationships between the elements are conventionally presented in the form of a series of structural cross-sections showing not only the fill of the basin but also the fluid flow therein, and, importantly, the progression of the petroleum migration front through time, and hence the limits of the system. An understanding of the limits of petroleum systems within basins is critical to an evaluation of their potential prospectivity, with exploration risk increasing in proportion to the distance from an established petroleum system. For example, in the North Sea basin, where migration is almost exclusively vertical, the limit of prospectivity is essentially controlled by the limit of the mature Kimmeridge Clay formation source-rock. An understanding of the certainty with which limits can be proscribed is also important. For example, in the Campos basin, offshore Brazil, the limit of the petroleum system and of prospectivity was recently significantly extended when advances in drilling technology enabled exploration in deep water for the first time. Petroleum systems are conventionally named after the source- rather than the reservoir-rock, since one source may charge only one or more than one reservoir. There may be only one petroleum system in a basin, or there may be more than one, in which case they may be stratigraphically or geographically discrete, or they may be overlapping. Identification of individual petroleum systems in basins where more than one is operative requires typing of reservoired petroleum to the source-rock. This is nowadays relatively easily achievable using a range of geochemical techniques including, for example, biomarker and diamondoid analysis.

There are certain empirical observations with regard to the distribution of oil and gas reserves within a basin that constitute general rules (although there are also specific exceptions, especially when more than one petroleum system is operative). As a general rule, oil gravity decreases with depth. In other words, heavy oil is found at shallow depths, light oil and condensate at intermediate depths, and gas at depth. This is a function of the combined effects of source-rock maturity and of biodegradation, or rather the lack thereof at depth, where temperature conditions are inimical to the Bacteria responsible. Also as a general rule, oil gravity decreases towards the centre of the basin. In other words, heavy oil is found at the basin margin, light oil and condensate in an intermediate position, and gas in the basin centre (where it can be stratigraphically rather than structurally trapped). This has been interpreted by Gussow as indicative of displacement of oil by gas along a so-called 'fill-spill chain'. 'Gussow's principle' relies on the existence of laterally extensive permeable carrier beds, in other words, low impedance to fluid flow (see the further reading list at the end of the chapter). Even in such a system, though, charge can be limited, and those structures most remote from the source kitchen in the centre of the basin run the risk of remaining uncharged.

### Palaeontological inputs into petroleum systems analysis

*Biostratigraphy and chronostratigraphy*   One key input in petroleum systems analysis is absolute age data, which are used to establish the ages of the source-rock and overburden, and, critically, to model the timing of migration as against that of trap formation. Chronostratigraphic data are in turn an output of routine palaeontological or biostratigraphic analysis, or of graphic correlation. Resolution is typically best – better than 1 Ma – in marine environments.

*Palaeobiology and palaeoenvironmental interpretation*   Another key input in petroleum systems analysis is palaeoenvironmental interpretation based on palaeobiology, used to infer palaeobathymetry and palaeotemperature for modelling purposes. Lithology away from areas of control is inferred from a descriptive and predictive sequence stratigraphic model into which palaeoenvironmental interpretation has been integrated.

*Palaeobathymetry.* Palaeobathymetric interpretation – essentially based on proxy living benthic foraminiferal distribution data – enables the characterisation of a range of depth environments. Resolution is typically sufficiently good to enable recognition of the following depth environments: non-marine (obviously, on the basis of fresh-water Algae, diatoms and ostracods, terrestrially derived spores and pollen, or land plants and animals, rather than Foraminifera); marginal marine; shallow marine, inner shelf or neritic (0–50 m); middle shelf (50–100 m); outer shelf (100–200 m); deep marine, upper slope or bathyal (200–1000 m); middle slope (1000–1500 m); lower slope (1500–2000 m); and abyssal plain (> 2000 m). (Note that the resolution is of the order of tens of metres in shallow marine environments, but only of the order of hundreds or even thousands of metres in deep marine environments.) Importantly, high-resolution palaeobathymetric curves can be constructed and imported into work-station environments through presently proprietary but hopefully ultimately publicly-available software packages containing links to bathymetric distribution data-sets or 'look-up' tables.

Additionally, dysoxic and anoxic environments can be characterised by foraminiferal assemblages, for example, Oxygen Minimum Zones (OMZs), typically on the upper slope, by infaunal Buliminida (see Sections 4.5.1 and 5.6.1).

*Palaeotemperature.* Palaeobiogeographic data can provide indications as to the palaeolatitude, palaeoclimate and palaeotemperature history of a basin. For example, the palaeotemperature history of the Middle East and north Africa has been established partly on the basis of biological proxies, such as essentially tropical Larger Benthic Foraminifera (LBFs).

*(Palaeo)biological controls on source-rock development*    There are (palaeo)biological controls on biogenic and thermogenic source-rock development, as well as the stratigraphic controls alluded to above.

Biogenic source-rock development is dependent on an excess of production over consumption of the organic carbon or OC that constitutes the feed-stock for the methanogenic 'Archaebacteria' that produce biogenic methane (see also Sections 4.5.1 and 5.6.1).

*Production of OC.* Production of OC is controlled by primary biological productivity in the terrestrial and marine realms. Note in this context that there is also a significant input of OC of terrigenous origin into the marine realm, associated not only with wind- and river-borne sediments, but also with turbidites, and with hyperpycnal turbidites generated by earthquakes and landslides on steep-sided oceanic islands, and by storms.

In the terrestrial realm, the production of OC is controlled by the productivity of land plants, and is typically highest in places or at times of optimal climate. In the case of the Indian subcontinent, it is significantly higher in the Ganges-Brahmaputra and Godavari drainage basins to the east, within which the vegetation is characterised by deciduous monsoon-forest and, in estuarine areas, mangrove-swamp or 'sundurban' elements, than in the Indus drainage basin to the west, within which the vegetation is characterised by montane-forest and semi-desert elements (in the Hindu Kush mountains and Thar desert, respectively). Moreover, the flux of terrigenous OC from the Ganges-Brahmaputra and Godavari rivers into the marine realm is significantly higher than that from the Indus river, and the concentration of terrigenous and total OC on the Bengal fan, fed by the Ganges-Brahmaputra and Godavari, is higher than that on the Indus fan, fed by the Indus (see also comments on 'palaeobiological de-risking of source presence' below).

In the marine realm, the production of OC is controlled by the productivity of marine Algae, and is highest in places or at times of optimal climate, or of enhanced nutrient supply associated with riverine input, or with upwelling (see Sections 4.5.1 and 5.6.1; see also the further reading list at the end of the chapter). Past upwelling can also be indicated by micropalaeontological evidence (see, for example, Kender *et al.*, 2009; see also Section 5.6.1.). It can be modelled using palaeoclimatological computer software.

*Consumption of OC*. Consumption and preservation of OC is essentially controlled by the availability of oxygen, and consumption is lowest, and preservation highest, in places of dysoxia or anoxia, such as the Black Sea or Indian Ocean, or at times of dysoxia or anoxia, for example the Oceanic Anoxic Events or OAEs of the Cretaceous (OAE1 in the Barremian–Aptian, OAE2 in the Cenomanian–Turonian, and OAE3 in the Senonian), such as have recently been simulated by a three-dimensional biogeochemical general circulation model (see Sections 4.5.1 and 5.6.1; see also the further reading list at the end of the chapter). In the case of the Indian Ocean, consumption of OC is particularly low, and preservation of OC particularly high, in the particularly well developed OMZ there (see also comments on 'palaeobiological de-risking of source presence' below).

*Palaeobiological de-risking of source presence*    In the case of the Krishna-Godavari basin off the east coast of India, where good-quality prior palaeobiological information is available, palynofacies indicates a significant flux of terrigenous OC into the marine realm, and microfacies indicates the development of an environment favourable for the preservation of OC, actual preservation of OC within the sediment, and even the possibility of active gas seepage, allowing a low risk to be assigned to the presence of a source-rock in the basin (see also Section 9.3.2). Benthic foraminiferal microfacies and palaeobathymetry based on modern analogy indicates an upper-slope environment of deposition for the Miocene–Pliocene Ravva formation of the Krishna-Godavari basin (Jones, 1994). A significant proportion of the species encountered (38%) exhibit internal functional morphological adaptations to low-oxygen conditions, and indicate the development of an OMZ, favourable for the preservation of OC (Charnock & Jones, 1997). A further significant proportion of the species encountered (46%) exhibit external functional morphological adaptations to infaunal habitats and detritivorous feeding strategies, and indicate actual preservation of OC within the sediment (Jones & Charnock, 1985). One species, *Arenomeandrospira glomerata*, previously described from regions of shallow gas occurrence, even indicates the possibility of active gas seepage (Jones & Wonders, in Hart *et al.*, 2000).

*Thermal alteration indices*   An independent measure of thermal maturity is provided by the spore, acritarch, chitinozoan, foraminiferal, ostracod and conodont colour indices, calibrated against vitrinite reflectance (see the further reading list at the end of the chapter).

### Reservoir-rocks

Reservoir rocks are those in which petroleum is reservoired. They require two properties, namely porosity and permeability. *Porosity* ($\varphi$) is the space between sedimentary particles that enables the reservoir to store pore fluid, including petroleum. It is measured by point counting of petrographic thin-sections or by petrophysical analysis of well logs, or by experimental analysis. Values are quoted as percentages or porosity units. Obviously, only open pores constitute effective porosity, that is, that which is capable of flowing hydrocarbons. Porosity can be either primary, and related to the original depositional facies or fabric, or secondary, and related to diagenesis. Primary porosity can be either enhanced or occluded by secondary diagenetic processes. Typically, dissolution, dolomitisation and fracturing are enhancive, and compaction and cementation occlusive, processes. Importantly, an early hydrocarbon charge can cause effective cessation of occlusive diagenetic processes. *Permeability* (k) is the ability of the rock to flow petroleum. It is measured by experimental analysis. Values are measured in darcies, or, more commonly, millidarcies (md). Permeability distribution within a reservoir is commonly characterised by anisotropy, with vertical permeability (kv) typically lower than horizontal permeability (kh), owing to bed effects. This has important consequences for production.

*Stratigraphic controls on reservoir-rock development*   There are both intrinsic and extrinsic stratigraphic controls on reservoir- as well as source-rock development. There is intrinsic control within the sequence stratigraphic framework, with preferential development of deep marine clastic reservoirs during low-stand systems tracts or LSTs, and shallow marine clastic and carbonate reservoirs during high-stand systems tracts or HSTs, and secondary enhancement of the quality of shallow marine carbonate reservoirs during LSTs in the case of dissolution, and transgressive systems tracts or TSTs in the case of – brine-reflux – dolomitisation (see the further reading list at the end of the chapter). There is also some extrinsic stratigraphic control on reservoir-rock development outwith this framework.

The primary reservoir properties of the chalk reservoirs of the North Sea and the nummulite bank reservoirs of the circum-Mediterranean are controlled in part by the evolution of the fossils that constitute them.

*Nummulite reservoirs*

Nummulite banks are locally of commercial significance as petroleum reservoirs: for example in the Bourri field, offshore Libya; in the Ashtart and Hasdrubal fields, offshore Tunisia; and in the Kerkennah and adjacent fields, onshore Tunisia (see Section 5.4.2; see also the further reading list at the end of the chapter). Unfortunately, the reservoirs are not easy to exploit, partly because the permeability distribution within, and therefore the producibility of, nummulite banks is difficult to describe and predict. Permeability distribution within nummulite bank reservoirs appears to be controlled by a complex combination of primary, depositional and secondary, diagenetic factors. Primary permeability is low (<10 md) in most nummulite bank reservoirs, in which the primary porosity is typically *within* individual nummulites (intra-particular), and therefore unconnected and ineffective; although it can also be high (>1000 md), for example in high-energy, well-sorted 'B'-form-dominated grainstones, in which the primary porosity is atypically *between* individual nummulites (inter-particular), and therefore connected and effective. Secondary enhancement of primary permeability has been described as having been brought about by dolomitisation, dissolution or leaching, and fracturing; secondary occlusion by compaction, pressure solution, stylolitisation and cementation. Importantly, the occlusive effects of cementation can be negated by an early emplacement of hydrocarbon, and actually reversed by dissolution by acidic basinal fluids associated with the emplacement of hydrocarbon.

*Bourri field, Libya*   Bourri field is situated in the Pelagian basin, offshore north-western Libya. The Early Eocene Jdeir formation part of the reservoir is approximately 125–175 m thick, and is interpreted as a nummulite bank, deposited in a rimmed shelf setting. Porosity within the nummulite bank ranges up to 21%, and averages 16%. It is interpreted as having been secondarily enhanced primarily by dissolution or leaching. Oil reserves for the Bourri field have been quoted as within the range 1–3 billion barrels.

*Ashtart field, Tunisia*   Ashtart field is situated in the Gulf of Gabes, offshore eastern Tunisia. The Early Eocene El Garia formation reservoir is approximately 90 m thick, and is interpreted as a re-deposited and allochthonous nummulite bank. Porosity within the cored part averages 17.2%. Permeability ranges from 0.2 to 1000 md, and averages 50 md. Both are interpreted as having been secondarily enhanced by dissolution and by fracturing. Oil production rates have been quoted as approximately 3000 barrels per day per well. Reserves have been quoted as within the range 350–400 million barrels.

*Hasdrubal field, Tunisia*   Hasdrubal field is also situated in the Gulf of Gabes, offshore eastern Tunisia. As in the case of the Ashtart field (see above), the Early Eocene El Garia formation reservoir is approximately 90 m thick, and is interpreted as a re-deposited and allochthonous nummulite bank. Porosity ranges from 0.5 to 23%, and averages 10.5%. Permeability ranges from 0.02 to 60 md, and averages 0.5 md. Both are interpreted as having been secondarily enhanced by dolomitisation and by fracturing, and also secondarily occluded by cementation. Gas production rates have been quoted on the BG Group website as within the range 5–20 million standard cubic feet per day per well. Reserves have been quoted also on the BG Group website as 250 billion cubic feet of gas and 20 million barrels of oil/condensate.

*Kerkennah and adjacent fields, Tunisia*   Kerkennah and adjacent fields are situated onshore eastern Tunisia. The Early Eocene El Garia formation reservoir is interpreted on the basis of analogy with equivalent three-dimensionally exposed outcrops elsewhere in eastern Tunisia as a nummulite bank, deposited in a mid-ramp setting, below fair-weather wave-base but above storm wave-base. The outcrops demonstrate that the grain-prone mid-ramp bank facies passes up-dip into a mud-prone inner-ramp back-bank facies, and down-dip into a mud-prone outer-ramp fore-bank and basinal facies, within less than 3 km (less than the conventional well spacing). Oil production rates for the Kerkennah field have been quoted as approximately 300 barrels per day per well. Reserves for the Kerkennah and adjacent fields have been quoted as < 100 million barrels of oil.

### Cap-rocks (seals)

Cap-rocks are those beneath which petroleum is trapped. They require a low permeability (see above) and a high sealing capacity, quantified using capillary entry pressure measurements. They can be of any lithology, although mudstones are the commonest, and evaporites such as anhydrite and halite the most effective.

*Stratigraphic controls on cap-rock development*   There is some intrinsic stratigraphic control on cap-rock development within the sequence stratigraphic framework, with preferential development during TSTs and maximum flooding surfaces or MFSs.

### Micropalaeontological characterisation of cap-rocks

*Introduction*   Various types of mudstones from the Oligo-Miocene sections of approximately 30 wells from offshore Angola have been analysed

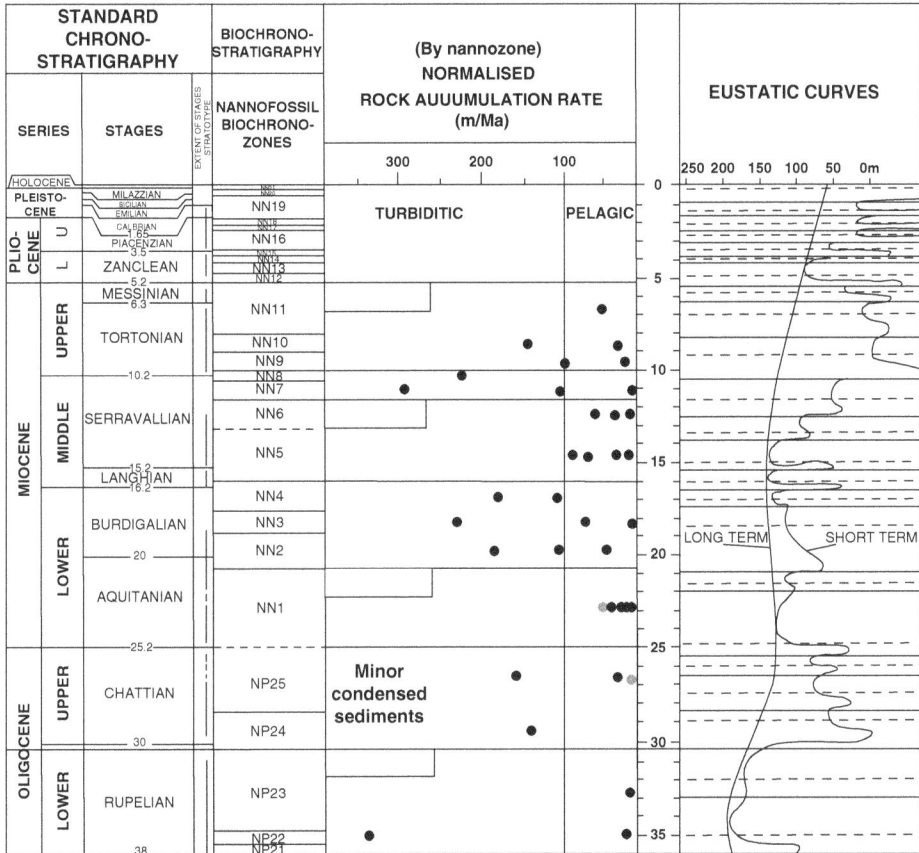

Fig. 9.1 **Mud-rock seal capacity of the Oligocene–Miocene of Angola**.
Shading indicates stratigraphic positions of CSs identified on bases of normalised
rock accumulation rates, and foraminiferal abundance and diversity and other
data, and interpreted as regional seals. Interpretation supported by capillary entry
pressure and other data. Note coincidence between CSs/seals and ?glacio-eusta-
tically mediated second-order sea-level high-stands. From Jones (2006).

micropalaeontologically, in order to establish whether those that constitute barriers
or baffles to fluid flow can be characterised (Jones, 2003b).

*Results* Condensed sections (CSs) in the Oligo-Miocene of offshore Angola,
interpreted as representing hemipelagites, have been distinguished from expanded
sections, interpreted as representing turbidites, on the bases of rock accumulation
rate data derived from thickness/age relationships and/or graphic correlation, and
of foraminiferal abundance and diversity (Fig. 9.1). The CSs have been distinguished essentially on the basis of rock accumulation rates of less than 100 m/Ma,
comparable to those exhibited by – uncompacted – Recent pelagites from

comparable depths in the North Atlantic. The CSs have also been distinguished on the basis of maximum benthic foraminiferal abundance and diversity, especially of infaunal morphotypes. High abundance and diversity is characteristic of CSs or MFSs. Note, however, that high abundance and diversity is not diagnostic of CSs, which can be characterised by low abundance and diversity, especially if affected by dissolution of calcareous species below the lysocline. Debris flows can be character-ised by high abundance and diversity, although they can still be distinguished from high abundance and diversity CSs on the basis of their allochthonous components, identified on the basis of bathymetric interpretation. (Proportionately) high abun-dance and diversity of infaunal 'morphogroup' C does appear to be diagnostic of CSs, reflecting colonisation of the sea floor under the comparatively tranquil condi-tions obtaining in between turbidite episodes, and over a comparatively long time.

*Discussion*     Importantly, the micropalaeontologically characterised CSs in the Oligo-Miocene of offshore west Africa correlate well with regional seals identified on seismic facies, wireline log and capillary entry pressure data. Interestingly, they are also coincident in time with tectono-eustatically- or glacio-eustatically medi-ated second-order sea-level high-stands in the Early Oligocene (global standard calcareous nannoplankton zones NP21–NP23), earliest Miocene (NN1), Middle Miocene (NN5–NN6) and latest Miocene (NN9–NN11). This suggests that the methodology has essentially global applicability.

Incidentally, ineffective as well as effective seals have also been characterised micropalaeontologically, elsewhere, on the basis of the similarity of their contained benthic foraminiferal assemblages to those associated with modern seeps. For example, the seal sequence overlying the platform carbonate reservoir target in the Miocene of the Nam Con Son basin, offshore Vietnam, is characterised by *Arenomeandrospira glomerata*, a species originally described from the Skagerak and Kattegat, regions of shallow gas occurrence and sea-bed 'pockmarks'.

*Traps*

Traps are configurations of reservoir and seal that do not allow the escape of petroleum. They are classified into three main types: structural, for example, anticlinal and diapiric traps; stratigraphic, for example, pinch-out, reef and sub-unconformity traps, with no structural component other than regional dip; and combination traps. Structural traps are by far the most common, not least because they are typically readily imaged on seismic data. Stratigraphic traps are typically much more subtle, and require a deliberate search.

There appears to be some extrinsic stratigraphic control on trap development outwith the sequence stratigraphic framework, with the Carboniferous, the 'Middle' Cretaceous and the 'late' Tertiary important times of plate reorganisation and associated structuration and structural trap formation in the Middle East and formerly contiguous north Africa, and the 'early' Tertiary an important time in the Indian subcontinent and adjacent areas of central and south-east Asia, and also in the Caribbean (see the further reading list at the end of the chapter).

### *9.1.2 Petroleum exploration*

Petroleum exploration involves a number of geological, geophysical and integrated techniques, and a number of drilling and petrophysical logging technologies (Jones, 1996, 2006, 2011a; see also Sections 9.2 and 9.3, and the further reading list at the end of the chapter).

#### *Geological techniques*

Geological techniques include not only field geological mapping and sampling (see below), but also office-based petroleum systems analysis or modelling (see above). Field geological mapping and sampling is undertaken principally to provide information on the distributions of source-, reservoir- and cap-rocks, on the structural configuration on the surface and in the subsurface, on potentially hydrocarbon-bearing structures, and on indications of hydrocarbons. It is particularly important in areas where the acquisition of geophysical data is difficult, for example, in mountainous areas, in areas of structural complexity and/or steep dips, or in areas where the surface outcrops are of limestones or evaporites.

#### *Geophysical techniques*

Geophysical techniques include seismic surveying, gravity and magnetics surveying, and other remote-sensing techniques.

*Seismic surveying* is undertaken to provide further data and interpretations on the structural configuration in the subsurface, to identify prospects suitable for drilling, to infer lithology, porosity and pore fluid content, and to identify 'Direct Hydrocarbon Indications' or DHIs in reservoirs. Acquisition of seismic data on land and at sea is achieved by detonating small explosive charges at the surface and recording the time taken for the shock waves generated by them to reflect back from the various subsurface layers. Processing of seismic data leads to a product that is essentially a structural cross-section through the subsurface, albeit in time rather than depth (until converted to depth using velocity data). Seismic rock property data and other attributes are used to infer lithology, porosity and pore fluid content, and to identify DHIs in reservoirs. Three-dimensional seismic data can be used to infer the geometry of individual sedimentary bodies within reservoirs.

*Gravity and magnetics surveying* is undertaken to provide further data and interpretations on the structural configuration on the surface and in the subsurface, and on the structure and composition of the basement. Acquistion of gravity and magnetic data is achieved by land, marine or airborne surveying. Processing leads to a product that is essentially a map of the local gravity or magnetic field (with the effects of the regional field filtered out). The map highlights density and magnetic contrasts, such as those between the basement and basin fill, and allows interpretations to be made as to the structure and composition of the basement.

*Other remote-sensing techniques* include high-resolution aerial photography, satellite, radar and multi-spectral. Processing of remote-sensing data leads to images or maps highlighting particular aspects of culture and geology, such as, for example, petroleum seeps or the alteration products thereof.

### Integrated techniques

Integrated techniques include basin analysis, and play fairway analysis or evaluation.

*Basin analysis* involves the integration of geological and geophysical data and interpretations (see above), and the development of exploration models and strategies. Interestingly, empirical observations in intensively explored basins suggest some relationship between basin type and contained petroleum reserves. The richest in terms of proven reserves are the 'Type IV' basins or 'continental borderland down-warps' of the Middle East. Note, though, that some of the largest reserves proven recently have been in 'Type VIII' basins or 'Tertiary deltas' such as those of the deep-water Gulf of Mexico, and Brazil and west Africa in the South Atlantic, that have only recently become drillable through advances in technology.

*Play fairway analysis or evaluation* is a methodology for mapping play component presence and effectiveness, play component presence and effectiveness risk, and composite play risk, in basin exploration (see the further reading list at the end of the chapter). It is becoming standard practice for the mapping and associated databasing to be undertaken using Geographic Information System (GIS) technology. Play component presence maps are essentially Gross Depositional Environment (GDE) – or palaeogeographic and facies – maps highlighting areas of proven or interpreted potential petroleum source-, reservoir- or cap-rock development (for example, maps of TSTs highlighting transgressive marine source- or reservoir-rock development). Play component effectiveness maps are those highlighting areas of proven or interpreted potential effective petroleum source-, reservoir- or cap-rock development, determined by burial depth, etc. Common Risk Segment (CRS) – or play component presence and effectiveness risk – maps colour code highlighted areas by perceived risk, conventionally green for low risk, yellow for moderate risk, and red for high risk. Composite CRS or CCRS – or play risk – maps are made by compositing play component presence and effectiveness risk maps (and, ideally, uncertainty or data confidence maps). The only segments that can be characterised as low risk on the composite risk map are those that are characterised as low risk on all of the individual play component risk maps. Similarly, segments characterised as moderate risk on the composite risk map are those that are characterised as no higher than moderate risk on all of the individual play component risk maps. Segments that are characterised as high risk on any of the individual play component risk maps are characterised as high risk on the composite risk map.

### Drilling technologies

Once a prospect has been matured, the technical and commercial risk of drilling evaluated, and the decision to drill approved, drilling can go ahead. The choice of rig will depend on a number of well-site and geological factors, including, for example, whether there is environmental legislation governing site planning or operations, and whether shallow gas pockets or over-pressured intervals are anticipated from seismic data, site investigation or basin modelling, which would necessitate fitting or upgrading of blow-out preventers. Incidentally, the sort of deep-water drilling currently being undertaken in the Gulf of Mexico and South Atlantic is close to the limits of available technology, and presents particular rig design, mobilisation and other logistical problems.

Put simply, drilling is effected by rotation of a diamond bit at the end of a drill string formed of lengths of pipe connected by collars, the weight of the apparatus providing the necessary downward force. Drilling can be either vertical, deviated or high-angle, or horizontal, with direction provided by various types of 'geosteering', including 'biosteering' (see below). Periodic casing of the borehole minimises down-hole contamination or caving, and generally maintains the condition of the hole, reducing the risk of the drill-pipe sticking and having to be fished out, which can be a time-consuming and costly exercise. Importantly, casing can also be used to contain over-pressured zones, reducing the risk of potentially extremely dangerous leaks or blow-outs.

Specially prepared drilling mud pumped down the drill-pipe lubricates the bit, and seals the walls of the bore. It returns to the surface between the drill-pipe and the walls of the bore, bringing with it loose chippings broken off by the action of the bit. These chippings, generally known as ditch cuttings, are collected in a trap or 'shale shaker' prior to recycling of the drilling mud. Ditch-cuttings samples are generally collected every 3 metres or 10 feet for the purposes of lithological description and micropalaeontological and other analysis. In the increasingly cost-conscious oil industry, in which the main aim of drilling is to reach the reservoir target as quickly and cheaply as possible, conventional coring, which is time-consuming and costly, is generally undertaken over critical sections of the reservoir, over which it allows extraction of invaluable data on depositional and diagenetic facies, and porosity and permeability characteristics. Side-wall coring is a useful adjunct to the conventional coring programme.

*Petrophysical logging technologies*

A wide range of petrophysical or wireline logging technologies are available for formation evaluation. (Note in this context that formation evaluation techniques practised by former eastern-bloc geologists differ somewhat from those practised by western geologists). The most commonly run petrophysical or wireline logs are gamma, sonic and resistivity logs (see below). Together, these provide an indication not only of lithology but also of porosity and pore fluid content. 'Logging-while-drilling' or LWD technologies have recently been developed that allow these logs to be run in real-time in the reservoir. Other logs that are becoming more widely used on account of their usefulness specifically in structural geological and sedimentological interpretation are dipmeter logs and image logs. Interestingly from the palaeontologist's point of view, individual fossils and trace fossils are occasionally identifiable on image logs.

*Gamma logs* measure radioactivity. Because in sedimentary rocks this is essentially a function of lithology, being high in fine clastics in which radioactive constituents tend to be concentrated, and low in coarse clastics and carbonates, gamma logs also provide a measure of this parameter. Note, though, that sandstones containing igneous clasts can yield anomalously high values, and non-radioactive clay minerals such as kaolinites anomalously low values. Also, that the so-called natural gamma-ray spectrometry tool can be used to identify individual clay minerals (incidentally, it is also useful in source-rock evaluation, since it can distinguish uranium from other radioactive elements).

*Sonic logs* measure sonic transit time. Because this is essentially a function of density and porosity, sonic logs provide measures of these parameters. Density values are highest in dense carbonates, and lowest in uncompacted clastics. Porosity values are highest in porous, and lowest in tight, carbonates and clastics. Porosity can also be measured using neutron logs. Porosity and permeability can be measured using nuclear magnetic resonance logs.

*Resistivity logs* measure resistivity (the inverse of electrical conductivity). Because this is essentially a function of pore fluid content, resistivity logs provide a measure of this parameter (and of hydrocarbon saturation). Values are highest in rocks whose pore spaces are occupied by hydrocarbons, and lowest in rocks whose pore spaces are occupied by water or brine.

### 9.1.3 Reservoir exploitation

Reservoir exploitation involves the processes of appraisal, development and production (Jones, 2006, 2011a; see also the further reading list at the end of the chapter). Appraisal is the process whereby the extent of the reservoir and its contained reserves of hydrocarbon is determined, and its economic viability evaluated. Development and production are the processes whereby the contained reserves are exploited as efficiently as possible, through the implementation on an integrated reservoir management strategy. (Development refers specifically to the installation of pipelines and associated infrastructure; production to the flow of hydrocarbon from the reservoir into the pipeline system.)

Reservoir exploitation also involves a number of geological techniques; geophysical techniques, including four-dimensional seismic (essentially 'time-lapse' three-dimensional, illustrating the movement of reservoir fluids through time); and integrated techniques (see also below, and Sections 9.4 and 9.5). It also involves a number of drilling and petrophysical logging technologies (see Section 9.1.2).

*Integrated techniques*

Integrated techniques include integrated reservoir characterisation, and reservoir modelling or simulation. Integrated reservoir characterisation is a process, implemented across appraisal, development and production, whereby general geological, geophysical and static and dynamic reservoir engineering data, and specialist micropalaeontological, reservoir geochemical and sedimentological data, are integrated and iterated; and reservoir layering and architecture, permeability distribution, and flow behaviour, described (and predicted).

Failure to integrate specialist micropalaeontological data can result in serious flaws in the reservoir model, and costly inefficiencies in the reservoir exploitation programme (see also the further reading list at the end of the chapter).

### 9.1.4 Well-site operations

Well-site operations involve a number of general and specialist geological, including micropalaeontological, techniques; and geophysical techniques, including borehole seismic (Jones, 1996, 2006, 2011a; see also Sections 9.6 and 9.7). They also involve a number of drilling and petrophysical logging technologies (see Section 9.1.2); and testing, and completion and production, technologies (including 'enhanced oil recovery' or EOR technologies). Testing is a process initiated on indication of hydrocarbons, whereby the bottom-hole formation is made to flow some of its contents into the drill-pipe for recovery and analysis. It provides valuable information on potential flow rate, or reservoir deliverability.

## 9.2 Applications in petroleum exploration

In petroleum exploration, micropalaeontology assists generally in the description and prediction of source-, reservoir- and cap-rock distributions in time and in space, in a high-resolution sequence stratigraphic framework (Jones, 1996, 2006, 2011a; see also the further reading list at the end of the chapter).

### 9.2.1  Chronostratigraphy and palaeoenvironmental interpretation

In this, micropalaeontology plays a key role in the absolute chronostratigraphic age-dating and palaeoenvironmental interpretation of samples acquired during field geological mapping and exploration drilling. Basic application calibrates well and seismic data, and facilitates correlation. More advanced application constrains the timing of structural geological events and enables the establishment of a tectono-sequence stratigraphic framework, and the construction of chronostratigraphic diagrams, for the basin, and also provides important inputs into basin and petroleum systems analysis, and play fairway analysis or evaluation. In play fairway evaluation, the principal input is in the identification of time-slices for the purposes of palaeogeographic and facies, and CRS and CCRS, mapping, and in the population of palaeogeographic and facies maps with pertinent palaeoenvironmental data and/or interpretations (made on the basis of a predictive sequence stratigraphic model).

## 9.3  Case studies of applications in petroleum exploration

### 9.3.1  Middle East

The geological history of the Middle East is long and complex (Eva *et al.*, 1990; Goff *et al.*, in Al-Husseini, 1995; Jones, 1996, 2000; Fraser *et al.*, 2003; Goff *et al.*, 2004; Ruiz *et al.*, 2004; Goff *et al.*, 2005; Fraser *et al.*, 2006; Jones, 2006; Fraser *et al.*, 2007; Jones, 2011a; and additional references cited therein; see also the further reading list at the end of the chapter).

It began essentially with extensional? tectonism and the formation of a series of salt basins, the evidence of which is most clearly seen in the present-day south Oman salt basin, in the area around the Strait of Hormuz in the southern Gulf, and in the Bandar Abbas and Fars provinces in Iran, in the Precambrian or 'Infracambrian' (Megasequence AP1 of Sharland *et al.*, 2001; Fig. 9.2). These basins underwent deformation, inversion, erosion and peneplanation at the 'Infracambrian'/Cambrian boundary. The entire Arabian plate then underwent passive margin subsidence in the Cambrian to Ordovician (Megasequence AP2). A glaciation took place in the latest Ordovician, Ashgillian (Hirnantian), and a deglaciation, resulting in the widespread deposition of transgressive marine sediments, in the Early Silurian, Llandoverian, and the subsequent deposition of regressive sediments in the Late Silurian–Devonian (Megasequence AP3). The plate then underwent compressional? tectonism and uplift, ultimately resulting in extensive erosion associated with the so-called 'Hercynian unconformity' or, in

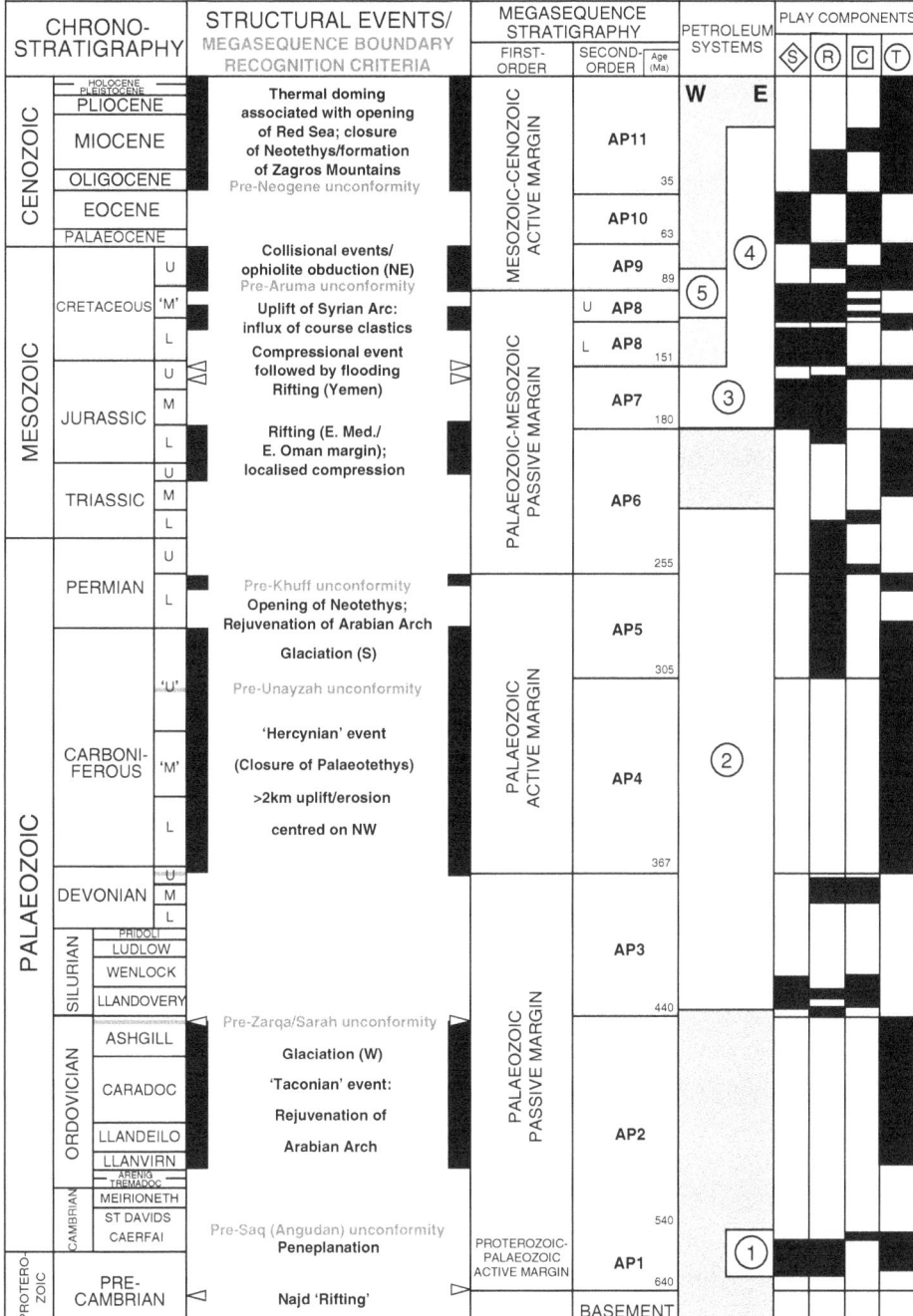

Fig. 9.2 **Stratigraphic framework of the Middle East**. Also showing petroleum resource distribution. S = source-rock; R = reservoir-rock; C = cap-rock; T = trap. Petroleum systems. 1 = Infracambrian; 2 = Silurian; 3 = Jurassic, Sargelu/Naokelekan and Diyab basins; 4 = Cretaceous-Tertiary, Garau, Kazhdumi and Pabdeh basins; 5 = 'Middle' Cretaceous, Bab/Shilaif basin. From Jones (2006).

Saudi Arabia, 'Pre-Unayzah unconformity', in the Early Carboniferous (Megasequence AP4). A second glaciation took place in the Late Carboniferous to Early Permian (Megasequence AP5).

Late Permian to Early Jurassic (Megasequence AP6) rifting and post-rift subsidence resulted in the break-up of the supercontinent of Gondwana, and, for the first time, extensive carbonate deposition. Middle to Late Jurassic (Megasequence AP7) rifting resulted in the break-up of Pangaea, and the formation of a series of intra-shelf basins, of which the most important were the Gotnia and Diyab basins. Carbonate deposition continued to dominate, with clastic deposition essentially confined to certain graben systems in the extreme south-west (Yemen). Extensive evaporite deposition took place at the end of the Jurassic. The Early Cretaceous, Neocomian to early Aptian, and 'Middle' Cretaceous, late Aptian to middle Turonian (Megasequence AP8) was characterised by post-rift subsidence and the formation of further of intra-shelf basins, of which the most important were the Bab basin in the Early Cretaceous and the Kazhdumi and Shilaif basins in the 'Middle' Cretaceous. At this time, carbonate sedimentation in the east became balanced by clastic deposition in the west. The Late Cretaceous, late Turonian to Senonian (Megasequence AP9) saw structuration and extensive erosion associated with the so-called, in Saudi Arabia, 'Pre-Aruma unconformity', generated by ophiolite obduction in the east.

The Cenozoic saw structuration associated with the collision between the Arabian and Eurasian plates and the closure of Tethys, beginning in the Late Cretaceous and effectively ending in the Miocene, and, to a lesser extent, with the opening of the Red Sea, beginning, with rift shoulder uplift, and the shedding of coarse clastic sediments from the 'Western Arabian Highlands' to the west, in the Oligocene. The initial phases involved the formation of a deep marine foredeep known as the Pabdeh basin in what were to become the Zagros Mountains in Iran in the Palaeocene–Eocene (Megasequence AP10). The later phases witnessed the widespread deposition of a typical 'molasse' sequence of shallow marine carbonates, marginal marine evaporites and non-marine clastics in the Oligocene–Holocene (Megasequence AP11).

*Outcrop geology*    The outcrop geology of the Middle East is well documented in the public-domain literature. Published field geological maps are available for the entire area on various scales.

Direct or indirect analogues for some of the source-rocks and reservoirs of the region can be studied at outcrop, for example around the Arabian Shield in Saudi Arabia, in the Oman Mountains of Oman and the United Arab Emirates, and in the Zagros Mountains of Iran and Iraq (see also the further reading list at the end of the chapter).

*Petroleum geology*   The Middle East is the site of 20% of the world's 'giant' oil and gas fields, that is, those containing $> 500$ million barrels of oil – or equivalent – recoverable (see the further reading list at the end of the chapter). The consensus view as to why the area is so rich in giant fields is that its large and on the whole relatively simple traps were able to drain large source kitchens by way of simple migration pathways. This process was facilitated by a peculiarly fortuitous juxta-position of source, reservoir and seal units in trap configurations, a function of the area's unique geological history. There is some stratigraphic control on play component distribution in the Middle East.

*Source-rocks.* Source-rocks occur at several stratigraphic horizons, the most important ones being Infracambrian, Silurian, Middle–Late Jurassic, Early Cret-aceous, 'Middle' Cretaceous and Palaeocene–Eocene (Fig. 9.2; see also the further reading list at the end of the chapter). The Infracambrian source-rock is apparently of comparatively limited regional extent, only proven to be present in the south Oman salt basin. However, the Silurian, Middle–Late Jurassic, Early Cretaceous, 'Middle' Cretaceous and Palaeocene–Early Miocene source-rocks are of at least sub-regional extent. Collectively, these prolific source-rock basins have generated a significant proportion of the world's known petroleum reserves (see the further reading list at the end of the chapter). The Silurian petroleum system accounts for all of the oil and gas reservoired in the Palaeozoic of the Middle East, including that in North Dome/South Pars, the world's largest gas field (Fraser *et al.*, 2003). The Middle–Late Jurassic petroleum system accounts for all of the oil and gas reservoired in the Jurassic of the Middle East, including that in Ghawar, the world's largest oil field (Fraser *et al.*, 2003).

*Reservoir-rocks.* Reservoir-rocks occur at practically every horizon (Fig. 9.2; see also the further reading list at the end of the chapter). Important Early Permian clastic reservoirs occur in Oman and Saudi Arabia; and Late Permian platform carbonate reservoirs occur across the Arabian Peninsula and Gulf and into Fars province in Iran. Jurassic platform carbonate reservoirs occur in Saudi Arabia and the United Arab Emirates. Cretaceous paralic and peri-deltaic clastic reservoirs occur in Kuwait and Iraq; and Cretaceous platformal, and locally peri-reefal, carbonate reservoirs in Kuwait, Iraq, Khuzestan and Lurestan provinces in Iran, Oman and the United Arab Emirates. Oligocene–Miocene paralic and peri-deltaic clastic reservoirs and platformal, and locally peri-reefal, reservoirs occur in in Kuwait, Iraq and Khuzestan and Lurestan provinces in Iran.

The quality of the carbonate reservoirs is controlled partly by primary depos-itional factors and partly by secondary diagenetic factors (fracturing also locally plays an important role in enhancing permeability, as, for example, in the Zagros Mountains of Iran and Iraq: see the further reading list at the end of the chapter). There is some sequence stratigraphic control on the quality of the carbonate reservoirs, with preferential deposition during HSTs, and preferential diagenetic enhancement during LSTs in the case of dissolution, and TSTs in the case of dolomitisation (Goff *et al.*, in Al-Husseini, 1995; Goff *et al.*, 2004; Lehmann *et al.*, 2004, 2006a–c; see also the further reading list at the end of the chapter).

*Cap-rocks (seals) and traps.* Cap-rocks (seals) occur at several horizons (Fig. 9.2). The most important are Infracambrian halites and anhydrites, Early Triassic shales, Late Jurassic halites and anhydrites, 'Middle' Cretaceous shales, Late Cretaceous shales, and Middle Miocene halites and anhydrites. Locally, as in parts of the Zagros Mountains, some of the seals have been breached by surface erosion, or rendered ineffective by fracturing (see the further reading list at the end of the chapter). Generally speaking: the Infracambrian cap-rocks delineate the top of the Infracambrian petroleum system; the Early Triassic cap-rocks the top of the Silurian petroleum system; the Late Jurassic cap-rocks the top of the Middle–Late Jurassic petroleum systems; the 'Middle' Cretaceous cap-rocks the top of the Early Cretaceous petroleum systems; the Late Cretaceous cap-rocks the top of the 'Middle' Cretaceous petroleum systems; and the Middle Miocene cap-rocks the top of the Palaeocene–Early Miocene petroleum system.

Traps are predominantly structural, the style varying from inverted basement-cored anticlinal in the foreland to thrust anticlinal in the fold-belt of the Zagros Mountains of Iran and Iraq (see the further reading list at the end of the chapter). Mobilisation of Infracambrian salt has played an important role in trap formation in Iran, the Gulf, and the eastern part of the Arabian Peninsula. Combination or straight stratigraphic traps include that of the Fateh field in Dubai in the United Arab Emirates.

## Biostratigraphy

A number of micropalaeontological, nannopalaeontological and palynological biostratigraphic zonation schemes are applicable in the Middle East, the scale ranging from global through regional to reservoir, and the resolution in inverse proportion to the scale (see below; see also Sections 9.5.3 and 9.7.1).

Quantitative graphic correlation (GC) techniques have been applied in the interpretation of biostratigraphic data from the Cretaceous (Scott, in Robertson *et al.*, 1990; Simmons, in Simmons, 1994; see also Section 7.6).

A range of non-biostratigraphic techniques have also been applied, including chemostratigraphy and cyclostratigraphy (see the further reading list at the end of the chapter).

*Palaeozoic* Local to regional Palaeo-Tethyan larger benthic foraminiferal and Gondwanan palynological biostratigraphic zonation schemes are applicable in the Palaeozoic of the Middle East, the former essentially in the Carboniferous–Permian only, the latter throughout (Douglas, 1936, 1950; Bozorgnia, 1964;

Setudehnia, 1972; Bozorgnia, 1973; Szabo & Kheradpir, 1978; Zaninetti *et al.*, 1978; Lys, 1980; Johnson, in Neale & Brasier, 1981; Kalantari, 1982; El-Bishlawy, 1985; Kalantari, 1986; Al-Aswad & Kamel, in Sadek, 1992; Martinez-Dias *et al.*, 1996; Al-Aswad, 1997; Angiolini *et al.*, 1998; Sharland *et al.*, 2001; Vachard *et al.*, 2002; Al-Husseini, 2004; Hughes, in Powell & Riding, 2005; Vachard *et al.*, 2005; Vaslet *et al.*, 2005; Insalaco *et al.*, 2006; Gaillot & Vachard, 2007; Weidlich, in Alvaro *et al.*, 2007; Hughes, in Demchuk & Gary, 2009; Forke *et al.*, 2010; Hughes, 2010a).

Key microfacies are illustrated by Sartorio & Venturini (1988).

*Mesozoic* Regional to modified global planktic foraminiferal and nannopalaeontological biostratigraphic zonation schemes are applicable in the Mesozoic of the Middle East, together with local to regional larger benthic and local smaller benthic schemes (Cox, 1937; Henson, 1948; Grimsdale, 1952; Dunnington, 1955; Eames & Smout, 1955; Smout, 1956; Owen & Nasr, in Weeks, 1958; Hudson & Chatton, 1959; Kavary & Frizzell, 1963; Bozorgnia, 1964; Redmond, 1964a, b; James & Wynd, 1965; Redmond, 1965; Derin & Reiss, 1966; Kalantari, 1969; Sampo, 1969; Loutfi & Jaber, 1970; El-Naggar & El-Rifaiy, 1972, 1973; Setudehnia, 1972; Gollesstaneh, 1974; Kalantari, 1976; Schroder & Darmoian, 1977; Kassab, 1978; Szabo & Kheradpir, 1978; Zaninetti *et al.*, 1978; Al-Shamlan, 1980; Kalantari, 1982; Harris *et al.*, in Schlee, 1984; El-Naggar & El-Nakhal, 1986; Kalantari, 1986; Shakib, 1987; Simmons & Hart, in Hart, 1987; El-Naggar & Kamel, 1988; Scott *et al.*, in Wilgus *et al.*, 1988; Banner & Highton, 1990; El-Nakhal, 1990; Scott, in Robertson *et al.*, 1990; Shakib, 1990; Smith *et al.*, 1990; Banner & Whittaker, 1991; Banner *et al.*, 1991; Kennedy & Simmons, 1991; Athersuch *et al.*, 1992; Rahaghi, 1992; Kaddouri & Al-Shaibani, 1993; Toland *et al.*, 1993; Banner & Simmons, in Simmons, 1994; de Matos, in Simmons, 1994; de Matos *et al.*, 1994; Shakib, in Simmons, 1994; Simmons, in Simmons, 1994; Simmons & Al-Thour, in Simmons, 1994; Witt & Gokdag, in Simmons, 1994; Alsharhan & Whittle, 1995; Al-Shdidi *et al.*, 1995; Al-Silwadi *et al.*, 1996; Hughes, 1996; Kuznetsova *et al.*, 1996; Aziz & Abd El-Sattar, 1997; Al-Fares *et al.*, 1998; Whittaker *et al.*, 1998; Granier, 2000; Hewaidy & Al-Saad, 2000; Hughes, 2000; Hughes, in Hart *et al.*, 2000; Simmons *et al.*, in Hart *et al.*, 2000; Sharland *et al.*, 2001; Abdelghany, 2003; Granier *et al.*, 2003; Sadooni & Alsharhan, 2003; Hughes, 2004a, b; Jones *et al.*, in Bubik & Kaminski, 2004; Abdelghany, 2006; Lehmann *et al.*, 2006a–c; Al-Saad, 2008; Hosseini & Conrad, 2008; Hughes *et al.*, 2008; Hughes, 2009; Hughes, in Demchuk & Gary, 2009; Hughes & Nani, 2009; Hughes *et al.*, 2009; Ghabeishavi *et al.*, 2010; Rashid, 2010; Busnardo & Granier, 2011; Granier *et al.*, in Grosheny & Granier, 2011; Cherchi & Schroder, 2013; Granier & Busnardo, 2013; Figs. 9.3–9.4).

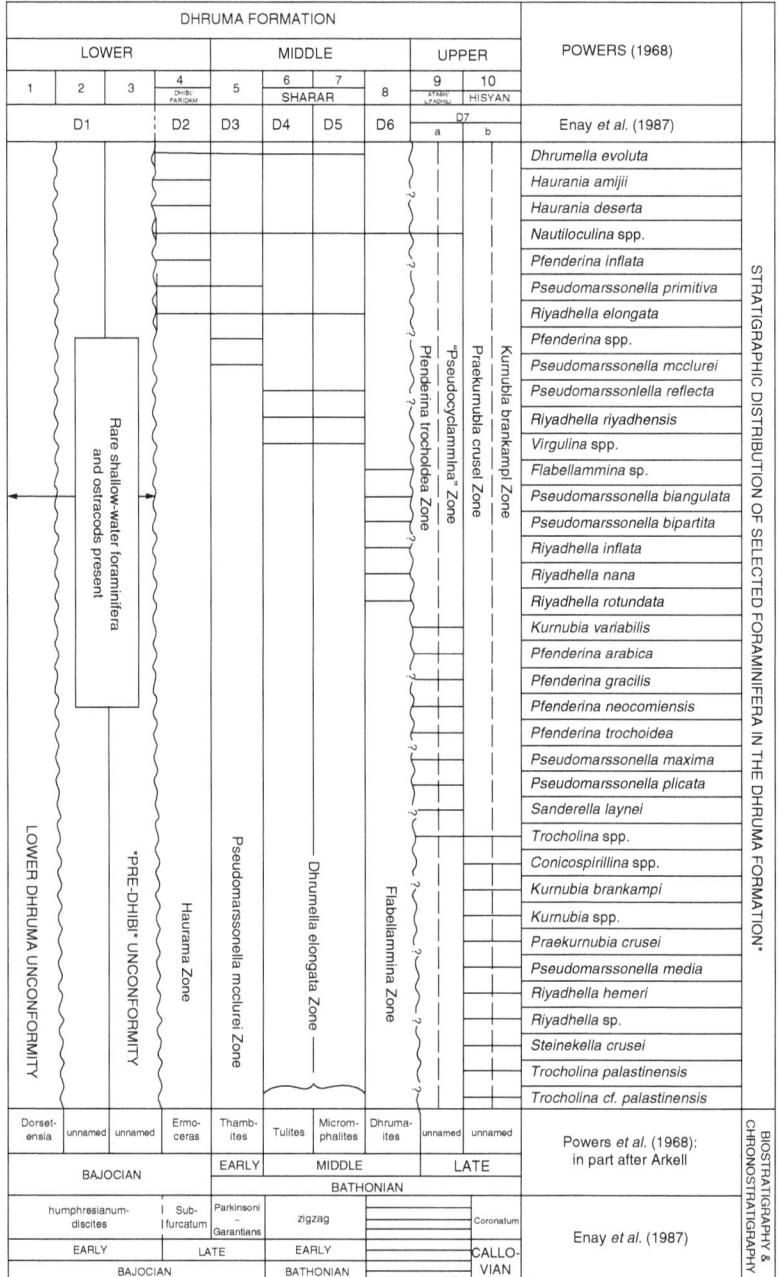

Fig. 9.3  **Stratigraphic distribution of Foraminifera in the Middle Jurassic Dhruma formation of Saudi Arabia**. Based on data in Powers (1968). Note that *Haurania amijii* is now referred to *Amijiella*, *Pseudomarssonella mcclurei* to *Riyadhoides*, *P. reflecta* to *P. plicata*, *P. biangulata* to *Riyadhella regularis*, *Riyadhella rotundata* to *Redmondoides*, and *R. hemeri* to *R. regularis* (Whittaker *et al.*, 1998).

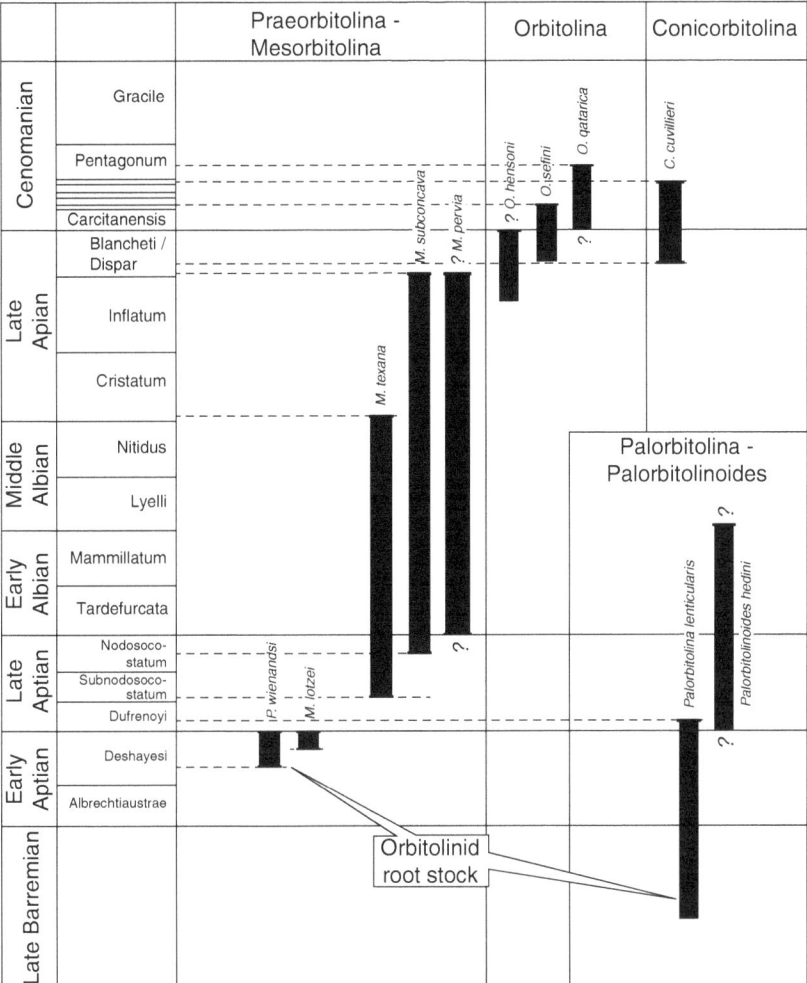

Fig. 9.4 **Stratigraphic Distribution of Middle Eastern orbitolinid Forami-nifera in the 'Middle' Cretaceous of the Middle East**. Reproduced with permission from Jones *et al.*, in Bubik & Kaminski (2004). Calibration against ammonite zones best-fit.

Dunnington, in Weeks (1958) established a Late Cretaceous planktic forami-niferal biozonation scheme based on variations in the proportions of globotrunca-nid species groups, the resolution of which was sufficiently high to enable it to be used in predictions of thicknesses of eroded Late Cretaceous sediment, and of unpenetrated Late Cretaceous sediment.

Some stratigraphically and palaeoenvironmentally significant species are illus-trated on Pls. 41–42 and 47; microfacies by Sartorio & Venturini (1988).

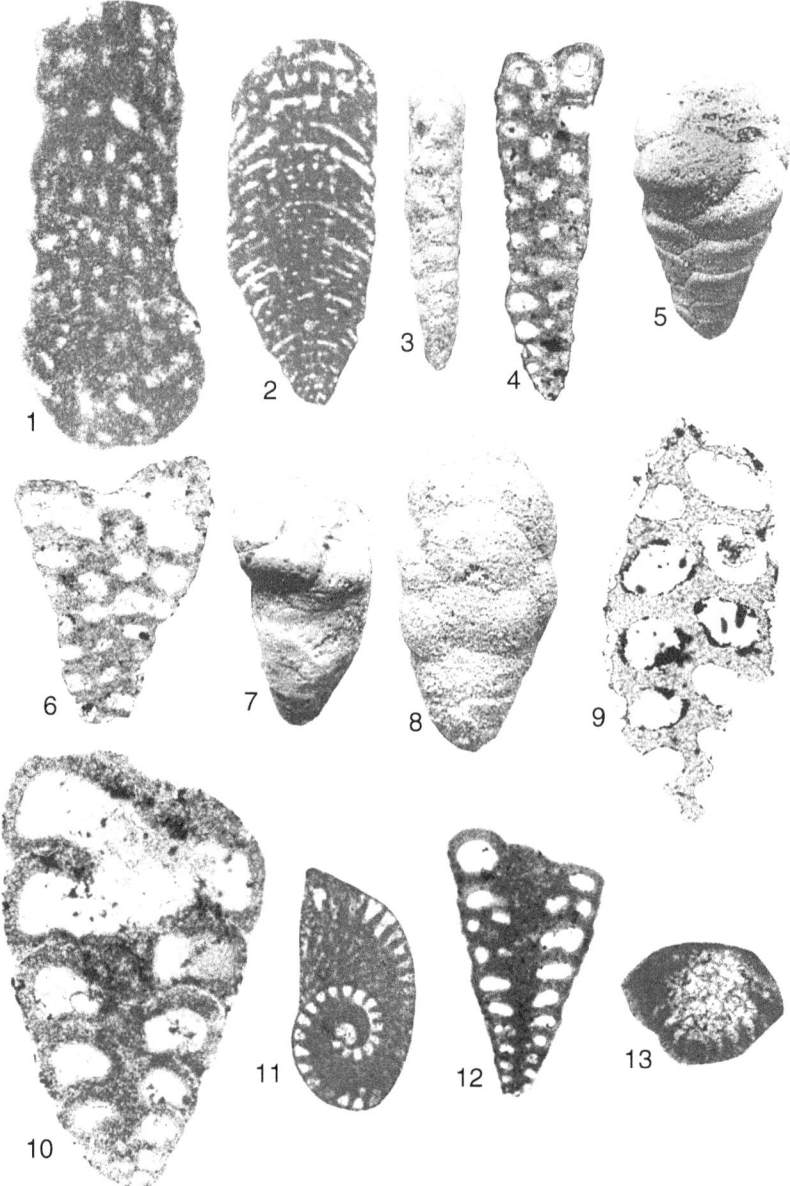

**Pl. 41  Some stratigraphically significant Foraminifera from the Jurassic of the Middle East.** Modified after Whittaker *et al.* (1998). 41.1 *Amijiella* [*Haurania*] *amijii*, ×145; 41.2 *Haurania deserta*, ×70; 41.3 *Riyadhella elongata*, ×165; 41.4 *Riyadhoides* [*Pseudomarssonella*] *mcclurei*, ×170; 41.5 *Pseudomarssonella bipartita* [*P. reflecta*], ×185; 41.6 *Redmondoides medius* [*Pseudomarssonella biangulata*], ×160; 41.7 *Pseudomarssonella bipartita*, ×90; 41.8 *Riyadhella inflata*, ×120; 41.9 *Riyadhella regularis* [*R. nana, R. hemeri*], ×350; 41.10 *Redmondoides* [*Riyadhella*] *rotundatus*, ×235; 41.11 *Pfenderina trochoidea*, ×55; 41.12 *Pseudomarssonella maxima*, ×110; 41.13 *Trocholina palastiniensis*, ×140.

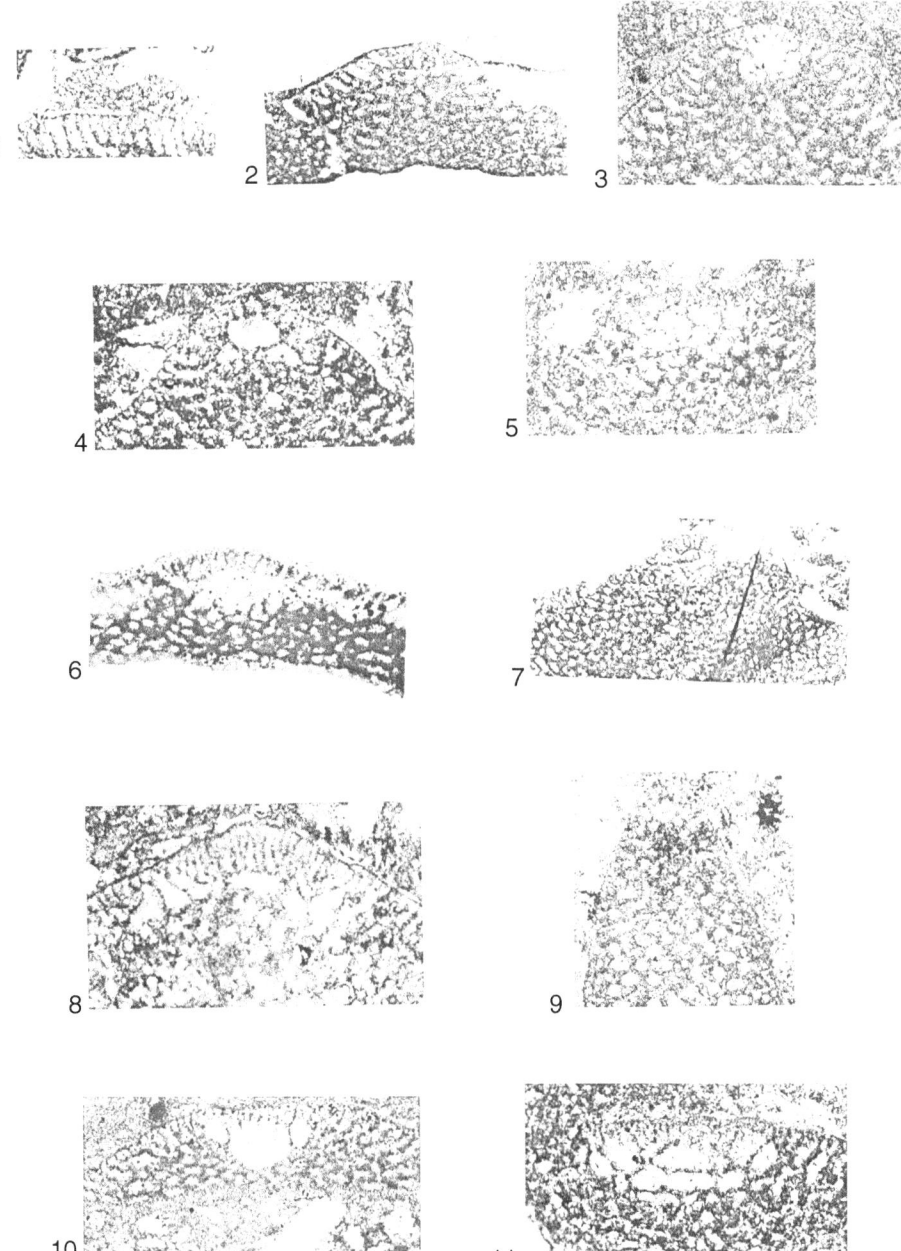

Pl. 42 **Some stratigraphically significant Foraminifera from the Cretaceous of the Middle East**. Modified after Simmons *et al.*, in Hart *et al.* (2000). 42.1 *Praeorbitolina wienandsi*; 42.2 *Mesorbitolina lotzei*; 42.3 *Mesorbitolina texana*; 42.4 *Mesorbitolina subconcava*; 42.5 *Mesorbitolina pervia*; 42.6 *Orbitolina hensoni*; 42.7 *Orbitolina sefini*; 42.8 *Orbitolina qatarica*; 42.9 *Conicorbitolina cuvillieri*; 42.10 *Palorbitolina lenticularis*; 42.11 *Palorbitolinodes hedini*. All scale bars 0.2 mm.

*Cenozoic*     Regional to modified global planktic foraminiferal and nannopalaeon-tological biostratigraphic zonation schemes are applicable in the Cenozoic of the Middle East, together with local to regional larger benthic and local smaller benthic schemes (Henson, 1948, 1950b; Grimsdale, 1952; Thomas, 1952a, b; Smout, 1954; van Bellen, 1956; Owen & Nasr, in Weeks, 1958; Sander, 1962; Kavary & Frizzell, 1963; Bozorgnia, 1964; Bozorgnia & Kalantari, 1965; James & Wynd, 1965; Setudehnia, 1972; El-Khayal, 1974; Kalantari, 1976; Kassab, 1978; Al-Hashimi, 1980; Al-Qayim *et al.*, 1980; Rahaghi, 1980; Fermont, 1982; Dullo *et al.*, 1983; Rahaghi, 1983; Al-Hashimi & Amer, 1985; Hasson, 1985; Majid & Veizer, 1986; Kalantari, 1986; El-Naggar & Kamel, 1988; Abawi, 1989; Hughes *et al.*, 1991; Abawi & Maroof, 1992; Hughes & Beydoun, 1992; Jones & Racey, in Simmons, 1994; Racey, in Simmons, 1994; White, in Simmons, 1994; Hughes & Filatoff, in Al-Husseini, 1995; Racey, 1995; Seyrafian & Hamedani, 1998; Hughes *et al.*, 1999; Seyrafian, 2000; Sharland *et al.*, 2001; Hughes & Johnson, 2005; Amirshahkarami *et al.*, 2007; Mossadegh *et al.*, 2009; Sander, 2012).

Some stratigraphically significant species are illustrated on Pls. 43–46; micro-facies by Sartorio & Venturini (1988).

### *Palaeoenvironmental interpretation*

Benthic Foraminifera are of use in the interpretation of the palaeoenvironments of the Middle East (Hughes, 2002).

*Palaeobathymetry*     Benthic Foraminifera are of use in the discrimination of marginal, shallow and deep marine environments, in part on the basis of modern analogy (see Section 4.2).

*Sedimentary environments*     Benthic Foraminifera are of especial use in the discrimination, on the bases of modern analogy, functional morphology and associated fossils and sedimentary facies, of peri-reefal, that is, back-reefal, reefal and fore-reefal sub-environments in shallow marine (platform) carbonate environments in the Palaeozoic (Sartorio & Venturini, 1988); similarly in the Mesozoic (Henson, 1950a; Stampfli *et al.*, 1976; Riche & Prestat, 1980; Mouty & Saint-Marc, 1982; Reulet, in Reeckmann & Friedmann, 1982; Frost *et al.*, in Harris, 1983; Burchette & Britton, in Brenchley & Williams, 1985; Jordan *et al.*, in Roehl & Choquette, 1985; Sartorio & Venturini, 1988; Burchette, in Simo *et al.*, 1993; Banner & Simmons, in Simmons, 1994; Jones, 1996; Lehmann *et al.*, 2006a–c); and also in the Cenozoic (Henson, 1950a; van Bellen, 1956; Fermont, 1982; Majid & Veizer, 1986; Jones & Racey, in Simmons, 1994; Sartorio & Venturini, 1988; Beavington-Penney *et al.*, 2006).

Back-reef sub-environments are or were characterised especially by smaller and some larger Miliolida and Fusulinida; reefal sub-environments by larger Miliolida,

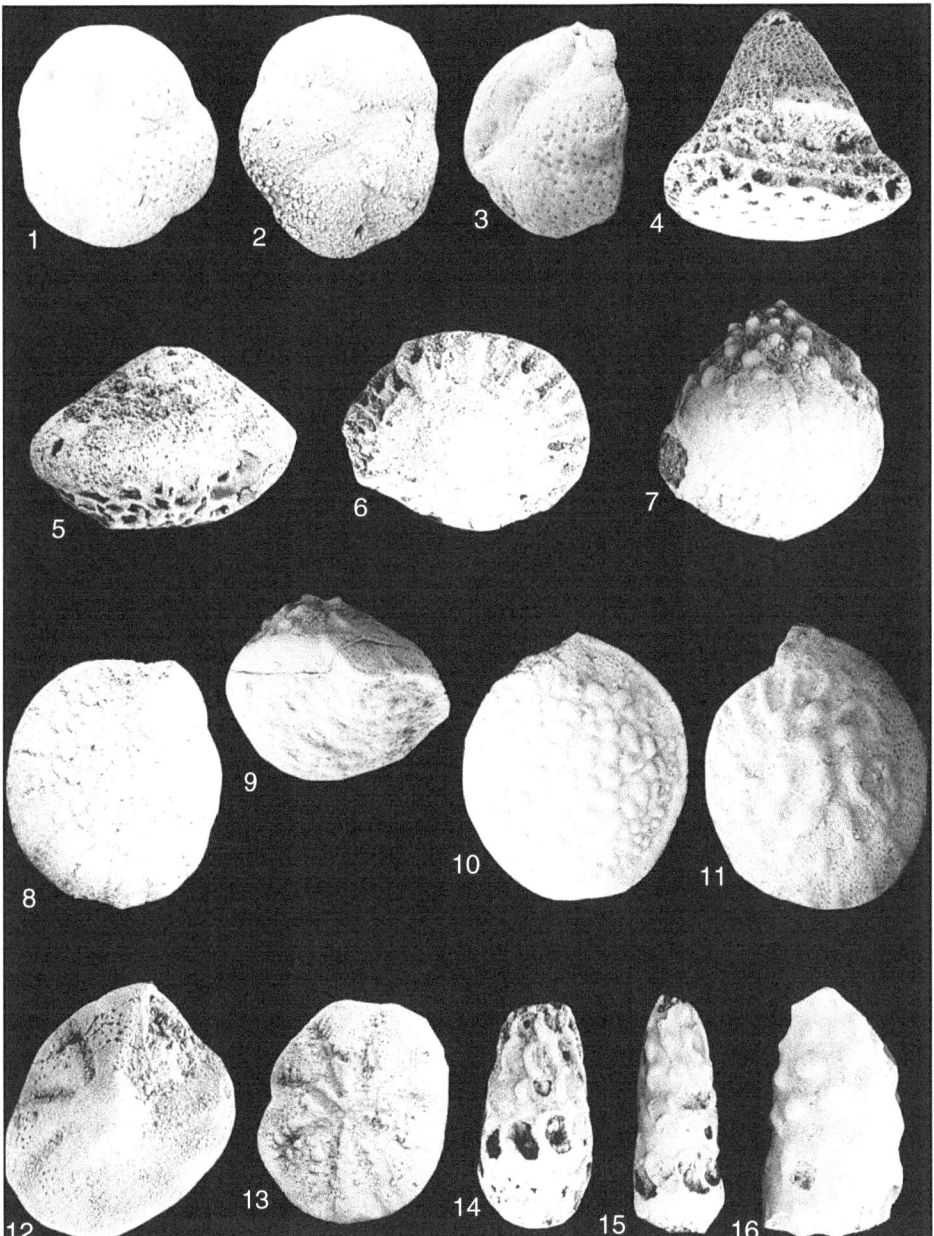

Pl. 43   **Some stratigraphically significant Foraminifera from the Cenozoic of the Middle East, 1. Palaeocene–Eocene**. 43.1–43.3 *Alabamina dubifera*; 43.4 *Chapmanina* cf. *gassinensis;* 43.5 *Dictyoconoides cooki*; 43.6 *Halkyardia minima*; 43.7 *Lockhartia altispira;* 43.8 *Lockhartia hunti pustulosa*; 43.9–43.11 *Lockhartia praehaimei;* 43.12 *Rotalia* cf. *trochidiformis;* 43.13 *Rotalia hensoni;* 43.14 *Sakesaria cotteri;* 43.15 *Sakesaria cordata;* 43.16 *Sakesaria ornata.*

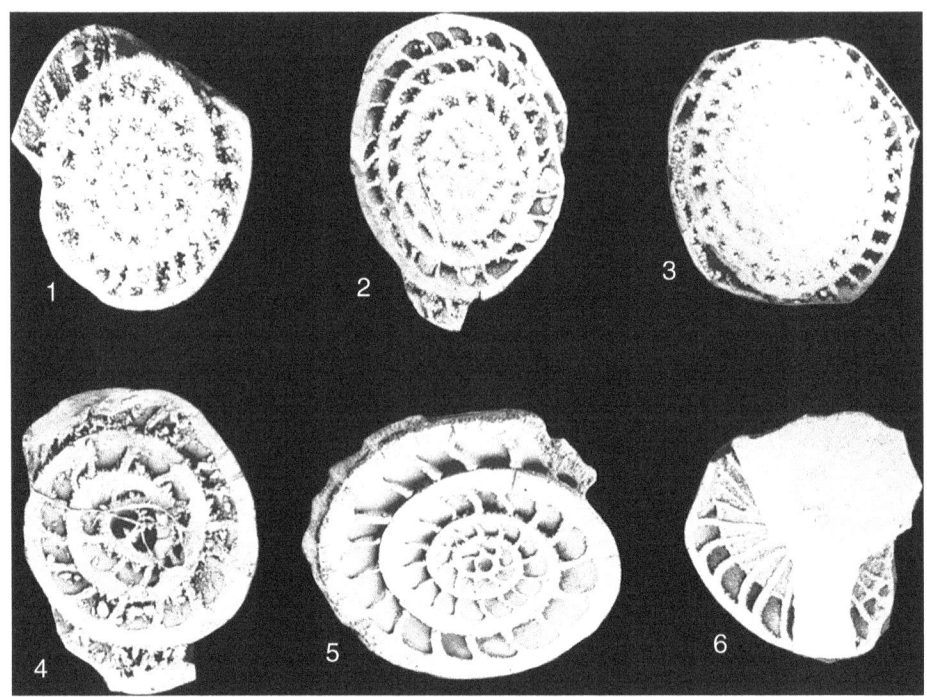

Pl. 44    **Some stratigraphically significant Foraminifera from the Cenozoic of the Middle East, 2. Palaeocene–Eocene contd**. 44.1 *Nummulites beaumonti*; 44.2 *Nummulites* cf. *cuvillieri*; 44.3 *Nummulites discorbinus*; 44.4 *Nummulites fabianii*; 44.5 *Nummulites ptukhiani*; 44.6 *Nummulites sorenbergensis*.

Fusulinida and Rotaliida; and fore-reef sub-environments by larger Rotaliida and Globigerinida.

The depth distributions of many ancient LBFs in peri-reefal sub-environments have been established by means of calibration against associated photosynthetic Algae with known light requirements and hence water depth preferences (Banner & Simmons, in Simmons, 1994; Jones, 1996; Jones *et al.*, in Bubik & Kaminski, 2004; Fig. 9.5). Using this information, high-resolution palaeobathymetric interpretations have been undertaken – and indeed detailed palaeobathymetric curves constructed – whenever and wherever sufficiently closely spaced samples are available for analysis. Integrated with sedimentology, this has facilitated reservoir characterisation, and even the identification of reservoir 'sweet spots' (see Section 9.4).

Some supplementary information as to depth is provided by morphological trends within certain groups of LBFs, such as the Orbitolinidae, with conical morphotypes characterising shallow sub-environments and discoidal or flattened types – generally – deep ones (cf. Reiss & Hottinger, 1984). Note, though, that as intimated above the primary control on distribution is water clarity rather rather than

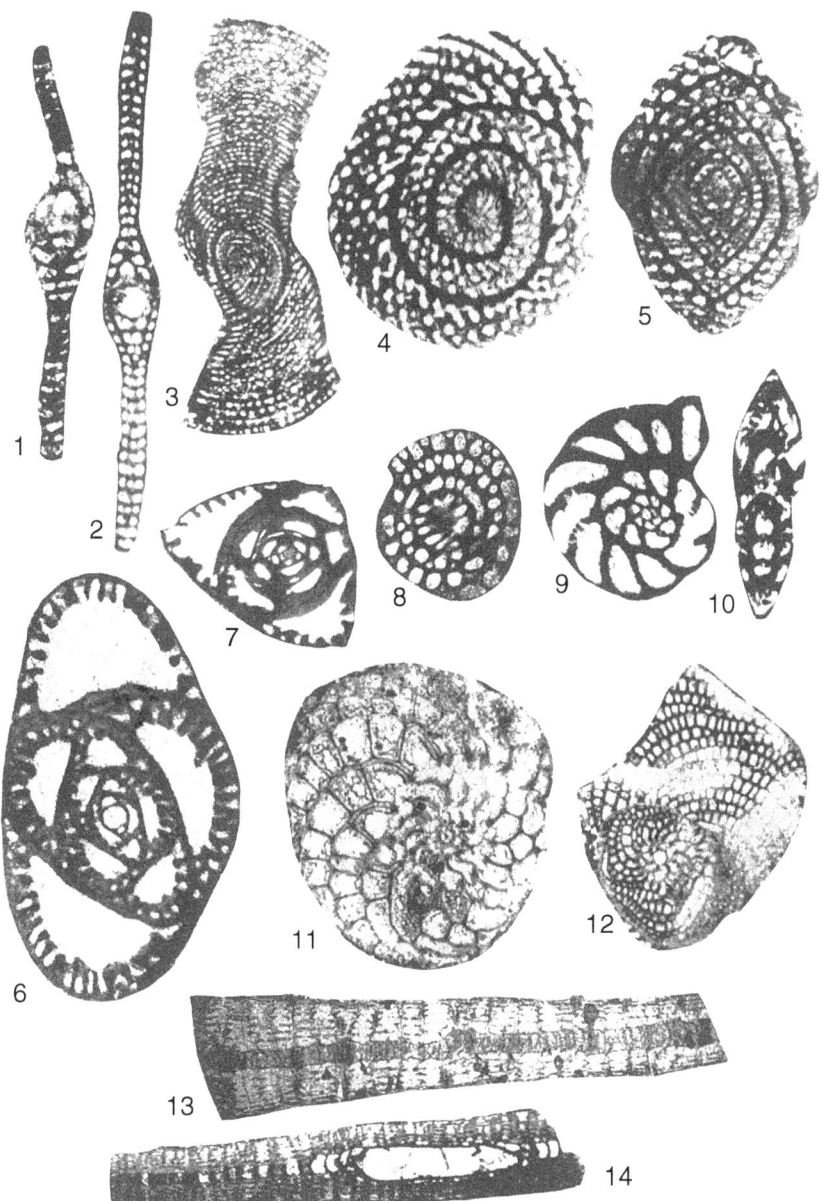

**Pl. 45    Some stratigraphically significant Foraminifera from the Cenozoic of the Middle East, 3. Oligocene–Miocene**. 45.1 *Archaias asmaricus* ×45; 45.2 *Archaias hensoni* ×35; 45.3 *Archaias kirkukensis* ×23; 45.4–45.5 *Archaias oper-culiniformis* ×35; 45.6 *Austrotrillina asmariensis* ×45; 45.7 *Austrotrillina pau-cialveolata* ×45; 45.8 *Borelis melo curdica*, ×56; 45.9–45.10 *Dendritina rangi* ×45; 45.11 *Heterostegina costata* ×88; 45.12 *Heterostegina praecursor* ×45; 45.13 *Lepidocyclina (Eulepidina) dilatata* ×35; 45.14 *Lepidocyclina (Eulepidina) elephantina* ×14.

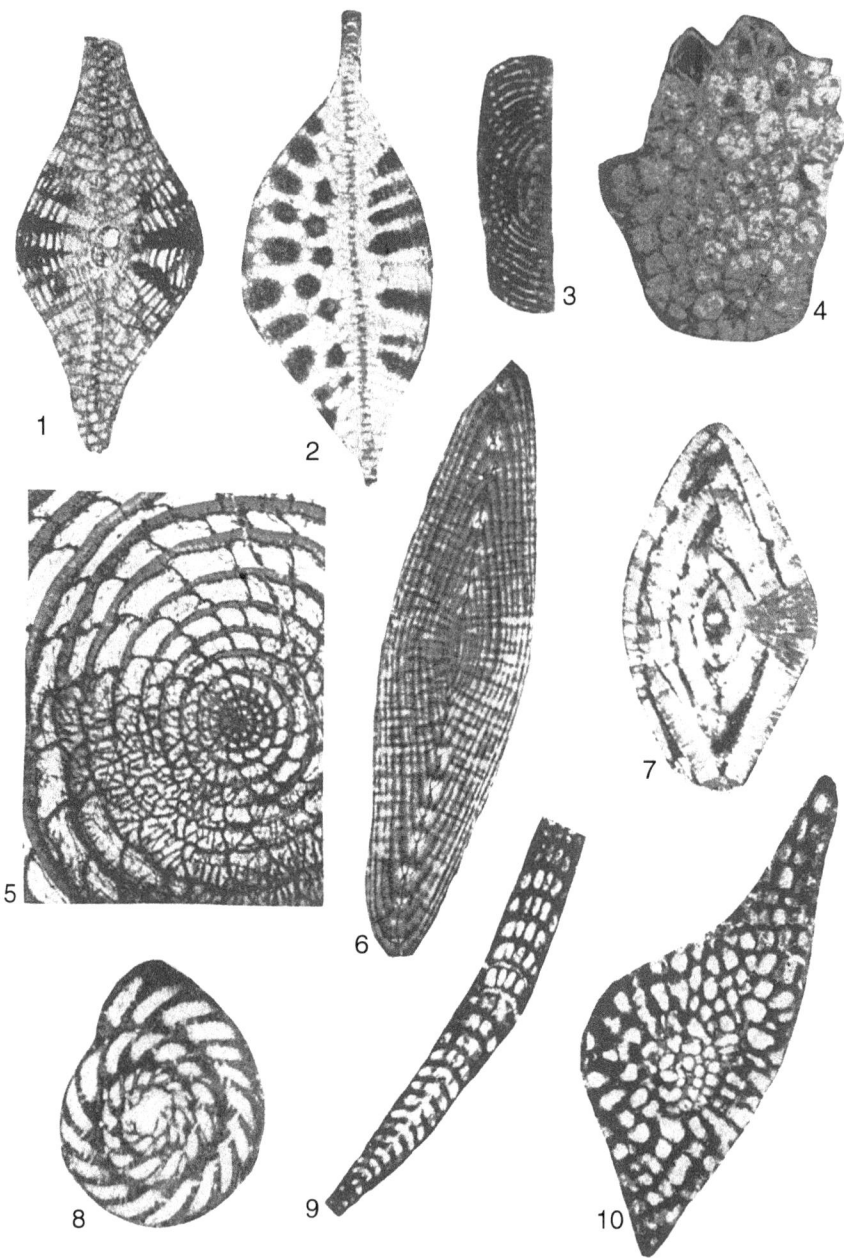

Pl. 46    **Some stratigraphically significant Foraminifera from the Cenozoic of the Middle East, 4. Oligocene–Miocene contd**. 46.1 *Lepidocyclina (Nephrolepidina) marginata* ×14; 46.2 *Lepidocyclina (Nephrolepidina) tournoueri* ×45; 46.3 *Meandropsina anahensis* ×45; 46.4 *Miogypsinoides complanatus* ×45; 46.5–46.6 *Nummulites fichteli/intermedius* ×35, ×14; 46.7 *Nummulites vascus* ×35; 46.8 *Peneroplis farsensis* ×35; 46.9 *Praerhapydionina delicata* ×35; 46.10 *Spiroclypeus ranjanae* ×45.

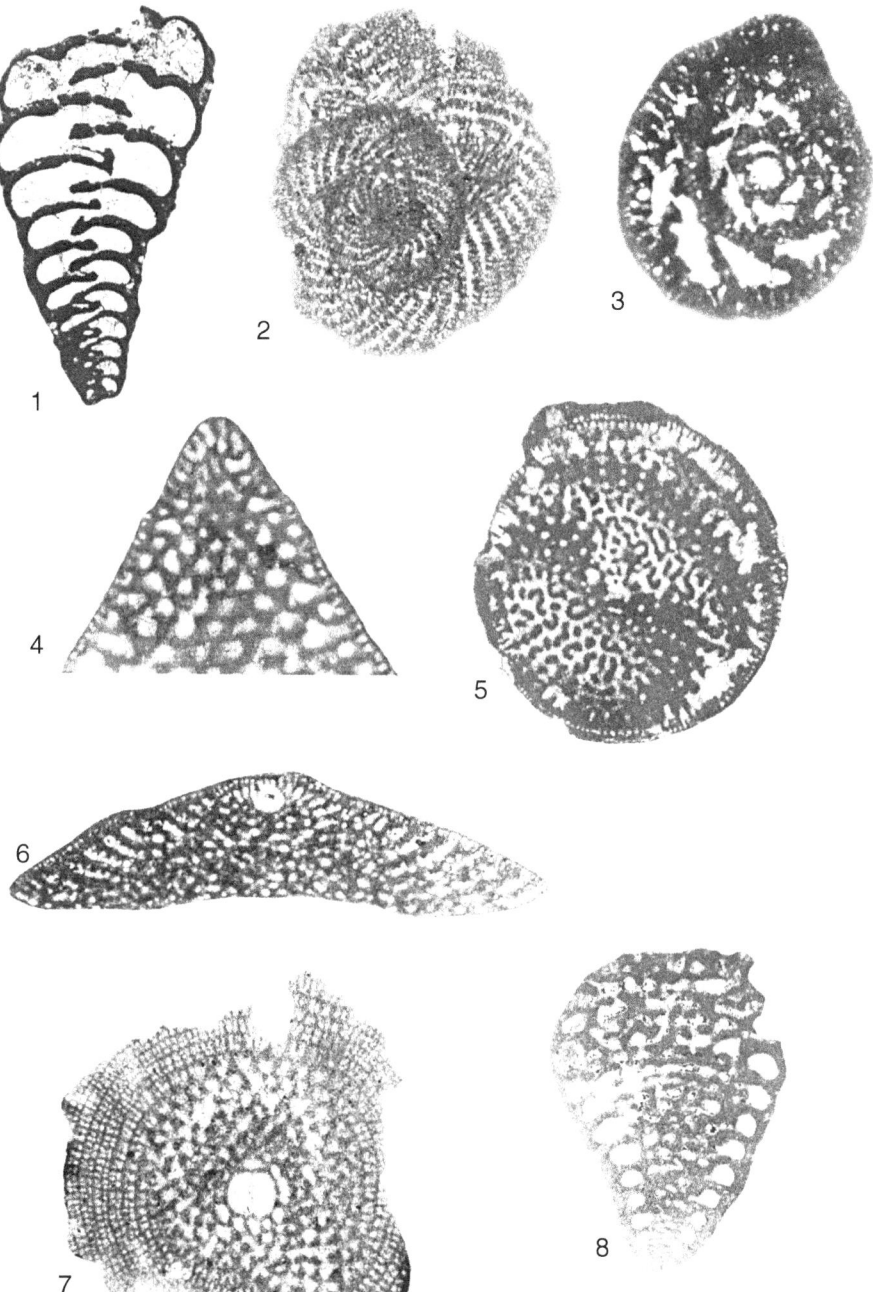

Pl. 47 **Some palaeoenvironmentally significant Foraminifera from the Cretaceous of the Middle East**. Modified after Whittaker *et al.* (1998). 47.1 *Praechrysalidina infracretacea*, ×40; 47.2 *Choffatella decipiens*, ×25; 47.3 *Pseudocyclammina lituus*, ×35; 47.4–47.5 *Palaeodictyoconus arabicus*, ×65, ×60; 47.6–47.7 *Palorbitolina lenticularis*, ×40, ×70; 47.8 *Paravalvulina arabica*, ×35.

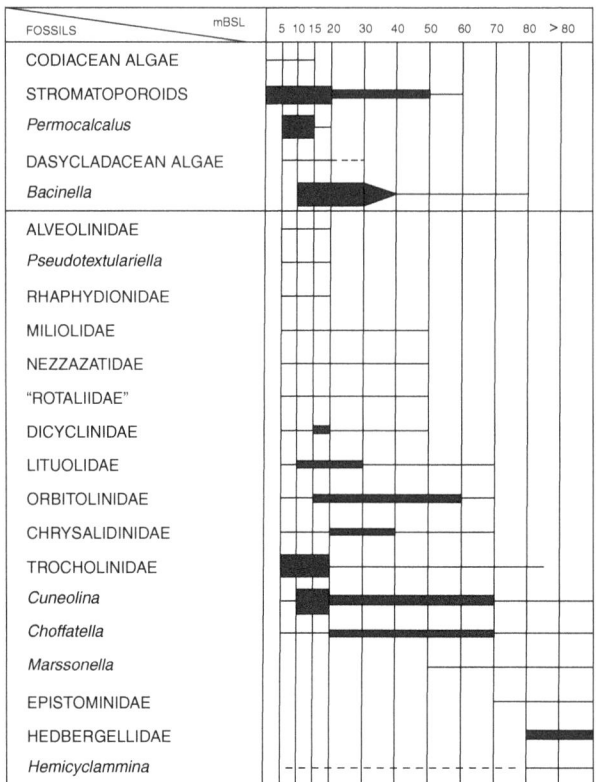

Fig. 9.5    **Palaeobathymetric distribution of Algae and Foraminifera in the 'Middle' Cretaceous of the Middle East**.

depth *per se*, and that discoidal or flattened morphotypes can – exceptionally – characterise poorly-lit shallow, as well as deep sub-environments (Pittet *et al.*, 2002).

### Integrated studies

A number of integrated biostratigraphic and sequence stratigraphic schemes for the Middle East have been published (Langdon & Malecek, 1987; Harris *et al.*, in Schlee, 1984; Scott *et al.*, in Wilgus *et al.*, 1988; le Nindre *et al.*, 1990; Alsharhan & Kendall, 1991; Cherif *et al.*, 1992; Simmons *et al.*, 1992; McGuire *et al.*, 1993; Jones & Racey, in Simmons, 1994; Goff *et al.*, in Al-Husseini, 1995; Koepnick *et al.*, in Al-Husseini, 1995; Kompanik *et al.*, in Al-Husseini, 1995; van Buchem *et al.*, 1996; Al-Husseini, 1997; Immenhauser *et al.*, 1999; Sharland *et al.*, 2001; Davies *et al.*, 2002; van Buchem *et al.*, 2002a, b; Pittet *et al.*, 2002; Al-Husseini, 2004; Sharland *et al.*, 2004; Haq & Al-Qahtani, 2005; Lehmann *et al.*, 2006a–c; Simmons *et al.*, 2007; Droste, in van Buchem *et al.*, 2010; Razin *et al.*, in van Buchem *et al.*, 2010; van Buchem *et al.*, in van Buchem *et al.*, 2010). The ages of sequence boundaries are constrained by conventional biostratigraphy, supplemented by GC techniques (Sharland *et al.*, 2001, 2004; Simmons *et al.*, 2007). The ages of MFSs are also constrained by conventional biostratigraphy: the Late Permian, Kazanian P20 MFS by *Neoschwagerina margaritae*; the Late Permian, Tatarian P30 MFS by *Colaniella parva*; the Late Jurassic, late Kimmeridgian J80 MFS by *Kurnubia jurassica*; the Tithonian J110 MFS by *Everticyclammina kelleri*; the Early Cretaceous, Barremian K60 MFS by *Palaeodictyoconus arabicus*; the early Aptian K70 MFS by *Palorbitolina lenticularis*; the 'middle' Aptian K80 Maximum Flooding Surface (MFS) by *Wondersella athersuchi*; the early Albian K90 MFS by *Mesorbitolina texana*; the middle Albian K100 MFS by *Mesorbitolina subconcava*; the late Albian K110 MFS by *Orbitolina sefini*; the Late Cretaceous, early Cenomanian K120 MFS by *Favusella washitensis*; the middle Cenomanian K130 MFS by *Praealveolina tenuis*; the early Turonian K140 MFS by *Whiteinella archaeocretacea*; the Campanian K170 MFS by *Globotruncana ventricosa*; the Maastrichtian K180 MFS by *Omphalocyclus macroporus*; the Late Palaeocene Pg10 MFS by *Planorotalites pseudomenardii*; the Early Eocene Pg20 MFS by *Nummulites globulus* and *Coskinolina balsillei*; the Early Oligocene Pg30 MFS by *Nummulites fichteli*; the 'Middle' Oligocene Pg40 MFS by *N. fichteli*, *Lepidocyclina* and *Praerhapidionina delicata*; the Late Oligocene Pg50 MFS by *Miogypsinoides complanatus*; the early Early Miocene Ng10 MFS by *Miogypsina globulina*; the late Early Miocene Ng20 MFS by *Borelis melo curdica*; and the late Middle Miocene Ng40 MFS by *Flosculinella bontangensis* and *Orbulina universa*) (Sharland *et al.*, 2001, 2004; Simmons *et al.*, 2007; see also Al-Husseini, 2007). The MFSs are characterised micropalaeontologically by maximum flooding, as indicated by foraminiferal palaeobathymetry, and also by maxima in abundance and diversity; and STs by palaeobathymetric trends (TSTs by upward deepening, and by upward thickening of parasequences, and HSTs by upward shallowing, and by upward thinning of parasequences) (Simmons *et al.*, 1992; Lehmann *et al.*, 2006a–c; see also Section 8.2).

In terms of play fairway evaluation, a number of palaeogeographic and facies maps for selected time-slices have also been published, the time-slices identified in part on the basis of biostratigraphy, and the palaeogeography and facies on the basis of palaeobiology (Saint-Marc, in Moullade & Nairn, 1978; Murris, 1980a, b; Jones & Racey, in Simmons, 1994; Goff *et al.*, in Al-Husseini, 1995; Sharland *et al.*, 2001; Ziegler, 2001; Jones, 2006, 2011a; Figs. 9.6–9.9). Source and/or reservoir distributions can be described on and predicted from maps such as these (Goff *et al.*, in Al-Husseini, 1995; Jones, 2006, 2011a). It is evident that the Middle–Late Jurassic, earliest Cretaceous and 'Middle' Cretaceous source facies are restricted to intra-shelf basins, and that the Palaeocene–Eocene source facies is restricted to the Zagros foredeep basin (Jones, 2006, 2011a). Oligocene and Early Miocene paralic and peri-deltaic, and platformal and peri-reefal, reservoirs are

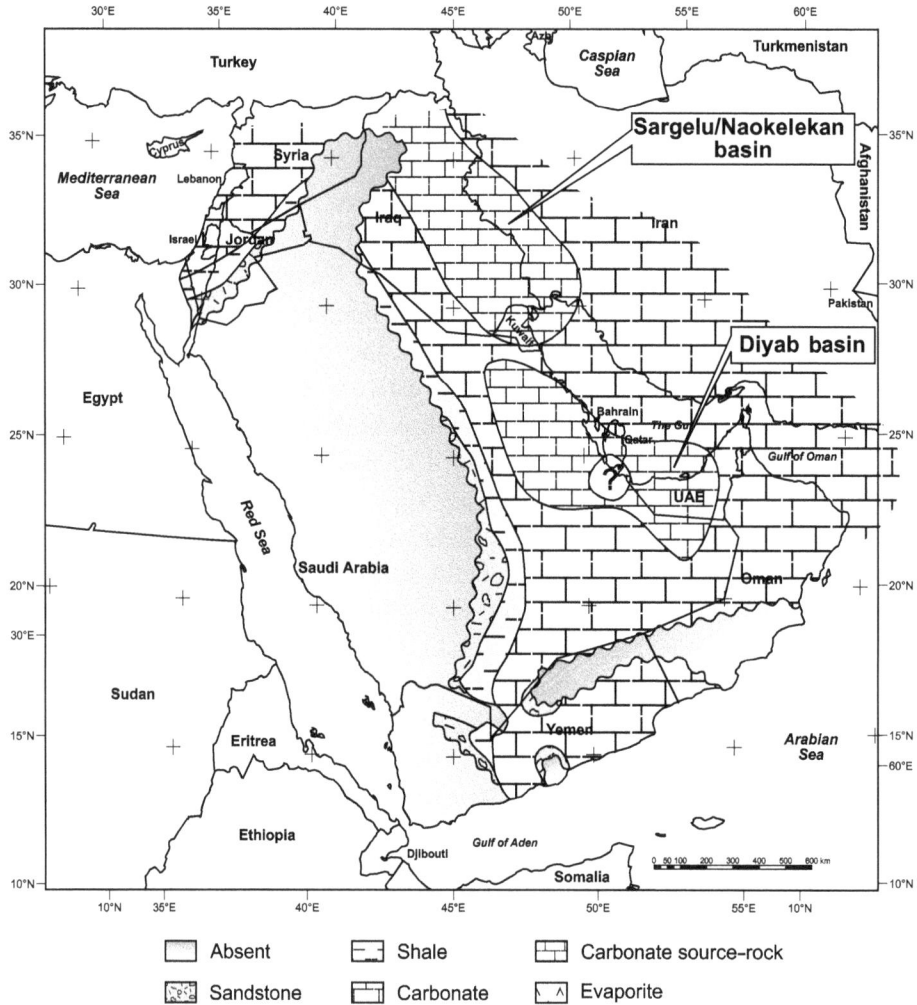

Fig. 9.6 **Middle–Late Jurassic palaeogeography of the Middle East**. Showing maximum extent of source facies (Sargelu/Naokelekan and Diyab basins). From Jones (2006).

essentially restricted to the margins of the Zagros foredeep basin in Khuzestan and Lurestan Provinces in Iran, and in contiguous Iraq (Goff *et al.*, in Al-Husseini, 1995). Note, though, that the clastic reservoirs, sourced from the 'Western Arabian Highlands', emerging in response to rifting in the Red Sea, extend further west.

### 9.3.2 *Indian subcontinent*

The geological history of the Indian subcontinent, like that of the Middle East and of other formerly contiguous parts of the supercontinent of Gondwana, with which

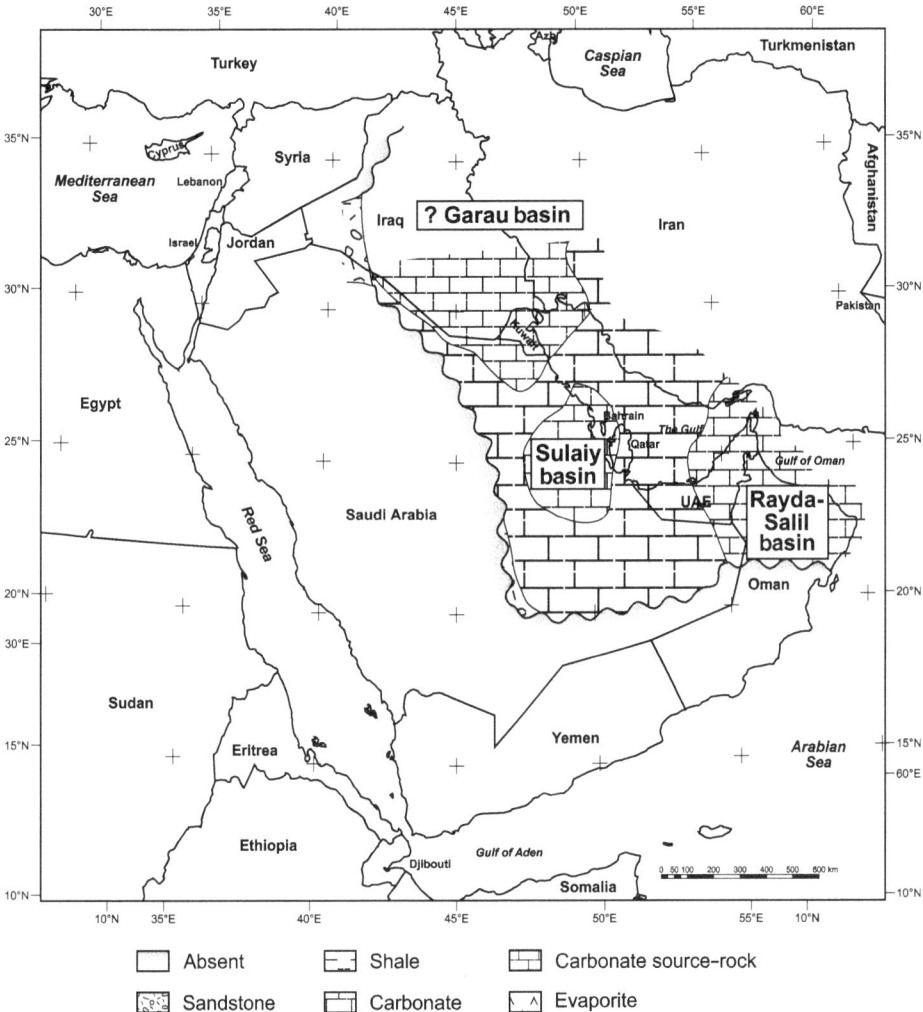

Fig. 9.7 **Early Cretaceous palaeogeography of the Middle East**. Showing maximum extent of source facies (Garau, Sulaiy and Rayda/Salil basins). From Jones (2006).

it shares certain common characteristics, is long and complex (Carmichael *et al.*, 2009; Jones, 2011a; see also comments on 'palaeobiogeography' below, and the further reading list at the end of the chapter).

It essentially began with extensional? tectonism and the formation of a series of basins, including salt basins, the best known of which is in the Salt Ranges in Pakistan, in the Precambrian–Infracambrian. Nothing remains other than in parts of the Himalaya of the later Cambrian to Carboniferous stratigraphy.

The preserved latest Carboniferous to Early Jurassic stratigraphy records a series of incipient rifting events related to the break-up of Gondwana.

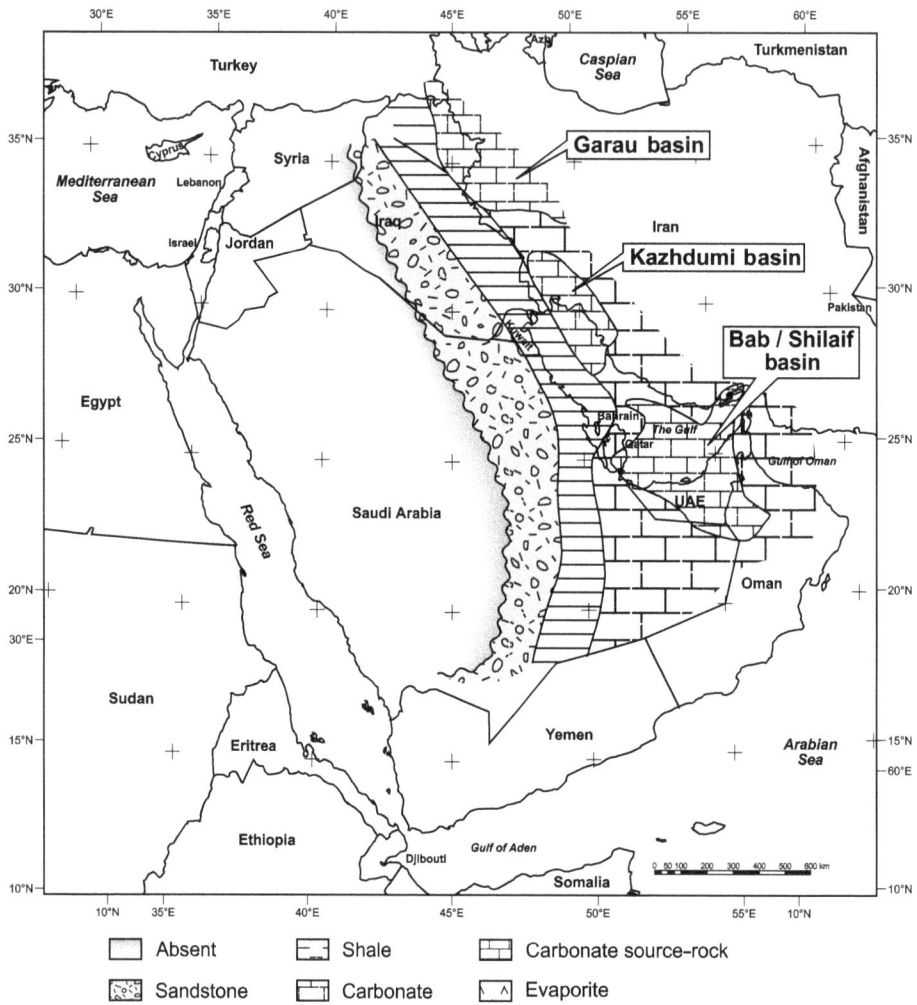

Fig. 9.8   **'Middle' Cretaceous palaeogeography of the Middle East**. Showing maximum extent of source facies (Garau, Kazhdumi and Bab/Shilaif basins). From Jones (2006).

Continental clastic sedimentation tended to dominate, although there is also evidence of intermittent marine incursion onto the continent. Glacial sediments, including tillites such as the widely developed Talchir Boulder Bed and its correlatives, characterise the latest Carboniferous to earliest Permian, Gzelian to Asselian; and deglacial sediments, including marine mudstones, the 'early' Early Permian, Sakmarian. The so-called Barakar 'Coal Measures' and their correlatives characterise the 'mid' Early Permian, Artinskian; the Barren Measures the 'late' Early Permian, Kungurian; and the Raniganj Coal Measures the Middle–Late Permian, Ufimian–Tatarian. Further continental clastics,

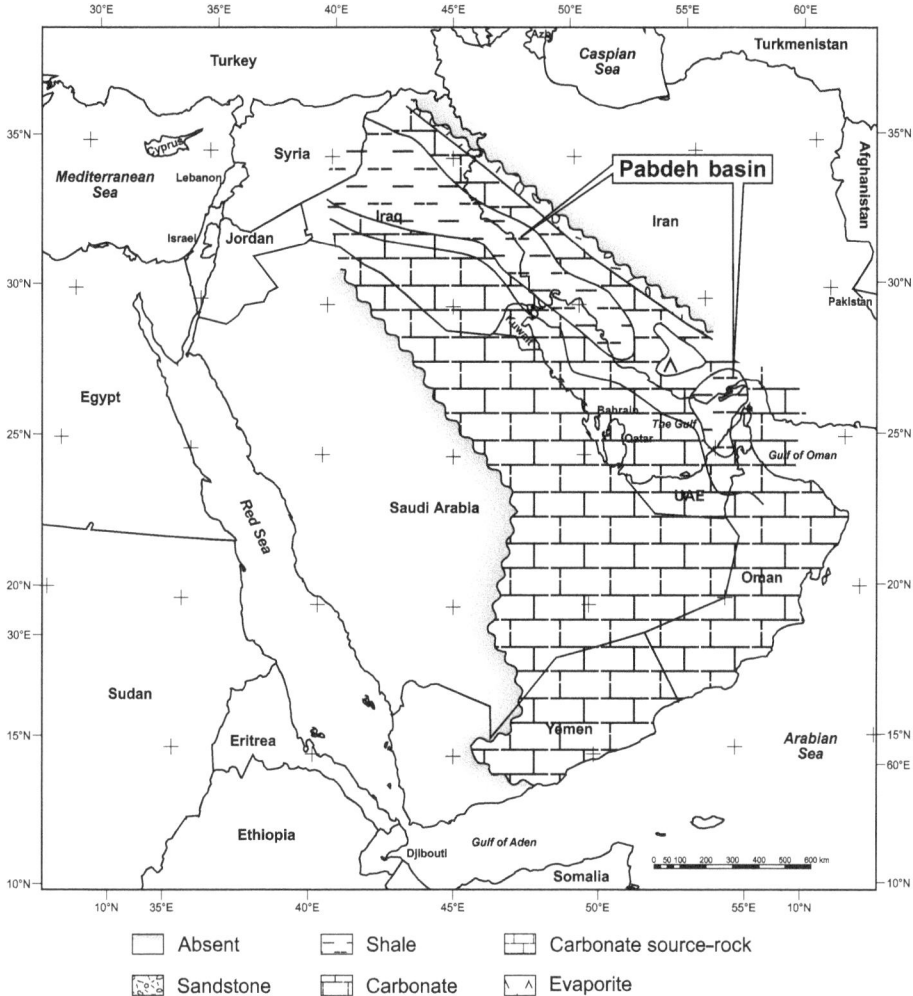

Fig. 9.9 **Palaeocene–Eocene palaeogeography of the Middle East**. Showing maximum extent of source facies (Pabdeh basin). From Jones (2006).

including the Panchet and supra-Panchet and their correlatives, characterise the Triassic–Early Jurassic.

Greater India, including Madagascar and the Seychelles, and Antarctica and Australasia, drifted away from east Africa in the Middle Jurassic, approximately 160 Ma. In the Middle–Late Jurassic, shallow marine carbonate sedimentation commenced along the west coast of the subcontinent, although continental clastic sedimentation continued to dominate in the interior and on what was to become the east coast, which was still attached to Antarctica and Australia, and there was also widespread volcanism associated with the development of the Karoo–Ferrar Large Igneous Province. Antarctica and Australasia drifted away in the earliest

Cretaceous, approximately 137 Ma, Madagascar and the Seychelles drifted away in the Late Cretaceous, approximately 90 Ma and 65 Ma, respectively. Continental to marine clastic and shallow to deep marine carbonate sedimentation were equally important in the Cretaceous, and again there was also widespread volcanism, this time associated with the Rajhamal hotspot event in the Early Cretaceous, the St Mary's Trap event in the Late Cretaceous and the Deccan Trap event and the development of the Deccan Large Igneous Province at the end of the Late Cretaceous.

India collided with Eurasia at the end of the Palaeocene, approximately 55 Ma. The initial elevation of the Himalaya and the associated initiation of southerly drainage was during the time the clastic Ghazij formation was deposited in Pakistan, in the Early Eocene, approximately 50 Ma. By the end of the Eocene, clastic sedimentation once more dominated over carbonate sedimentation across the entire subcontinent. By the end of the Miocene, approximately 9–7 Ma, the Himalaya was sufficiently elevated to affect atmospheric circulation, and to cause the climate to become markedly seasonal, or monsoonal. Himalayan 'molasse' and associated continental clastic sedimentation commenced across the entire subcontinent, exemplified by the Siwalik group and its correlatives.

*Outcrop geology*   The outcrop geology of the Indian subcontinent is well documented in the public-domain literature. Published field geological maps are available for the entire area on various scales.

Direct or indirect analogues for some of the source-rocks and reservoirs of the region can be studied at outcrop, for example on the margins of the Cambay basin in India and in the Sulaiman ranges on the margins of the Indus basin in Pakistan (see the further reading list at the end of the chapter).

*Petroleum geology*   The principal petroliferous basins of the Indian subcontinent, are, anti-clockwise from west to east, the Indus basin of Pakistan, the Cambay basin on the west coast of India, the Bombay or Mumbai basin also on west coast of India, the Cauvery basin on the east coast of India, the Krishna-Godavari basin also on the east coast of India, and the Assam basin of India and contiguous Surma basin of Bangladesh.

*Source-rocks.* A range of peri-deltaic, coaly, prodeltaic and marine thermogenic – and terrestrially influenced marine biogenic – source-rocks are developed in the Indian subcontinent (see comments on '(palaeo)biological controls on source-rock development' in Section 9.1.1; see also the further reading list at the end of the chapter).

Predominantly prodeltaic to marine source-rocks are developed in the Mesozoic, for example in the Mianwali and Datta formations in the Potwar basin, and the Sembar formation in the Indus basin, in Pakistan; in the Jhurio and Jumara formations in the Kutch basin, and the Baisakhi-Bedesir and Pariwar formations in the Jaisalmer basin, in western India; in the Mannar basin in Sri Lanka; and in the Andimadam and Sattapadi formations in the Cauvery basin, and the Mandapeta, Krishna, Gajulapadu and Raghavapuram formations in the Krishna-Godavari basin, in eastern India. Peri-deltaic, coaly, prodeltaic and marine source-rocks are developed in the Cenozoic, for example in the Lockhart, Patala and Sakesar formations in the Potwar basin, and in the Indus basin, in Pakistan; in the Olpad, Cambay Shale and Kalol formations in the Cambay basin, and the Panna formation in the Bombay basin, in western India; in the Palakollu, Vadaparru and Ravva formations in the Krishna-Godavari basin, in the Mahanadi-Bengal basin, and in the Kopili and Jenam formations in Assam, in eastern India; and in the Tura, Kopili, Jenam and Bhuban formations in Bangladesh.

Some of the more important of these, that have been identified on the basis of typing as contributing to important working petroleum systems, are the Cambay Shale formation in the Cambay basin; the Panna formation in the Bombay basin; the Andimadam formation in the Cauvery basin; the Kommugedem, Gajulapadu, Raghavapuram, Palakollu and Vadaparru formations in the Krishna-Godavari basin; and the Kopili and Jenam formations in Assam.

*Reservoir-rocks.* A range of continental to marginal marine, peri-deltaic to deep marine, submarine fan clastic; and shallow marine, platformal carbonate, reservoir-rocks are developed in the Indian subcontinent (see the further reading list at the end of the chapter).

Predominantly continental to marginal marine, peri-deltaic to deep marine, submarine fan clastic reservoir-rocks are developed in the Mesozoic, for example in the Datta, Shinawari, Samana Suk, Chichali and Lumshiwal formations in the Potwar basin, and the Lower Goru formation in the Indus basin, in Pakistan; in the Kutch basin in western India; and in the Bhuvanagiri and Nannilam formations in the Cauvery basin, and the Golapalli, Kanakollu and Tirupati formations in the Krishna-Godavari basin, in eastern India. Clastic and carbonate reservoir-rocks are developed in the Cenozoic, for example in the Potwar basin and in the Kirthar and Sulaiman ranges in Pakistan; in the Fatehgarh formation in the Barmer basin, the Anklesvar, Kadi and Kalol formations in the Cambay basin, and the Bombay formation in the Bombay basin, in western India; in the Niravi formation in the Cauvery basin, the Pasarlapudi and Ravva formations in the Krishna-Godavari basin, and the Tura, Sylhet and Kopili formations and Barail and Tipam groups in Assam, in eastern India; and the Jaintia, Barail and Surma groups in Bangladesh.

Current exploration is targetted in part at deep marine, submarine fan clastic plays in the deep offshore, for example in the Krishna-Godavari basin.

*Cap-rocks and traps.* A range of continental to marine predominantly fine-grained clastic cap-rocks are developed in the Indian subcontinent (see the further reading list at the end of the chapter). The Upper Goru formation seals the Lower Goru formation reservoir in the Indus basin; the Tarapur seals the Anklesvar, Kadi and Kalol reservoirs in the Cambay basin; the Bandra seals the Bombay reservoir in the Bombay basin; the Kudavasal seals the Bhuvanagiri reservoir, the Komarakshi, the Nannilam reservoir, and the Shiyali, the Niravi reservoir, in the Cauvery basin; the Raghavapuram seals the Kanakollu reservoir, the Razole, the Tirupati reservoir, the Vadaparru, the Pasarlapudi reservoir, and the Godavari, the Ravva reservoir, in the Krishna-Godavari basin in India; and the Upper Manna seals the Surma reservoir in Bangladesh.

Traps are predominantly structural, although a wide range of structural styles is known. Some recent discoveries in deep marine, submarine fan clastic reservoirs appear to have a stratigraphic component.

## *Biostratigraphy*

A wide range of biostratigraphically and/or palaeoenvironmentally useful fossil groups have been recorded in the surface and subsurface sections of the Indian subcontinent, including dinoflagellates, diatoms, calcareous nannoplankton, calcareous Algae, acritarchs, Foraminifera, plants, ammonoids, ostracods, branchiopods, fish, amphibians, reptiles and birds, and mammals. A similarly wide range of biostratigraphic zonation schemes is applicable, the scale ranging from regional to reservoir, the resolution in inverse proportion to the scale.

Various quantitative techniques have been applied in the interpretation of biostratigraphic data from the Cretaceous to the Cenozoic of the Indus basin in Pakistan (Wakefield & Monteil, 2002; see also Section 7.6).

*Palaeozoic*   Local to regional, Palaeo-Tethyan larger benthic foraminiferal – and Gondwanan palynological – biostratigraphic zonation schemes are applicable in the Palaeozoic of the Indian Subcontinent (Kalia & Sharma, in Samanta, 1985; Jones, 2011a; see also Toriyama *et al.*, 1975; Brönnimannn *et al.*, 1978; Toriyama, 1984; Martinez-Dias *et al.*, 1985; Dawson, 1993; Dawson & Racey, 1993; Dawson *et al.*, 1993).

*Mesozoic*   Global standard planktic foraminiferal and regional larger benthic foraminiferal schemes, or modifications thereof, are applicable in the marine Mesozoic of the Indian subcontinent (Nagappa, 1959; Kureshy, 1977a, b, 1978a; Watkinson *et al.*, 2007; Nagendra *et al.*, 2010). Smaller benthics are also locally biostratigraphically and/or palaeoenvironmentally useful (Bhalla, 1969; Govindan & Sastri, in Verdenius *et al.*, 1983; Bhalla & Talib, 1991; Rajshekar, 1995; Rajshekar & Atpalkar, 1995; Abbas, in Pandey *et al.*, 1996; Wakefield & Monteil, 2002; Talib *et al.*, 2007; Singh *et al.*, 2008; Talib & & Gaur, 2008; Gaur & Talib, 2009).

Many of the key species found in the shallow-water environments of the Jurassic–Cretaceous of the Indian subcontinent are also found in the Middle East (see Section 9.3.1).

*Cenozoic* Again, global standard planktic foraminiferal and regional larger benthic foraminiferal schemes, or modifications thereof, are applicable in the marine Cenozoic of the Indian subcontinent (Davies & Pinfold, 1937; Eames, 1952; Nagappa, 1959; Latif, 1961; Samanta, 1969; Raju *et al.*, 1970; Samanta, 1972, 1973; Raju, 1974; Kureshy, 1978a, b, c; Mohan, 1982; Kureshy, 1984; Singh, 1984; Samanta & Lahiri, in Samanta, 1985; Mishra, 1996; Jones, 1997; Dave, in Nanda & Mathur, 2000; Warraich *et al.*, 2000; Warraich & Nishi, 2003; Wakefield & Monteil, 2002; Afzal *et al.*, 2005; Raju, in Raju *et al.*, 2005; Raju & Uppal, in Raju *et al.*, 2005; Raju *et al.*, in Raju *et al.*, 2005; Rau, Ramesh *et al.*, in Raju *et al.*, 2005; Afzal *et al.*, 2009; Keller *et al.*, 2009; Fig. 9.10). And smaller benthics are also locally biostratigraphically and/or palaeoenvironmentally useful (Haque, 1956; Abbas, in Pandey *et al.*, 1996; Pandey & Dave, in Pandey *et al.*, 1996; Govindan, 2004).

Importantly, LBFs are applicable in reservoir characterisation, for example in the Oligocene–Miocene shallow-water carbonate reservoir in the Bombay High field in the offshore Bombay or Mumbai basin on the west coast of India (Raju & Uppal, in Raju *et al.*, 2005; Rau, Ramesh *et al.*, in Raju *et al.*, 2005).

Many of the key species found in the shallow-water environments and reservoirs of the Palaeogene of the Indian subcontinent are also found in the Middle East (see Section 9.3.1).

### *Palaeoenvironmental interpretation*

Benthic foraminifera are the most important groups of fossils in palaeobiological, palaeoecological and palaeoenvironmental interpretation in the Indian subcontinent (Kumar, 1984; Pandey & Choubey, 1984; Singh, 1984; Govindan & Bhandari, in Rogl & Gradstein, 1988; Raju & Dave, in Pandey *et al.*, 1996; Jones, 1997; Raju *et al.*, 1999; Warraich *et al.*, 2000; Wakefield & Monteil, 2002; Warraich & Nishi, 2003; Govindan, 2004; Afzal *et al.*, 2005; Raju, in Raju *et al.*, 2005; Raju & Naidu, in Raju *et al.*, 2005; Raju, Satyanarayana *et al.*, in Raju *et al.*, 2005; Ramesh & Peters, in Raju *et al.*, 2005; Ramesh & Raju, in Raju *et al.*, 2005). Interpretation is based on analogy with their living counterparts, on functional morphology, and on associated fossils and sedimentary environments (Jannink *et al.*, 1998; Bhalla *et al.*, 2007; Bhattacharjee *et al.*, in Sinha, 2007; Nigam *et al.*, 2007; Schumacher *et al.*, 2007; De & Gupta, 2010; Nagendra *et al.*, 2010; Schumacher *et al.*, 2010).

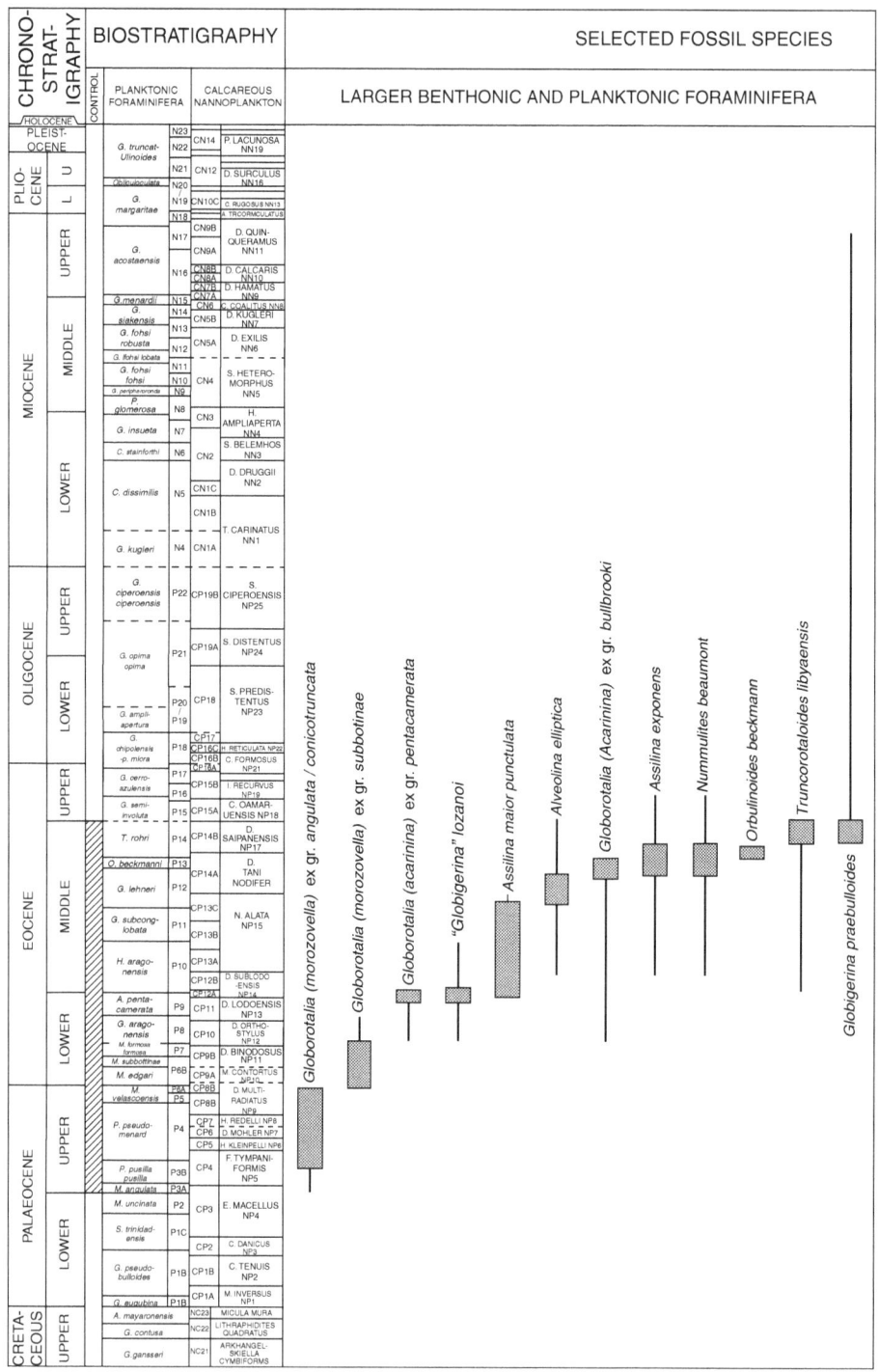

Fig. 9.10 **Stratigraphic distribution of Foraminifera and calcareous nanno-fossils in the Palaeogene of the Sulaiman Ranges, Pakistan**.

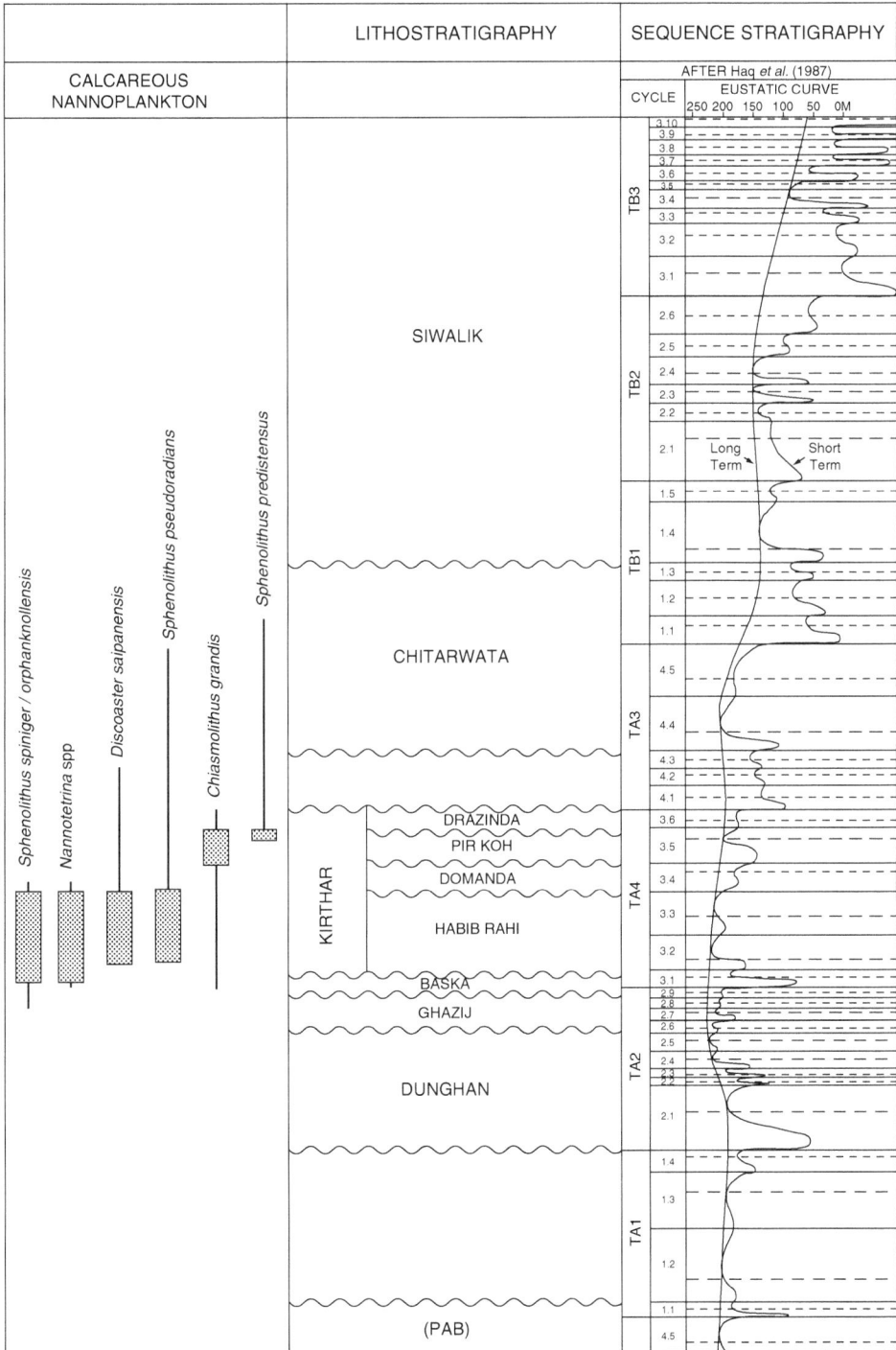

Fig. 9.10 (*cont.*)

Various quantitative techniques have been applied in the interpretation of palaeoenvironmental data from the Cretaceous to the Cenozoic of the Indus basin in Pakistan (Wakefield & Monteil, 2002; see also Section 7.7).

*Palaeobathymetry*    Benthic foraminifera are the most important group of fossils in palaeobathymetric interpretation in the Indian subcontinent. Those from the Palaeogene formations of the Sulaiman ranges of Pakistan represent a range of predominantly shallow marine carbonate environments and palaeobathymetries (Jones, 1997; Warraich *et al.*, 2000; Warraich & Nishi, 2003; Afzal *et al.*, 2009). Those from the Neogene Ravva formation of the Krishna-Godavari basin off the east coast of India represent a range of predominantly deep marine clastic environments and palaeobathymetries (Govindan, 2004). Here, palaeoenvironmental and palaeobathymetric interpretation has been undertaken on the basis of analogy between contained fossil Deep-Water Agglutinated Foraminifera or DWAFs and their living counterparts. Three biofacies groupings have been recognised: A, interpreted as representing bathyal to abyssal bathymetries; B, interpreted as representing upper bathyal bathymetries; and C, interpreted as representing outer neritic to upper bathyal bathymetries. The observed succession from biofacies C to B to A is interpreted as representing shallowing from abyssal to lower to middle bathyal, through upper bathyal, to upper bathyal to outer neritic. The ancient OMZ in upper bathyal bathymetries is characterised by infaunal agglutinated Foraminifera (see also Section 9.1.1).

*Palaeobiogeography*    In the Mesozoic, the continuing occurrence of essentially identical suites of land plants in the Jurassic–Early Cretaceous of India and Australia indicates that at this time these areas continued to constitute part of Gondwana, although the increasing difference between Indian and Australian land plants from the 'Middle' Cretaceous onwards indicates that by this time Gondwana had started to rift apart and its former constituent parts to drift apart, and the resulting proto-eastern Indian Ocean had started to form a barrier to the dispersal of terrestrial organisms (Jones, 2011a; and additional references cited therein). In the Cenozoic, the occurrence of similar suites of land vertebrates in the Eocene of Europe, west Asia and India, and different suites in the rest of south and east Asia, indicates that even at this time the rising mountain range of the proto-Himalaya in between had started to form a barrier to the dispersal of terrestrial organisms.

### Integrated studies

A number of integrated biostratigraphic and sequence stratigraphic schemes for the Indian subcontinent have been published (Lindsay *et al.*, 1991; Copestake *et al.*, in Caughey *et al.*, 1996; Jones, 1997; Das *et al.*, 1999; Daley & Alem, in Clift *et al.*,

2002; Smewing *et al.*, in Clift *et al.*, 2002; Aswal *et al.*, in Swamy & Kapoor, 2002; Satyanarayana *et al.*, in Swamy & Kapoor, 2002; Singh *et al.*, in Swamy & Kapoor, 2002; Wakefield & Monteil, 2002; Raju, in Raju *et al.*, 2005; Reddy *et al.*, 2005; Bastia *et al.*, 2007; Singh *et al.*, 2008; Carmichael *et al.*, 2009; Nagendra *et al.*, 2010; Kale, 2011; Raju, 2011; Ahmad *et al.*, 2012). The ages of sequence boundaries and MFSs are constrained by conventional biostratigraphy. MFSs are characterised micropalaeontologically by maximum flooding, as indicated by for-aminiferal palaeobathymetry, and also by maxima in abundance and diversity. Systems tracts are characterised micropalaeontologically by palaeobathymetric trends: TSTs by upward deepening; HSTs by upward shallowing (Raju, in Raju *et al.*, 2005; see also Section 8.2).

Palaeogeographic and facies maps of sequences have also been published (Smewing *et al.*, in Clift *et al.*, 2002; Carmichael *et al.*, 2009; Ahmad *et al.*, 2012; Mishra *et al.*, 2012). Source and reservoir distributions can be predicted from maps such as these.

### 9.3.3  Central and northern North Sea

The North Sea basin essentially formed in response to extensional episodes associated with rifting in the North Atlantic in the Permo-Triassic and Jurassic (Jones, 1996, 2006, 2011a; see also the further reading list at the end of the chapter). (Regional evidence from the onshore United Kingdom also points towards an earlier – Variscan or Hercynian – compressional episode, during which back-arc rift systems formed and deformed in response to the closure of the Rheic Ocean.)

The Permo-Triassic witnessed multiple phases of rifting and post-rift thermal subsidence, and was characterised by the deposition of continental clastics and evaporites. The Early–Middle Jurassic was a time of thermal doming prior to another extensional phase in the Late Jurassic, and was characterised by deposition of volcanics, volcaniclastics and marginal marine, peri-deltaic clastics, and by locally significant erosion at the 'mid Cimmerian unconformity'. The – diachronous – Late Jurassic syn-rift sequence was characterised by deposition of deep marine shales and submarine fan sandstones in the basin centre, and of shallow marine, coastal plain sandstones at the basin margins. Significant erosion took place at or near the Jurassic/Cretaceous boundary 'late Cimmerian unconformity'. The post-rift package was characterised initially, in the Early Cretaceous, by deposition of deep marine marls and shales and submarine fan sandstones, and later, in the Late Cretaceous to Early Palaeocene, by deposition of marls in the northern North Sea and chalks in the central North Sea.

The Late Palaeocene saw the onset of sea-floor spreading in the North Atlantic, and attendant uplift of hinterland areas in Scotland and northern England. The amount of uplift has been estimated to have been as much as 300–400 m at the time of the 'Chron-25N Hiatus' or 'Forties low-stand'. Rejuvenation of sediment supply resulted in the progradation of successive sequences of deltaic and associated submarine fan clastics into the basin in the Late Palaeocene and Eocene. The morphology of the fan units was controlled by the evolving structural configuration of the basin. Inversion at the end of the Eocene resulted in the generation of the 'Pyrenean unconformity'. The succeeding Oligocene to Holocene was characterised by renewed basin fill.

*Outcrop geology*    The outcrop geology of the margins of the North Sea basin is well documented in the public-domain literature (see, for example, the British Geological Survey (BGS)'s 'British Regional Geology' guides; see also their 'United Kingdom Offshore Regional Reports'). Published field geological maps are available for the entire area on various scales (see the BGS's Sheets, and accompanying 'Sheet Memoirs').

Direct or indirect analogues for some of the source-rocks and reservoirs of the region can be studied at outcrop. The principal source-rock, the Kimmeridge Clay, can be studied at outcrop in the Wessex basin, and one of the principal reservoirs, the Chalk, on the east coast as well as the south (see the further reading list at the end of the chapter). Analogues for the submarine fan reservoirs can be studied in County Clare in the Republic of Ireland, in the Austrian and Swiss Alps, Polish Carpathians and Spanish Pyrenees.

*Petroleum geology    Source-rocks.* The most important source-rock – and also incidentally one of the most important regional cap-rocks – in the central and northern North Sea is the latest Jurassic to earliest Cretaceous Kimmeridge Clay formation (see the further reading list at the end of the chapter). The Kimmeridge Clay is modelled as having been deposited in a basin characterised by at least intermittently anoxic bottom conditions brought about by salinity stratification, and inimical to most life forms. These conditions favoured the low consumption and hence high preservation of organic material observed in palynological preparations.

*Reservoir-rocks.* Reservoir rocks are developed at several stratigraphic horizons (see the further reading list at the end of the chapter). The most important are: Middle Jurassic to Early Cretaceous marginal marine and shallow marine, including peri-deltaic, sandstones (see Section 9.5.1); Late Jurassic to Early Cretaceous and Late Palaeocene to Eocene deep marine, submarine fan sandstones (see Sections 9.5.4. and 9.7.3); and Late Cretaceous to Early Palaeocene deep marine chalks (see Section 9.7.2). The quality of the chalk reservoirs is controlled in part by primary porosity and in part by secondary porosity and permeability enhancement associated with fracturing, both in turn controlled in part by the nature of the constituent micro- and nanno-fossils, or micro- and nanno-fabric.

*Cap-rocks and traps.* Cap-rocks are predominantly intra-formational. Traps are predominantly structural, although combination and stratigraphic traps are becoming increasingly important exploration targets (see the further reading list at the end of the chapter). Mobilisation of Permian salt has locally played an important role in trap development.

## Biostratigraphy

A number of micropalaeontological, nannopalaeontological and palynological biostratigraphic zonation schemes are applicable in the central and northern North Sea and its borderland, the scale ranging from global through regional to reservoir, and the resolution in inverse proportion to the scale (see below, and Sections 9.5.1, 9.5.4, 9.7.2 and 9.7.3). Unfortunately, on account of their commercial sensitivity, some such schemes remain either unpublished or published only in coded or otherwise unusable form.

A range of quantitative techniques and visualisation tools has been applied in the interpretation of biostratigraphic data from the North Sea (Gradstein *et al.*, in Rogl & Gradstein, 1988; Gradstein *et al.*, 1992; Gradstein *et al.*, 1994; Neal *et al.*, 1994; Neal *et al.*, in Mann & Lane, 1995; Neal, in Knox *et al.*, 1996; Agterberg & Gradstein, 1999; Watson & Nairn, 2005; Gradstein *et al.*, 2008; Watson *et al.*, in Ratcliffe & Zaitlin, 2010; see also Section 7.6).

Chemostratigraphic techniques have been employed in the correlation of the biostratigraphically barren Triassic.

*Mesozoic*   Generally globally applicable micropalaeontological and nannopalaeontological biostratigraphic zonation schemes are essentially inapplicable in the Triassic–Jurassic of the North Sea, owing to the prevalence at this time of marginal marine to non-marine facies unfavourable for the essentially open oceanic planktic Foraminifera and calcareous nannoplankton (see above). Instead, local micropalaeontological (benthic foraminiferal, ostracod and radiolarian) and palynological (palynomorph, including terrestrially derived spore and pollen and marine dinoflagellate cyst) schemes or datums are used, some of which are based on the offshore and some on the onshore (Copestake, in Jenkins & Murray, 1989; Morris & Coleman, in Jenkins & Murray, 1989; Shipp, in Jenkins & Murray, 1989; Morris & Dyer, in Hemleben *et al.*, 1990; Copestake, in Jenkins, 1993; Partington, Copestake *et al.*, in Parker, 1993; Carruthers *et al.*, in Hurst *et al.*, 1996; Ainsworth, in Underhill, 1998; Fritsen *et al.*, 1999; Morris *et al.*, in Jones & Simmons, 1999). (The radiolarian datums, in particular, are of considerable chronostratigraphic and correlative value in the 'Central Graben', where sediments are deeply buried, and where contained palynomorphs are often rendered unidentifiable by thermal alteration.)

Generally globally applicable micropalaeontological and nannopalaeontological biozonation schemes, or local modifications thereof, are applicable in the Cretaceous, especially in the Late Cretaceous (to Early Palaeocene), on account of the more widespread development at this time of marine facies favourable for the planktic Foraminifera and calcareous nannoplankton (see above). Local micropalaeontological (principally benthic foraminiferal) and palynological schemes or datums are also used, some of which are based on the offshore and some on the onshore (Hart & Bailey, in Wiedmann, 1979; Carter & Hart, 1977; Crittenden, 1986; Banner & Desai, 1988; Copestake, in Jenkins & Murray, 1989; Hart *et al.*, in Jenkins & Murray, 1989; King *et al.*, in Jenkins & Murray, 1989; Morris & Coleman, in Jenkins & Murray, 1989; Shipp, in Jenkins & Murray, 1989; Bang, 1990; Banner *et al.*, 1993; Copestake, in Jenkins, 1993; Bergen & Sikora, in Jones & Simmons, 1999; Hart *et al.*, in Demchuk & Gary, 2009; Wilkinson, 2011).

The Early Cretaceous, Ryazanian of the central and northern North Sea is marked by the *Haplophragmoides* spp. Zone (FCN1); the late Ryazanian–early Valanginian by the *Trocholina infragranulata* Zone (FCN2); the late Valanginian–early Hauterivian by the *Marssonella kummi* Zone (FCN3); the late Hauterivian by the *Falsogaudryinella moesiana* Zone (FCN4); the early Barremian by the *Falsogaudryinella* sp. X Zone (FCN5); the late Barremian by the *Gavelinella barremiana* Zone (FCN6); the early Aptian by the *Hedbergella infracretacea* Zone (FCN7); the late Aptian by the *Verneuilinoides chapmani* Zone (FCN8); the early Albian by the '*Globigerinelloides*' *gyroidinaeformis* Zone (FCN9); the middle Albian by the *Recurvoides* sp. and *Falsogaudryinella* sp. 1 zones (FCN10a and FCN10b); and the late Albian by the *Globigerinelloides bentonensis* and *Osangularia schloenbachi* zones (FCN11 and FCN12a) (King *et al.*, in Jenkins & Murray, 1989).

The Late Cretaceous, early Cenomanian of the central and northern North Sea is marked by the *Sigmoilina antiqua* Zone (FCN12b); the middle Cenomanian by the *Hedbergella brittonensis* Zone (FCN13); the late Cenomanian–early Turonian by the *Dicarinella* spp. Zone (FCN14); the late Turonian–early Coniacian by the *Marginotruncana marginata* Zone (FCN15); the late Coniacian–early Santonian by the *Stensioeina granulata polonica* Zone (FCN16); the late Santonian by the Spherical Radiomaria Zone (FCN17); the early Campanian by the Unnamed Zone (FCN18); the late Campanian by the *Tritaxia capitosa* Zone (FCN19); the early Maastrichtian by the *Reussella szajnochae* Zone (FCN20); and the late Maastrichtian by the *Pseudotextularia elegans* Zone (FCN21) (King *et al.*, in Jenkins & Murray, 1989).

The Cretaceous of the southern North Sea is marked, in chalky facies, by the parallel *Ammovertella cellensis* to *Pseudotextularia elegans* zones (FCS2 to FCS23).

Some stratigraphically significant species are illustrated on Pl. 48.

*Cenozoic* Local micropalaeontological (principally benthic foraminiferal) and palynological biostratigraphic zonation schemes or datums are used in the Cenozoic of the North Sea, some of which are based on the offshore and some on the onshore (Gradstein & Berggren, 1981; King, 1983; Gradstein *et al.*, in Rogl & Gradstein, 1988; Vinken, 1988; Gradstein & Kaminski, 1989; King, in Jenkins & Murray, 1989; Charnock & Jones, in Hemleben *et al.*, 1990; Gradstein *et al.*, 1992; Schroder, 1992; Copestake, in Jenkins, 1993; Bujak & Mudge, 1994; Gradstein *et al.*, 1994; Neal *et al.*, 1994; Neal *et al.*, in Mann & Lane, 1995; Gradstein & Backstrom, 1996; Mudge & Bujak, in Knox *et al.*, 1996; Mudge & Bujak, 1996;

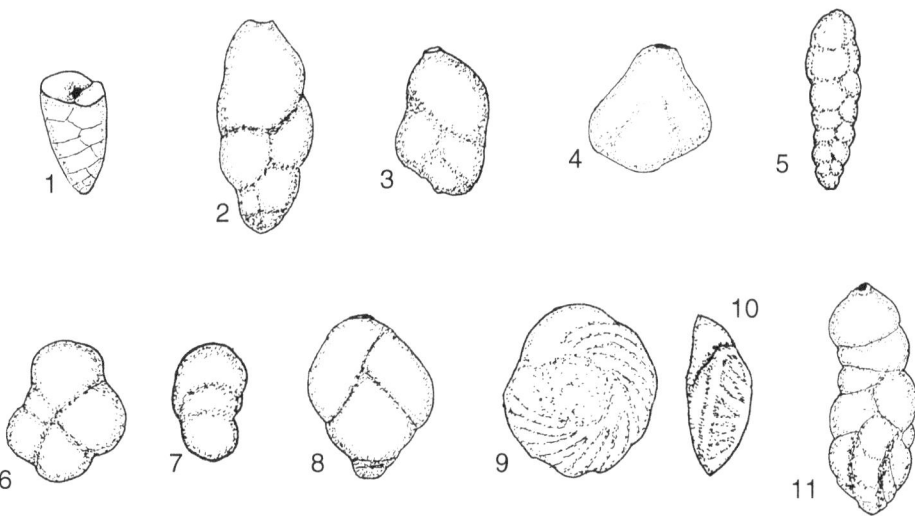

Pl. 48 **Some stratigraphically significant Foraminifera from the Cretaceous of the North Sea**. Modified after King *et al.*, in Jenkins & Murray (1989). 48.1 *Marssonella kummi*, nominate taxon of *Marssonella kummi* Zone (FCN3), ×55; 48.2–48.3 *Falsogaudryinella moesiana*, nominate taxon of *Falsogaudryinella moesiana* Zone (FCN4), ×120; 48.4 *Falsogaudryinella* sp. X, nominate taxon of *Falsogaudryinella* sp. X Zone (FCN5), ×100; 48.5 *Verneuilinoides chapmani*, nominate taxon of *Verneuilinoides chapmani* Zone (FCN8), ×35; 48.6–48.7 '*Globigerinelloides*' *gyroidinaeformis*, nominate taxon of '*Globigerinelloides*' *gyroidinaeformis* Zone (FCN9), ×60; 48.8 *Falsogaudryinella* sp. 1, nominate taxon of *Falsogaudryinella* sp. 1 Sub-Zone (FCN10b), ×80; 48.9–48.10 *Osangularia schloenbachi*, nominate taxon of *Osangularia schloenbachi* Sub-Zone (FCN12a), ×75; 48.11 *Tritaxia capitosa*, nominate taxon of *Tritaxia capitosa* Zone (FCN19), ×45.

Neal, in Knox *et al.*, 1996; Charnock & Jones, 1997; Gradstein & Kaminski, 1997; Agterberg & Gradstein, 1999; Gradstein *et al.*, 2008).

The Early Palaeocene of the North Sea is marked by the calcareous benthic foraminiferal *Tappanina selmensis* Zone (NSB1); the Late Palaeocene by the *Stensioeina beccariiformis* and *Bulimina trigonalis* Zones (NSB1b and NSB1c); the Late Palaeocene–Early Eocene, by the Unnamed Zone (NSB2); the Early Eocene by the *Gaudryina hiltermanni* and *Bulimina* sp. A zones (NSB3a and NSB3b); the Middle Eocene by the Unnamed, *Neoeponides karsteni* and *Lenticulina gutticostata* zones (NSB4, NSB5a and NSB5b); the Late Eocene by the *Planulina costata* and *Cibicidoides truncanus* zones (NSB5c and NSB6a); the Early Oligocene by the *Uvigerina germanica, Cassidulina carapitana* and *Rotaliatina buliminoides* zones (NSB6b, NSB7a and NSB7b); the Late Oligocene by the *Asterigerina guerichi guerichi, Elphidium subnodosum* and *Bolivina antiqua* zones (NSB8a, NSB8b and NSB8c); the Early Miocene by the *Plectofrondicularia seminuda* and *Uvigerina tenuipustulata* zones (NSB9 and NSB10); the Middle Miocene by the *Asterigerina guerichi staeschei, Uvigerina semiornata saprophila, Elphidium antoninum* and *Uvigerina* sp. A zones (NSB11, NSB12a, NSB12b and NSB12c); the Late Miocene by the *Uvigerina pigmea* and *Uvigerina venusta saxonica* zones (NSB13a and NSB13b); the Early Pliocene by the *Cibicidoides limbatosuturalis* and *Monspeliensina pseudotepida* zones (NSB14a and NSB14b); the Late Pliocene by the *Cibicidoides pachyderma* and *Cibicides grossus* zones (NSB15a and NSB15b); and the Pleistocene by the *Nonionellina* [*Nonion*] *labradorica* zone (NSB16) (King, in Jenkins & Murray, 1989).

The Cenozoic of the North Sea is also marked, in deep-water, turbiditic facies, by the parallel agglutinated foraminiferal *Spiroplectammina spectabilis* to Unnamed zones (NSA1 to NSA12); and in deep-water, pelagic facies by the parallel planktic foraminiferal *Globoconusa daubjergensis* to *Neogloboquadrina pachyderma* zones (NSP1 to NSP16) (King, in Jenkins & Murray, 1989).

King (1983, and, in Jenkins & Murray, 1989) has elegantly demonstrated that the apparent extinction of DWAFs in the Cenozoic of the North Sea is facies-controlled and measurably diachronous, younging from the margin of the basin to the centre. The same holds true for individual DWAF species, including *Cyclammina* (*Reticulophragmium*) *amplectens* and *Reticulophragmoides jarvisi*, whose extinctions are in the Eocene on the margin of the basin and in the Oligocene in the centre, and *Cyclammina* (*Cyclammina*) *placenta*, whose extinction is in the Oligocene on the margin of the basin and in the Miocene in the centre (Charnock & Jones, in Hemleben *et al.*, 1990; Charnock & Jones, 1997).

Some stratigraphically and palaeoenvironmentally significant species are illustrated on Pls. 49–52.

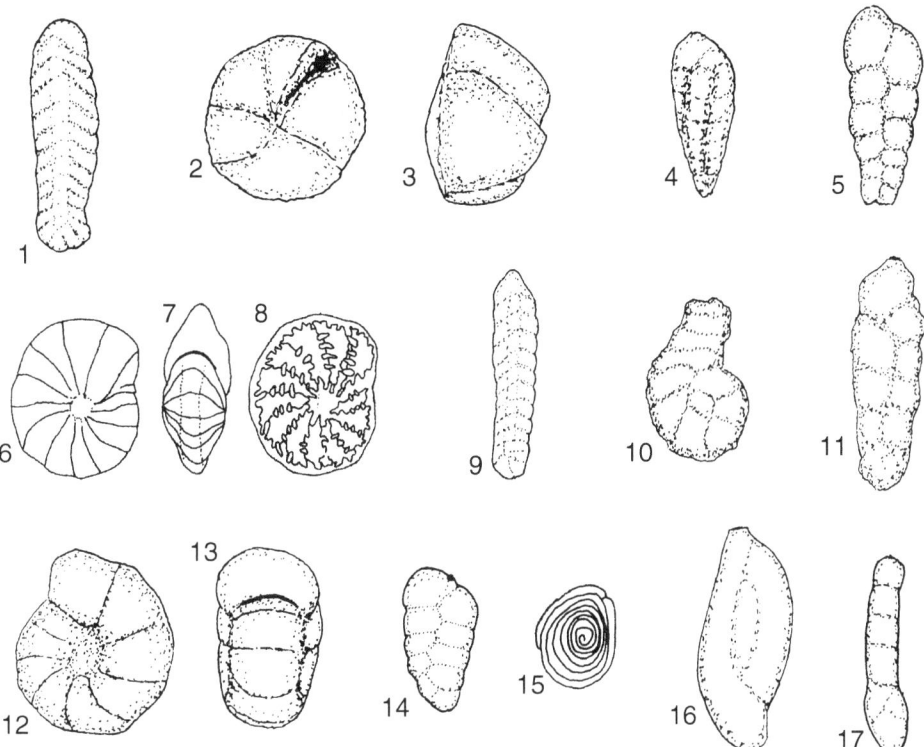

**Pl. 49 Some stratigraphically significant Foraminifera from the Cenozoic of the North Sea, 1**. Modified after King, in Jenkins & Murray (1989) and Charnock & Jones (1997). 49.1 = *Spiroplectammina spectabilis*, nominate taxon of *Spiroplectammina spectabilis* Sub-Zone (NSA1a), ×38; 49.2–49.3 *Trochammina ruthvenmurrayi*, nominate taxon of *Trochammina ruthvenmurrayi* Sub-Zone (NSA1b), ×49; 49.4 *Verneuilinoides subeocaenus*, nominate taxon of *Verneuilinoides subeocaenus* Zone (NSA2), ×45; 49.5 *Textularia plummerae*, nominate taxon of *Textularia plummerae* Sub-Zone (NSA4a), ×48; 49.6–49.8 *Reticulophragmium amplectens*, nominate taxon of *Reticulophragmium amplectens* Sub-Zone (NSA4b), ×56; 49.9 *Spiroplectammina* aff. *spectabilis*, nominate taxon of *Spiroplectammina* aff. *spectabilis* Zone (NSA5), ×32; 49.10 *Ammomarginulina macrospira*, nominate taxon of *Ammomarginulina macrospira* Sub-Zone (NSA6a), ×130; 49.11 *Karrerulina conversa*, nominate taxon *Karrerulina conversa* of Sub-Zone (NSA6b), ×64; 49.12–49.13 *Veleroninoides* [*Cribrostomoides*] *scitulus*, nominate taxon of *Cribrostomoides scitulus* Zone (NSA7), ×38; 49.14 *Karreriella chilostoma*, nominate taxon of *Karreriella chilostoma* Zone (NSA8), ×21; 49.15 *Glomospirella biedae* [*Ammodiscus* sp. B], nominate taxon of *Ammodiscus* sp. B Zone (NSA9), ×27; 49.16 *Spirosigmoilinella* sp. A, nominate taxon of *Spirosigmoilinella* sp. A Zone (NSA10), ×68; 49.17 *Martinottiella bradyana*, nominate taxon of *Martinottiella bradyana* Zone (NSA11), ×26.

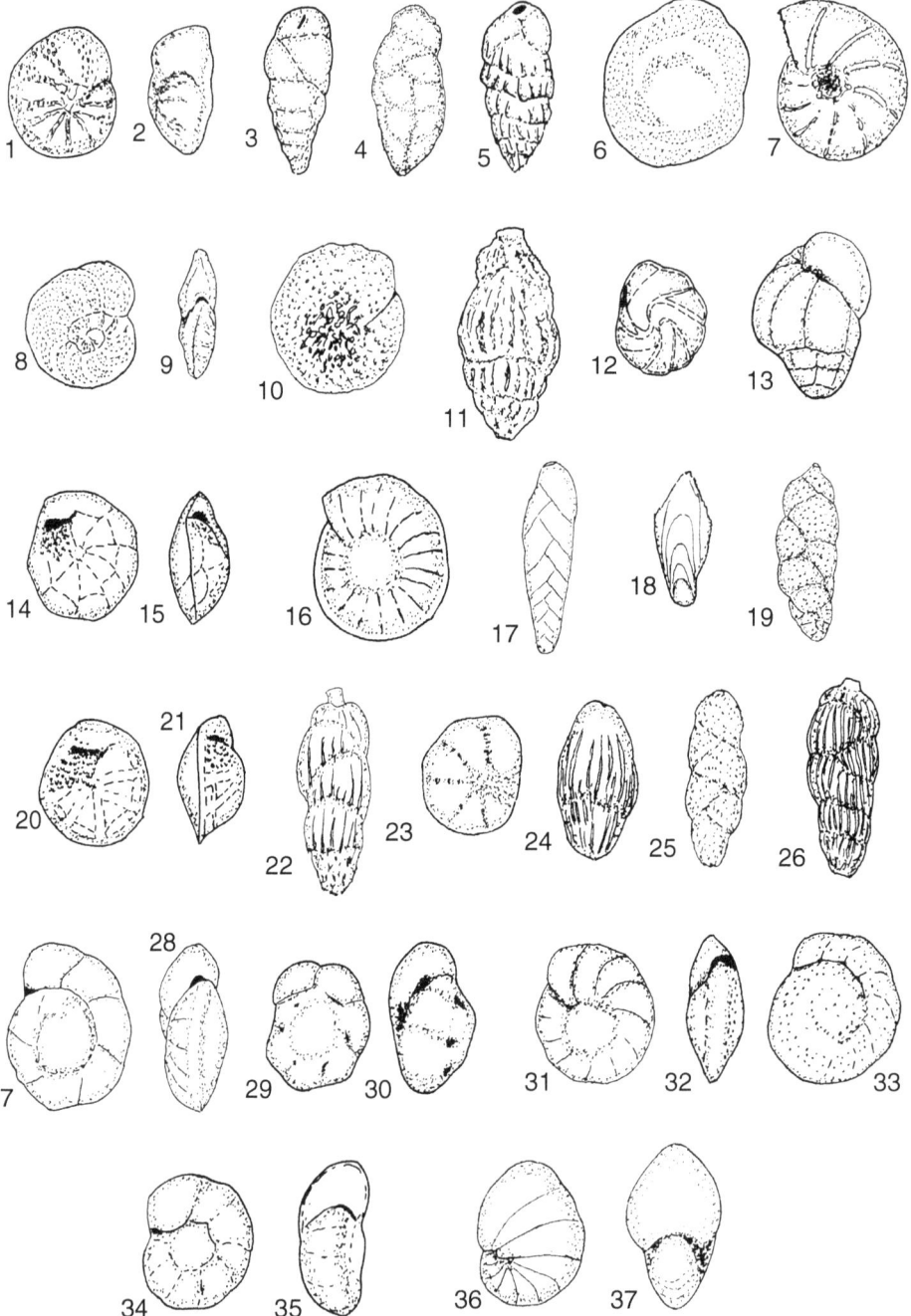

Pl. 50 **Some stratigraphically significant Foraminifera from the Cenozoic of the North Sea, 2**. Modified after King, in Jenkins & Murray (1989). 50.1-50.2 *Stensioeina beccariiformis*, nominate taxon of *Stensioeina beccariiformis*

*Palaeoenvironmental interpretation*

Benthic foraminiferal microfacies and palynofacies techniques are both of use in the interpretation of the palaeoenvironments of the North Sea.

Quantitative 'fuzzy C means' (FCM) clustering techniques have been applied in the interpretation of (palynological) palaeoenvironmental data from the Jurassic reservoir of Hawkins field (Gary *et al.*, in Demchuk & Gary, 2009; see also Section 5.9).

*Palaeobathymetry*   Microfacies is of considerable use in the interpretation of palaeobathymetry, with marginal, shallow and deep marine environments generally identifiable, and with inner-, middle- and outer-shelf shallow marine, and

Pl. 50 caption (*cont.*)   Sub-Zone (NSB1b), ×72; 50.3 *Bulimina trigonalis*, nominate taxon of *Bulimina trigonalis* Sub-Zone (NSB1c), ×60; 50.4 *Gaudryina hiltermanni*, nominate taxon of *Gaudryina hiltermanni* Sub-Zone (NSB3a), ×29; 50.5 *Bulimina* sp. A, nominate taxon of *Bulimina* sp. A Sub-Zone (NSB3b), ×56; 50.6 *Neoeponides karsteni*, nominate taxon of *Neoeponides karsteni* Sub-Zone (NSB5a), ×43; 50.7 *Lenticulina gutticostata*, nominate taxon of *Lenticulina gutticostata* Sub-Zone (NSB5b), ×34; 50.8–50.9 *Planulina costata*, nominate taxon of *Planulina costata* Sub-Zone (NSB5c), ×32; 50.10 *Cibicidoides truncanus*, nominate taxon of *Cibicidoides truncanus* Sub-Zone (NSB6a), ×38; 50.11 *Uvigerina germanica*, nominate taxon of *Uvigerina germanica* Sub-Zone (NSB6b), ×110; 50.12 *Cassidulina carapitana*, nominate taxon of *Cassidulina carapitana* Sub-Zone (NSB7a), ×45; 50.13 *Rotaliatina bulimoides*, nominate taxon of *Rotaliatina bulimoides* Sub-Zone (NSB7b), ×90; 50.14–50.15 *Asterigerina guerichi guerichi*, nominate taxon of *Asterigerina guerichi guerichi* Sub-Zone (NSB8a), ×56; 50.16 *Elphidium subnodosum*, nominate taxon of *Elphidium subnodosum* Sub-Zone (NSB8b), ×22; 50.17 *Bolivina antiqua*, nominate taxon of *Bolivina antiqua* Sub-Zone (NSB8c), ×78; 50.18 *Plectofrondicularia seminuda*, nominate taxon of *Plectofrondicularia seminuda* Zone (NSB9), ×30; 50.19 *Uvigerina tenuipustulosa*, nominate taxon of *Uvigerina tenuipustulosa* Zone (NSB10), ×45; 50.20-50.21 *Asterigerina guerichi staeschei*, nominate taxon of *Asterigerina guerichi staeschei* Zone (NSB11), ×56; 50.22 *Uvigerina semiornata saprophila*, nominate taxon of *Uvigerina semiornata saprophila* Sub-Zone (NSB12a), ×62; 50.23 *Elphidium antoninum*, nominate taxon of *Elphidium antoninum* Sub-Zone (NSB12b), ×75; 50.24 *Uvigerina* sp. A, nominate taxon of *Uvigerina* sp. A Sub-Zone (NSB12c), ×33; 50.25 *Uvigerina pygmea langeri*, nominate taxon of *Uvigerina pygmea langeri* Sub-Zone (NSB13a), ×55; 50.26 *Uvigerina venusta saxonica*, nominate taxon of *Uvigerina venusta saxonica* Sub-Zone (NSB13b), ×60; 50.27-50.28 *Cibicidoides limbatosuturalis*, nominate taxon of *Cibicidoides limbatosuturalis* Sub-Zone (NSB14a), ×56; 50.29–50.30 *Monspeliensina pseudotepida*, nominate taxon of *Monspeliensina pseudotepida* Sub-Zone (NSB14b), ×44, ×50; 50.31–50.33 *Cibicidoides pachyderma*, nominate taxon of *Cibicidoides pachyderma* Sub-Zone (NSB15a), ×42; 50.34–50.35 *Cibicides grossus*, nominate taxon of *Cibicides grossus* Sub-Zone (NSB15b), ×30; 50.36–50.37 *Nonionellina labradorica* [*Nonion labradoricum*], nominate taxon of *Nonion labradoricum* Zone (NSB16).

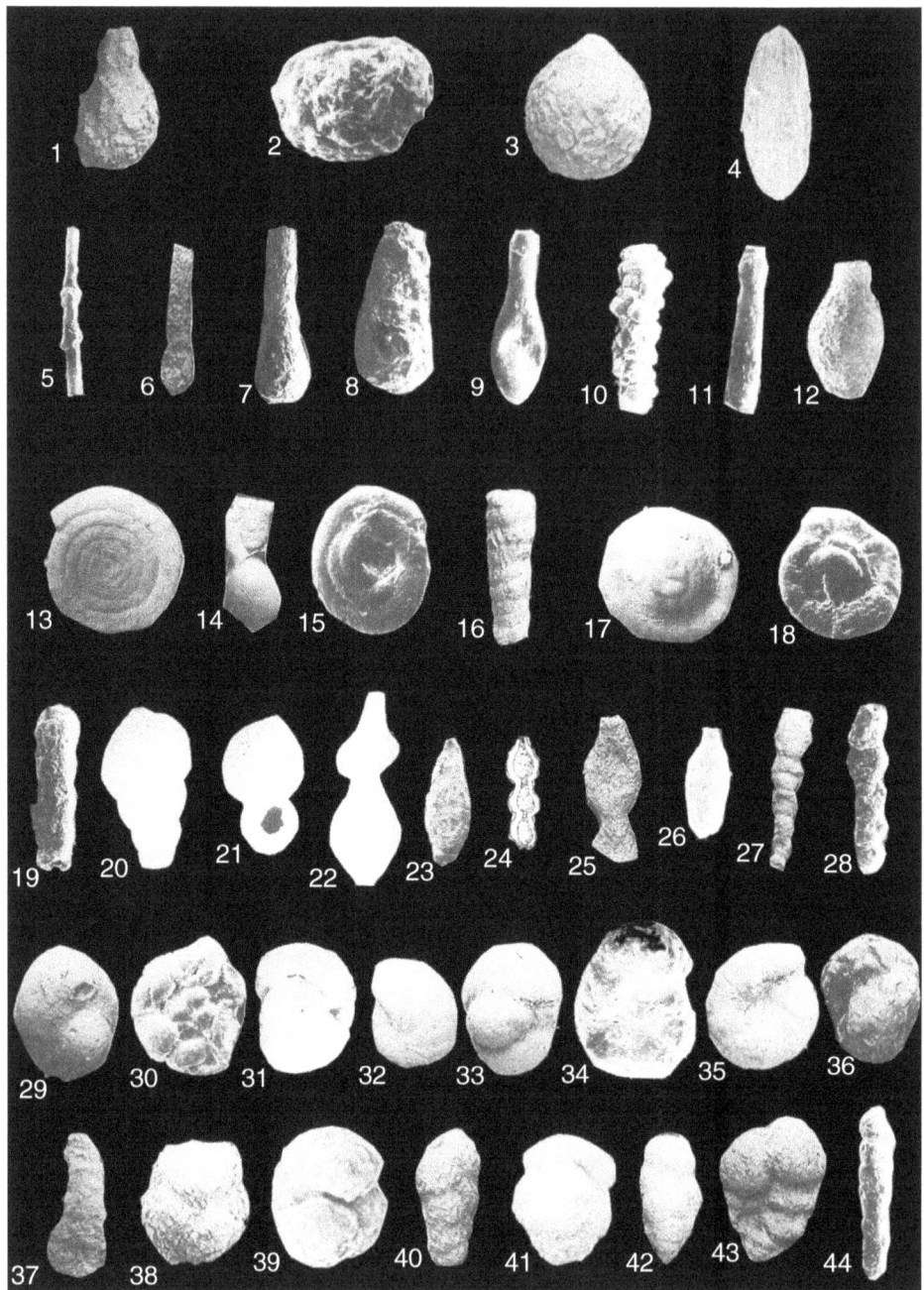

Pl. 51 **Some palaeoenvironmentally significant Foraminifera from the Cenozoic of the North Sea**. Modified after Charnock & Jones, in Hemleben *et al.*, (1990). 51.1 *Lagenammina ampullacea*, ×100; 51.2 *Psammosphaera fusca*, ×30; 51.3 *Saccammina sphaerica*, ×60; 51.4 *Technitella legumen*, ×30; 51.5 *Bathysiphon capillaris*, ×30; 51.6 *Hyperammina cylindrica*, ×65; 51.7

upper-, middle- and lower-slope deep marine environments often also identifiable, in part on the basis of modern analogy (King, 1983; King, in Jenkins & Murray, 1989; Jones, 1994; Charnock & Jones, in Hemleben *et al.*, 1990; Gradstein & Backstrom, 1996; Fig. 9.11; see also Section 4.2).

Benthic foraminifera from the Palaeogene are interpreted on the basis of modern analogy as indicative of deep marine, submarine fan rather than marginal to shallow marine, peri-deltaic environments (Charnock & Jones, in Hemleben *et al.*, 1990).

*Sedimentary environments*    Various marginal to shallow marine, including peri-deltaic, and deep marine, including submarine fan, sedimentary sub-environments are identifiable on the basis of microfacies or palynofacies (see Sections 4.2.1. and 4.2.3).

*Marginal to shallow marine, including peri-deltaic, sub-environments.* Microfacies and palynofacies techniques are of equal use in the interpretation of the marginal to shallow marine, including peri-deltaic, sub-environments of the North Sea. A range of peri-deltaic sub-environments, including delta-top, delta-front and prodelta, have been identified on the basis of cross-plots of epifaunal, surficial and infaunal 'morphogroups' of agglutinating Foraminifera (Nagy *et al.*, 1984; Nagy, 1985a, b; Nagy & Johansen, 1991; Nagy, 1992; Nagy *et al.*, 2001; Nagy & Seidenkrantz, 2003; Nagy *et al.*, 2010; Murray *et al.*, 2011; see also Sections 3.2.2, and 9.5.1). Delta-top and, to a lesser extent, delta-front sub-environments are characterised by epifaunal, including epiphytic, morphotypes; and prodelta sub-environments by surficial and infaunal morphotypes.

Pl. 51 caption (*cont.*)    *Hyperammina elongatea*, ×60; 51.8 *Hyperammina friabilis*, ×65; 51.9 *Hyperammina laevigata*, ×85; 51.10 *Rhabdammina abyssorum*, ×15; 51.11 *Rhabdammina discreta*, ×25; 51.12 *Rhabdammina linearis*, ×25; 51.13 *Ammodiscus tenuis*, ×60; 51.14 *Ammolagena clavata*, ×75; 51.15 *Glomospira gordialis*, ×130; 51.16 *Turritellella shoneana*, ×210; 51.17 *Usbekistania charoides*, ×110; 51.18 *Lituotuba lituiformis*, ×60; 51.19 *Hormosina bacillaris*, ×60; 51.20 *Hormosina globulifera*, ×30; 51.21 *Hormosina pilulifera*, ×90; 51.22 *Hormosinella carpenteri*, ×8; 51.23 *Hormosinella distans*, ×40; 51.24 *Hormosinella guttifera*, ×50; 51.25 *Hormosinella ovicula*, ×45; 51.26 *Reophax fusiformis*, ×65; 51.27 *Reophax nodulosus*, ×30; 51.28 *Subreophax aduncus*, ×45; 51.29 *Buzasina galeata*, ×100; 51.30 *Conglophragmium coronatum*, ×30; 51.31 *Cribrostomoides subglobosus*, ×40; 51.32 *Cyclammina rotundidorsata*, ×30; 51.33 *Cystammina pauciloculata*, ×100; 51.34 *Evolutinella rotulata*, ×80; 51.35 *Veleroninoides scitulus* [*Labrospira scitula*], ×30; 51.36 *Recurvoides turbinatus*, ×30; 51.37 *Eratidus foliaceus*, ×110; 51.38 *Adercotryma glomerata*; ×105; 51.39 *Trochamminopsis challenger*, ×65; 51.40 *Dorothia scabra*, ×55; 51.41 *Verneuilinulla propinqua*, ×30; 51.42 *Martinottiella communis*, ×30; 51.43 *Karreriella bradyi*, ×65; 51.44 *Karrerulina conversa*, ×70.

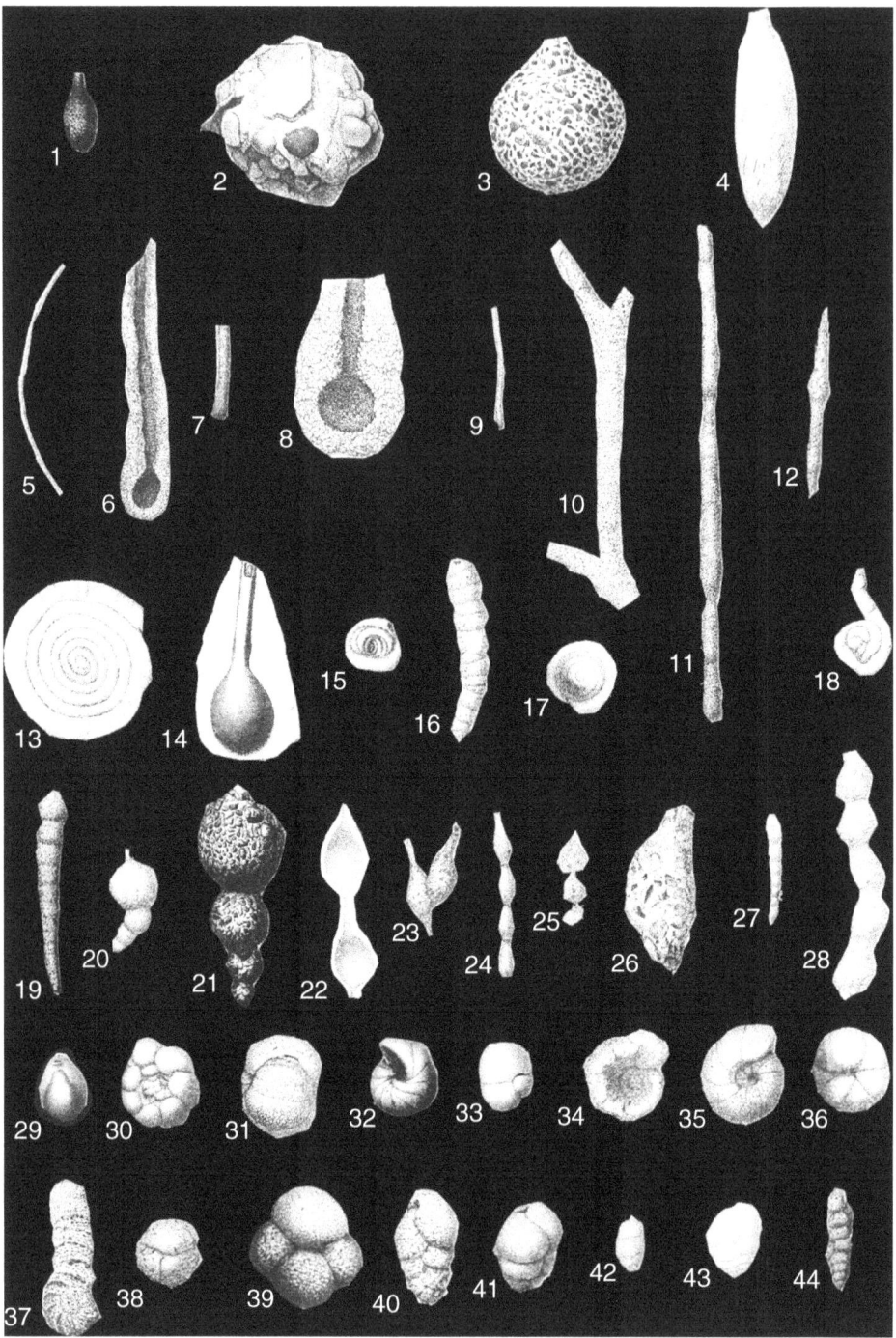

Pl. 52  **Living counterparts of palaeoenvironmentally significant Foramini-**
**fera from the Cenozoic of the North Sea.** Modified after Jones (1994).

*Deep marine, including submarine fan, sub-environments.* Microfacies techniques are of most use in the interpretation of the deep marine, including submarine fan, sub-environments of the North Sea. A range of deep marine, submarine fan sub-environments, including channel, levee and overbank, have been identified on the basis of cross-plots of epifaunal suspension-feeding, epifaunal detritus-feeding and infaunal detritus-feeding morphogroups of agglutinating Foraminifera (Jones & Charnock, 1985; Jones, in Jones & Simmons, 1999; Jones *et al.*, in Koutsoukos, 2005; Jones *et al.*, in Powell & Riding, 2005; Murray *et al.*, 2011; see also Sections 3.2.2, and 9.5.4. and 9.7.3). Channel sub-environments are characterised by epifaunal suspension-feeders; levee sub-environments by epifaunal detritus-feeders; and overbank sub-environments by infaunal detritus-feeders.

## Integrated studies

A number of integrated biostratigraphic and sequence stratigraphic schemes for the North Sea have been published (Stewart, in Brooks & Glennie, 1987; Mitchener *et al.*, in Morton, 1992; Armentrout *et al.*, in Parker, 1993; Fraser, in Parker, 1993; Partington, Copestake *et al.*, in Parker, 1993; Partington, Mitchener *et al.*, in Parker, 1993; Jones & Milton, 1994; Neal *et al.*, 1994; Jordt *et al.*, 1995; Carruthers *et al.*, in Hurst *et al.*, 1996; Neal, in Knox *et al.*, 1996; Liu & Galloway, 1997; Veeken, 1997; Michelsen *et al.*, in de Graciansky *et al.*, 1998; Neal *et al.*, in de Graciansky *et al.*, 1998; Charnock *et al.*, in Martinsen & Dreyer, 2001; Evans *et al.*, 2003). The ages of sequence boundaries are constrained by conventional biostratigraphy, supplemented by GC. The ages of MFSs are also constrained by

Pl. 52 caption (*cont.*) 52.1 *Lagenammina ampullacea*; 52.2 *Psammosphaera fusca*; 52.3 *Saccammina sphaerica*; 52.4 *Technitella legumen*; 52.5 *Bathysiphon capillaris*; 52.6 *Hyperammina cylindrica*; 52.7 *Hyperammina elongata*; 52.8 *Hyperammina friabilis*; 52.9 *Hyperammina laevigata*; 52.10 *Rhabdammina abyssorum*; 52.11 *Rhabdammina discreta*; 52.12 *Rhabdammina linearis*; 52.13 *Ammodiscus tenuis*; 52.14 *Ammolagena clavata*; 52.15 *Glomospira gordialis*; 52.16 *Turritellella shoneana*; 52.17 *Usbekistania charoides*; 52.18 *Lituotuba lituiformis*; 52.19 *Hormosina bacillaris*; 52.20 *Hormosina globulifera*; 52.21 *Hormosina pilulifera*; 52.22 *Hormosinella carpenteri*; 52.23 *Hormosinella distans*; 52.24 *Hormosinella guttifera*; 52.25 *Hormosinella ovicula*; 52.26 *Reophax fusiformis*; 52.27 *Reophax nodulosus*; 52.28 *Subreophax aduncus*; 52.29 *Buzasina galeata*; 52.30 *Conglophragmium coronatum*; 52.31 *Cribrostomoides subglobosus*; 52.32 *Cyclammina rotundidorsata*; 52.33 *Cystammina pauciloculata*; 52.34 *Evolutinella rotulata*; 52.35 *Veleroninoides [Labrospira] scitulus*; 52.36 *Recurvoides turbinatus*; 52.37 *Eratidus foliaceus*; 52.38 *Adercotryma glomerata*; 52.39 *Trochamminopsis challengeri*; 52.40 *Dorothia scabra*; 52.41 *Verneuilinulla propinqua*; 52.42 *Martinottiella communis*; 52.43 *Karreriella bradyi*; 52.44 *Karrerulina conversa*.

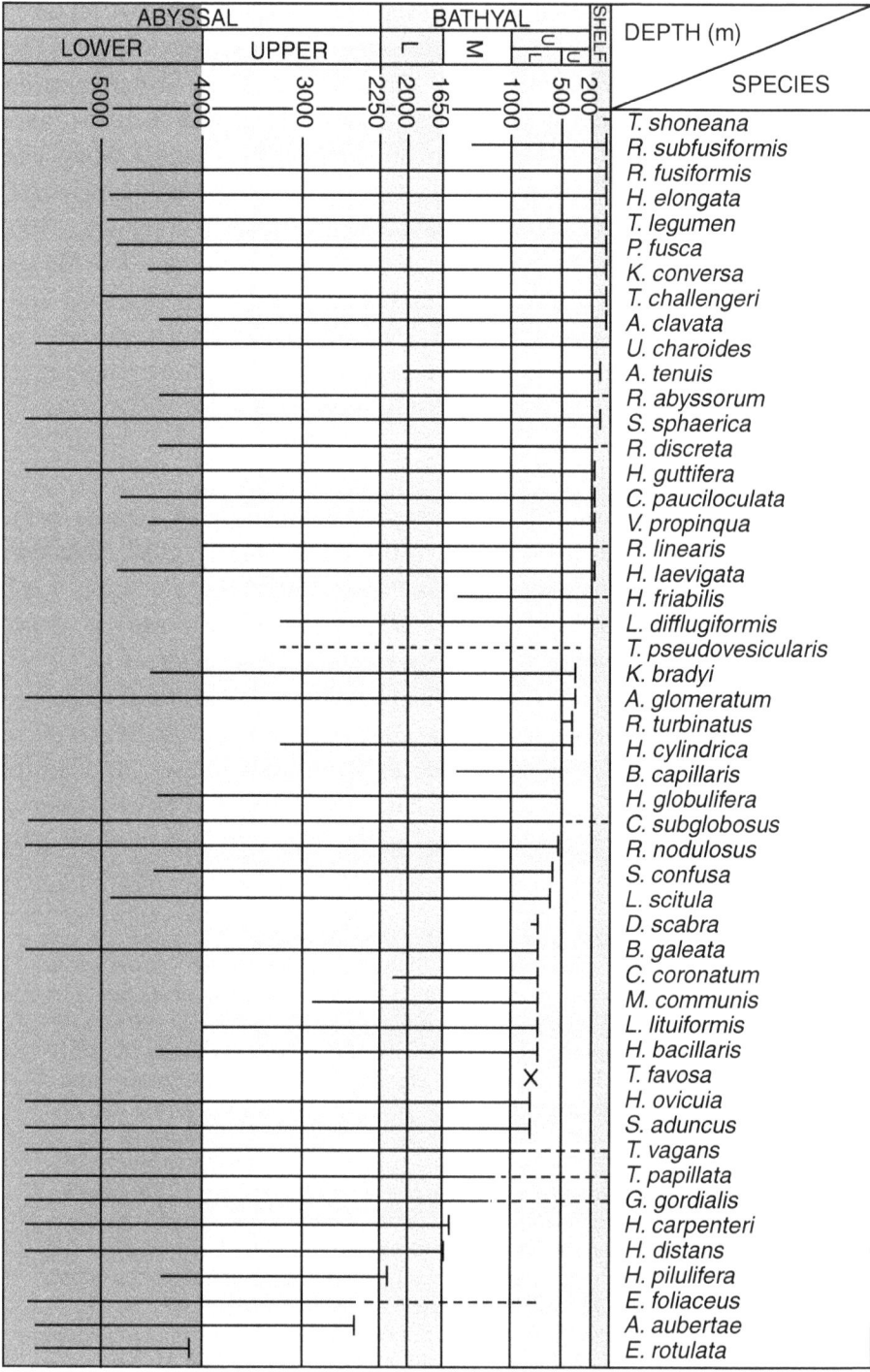

Fig. 9.11 **Bathymetric distribution of selected Foraminifera from the Cenozoic of the North Sea**. *T. shoneana = Turritellella*; *R. subfusiformis*,

conventional biostratigraphy. MFSs are characterised micropalaeontologically by maximum flooding, as indicated by foraminiferal palaeobathymetry, and also by maxima in abundance and diversity. Systems tracts are characterised micropalaeontologically by palaeobathymetric trends: TSTs by upward deepening; HSTs by upward shallowing.

Palaeogeographic and facies maps for selected time-slices have also been published, the time-slices identified in part on the basis of biostratigraphy, and the palaeogeography and facies on the basis of palaeobiology (Fraser *et al.*, in Evans *et al.*, 2003; Jones *et al.*, in Evans *et al.*, 2003; Jones, 2006, 2011a; Fig. 9.12). Source and/or reservoir distributions can be described on and predicted from maps such as these (Kubala *et al.*, in Evans *et al.*, 2003).

### *9.3.4 South Atlantic salt basins*

The South Atlantic salt basins, that is, those situated in eastern South America and west Africa between the Equatorial Fracture Zone to the north, and the Rio Grande Rise and Walvis Ridge to the south, owe their existence to South Atlantic rifting in the Early Cretaceous, Neocomian–'early' Aptian, and drifting from the 'middle' Aptian–Recent, some of the palaeobiogeographic evidence for which is outlined under 'palaeobiogeography' below (Jones, 1996; Belopolsky *et al.*, 2006, 2007; Jones, 2011a; see also comments on 'palaeobiogeography' below, and the further reading list at the end of the chapter).

Fig. 9.11 caption (*cont.*) *R. fusiformis = Reophax; H. elongata = Hyperammina; T. legumen = Technitella; P. fusca = Psammosphaera; K. conversa = Karrerulina; T. challengeri = Trochamminopsis; A. clavata = Ammolagena; U. charoides = Usbekistania; A. tenuis = Ammodiscus; R. abyssorum = Rhabdammina; S. sphaerica = Saccammina; R. discreta = Rhabdammina; H. guttifera = Hormosinella; C. pauciloculata = Cystammina; V. propinqua = Verneuilinulla; R. linearis = Rhabdammina; H. laevigata, H. friabilis = Hyperammina; L. difflugiformis = Lagenammina; T. pseudovesicularis = Trochamminopsis; K. bradyi = Karreriella; A. glomeratum = Adercotryma; R. turbinatus = Recurvoides; H. cylindrica = Hyperammina; B. capillaris = Bathysiphon; H. globulifera = Hormosina; C. subglobosus = Cribrostomoides; R. nodulosus = Reophax; S. confusa = Sorosphaera; L. scitula = Labrospira [Veleroninoides]; D. scabra = Dorothia; B. galeata = Buzasina; C. coronatum = Conglophragmium; M. communis = Martinottiella; L. lituiformis = Lituotuba; H. bacillaris = Hormosina; T. favosa = Thurammina; H. ovicula = Hormosinella; S. aduncus = Subreophax; T. vagans = Tolypammina; T. papillata = Thurammina; G. gordialis = Glomospira; H. carpenteri, H. distans = Hormosinella; H. pilulifera = Hormosina; E. foliaceus = Eratidus; A. aubertae = Ammomarginulina; E. rotulata = Evolutinella.* Reproduced with permission from Charnock & Jones, in Hemleben *et al.* (1990).

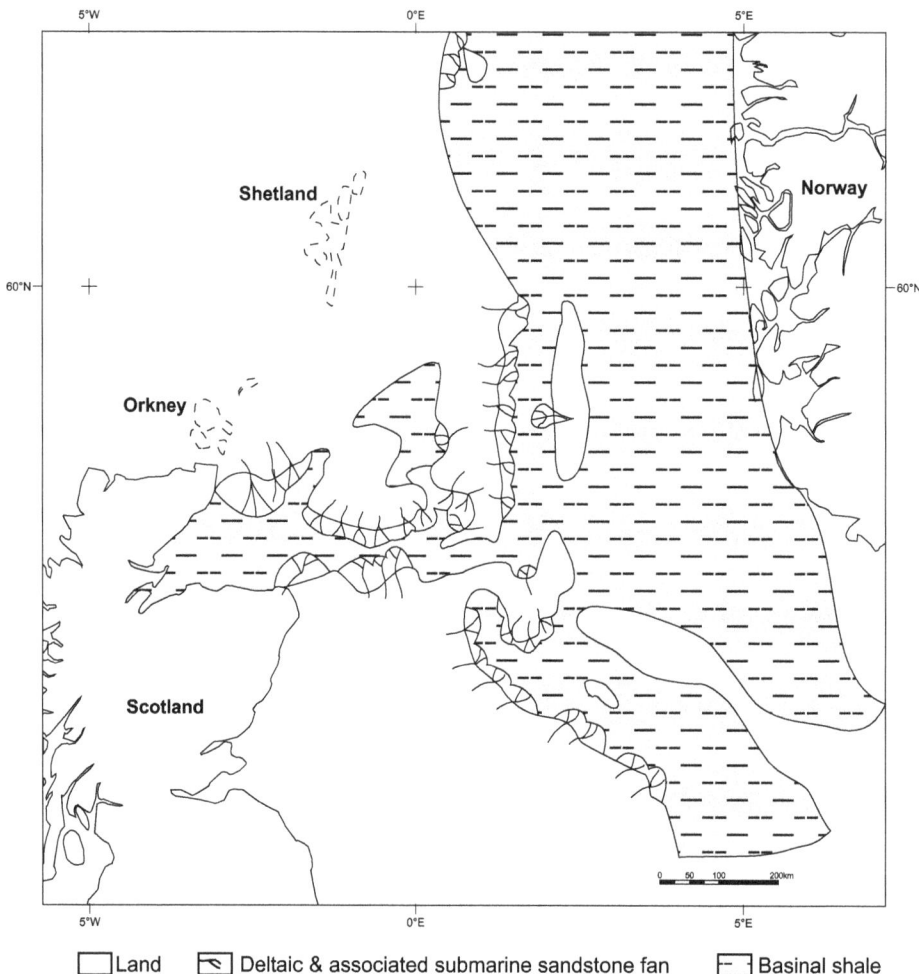

Fig. 9.12  **Late Jurassic palaeogeography of the North Sea**. Showing maximum extent of Kimmeridge Clay source facies (basin). From Jones (2006).

The nature of the pre-rift sequence is poorly understood. The Neocomian–'early' Aptian syn-rift sequence is restricted to grabens oriented sub-parallel to the modern coastlines of the formerly conjugate eastern South American and west African margins. It is characterised by predominantly non-marine facies, although also by marked lateral and vertical changes in facies. The boundary between the syn- and post-rift sequences is marked by a major – 'break-up' – unconformity. The 'middle' Aptian–Recent post-rift sequence is characterised by predominantly marine facies, and also by an upward transition from marginal through shallow to deep marine facies. It comprises, in ascending stratigraphic order, 'middle' Aptian marginal marine clastics and evaporites, 'late' Aptian–Albian shallow marine carbonates, and Late Cretaceous–Cenozoic deep marine clastics. The Aptian evaporites, including salts, were mobilised by sediment loading in the Albian to produce a variety of structures that controlled subsequent sedimentation. The deep marine clastics were predominantly fine-grained in the retrogradational Late Cretaceous–Eocene interval; and predominantly coarse-grained in the progradational Oligocene–Recent (following major tectono- or glacio-eustatically mediated sea-level falls in the Late Eocene and Early Oligocene).

*Outcrop geology*   The outcrop geology of the margins of the South Atlantic salt basins is reasonably well documented in the public-domain literature. Published field geological maps are available (see the further reading list at the end of the chapter).

Direct or indirect analogues for some of the source-rocks and reservoirs of the region can be studied at outcrop. Analogues for the submarine fan sandstone reservoirs can be studied at outcrop in Chile and in South Africa. Modern analogues can also be studied in the Almirante Camara fan system offshore Brazil (by means of side-scan sonar and shallow seismic and coring techniques).

*Petroleum geology*   The South Atlantic salt basins are interpreted as having remaining reserves in excess of 30 billion barrels of oil equivalent (see the further reading list at the end of the chapter).

*Source-rocks*. Non-marine, lacustrine source-rocks are developed in the pre-salt; and marine source-rocks at least locally at several horizons in the post-salt, including the Apto–Albian, Cenomanian–Turonian, Senonian, Palaeogene and Neogene (see the further reading list at the end of the chapter). The marine source-rocks in the Apto–Albian, Cenomanian–Turonian and Senonian appear to coincide with Cretaceous Oceanic Anoxic Events or OAEs 1, 2 and 3.

*Reservoir-rocks*. Non-marine carbonate and locally also clastic reservoir-rocks are developed in the pre-salt; and marine carbonate and clastic reservoirs at several horizons in the post-salt, including the Albian (marginal to shallow marine, peri-deltaic clastics and shallow marine, platform carbonates), and Late Cretaceous and Cenozoic (deep marine, submarine fan clastics and contourites) (see the further reading list at the end of the chapter).

Current exploration is targetted at pre-salt carbonate and post-salt deep marine submarine fan clastic plays in the deep offshore.

*Cap-rocks and traps*. The ultimate cap-rock to the pre-salt reservoirs is the salt. The principal cap-rocks to the post-salt turbidite reservoirs are associated hemipelagites.

Traps are predominantly structural, although stratigraphic traps are locally developed, for example in the Lagoa Parda field in the Espirito Santo basin in Brazil (see the further reading list at the end of the chapter).

## Biostratigraphy

A wide range of biostratigraphically and/or palaeoenvironmentally useful fossil groups have been recorded in the South Atlantic salt basins, including Cyanobacteria, dinoflagellates, diatoms, calcareous nannoplankton, Foraminifera, radiolarians, plants, fungi, bivalves, gastropods, ammonoids, ostracods, branchiopods, crinoids, fish, and reptiles and birds. A similarly wide range of biostratigraphic zonation schemes is applicable, the scale ranging from regional to reservoir, the resolution in inverse proportion to the scale.

A range of quantitative techniques and visualisation tools has been applied in the interpretation of biostratigraphic data from the South Atlantic salt basins (Watson & Nairn, 2005; Kender *et al.*, 2008; Kender *et al.*, in Kaminski & Coccioni, 2008; Watson *et al.*, in Ratcliffe & Zaitlin, 2010; see also Section 7.6, and Fig. 7.2).

*Pre-salt Mesozoic*   Only local terrestrially derived spore and pollen and freshwater ostracod biostratigraphic zonation schemes are applicable in the predominantly non-marine Neocomian–'early' Aptian, 'Gondwana Wealden' sequences of the pre-salt (Jones, 1996, 2011; and additional references cited therein).

*Post-salt Mesozoic and Cenozoic*   Global standard calcareous nannoplankton and planktic foraminiferal schemes, or modifications thereof, are applicable in the predominantly marine 'late' Aptian–Recent sequences of the post-salt, together with local smaller benthic schemes (de Klasz & Rerat, 1961, 1963; de Klasz & Gageonnet, 1965; Graham *et al.*, 1966; Hanse, 1965; Belmont, in van Hinte, 1966; Closs, in van Hinte, 1966; Reyment, in van Hinte, 1966; le Calvez *et al.*, 1971, 1974; de Klasz & van Hinte, 1977; Seiglie & Frost, 1979; Anglada *et al.*, 1981b; Reyment, in Reyment & Bengtson, 1981; Tronchetti, 1981; Koutsoukos, 1982; Seiglie & Baker, in Watkins, 1982; Seiglie & Baker, 1983; Brun *et al.*, in Oertli, 1984; Koutsoukos, 1984; Seiglie & Baker, in Schlee, 1984; Kogbe & Me'hes, 1986; Seiglie *et al.*, 1986; Carminatti & Scarton, in Weimer & Link, 1991; Koutsoukos *et al.*, 1991; Koutsoukos *et al.*, in Simo *et al.*, 1993; Preece *et al.*, in Cameron *et al.*, 1999; Grant *et al.*, 2000; Jones, 2001, 2003a, b; Kender *et al.*, 2006, 2007a, b, 2008; Kender *et al.*, in Kaminski & Coccioni, 2008; Coimbra *et al.*, 2009; Cetean & Kaminski, 2011). Larger benthics are also locally stratigraphically significant, for example in the Cenozoic of the Cameroon in west Africa (Kupper, 1960; Blondeau, in van Hinte, 1966; Drooger, in van Hinte, 1966) and of the Foz do Amazonas basin in Brazil (de Mello e Sousa *et al.*, 2003, 2009).

Some stratigraphically and palaeoenvironmentally significant species are illustrated on Pls. 53–55. Note that many of those encountered in the deep water

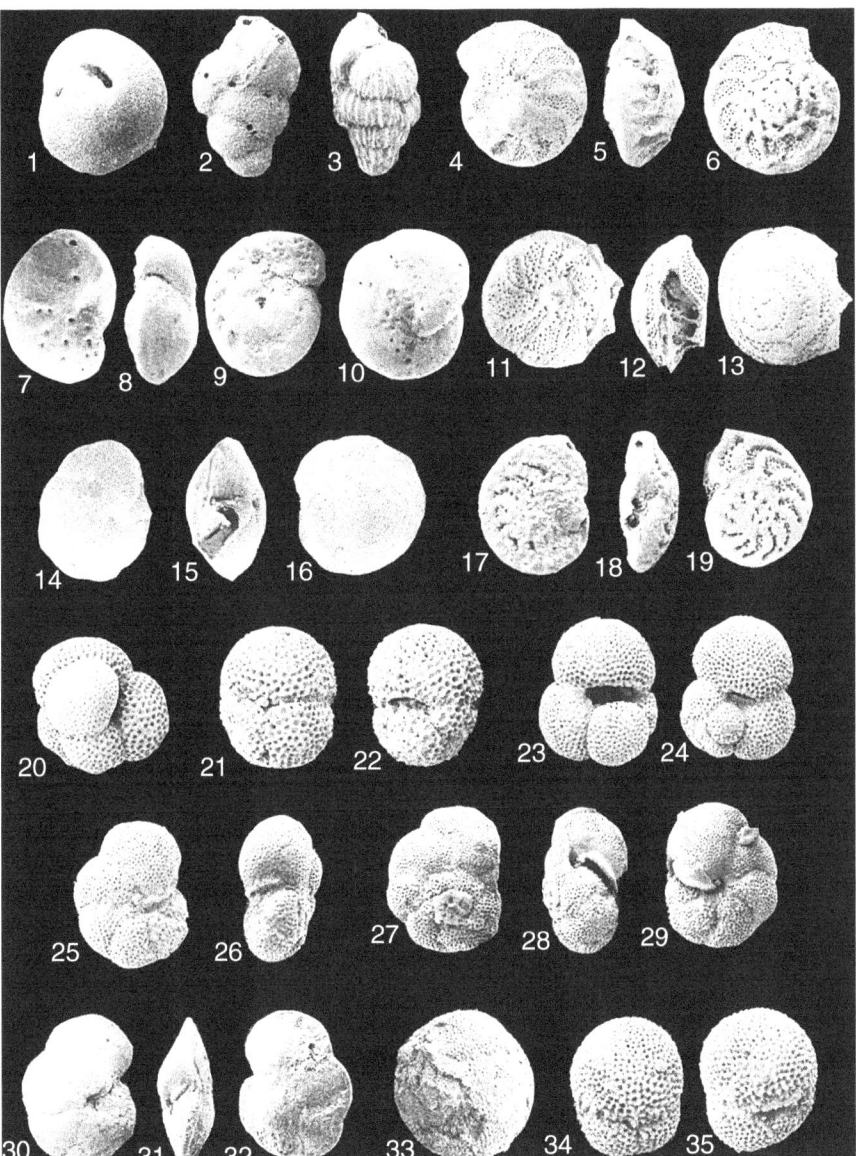

Pl. 53　**Some stratigraphically significant Foraminifera from the Cenozoic of the South Atlantic**. 53.1 *Cassidulina [Globocassidulina] punctata*; 53.2 *Uvigerina proboscidea*; 53.3 *Uvigerina spinulosa*; 53.4–53.6 *Cibicidoides crebbsi*; 53.7–53.9 *Cibicidoides dohmi*; 53.10 *Cibicidoides grimsdalei*; 53.11–53.13 *Cibicidoides havanensis*; 53.14–53.16 *Neoeponides campester*; 53.17–53.19 *Planulina renzi*; 53.20 *Catapsydrax unicavus*; 53.21–53.22 *Globigerinoides bisphericus*; 53.23–53.24 *Globigerinoides immaturus*; 53.25–53.26 *Globorotalia [Paraglobor-otalia] mayeri*; 53.27–53.29 *Globorotalia [Fohsella] peripheroronda*; 53.30–53.32 *Globorotalia [Menardella] praemenardii*; 53.33 *Praeorbulina glomerosa glomerosa*; 53.34–53.35 *Praeorbulina sicana*.

Pl. 54 **Some palaeoenvironmentally significant Foraminifera from the Cenozoic of the South Atlantic, 1**. Agglutinating Foraminifera. Modified after Kender *et al.*, in Kaminski & Coccioni (2008). 54.1 *Rhabdammina linearis*; 54.2 *Hyperammina cylindrica*; 54.3 *Glomospira gordialis*; 54.4 *Usbekistania charoides*; 54.5 *Hormosina globulifera*; 54.6 *Hormosina pilulifera*; 54.7 *Hormosinella carpenteri*; 54.8 *Hormosinella guttifera*; 54.9 *Lituotuba lituiformis*; 54.10 *Cribrostomoides subglobosus*; 54.11 *Cyclammina rotundidorsata*; 54.12. *Evolutinella rotulata*; 54.13 *Karrerulina conversa*.

sections in the Cenozoic of the South Atlantic salt basins also occur in deep water sections in the Cenozoic in the eastern Venezuelan basin (see Section 9.3.5).

*Palaeoenvironmental interpretation* Benthic Foraminifera are perhaps the most important group of fossils in palaeoenvironmental interpretation in the South Atlantic salt basins (Brun *et al.*, in Oertli, 1984; Koutsoukos, 1982, 1985a–c; Koutsoukos *et al.*, 1989, 1990, 1991; Koutsoukos *et al.*, in Simo *et al.*, 1993; Scarparo Cunha & Koutsoukos, 1998; Preece *et al.*, in Cameron *et al.*, 1999; Koutsoukos, in Hart *et al.*, 2000; Jones, 2001, 2003a, b; Kender *et al.*, 2006; Kender *et al.*, in Kaminski & Coccioni, 2008; Kender *et al.*, 2008, 2009; Cetean & Kaminski, 2011); although other micro- and macro-fossil groups are also important in the field of palaeobiogeography (see below). Interpretation is based in part on analogy with their living counterparts (Lohmann, 1978; Brun *et al.*, in Oertli, 1984; van

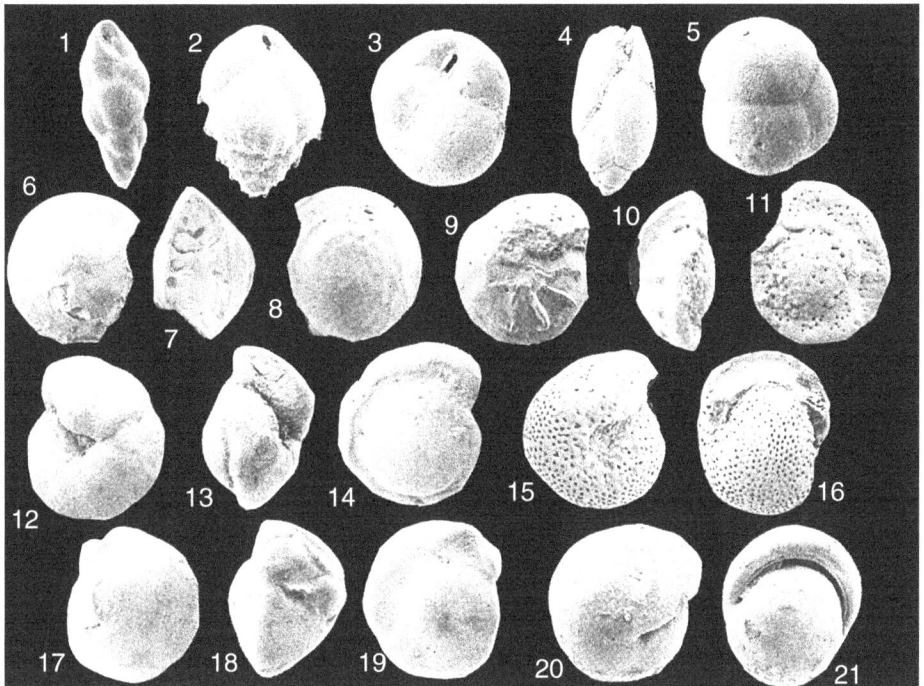

Pl. 55  **Some palaeoenvironmentally significant Foraminifera from the Cenozoic of the South Atlantic, 2. Buliminida (55.1–55.5), Robertinida (55.6–55.8) and Rotaliida (55.9–55.21)**. 55.1 *Bulimina elongata*; 55.2 *Bulimina marginata*; 55.3 *Cassidulina* [*Globocassidulina*] *subglobosa*; 55.4 *Praeglobobulimina ovata*; 55.5 *Sphaeroidina bulloides*; 55.6–55.8 *Hoeglundina elegans*; 55.9–55.11 *Cibicidoides mundulus*; 55.12–55.14 *Hansenisca soldanii/neosoldanii*; 55.15–55.16 *Melonis pompilioides*; 55.17–55.19 *Oridorsalis umbonatus*; 55.20–55.21 *Pullenia bulloides*.

Leeuwen, 1989; Debenay *et al.*, 1987; Kouyoumontzakis, 1987; Debenay & Basov, 1993; Debenay *et al.*, 1996; de Mello e Sousa *et al.*, 2006), in part on functional morphology (Preece *et al.*, in Cameron *et al.*, 1999; Kender *et al.*, 2006; Kender *et al.*, in Kaminski & Coccioni, 2008; Kender *et al.*, 2008; Kender *et al.*, 2009; Cetean & Kaminski, 2011), and in part on associated fossils and sedimentary environments.

*Palaeobathymetry*  Marginal, shallow and deep marine environments can be differentiated on the basis of benthic Foraminifera. Shallow marine, inner-, middle- and outer-neritic or -shelf; and deep marine, upper-, middle- and lower-bathyal or -slope, and abyssal sub-environments can at least locally also be differentiated (see, for example, Kender *et al.*, 2008).

Some measure of depth or distance from shoreline can also be derived from the ratio of planktic to benthic Foraminifera (although the trend to a higher ratio in deep marine environments is non-linear and indeed is locally reversed in upper-slope sub-environments).

*Sedimentary environments*   Various deep marine, submarine fan, sedimentary sub-environments are identifiable on the basis of benthic foraminiferal microfacies (Jones, 2001, 2003a, b; see also Section 5.4.3, and the comments on 'micro-palaeontological characterisation of cap-rocks' in Section 9.1.1).

*Palaeobiogeography*   The occurrences of riverine crocodilians in the Araripe basin in Brazil in eastern South America and in Niger in west Africa, and of shallow marine coelacanth fish also in Brazil in eastern South America and in Congo, Niger, Algeria, Morocco and Egypt in west, north-west and north Africa, in the immediately post-salt, 'late' Aptian–Cenomanian, indicates that the ocean had still not become established to such an extent as to constitute a significant barrier to the dispersal of non-marine or shallow marine animals from one side to the other by this time (Jones, 1996, 2011a, and additional references cited therein). However, the occurrence of Tethyan marine planktic Foraminifera and other organisms essentially throughout the South Atlantic by the Albian–Turonian indicates, in context, that a throughgoing ocean had become established by this time.

*Integrated studies*

A number of integrated biostratigraphic and sequence stratigraphic schemes for the South Atlantic basins have been published (Seiglie & Baker, in Watkins, 1982; Seiglie & Baker, in Schlee, 1984; Della Favera & Possatto, in Al-Hashimi & Sadooni, 1985; Carminatti & Scarton, in Weimer & Link, 1991; Brown *et al.*, 1995; Grant *et al.*, 2000; Moreira *et al.*, 2007; de Mello e Sousa *et al.*, 2009). The ages of sequence boundaries and MFSs are constrained by conventional biostratigraphy. MFSs are characterised micropalaeontologically by maximum flooding, as indicated by foraminiferal palaeobathymetry, and also by maxima in abundance and diversity; STs by palaeobathymetric trends (TSTs by upward deepening, and HSTs by upward shallowing) (Grant *et al.*, 2000; Jones, 2001, 2003a,b).

Palaeogeographic and facies maps of sequences have also been published (Melguen *et al.*, 1975; Anglada *et al.*, 1981a; Koutsoukos, 1985a; Zimmermann *et al.*, in Brooks & Fleet, 1987; Carminatti & Scarton, in Weimer & Link, 1991). Source and/or reservoir distributions can be predicted from maps such as these.

### 9.3.5 *Eastern Venezuelan basin*

The geological history of the eastern Venezuelan basin of eastern Venezuala and Trinidad is long and complex, with much of the complexity associated with the convergence and collision of the Caribbean and South American plates, and the uplift of the Andes, from the Oligocene onwards (Jones, in Ali *et al.*, 1998; Jones *et al.*, in Jones & Simmons, 1999; Jones, 2009, 2011a; see also comments on 'palaeobiogeography' below, and the further reading list at the end of the chapter).

To (over-)simplify, before the collision, foreland basin sedimentation character-ised the – locally preserved – Palaeozoic, and rifting and passive margin sedimen-tation the Jurassic, to the earlier part of the Oligocene (Fig. 9.13). (The best evidence for rifting comes from surface outcrop and subsurface seismic and well data from the Serrania del Interior and Espino graben in eastern Venezuela, and the Gulf of Paria between Venezuela and Trinidad). After the collision, foreland basin sedimentation came to dominate the later part of the Oligocene to the present. The collision caused the uplift of the Andes and the diversion of the Orinoco to its present-day position, and the shoaling and closure of the former Strait of Panama and the creation of the Isthmus of Panama, some of the palaeobiogeographic evidence for which is outlined under 'palaeobiogeography' below. Orinoco delta loading triggered extensive growth-faulting in the Columbus basin, offshore Trini-dad, in the Plio-Pleistocene. The entire area remains extremely tectonically active. There is abundant evidence of extensional and transtensional as well as compres-sional and transpressional tectonism.

The passive margin phase is recorded in eastern Venezuela by the Barranquin, Garcia, El Cantil, Chimana, Querecual, San Antonio, San Juan, Vidoño, Caratas, Tinajitas, Los Jabillos and Areo formations, and in Trinidad by the Naparima Hill, Guayaguayare, Lizard Springs and Navet. Associated predominantly clastic sedi-ments, derived from the Guyana Shield to the south, range from marginal marine sandstones through shallow marine sandstones to deep marine shales. Lithology varies with eustatically mediated sea level, with the sandstones associated with regressions and the shales with transgressions. Carbonate sediments (platform and pelagic limestones) are only locally volumetrically significant.

The foreland basin phase is recorded in eastern Venezuela by the seismically demonstrably syn-tectonic Naricual, Carapita, La Pica, Las Piedras and Mesa/Paria formations, and in Trinidad by the Cipero, Lengua, Cruse, Forest and equivalents. Associated clastic sediments, derived not only from the Guyana Shield to the south but also from the rising Serrania del Interior and Northern Ranges to the north, range from alluvial and fluvial conglomerates and sandstones through peri-deltaic sand-stones and coals and prodeltaic shales to turbiditic sandstones and basinal shales. On a basinal scale, lithology varies with tectonically mediated basin fill, with the turbiditic lithologies ('flysch') associated with early stages, and the prodeltaic, peri-deltaic, fluvial and alluvial lithologies ('molasse') associated with late stages. On a smaller scale, lithology varies with eustatically mediated sea level, with the peri-deltaic sandstones associated with regressions and the prodeltaic shales with transgressions.

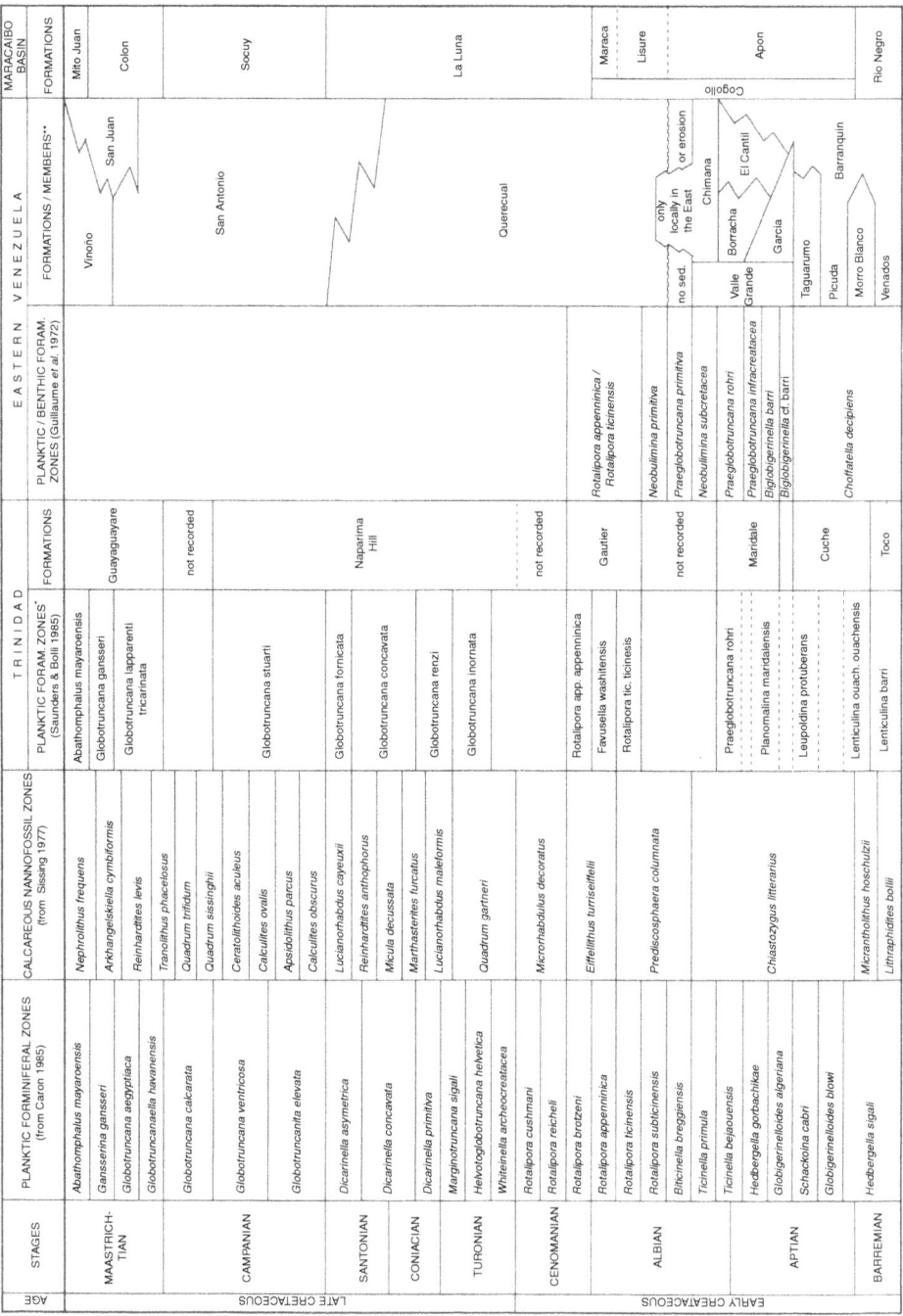

Fig. 9.13 **Stratigraphic framework for the eastern Venezuelan basin.**
(a) Cretaceous; (b) Cenozoic. Reproduced with permission from Bolli *et al.* (1994).

(b)

| Age | | Planktic foraminiferal zones (from Bolli et al 1985) | | Calcareous nannofossil zones (from Bolli et al 1985) | | South and South-east Trinidad rich in planktic Foraminifera | South and South-east trinidad predom. benthic Foraminifera | Venezuela — Eastern | Venezuela — South-east falcon | Venezuela — Central falcon | Venezuela — Maracaibo basin | Barbados |
|---|---|---|---|---|---|---|---|---|---|---|---|---|
| Miocene | Late | N17 | *Globorotalia humerosa* | NN11 | *Discoaster quinqueramus* | | Cruse | La Pica | Ojo de Agua | La Vela | Betijoque | |
| | | N16 | *Globorotalia acostaensis* | NN10 | *Discoaster calcaris* | | | | | | | |
| | | N15 | *Globorotalia menardii* | NN9 | *Discoaster hamatus* | | | Carapita | Pozon | Caujarao | Isnotú | Conset |
| | | N14 | *Globorotalia mayeri* | NN8 | *Catinaster coalitus* | Lengua | Karamat | | | | | |
| | middle | N13 | *Globigerinoides ruber* | NN7 | *Discoaster kugleri* | | | | | Socorro | | |
| | | N12 | *Globorotalia fohsi robusta* | | | Hiatus | | | | | Lagunillas | |
| | | N11 | *Globorotalia fohsi lobata* | NN6 | *Discoaster exilis* | | Herrera | | | | | |
| | | N10 | *Globorotalia fohsi fohsi* | | | | Ste. Croix | Chapapotal | | | | |
| | | N9 | *Globorotalia fohsi peripheroronda* | NN5 | *Sphenolithus heteromorphus* | | Brasso | | | Cerro Pelado | | |
| | Early | N8 | *Praeorbulina glomerosa* | | | Cipero | | | | | | Bissex Hill |
| | | N7 | *Globigerinatella insueta* | NN4 | *Helicosphaera ampliaperta* | | | Carapita | San Lorenzo | Agua Clara | | |
| | | N6 | *Catapsydrax stainforthi* | NN3 | *Sphenolithus belemnos* | | | | | | La Rosa | |
| | | N5 | *Catapsydrax dissimilis* | NN2 | *Discoaster druggii* | | Nariva | Naricual | | | | |
| | | N4 | *Globigerinoides primordius* | NN1 | *Triquetrorhabdulus carinatus* | | | | | Pedregoso | | |
| Oligocene | Late | P22 | *Globorotalia kugleri* | NP25 | *Sphenolithus ciperoensis* | | | Areo | Guacharaca | Pecaya | Icotea | |
| | | P21 | *Globigerina ciperoensis ciperoensis* | NP24 | *Sphenolithus distentus* | | | | | | | |
| | | P21 | *Globorotalia opima opima* | | | | | | | | | |
| | Early | P19/20 | *Globigerina ampliapertura* | NP23 | *Sphenolithus predistentus* | | | | | | | |
| | | P18 | *Cassig. chipolensis / Pseud. micra* | NP22 | *Helicosphaera reticulata* | Hiatus | | Los Jabillos | | El Paraiso | Mene Grande | Oceanic |
| Eocene | Late | P17 | *Turborotalia cerroazulensis s.l.* | NP21 | *Ericsonia subdisticha* | San Fernando | | | | | | |
| | | P16 | *Globigerinatheka semiinvoluta* | NP19/20 | *Isthmolithus recurves* | | | | Cerro Mision | Jarillal / Santa Rita | | |
| | | P15 | *Truncorotaloides rohri* | NP18 | *Chiasmolithus oamaruensis* | Navet | | Tinajitas | | | | |
| | middle | P14 | *Orbulinoides beckmanni* | NP17 | *Discoaster saipanensis* | | Pointe-à-Pierre | | | | Pauji | |
| | | P13 | *Morozovella lehneri* | NP16 | *Discoaster nodifer* | | | | | | | |
| | | P12 | *Globigerinatheka s. subconglobata* | NP15 | *Nannotetrina fulgens* | | | Caratas | | | | |
| | | P11 | *Hantkenina nuttalli* | NP15 | | | | | | | Misoa | |
| | Early | P10 | *Acarinina pentacamerata* | NP14 | *Discoaster sublodoensis* | | | | | | | |
| | | P9 | *Morozovella aragonensis* | NP13 | *Discoaster lodoensis* | Upper Lizard Springs | | | | | | |
| | | P8 | *Morozovella formosa formosa* | NP12 | *Tribrachiatus orthostylus* | | | | | | | |
| | | P7 | *Morozovella subbotinae* | NP11 | *Discoaster binodosus* | | | | | | Trujillo | |
| | | P6 | *Morozovella edgari* | NP10 | *Tribrachiatus contortus* | | | Vidoño | | | | |
| Paleocene | Late | P5 | *Morozovella velascoensis* | NP9 | *Discoaster multiradiatus* | Lower Lizard Springs | Chaudière | | | | | |
| | | P4 | *Planorotalites pseudomenardii* | NP8 | *Heliolithus riedeli* | | | | | | | |
| | | | | NP7 | *Discoaster mohleri* | | | | | | | |
| | | | | NP6 | *Heliolithus kleinpelli* | | | | | | | |
| | Middle | P3 | *Planorotalites pusilla pusilla* | NP5 | *Fasciculithus tympaniformis* | | | | | | Guasare | |
| | | | *Morozovella angulata* | NP4 | *Ellipsolithus macellus* | | | | | | | |
| | | P2 | *Morozovella uncinata* | NP3 | *Chiasmolithus danicus* | | | | | | | |
| | Early | P1 c | *Morozovella trinidadensis* | NP2 | *Cruciplacolithus tenuis* | not recorded or Hiatus | not recorded or Hiatus | | | | | |
| | | P1 b | *Morozovella pseudobulloides* | | | | | | | | | |
| | | P1 a | *Globigerina eugubina* | NP1 | *Markalius inversus* | | | | | | | |

Fig. 9.13 (*cont.*)

*Outcrop geology*    The outcrop geology of the margins of the eastern Venezuelan basin is reasonably well documented in the public-domain literature (Fig. 9.13; see also the further reading list at the end of the chapter). Published field geological maps are available for the entire area on various scales.

Direct or indirect analogues for some of the source-rocks and reservoirs of the region can be studied at outcrop. The principal source-rock(s), the Querecual and Naparima Hill formations, can be studied at outcrop in the Serrania del Interior of eastern Venezuela, and at Naparima Hill near San Fernando on the west coast of Trinidad, respectively (and the equivalent La Luna in Colombia and western Venezuela). Analogues for the marginal to shallow marine, including peri-deltaic, clastic reservoirs can be studied in the Serrania del Interior in eastern Venezuela, and on the south (Moruga) and east (Mayaro) coasts of Trinidad; analogues for the deep marine, submarine fan, clastic reservoirs, on Barbados (Jones, 2009).

*Petroleum geology*    Latin America has the second largest oil reserves in the world, after the Middle East (see the further reading list at the end of the chapter). The eastern Venezuelan basin of eastern Venezuela and Trinidad has been produc-ing significant quantities of oil – and gas – for over 100 years, and still has significant remaining reserves, including those of the so-called 'Faja Petrolifera Orinoco' or 'Orinoco Tar Belt'.

*Source-rocks.* The restricted deep marine Querecual formation of eastern Vene-zuela and Naparima Hill of Trinidad (equivalent to the La Luna of Colombia and western Venezuela), of 'Middle' Cretaceous age, constitute the most important source-rock(s) in the region (Jones *et al.*, in Jones & Simmons, 1999; Fig. 9.13; see also the further reading list at the end of the chapter). The preferred depositional model for the source-rock(s) invokes enhanced production of organic matter as a function of upwelling, and enhanced preservation as a function of at least intermit-tent anoxia.

*Reservoir-rocks.* Marginal to shallow marine, and, locally, deep marine, clastics of Late Cretaceous, Palaeogene and Neogene age constitute the most important reservoirs in eastern Venezuela, for example in the 'Orinoco Tar Belt', along the Oficina, Temblador and Furrial trends, and in the Quiriquire and Pedernales fields (Jones *et al.*, in Jones & Simmons, 1999; Fig. 9.13; see also the further reading list at the end of the chapter). Marginal to shallow marine, and, locally, deep marine, clastics of Neogene and Pleistogene constitute the most important reservoirs onshore and offshore Trinidad (Jones *et al.*, in Ali *et al.*, 1998; Fig. 9.13; see also the further reading list at the end of the chapter).

Current exploration is targetted in part at deep marine, submarine fan clastic plays in the deep offshore.

*Cap-rocks and traps.* The cap-rock to the Late Cretaceous and Palaeogene Naricual formation *s.l.* reservoirs along the Furrial trend in eastern Venezuela is the Areo or Carapita, a significant regional flooding surface at the base of the foreland basin megasequence (Jones *et al.*, in Jones & Simmons, 1999; Fig. 9.13). That to the Neogene La Pica formation, Pedernales member reservoir in Pedernales field is the Cotorra mudstone, an intra-formational flooding surface (Jones *et al.*, in Jones & Simmons, 1999; Fig. 9.13). The cap-rocks to the Neogene and Pleistogene reservoirs onshore and offshore Trinidad are similarly intra-formational (Jones, in Ali *et al.*, 1998).

Traps in the region are predominantly structural, although a wide range of structural styles is known. That to the Pedernales field may have a stratigraphic component to it, although it is essentially a mud-cored anticline (Jones *et al.*, in Jones & Simmons, 1999).

### Biostratigraphy

A wide range of biostratigraphically and/or palaeobiologically useful fossil groups have been recorded in the surface and subsurface sections of the eastern Venezuelan basin, including dinoflagellates, calcareous nannoplankton, Foraminifera, radiolarians, plants, corals, bivalves, gastropods, ammonoids, ostracods, fish, amphibians, reptiles, birds, and mammals. A similarly wide range of biostratigraphic zonation schemes is applicable, the scale ranging from regional to reservoir, the resolution in inverse proportion to the scale (see below; see also Section 9.5.2).

A range of quantitative techniques and visualisation tools has been applied in the interpretation of biostratigraphic data from Trinidad (Wakefield, in Olson & Leckie, 2003; Crux *et al.*, 2005; Crux *et al.*, in Ratcliffe & Zaitlin, 2010).

*Mesozoic* Global standard calcareous nannoplankton and planktic foraminiferal schemes, or modifications thereof, are applicable, together with local smaller benthic schemes (Renz, 1962; Lamb, 1964; Metz, in Saunders, 1968; Premoli-Silva & Bolli, 1973; Bartenstein, 1985; Saunders & Bolli, 1985; Koutsoukos & Merrick, 1986; Bartenstein, 1987; Bolli *et al.*, 1994; Carillo *et al.*, in Langer, 1995; de Romero & Galea-Alvarez, in Langer, 1995; Natural History Museum, Basel, 1996; Crespo de Cabrera *et al.*, in Huber *et al.*, 1999; Carr-Brown, in Ministry of Energy and Energy Industries, 2009; Fig. 9.13). Many of the now global standard planktic foraminiferal zones were first recognised in Trinidad (Saunders & Bolli, 1985; Carr-Brown, in Ministry of Energy and Energy Industries, 2009).

Key species are illustrated by Bolli *et al.* (1994).

*Cenozoic* Global standard calcareous nannoplankton and planktic foraminiferal schemes, or modifications thereof, are applicable, together with local smaller and larger benthic schemes (Hedberg, 1937; de Cizancourt, 1951; Stanley, 1960; Bermudez, 1962; Renz, 1962; Lamb, 1964; Bermudez, 1966; Bower & Hunter, in Saunders, 1968; Lamb & Sulek, in Saunders, 1968; Metz, in Saunders, 1968; Robinson, in Saunders, 1968; Stainforth, in Saunders, 1968; Premoli-Silva & Bolli, 1973; Bermudez & Stainforth, 1975; Hunter, 1976, 1978; Seiglie & Frost, 1979; Bolli & Saunders, in Bolli *et al.*, 1985; Saunders & Bolli, 1985; Seiglie *et al.*, 1986; Kaminski & Geroch, 1987; Kaminski *et al.*, in Rogl & Gradstein, 1988; Bolli *et al.*, 1994; Crespo de Cabrera & di Gianni Canudas, 1994; Galea-Alvarez & Moreno-Vasquez, 1994; Moreno-Vasquez, in Langer, 1995; Caudri, 1996; Jones, in Ali *et al.*, 1998; Jones *et al.*, in Jones & Simmons, 1999; Kaminski & Crespo de Cabrera, 1999; Carr-Brown, in Ministry of Energy and Energy Industries, 2009; Fig. 9.13). Again, many of the now global standard planktic foraminiferal zones were first recognised in Trinidad (Saunders & Bolli, 1985; Carr-Brown, in Ministry of Energy and Energy Industries, 2009). Interestingly, LBFs, although of shallow marine origin, are sufficiently common – contemporaneously – reworked into the evidently deep marine deposits of the Scotland group in Barbados so as to form the basis of a workable stratigraphy there (Jones, 2009). The lower part of the Scotland group, comprising, in ascending stratigraphic order, the Mount All, Chalky Mount and Murphy's formations, contains contemporaneously and/or non-contemporaneously reworked Palaeocene–Early Eocene species. The upper part, comprising the Morgan Lewis and Walkers formations, contains interpreted contemporaneously reworked Middle Eocene as well as non-contemporaneously reworked Middle Eocene species. The interpreted contemporaneously reworked Middle Eocene species include *Asterocyclina barbadensis*, *Eoconuloides senni* and *Neodiscocyclina anconensis*.

Key species are illustrated by Bolli & Saunders, in Bolli *et al.* (1985) (planktics) and Bolli *et al.* (1994) (see also Whittaker, 1988) (benthics).

### *Palaeoenvironmental interpretation*

Benthic foraminifera are perhaps the most important group of fossils in palaeoenvironmental interpretation in the eastern Venezuelan basin (Batjes, in Saunders, 1968; Bower & Hunter, in Saunders, 1968; Hunter, 1976; Koutsoukos & Merrick, 1986; Galea-Alvarez & Moreno-Vasquez, 1994; Carillo *et al.*, in Langer, 1995; de Romero & Galea-Alvarez, in Langer, 1995; Moreno-Vasquez, in Langer, 1995; Bornmalm, 1997; Jones, in Ali *et al.*, 1998; Crespo de Cabrera *et al.*, in Huber *et al.*, 1999; Wilson, 2003, 2004, 2007, 2008a, b; Jones, 2009; Wilson, 2010; Wilson *et al.*, 2010; Wilson, 2011; Wilson & Costelloe, 2011a, b; Wilson *et al.*, 2012); although again other micro- and macro-fossil groups are also important in the field of palaeobiogeography (see below). Interpretation is based on analogy with their living counterparts, and on associated fossils and sedimentary environments (Saunders, 1957; van Andel

& Postma, 1954; Todd & Brönnimann, 1957; Drooger & Kaasschieter, 1958; Saunders, 1958; Bermudez, 1966; Carr-Brown, 1972; Hofker, 1976; Culver & Buzas, 1982a; Jones, in Ali *et al.*, 1998; van der Zwaan *et al.*, in Tyson & Pearson, 1991; Jones, in Jones & Simmons, 1999; Wilson *et al.*, 2008; Fiorini, 2010).

Various quantitative techniques have been applied in the interpretation of palaeoenvironmental data from the Cenozoic of Trinidad, including the Cenozoic reservoir of Dolphin field (Wakefield, in Olson & Leckie, 2003; Wilson, 2008b; Wilson *et al.*, 2010; Wilson, 2011, Wilson & Costelloe, 2011a; Wilson *et al.*, 2012; see also Section 5.9).

*Palaeobathymetry*   Benthic foraminifera are the principal group of use in palaeo-bathymetric interpretation in the region (see, for example, Jones, 2009 (uplift history of Barbados)).

*Sedimentary environments*   These are of particular use in the discrimination of sedimentary sub-environments in marginal to shallow marine, peri-deltaic, environments and associated reservoirs. In the case of the modern and ancient Orinoco and other rivers in the region, fluvially dominated delta-top sub-environments are characterised by the testae amoeban *Centropyxis*; tidally dominated delta-top sub-environments by *Miliammina* and *Trochammina* foraminiferal assemblages; delta-front sub-environments are characterised by *Buliminella* foraminiferal assemblages; proximal prodelta sub-environments by *Eggerella* assemblages; and distal prodelta sub-environments by *Glomospira* and *Alveovalvulina/Cyclammina* assemblages (Batjes, in Saunders, 1968; Jones, in Ali *et al.*, 1998; Jones *et al.*, in Jones & Simmons, 1999; Fig. 9.14).

*Palaeobiogeography*   Plants, molluscs, fish, amphibians, reptiles and mammals record the uplift of the Andes from the Oligocene onwards, and the associated diversion of the debouchment of the Orinoco from western to eastern Venezuela (see Jones, 2011a; see also Hoorn *et al.*, 1995, and Hoorn & Wesselingh, 2010). Foraminifera and marine molluscs record the shoaling and closure of the Strait of Panama in the Early Pliocene (molluscs from the Pacific and Caribbean becoming isolated and beginning to evolve separately at this time). Terrestrial mammals record the creation of the Isthmus of Panama in the Middle Pliocene (those from North and South America beginning to interchange freely at this time).

### Integrated studies

A number of integrated biostratigraphic and sequence stratigraphic schemes for the eastern Venezuelan basin have been published (Galea-Alvarez & Moreno-Vasquez,

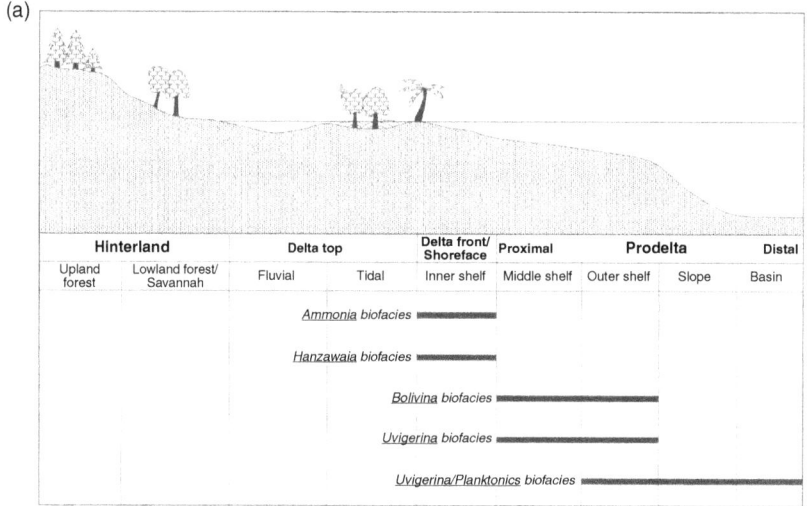

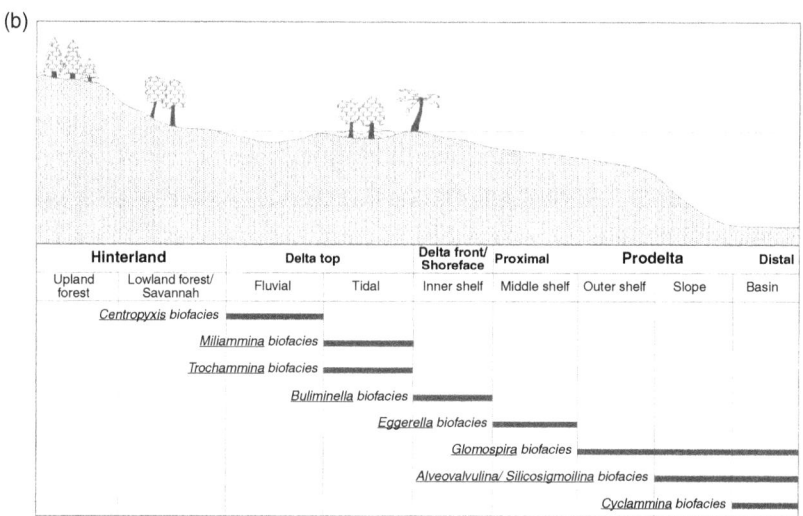

Fig. 9.14 **Palaeoenvironmental distribution of Foraminifera in the late Cenozoic of the eastern Venezuelan basin**. In (a) clear-water environments (outwith influence of the Orinoco delta); and (b) turbid-water environments (within influence of the Orinoco delta). Reproduced with permission from Jones, in Jones & Simmons (1999).

1994; Crespo de Cabrera & di Gianni Canudas, 1994; Carillo *et al.*, in Langer, 1995; Moreno-Vasquez, in Langer, 1995; Sams, 1995; Jones, in Ali *et al.*, 1998; Jones *et al.*, in Jones & Simmons, 1999; Wood, 2000). The ages of sequence boundaries and MFSs are constrained by conventional biostratigraphy. MFSs are characterised micropalaeontologically by maximum flooding, as indicated by foraminiferal palaeobathymetry, and also by maxima in abundance and diversity

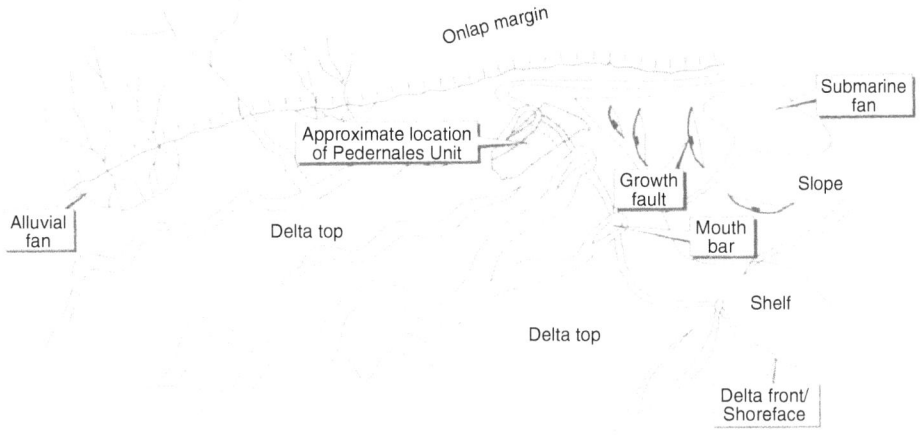

Fig. 9.15   **Pliocene palaeogeography of the eastern Venezuelan basin**. Reproduced with permission from Jones, in Jones & Simmons (1999).

(Jones *et al.*, in Jones & Simmons, 1999). Systems tracts are characterised micropalaeontologically by palaeobathymetric trends: TSTs by upward deepening; HSTs by upward shallowing.

Palaeogeographic and facies maps of sequences have also been published (Rohr, 1991; Crespo de Cabrera & di Gianni Canudas, 1994; Jones, in Ali *et al.*, 1998; Jones *et al.*, in Jones & Simmons, 1999; Erlich & Keens-Dumas, 2007). Source and/or reservoir distributions can be predicted from maps such as these (Fig. 9.15).

### 9.3.6  Gulf of Mexico

The Gulf of Mexico formed as a result of rifting between the Yucatan micro-plate and the North American craton associated with the opening of the central Atlantic Ocean (Jones, 1996; see also the further reading list at the end of the chapter).

The Late Triassic–Early Jurassic syn-rift was characterised by continental clastics. The succeeding post-rift was characterised initially (Middle Jurassic–Early Cretaceous) by carbonates (bank build-ups on rift shoulders on the basin margins, restricted muds in the starved basin centre) and associated evaporites, and by subordinate clastics (on the margins). The 'Middle' Cretaceous was characterised by initiation of significant coarse clastic sedimentation associated with a major regional unconformity. The succeeding Late Cretaceous was characterised by pelagic carbonate sedimentation associated with the continuing opening of the central Atlantic. The Cenozoic was and is characterised by coarse clastic sedimentation associated with the Laramide orogeny. The proto-Mississippi, now with an immense drainage area, built a series of – partly eustatically mediated – increasingly progradational river-dominated deltaic and associated submarine fan wedges out into the Gulf of Mexico. Asssociated loading caused salt mobilisation and growth faulting. The structures thus formed significantly influenced subsequent sedimentation.

*Outcrop geology*   Direct or indirect analogues for some of the source- and reservoir-rocks of the region can be studied at outcrop. For example, analogues for the deep marine, submarine fan reservoirs can be studied in the Pennsylvanian - Jackfork group of Baumgartner quarry in Arkansas (see the further reading list at the end of the chapter).

*Petroleum geology*   Note that further information on source- and reservoir-rocks is available through the United States Geological Survey (USGS) website: www.usgs.gov.

*Source-rocks.* Source-rocks in the onshore and offshore, some of them also shale-gas reservoirs, are developed in the Jurassic (the micro-laminated zone (MLZ) of the Smackover formation), the Late Jurassic to Early Cretaceous (the Haynesville formation), the Early Cretaceous (the Pearsall member of the Glenrose formation), the 'Middle' Cretaceous (the Eagle Ford formation), the Late Cretaceous (laminated carbonate muds at the base of the Austin Chalk), and the Eocene (unnamed) (see the further reading list at the end of the chapter). The Late Jurassic–Early Cretaceous carbonate muds are of regional extent (and probably also charged the accumulations in the Campeche, Golden Lane/Poza Rica and Reforma trends in Mexico).

*Reservoir-rocks.* Reservoir-rocks onshore and offshore are developed in Jurassic to Early Cretaceous shallow marine, peri-reefal carbonates (Smackover, Edwards and Sligo formations), fractured Late Cretaceous deep marine, pelagic carbonates (Austin and Taylor Chalk formations), and Cretaceous and Cenozoic marginal to shallow marine, peri-deltaic, and deep marine, submarine fan clastics (Woodbine and unnamed formations) (see the further reading list at the end of the chapter). Current exploration is targetted at deep marine, submarine fan clastic plays in the deep offshore. Current exploration is targeted at deep marine submarine fan clastic plays in the deep offshore.

*Cap-rocks and traps.* Cap-rocks to the submarine fan reservoirs in the outboard areas of the offshore are essentially intra-formational.

Traps are predominantly structural or combination structural–stratigraphic, many formed by salt movement, which may itself have been subject to some stratigraphic control (see the further reading list at the end of the chapter).

## Biostratigraphy

A range of quantitative techniques and visualisation tools has been applied in the interpretation of biostratigraphic data from the Gulf of Mexico (Martin & Fletcher, in Mann & Lane, 1995; Crux *et al.*, 2005; Gradstein *et al.*, 2008; Crux *et al.*, in Ratcliffe & Zaitlin, 2010; see also Section 7.6).

*Mesozoic* Global standard calcareous nannoplankton and planktic foraminiferal schemes, or modifications thereof, are applicable in the Mesozoic of the Gulf of Mexico (Clark & Bird, 1966; Pessagno, 1969; Poag, in Kaufmann & Hazel, 1977; Longoria, 1984; McNulty, in Perkins & Martin, 1985; Scott, 1990; Mancini *et al.*, 1996; Scott, 2002; Rosen & Rosen, 2008). The resolution of the schemes is sufficiently high to enable confident correlation across faults.

*Cenozoic* Again, global standard calcareous nannoplankton and planktic foraminiferal schemes, or modifications thereof, are applicable in the Cenozoic of the Gulf of Mexico, as are local benthic foraminiferal schemes (Rainwater, in van Hinte, 1966; Leutze, 1972; Skinner & Steinkraus, 1972; Poag, in Kaufmann & Hazel, 1977; Beard *et al.*, 1982; Bolli & Saunders, in Bolli *et al.*, 1985; Kohl, 1985; Nunn, 1986; van Morkhoven *et al.*, 1986; Akers, in Anon., 1987; Albers, in Anon., 1987; Armentrout, in Anon., 1987; Lamb *et al.*, in Ross & Haman, 1987; Mancini *et al.* in Anon., 1987; Allen *et al.*, in Armentrout & Perkins, 1990; Armentrout *et al.*, in Armentrout & Perkins, 1990; Hill & Rosen, in Armentrout & Perkins, 1990; Martin *et al.*, in Armentrout & Perkins, 1990; Pacht *et al.*, in Armentrout & Perkins, 1990; Taylor, in Armentrout & Perkins 1990; Vail & Wornardt, in Armentrout & Perkins, 1990; Wornardt & Vail, in Armentrout & Perkins, 1990; Flugelman *et al.*, 1990; Rosen & Hill, 1990; Shaffer, 1990; Armentrout, in Weimar and Link, 1991; Mancini & Tew, 1991; Pulham *et al.*, 1991; Pacht *et al.*, in Rhodes & Moslow, 1992; Aubry, 1993; Schnitker, 1993; Zhang *et al.*, 1993; Orndorff & Culver, 1998; Villamil *et al.*, 1998; Picou *et al.*, 1999; Rosen *et al.*, 2001; Denne & Sen Gupta, in Olson & Leckie, 2003). Again, the resolution of the schemes is sufficiently high to enable confident correlation across faults.

Some stratigraphically and palaeoenvironmentally significant species are illustrated by Bolli & Saunders, in Bolli *et al.* (1985) (planktics), and by van Morkhoven *et al.* (1986) (see also Picou *et al.*, 1999, and Rosen *et al.*, 2001) and on Pls. 56–70 below (benthics). The stratigraphic significance of many benthic species is enhanced by calibration against planktic zones. *Cibicidoides grosseperforatus*, *Cibicorbis herricki* and *Hanzawaia mantaensis*; range no younger than global standard planktic foraminiferal Zone N21 (Pliocene); *Rectuvigerina striata*, *Heterolepa dutemplei* and *Rectuvigerina multicostata* no younger than Zone N21;

Pl. 56   **Some stratigraphically significant Foraminifera from the Cenozoic of the Gulf of Mexico, 1 (Deep-Water Agglutinating Foraminifera)**. Modified after Green *et al.*, in Bubik & Kaminski (2004). 56.1 *Alveovalvulina suteri*; 56.2 *Alveovalvulinella pozonensis*; 56.3 *Ammosphaeroidina pseudopauciloculata* [?*Cystammina pauciloculata*]; 56.4 *Cyclammina acutidorsata*; 56.5 *Haplophragmoides carinatus*; 56.7 *Jarvisella karamatensis*; 56.7 *Martinottiella antillarum*; 56.8 *Praesphaerammina subgaleata* [?*Buzasina galeata*]; 56.9–56.10 *Textularia tatum*; 56.11–56.12 *Veleroniniodes veleronis*; 56.13 *Valvulina flexilis*.

*Cassidulina* [*Globocassidulina*] *punctata* and *Cancris nuttalli* no younger than N19; *Planulina dohertyi* no younger than N18 (Miocene); *Uvigerina carapitana*, *Plectofrondicularia parri*, *Plectofrondicularia vaughani* and *Planulina renzi* no younger than N17; *Cibicidoides compressus* no younger than N16; *Cibicidoides crebbsi* no younger than N15; *Cibicidoides guazumalensis*, *Siphonina pozonensis*, *Anomalinoides semicribratus* and *Rectuvigerina mexicana* no younger than N14; *Neoeponides campester* and *Bulimina tuxpamensis* no younger than N13; *Cibicidoides matanzasensis*, *Planulina mexicana* and *Buliminella grata* no younger than N12; *Matanzia bermudezi*, *Hanzawaia ammophila* and *Rectuvigerina transversa* no younger than N11; *Rectuvigerina sennii*, *Cibicidoides* sp. 11, *Cibicidina walli*, *Uvigerina adelinensis*, *Bulimina jarvisi*, *Cibicidoides barnetti* and *Cibicidoides havanensis* no younger than N10; *Uvigerina basicordata*, *Cibicidoides dohmi*,

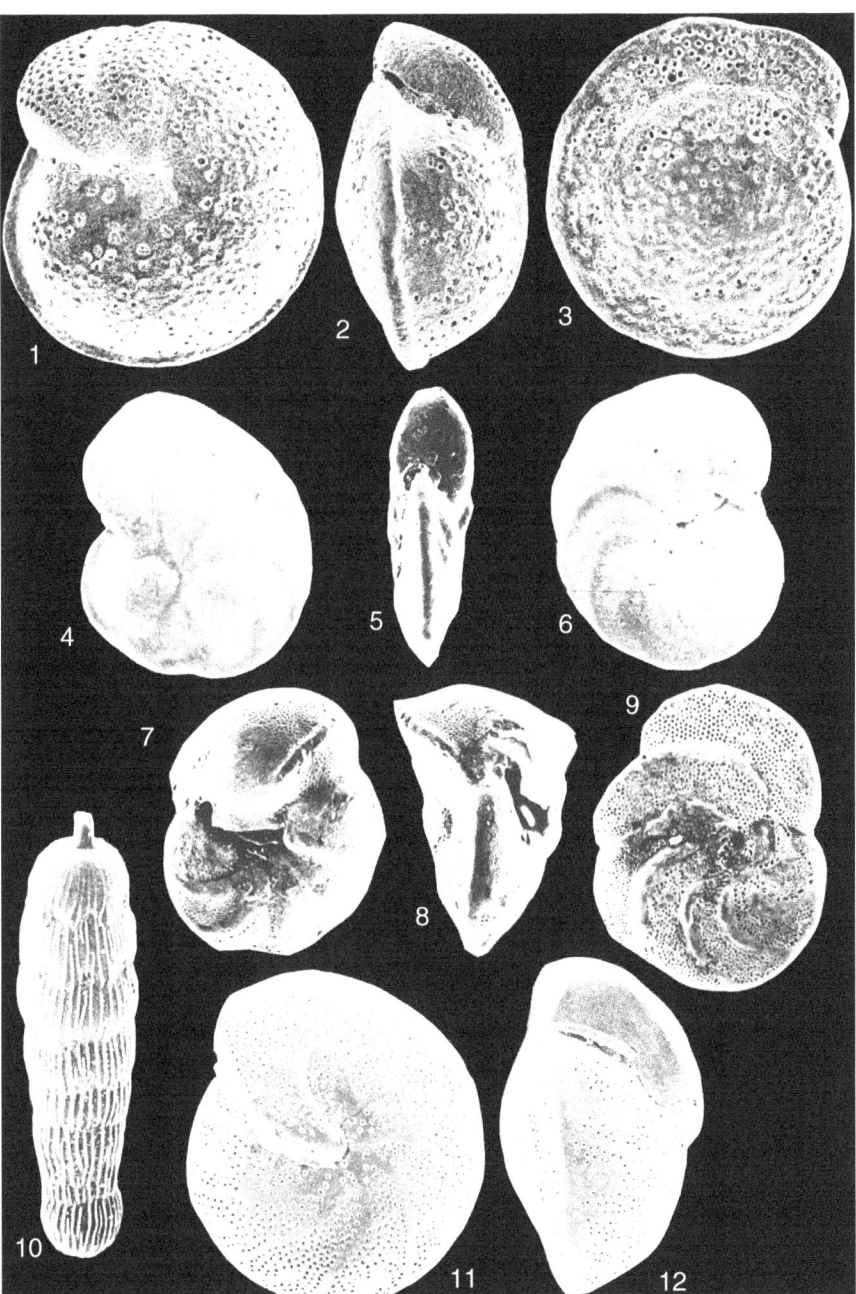

Pl. 57    **Some stratigraphically significant Foraminifera from the Cenozoic of the Gulf of Mexico, 2.** Modified after van Morkhoven *et al.* (1986). 57.1–57.3 *Cibicidoides grosseperforatus*; 57.4–57.6 *Hanzawaia mantaensis*; 57.7–57.9 *Cibicorbis herricki*; 57.10 *Rectuvigerina striata*; 57.11–57.12 *Heterolepa dutemplei.*

Pl. 58   **Some stratigraphically significant Foraminifera from the Cenozoic of the Gulf of Mexico, 3**. Modified after van Morkhoven *et al.* (1986). 58.1 *Rectuvigerina multicostata*; 58.2 *Cassidulina* [*Globocassidulina*] *punctata*; 58.3–58.4 *Cancris nuttalli*; 58.5–58.7 *Planulina dohertyi*; 58.8 *Uvigerina carapitana*; 58.9 *Plectofrondicularia parri*; 58.10 *Plectofrondicularia vaughani*; 58.11–58.12 *Planulina renzi*.

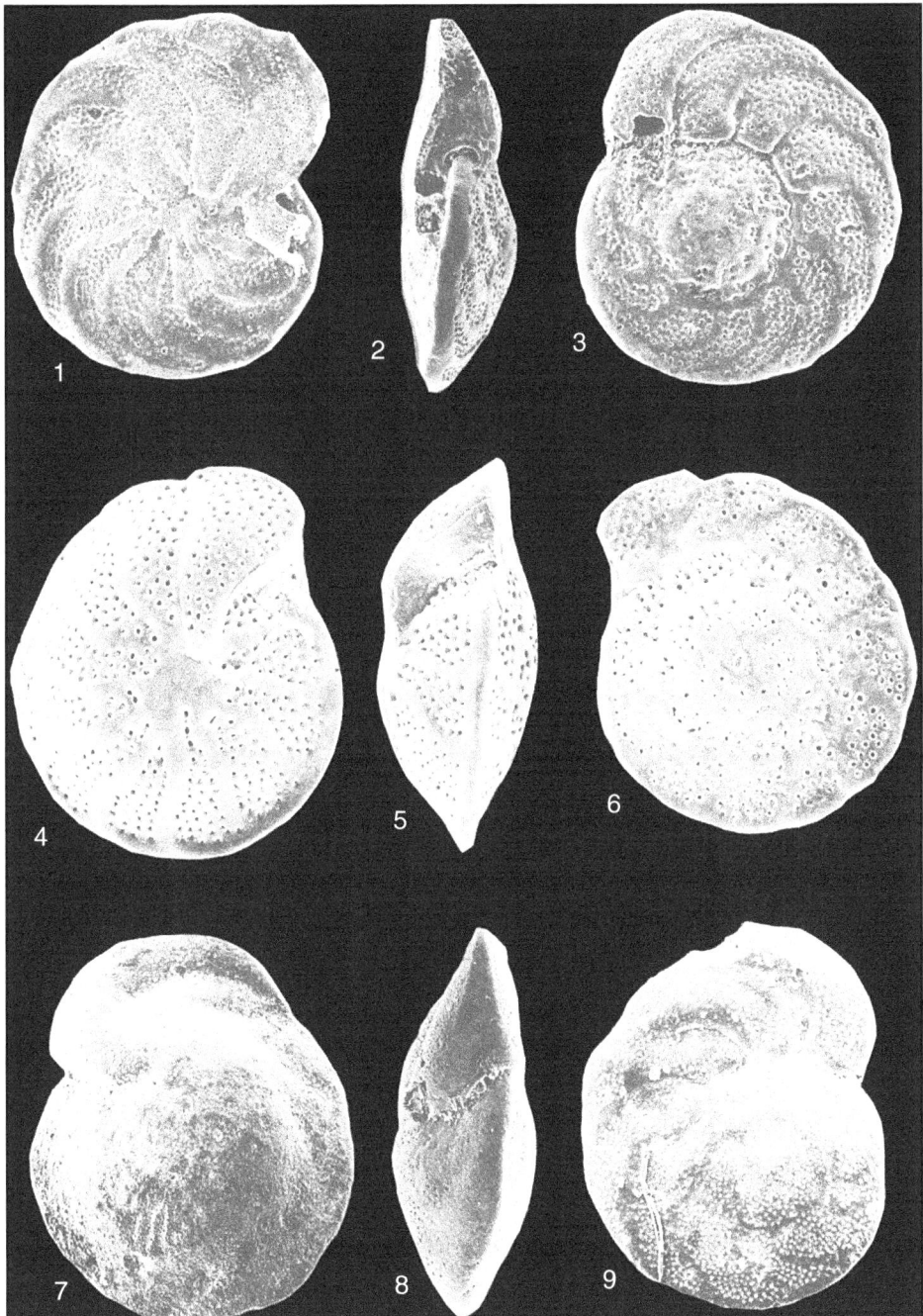

Pl. 59   **Some stratigraphically significant Foraminifera from the Cenozoic of the Gulf of Mexico, 4**. Modified after van Morkhoven *et al.* (1986). 59.1–59.3 *Cibicidoides compressus*; 59.4–59.6 *Cibicidoides crebbsi*; 59.7–59.9 *Cibicidoides guazumalensis*.

Pl. 60 **Some stratigraphically significant Foraminifera from the Cenozoic of the Gulf of Mexico, 5**. Modified after van Morkhoven *et al.* (1986). 60.1 *Siphonina pozonensis*; 60.2 *Anomalinoides semicribratus*; 60.3 *Rectuvigerina mexicana*; 60.4–60.6 *Neoepondes campester*; 60.7 *Bulimina tuxpamensis*; 60.8–60.10 *Cibicidoides matanzasensis*.

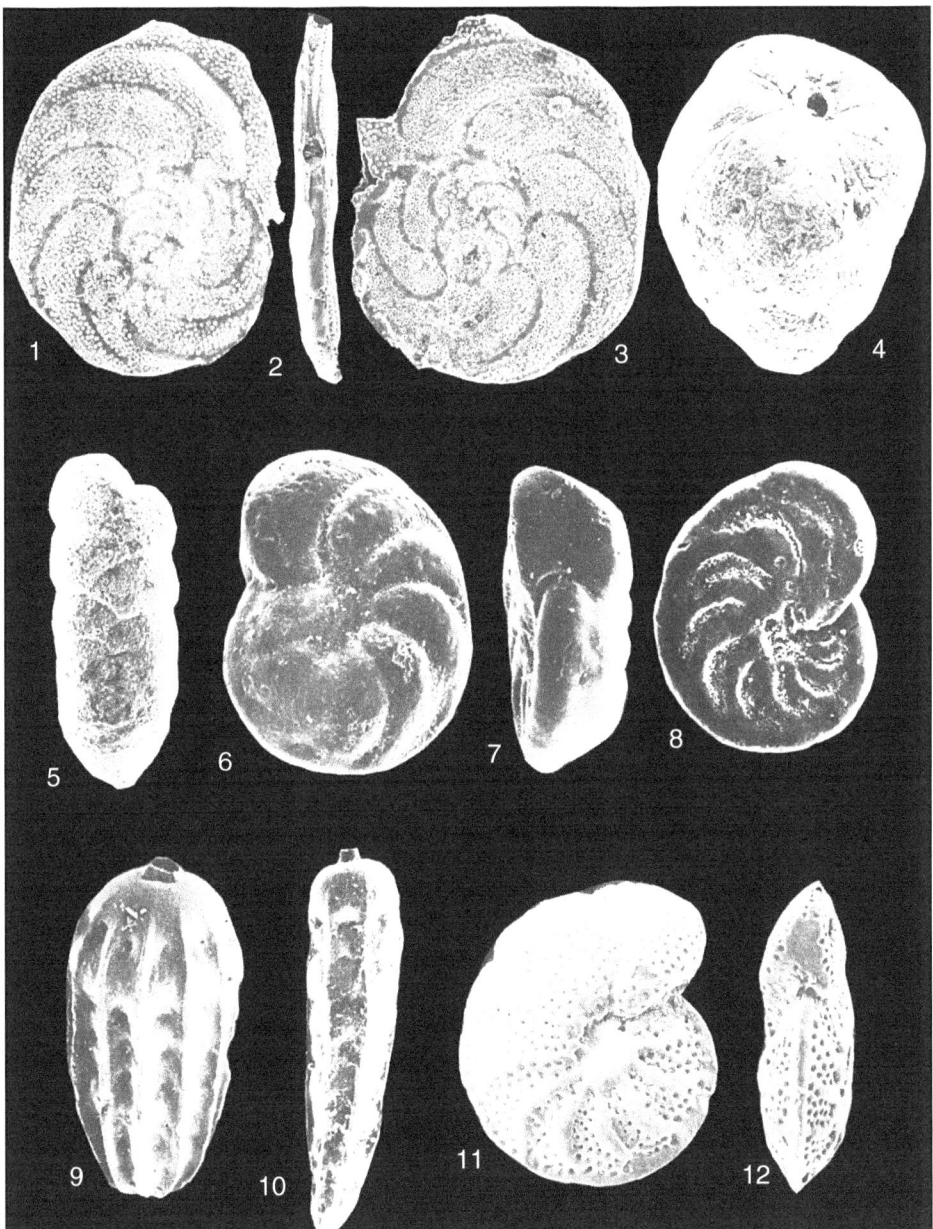

Pl. 61   **Some stratigraphically significant Foraminifera from the Cenozoic of the Gulf of Mexico, 6**. Modified after van Morkhoven *et al.* (1986). 61.1–61.3 *Planulina mexicana*; 61.4 *Buliminella grata*; 61.5 *Matanzia bermudezi*; 61.6–61.7 *Hanzawaia ammophila*; 61.8–61.9 *Rectuvigerina transversa*; 61.10 *Rectuvigerina sennii*; 61.11–61.12 *Cibicidoides* sp. 11.

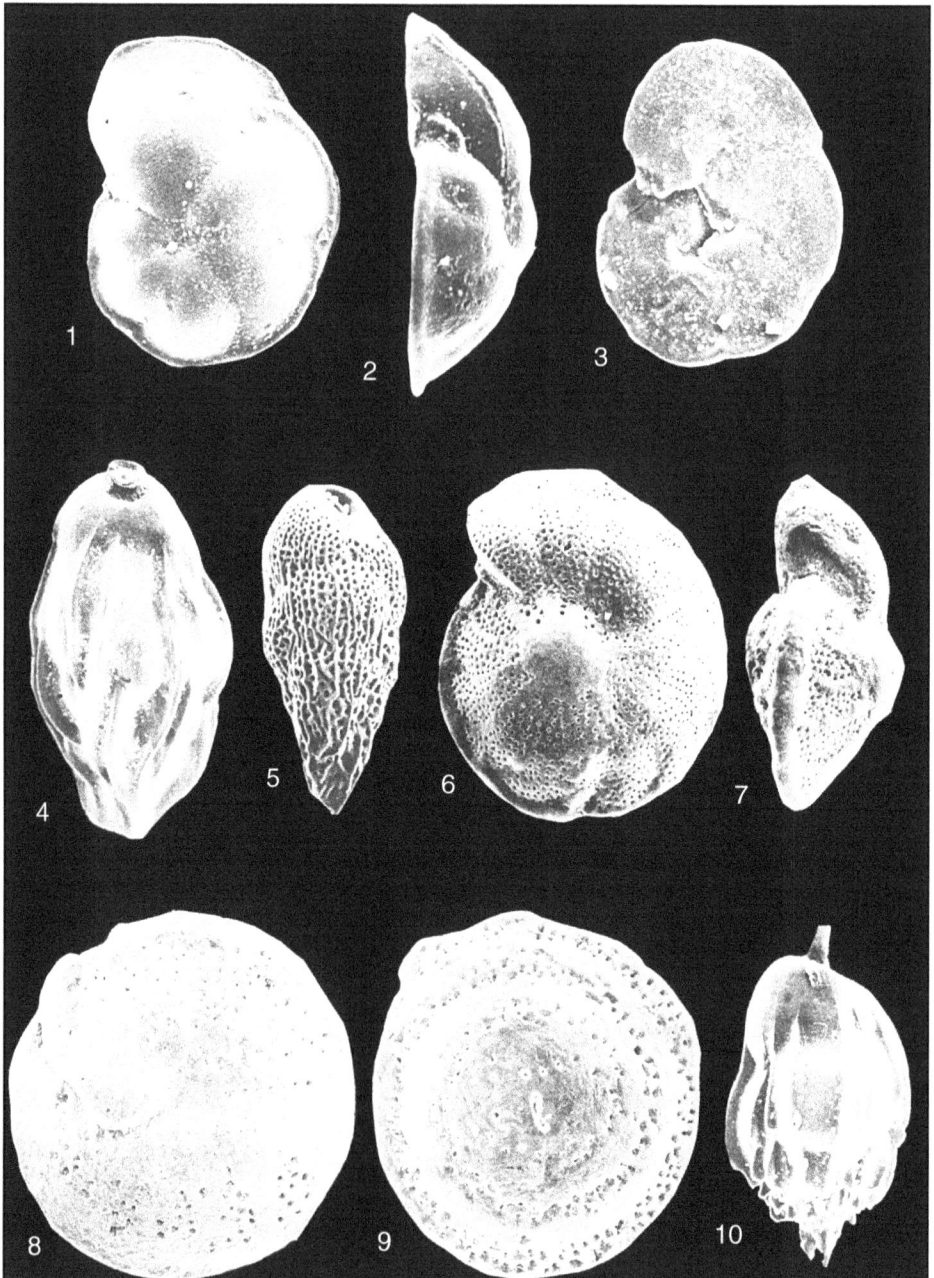

Pl. 62   **Some stratigraphically significant Foraminifera from the Cenozoic of the Gulf of Mexico, 7**. Modified after van Morkhoven *et al.* (1986). 62.1–62.3 *Cibicidina walli*; 62.4 *Uvigerina adelinensis*; 62.5 *Bulimina jarvisi*; 62.6–62.7 *Cibicidoides barnetti*; 62.8–62.9 *Cibicidoides havanensis*; 62.10 *Uvigerina basicordata*.

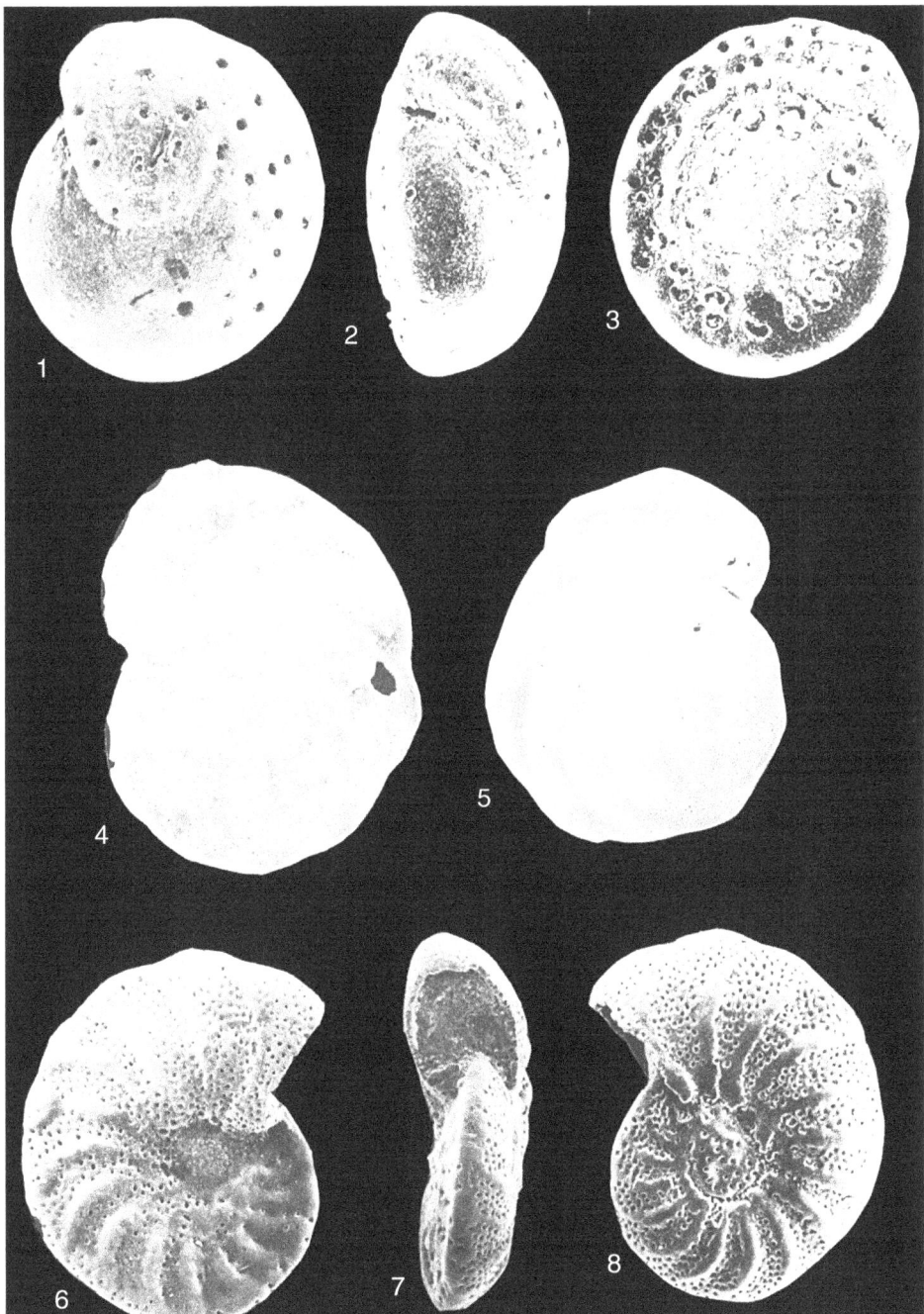

Pl. 63  **Some stratigraphically significant Foraminifera from the Cenozoic of the Gulf of Mexico, 8.** Modified after van Morkhoven *et al.* (1986). 63.1–63.3 *Cibicidoides dohmi*; 63.4–63.5 *Planulina subtenuissimima*; 63.6-63.8 *Cibicidoides alazanensis*.

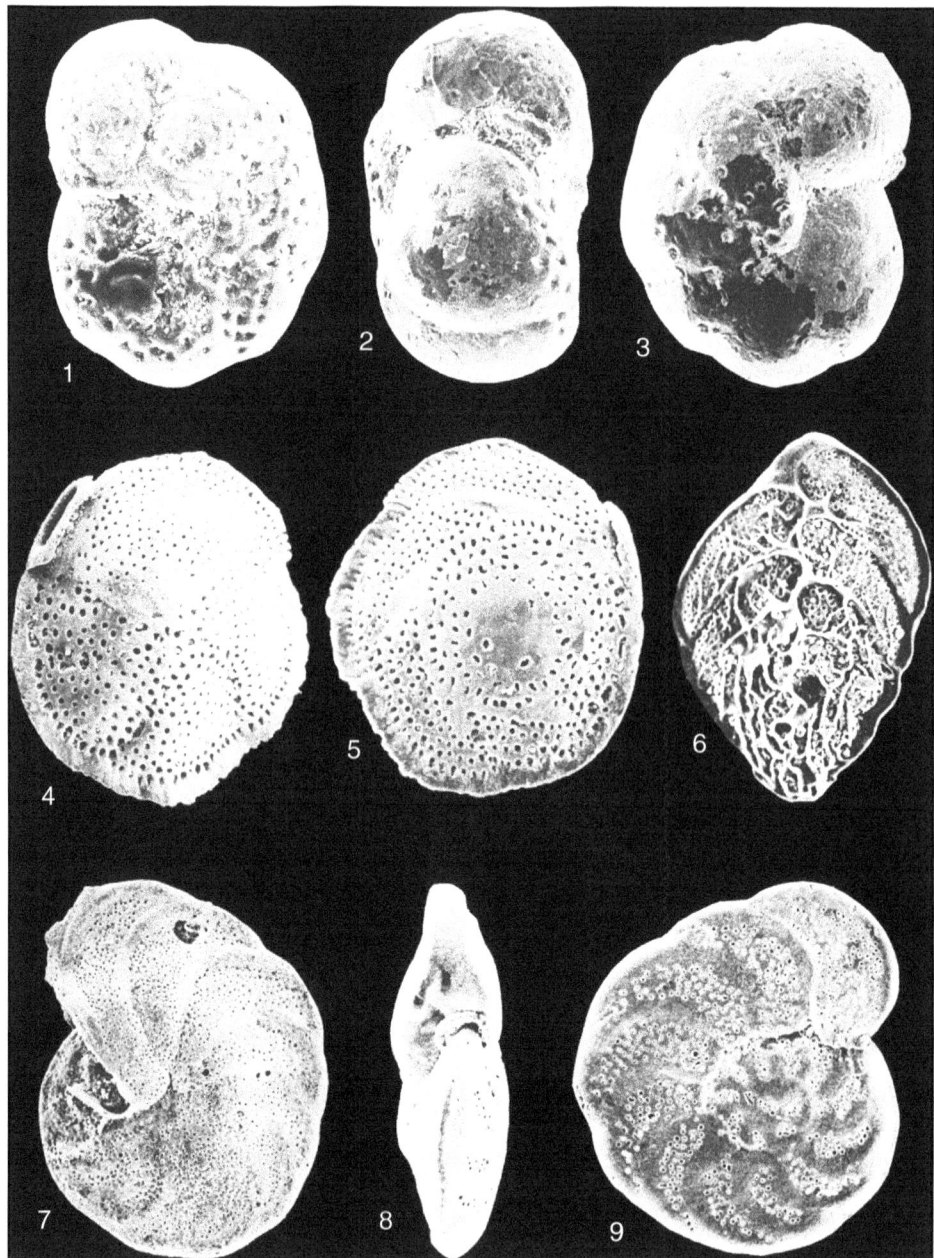

Pl. 64    **Some stratigraphically significant Foraminifera from the Cenozoic of the Gulf of Mexico, 9**. Modified after van Morkhoven *et al.* (1986). 64.1–64.3 *Anomalinoides pseudogrosserugosus*; 64.4–64.5 *Siphonina tenuicarinata*; 64.6 *Bolivina byramensis*; 64.7–64.9 *Planulina costata*.

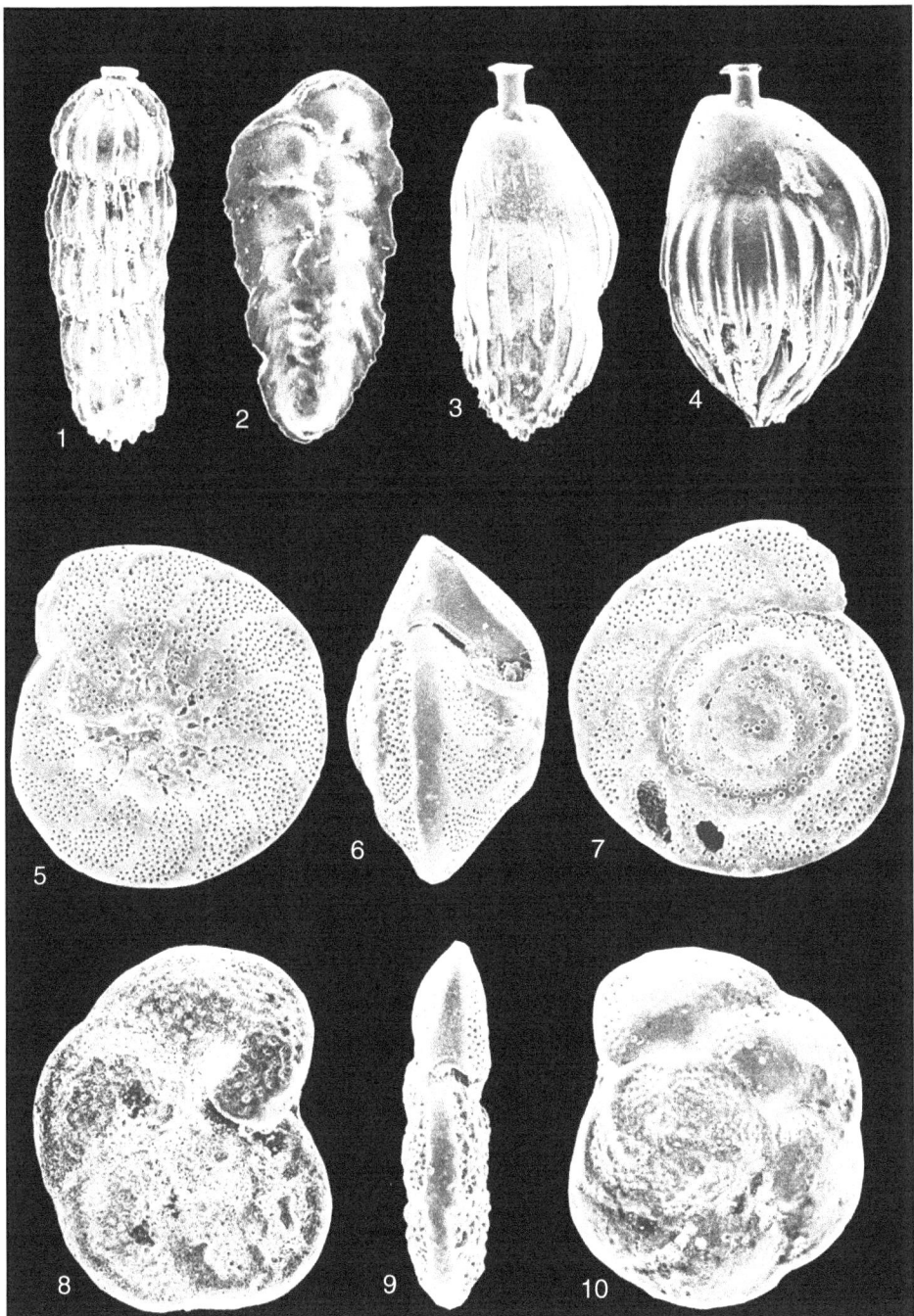

Pl. 65    **Some stratigraphically significant Foraminifera from the Cenozoic of the Gulf of Mexico, 10**. Modified after van Morkhoven *et al.* (1986). 65.1 *Rectuvigerina stonei*; 65.2 *Bolivina aliformis*; 65.3 *Uvigerina spinulosa*; 65.4 *Uvigerina mexicana*; 65.5–65.7 *Cibicidoides mexicanus*; 65.8–65.10 *Planulina ambigua*.

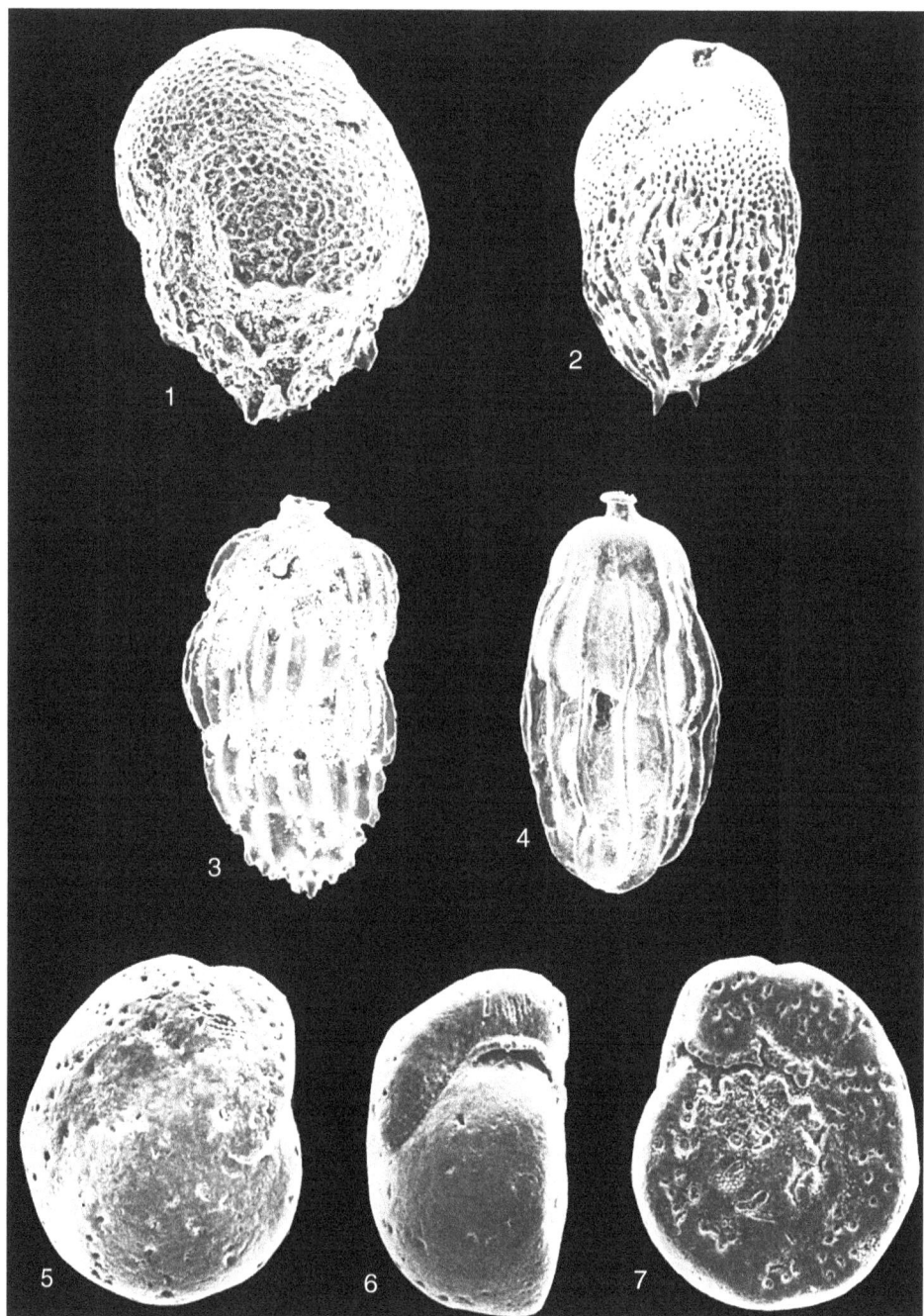

Pl. 66    **Some stratigraphically significant Foraminifera from the Cenozoic of the Gulf of Mexico, 11**. Modified after van Morkhoven *et al.* (1986). 66.1 *Bulimina impedens*; 66.2 *Bulimina glomarchallengeri*; 66.3 *Uvigerina havanensis*; 66.4 *Rectuvigerina nodifera*; 66.5–66.7 *Cibicidoides grimsdalei*.

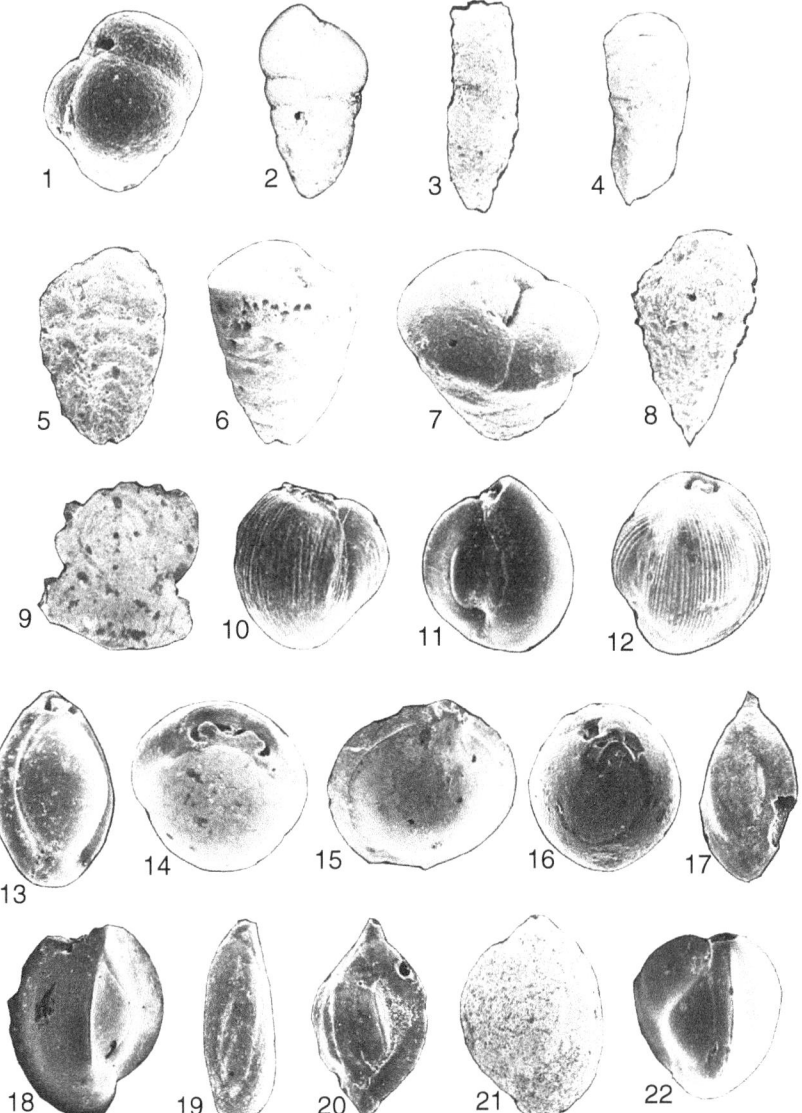

**Pl. 67 Some palaeoenvironmentally significant Foraminifera from the Cenozoic of the Gulf of Mexico, 1**. Agglutinating Foraminifera (67.1–67.8) and Miliolida (67.9–67.22). 67.1 *Eggerella bradyi*; 67.2 *Karreriella bradyi*; 67.3 *Martinottiella occidentalis*; 67.4 *Pseudogaudryina atlantica*; 67.5 *Textularia mexicana*; 67.6–67.7 *Textulariella barretti* (microspheric and macrospheric generations); 67.8 *Verneulinulla affixa*; 67.9 *Cornuspira foliacea*; 67.10 *Flintina bradyana*; 67.11 *Miliolinella subrotunda*; 67.12 *Pyrgo comata*; 67.13 *Pyrgo elongata*; 67.14 *Pyrgo laevis*; 67.15 *Pyrgo murrhina*; 67.16 *Pyrgoella sphaera*; 67.17 *Quinqueloculina compta*; 67.18 *Quinqueloculina lamarckiana*; 67.19 *Quinqueloculina tropicalis*; 67.20 *Sigmoilina distorta*; 67.21 *Sigmoilopsis schlumbergeri*; 67.22 *Triloculina tricarinata*.

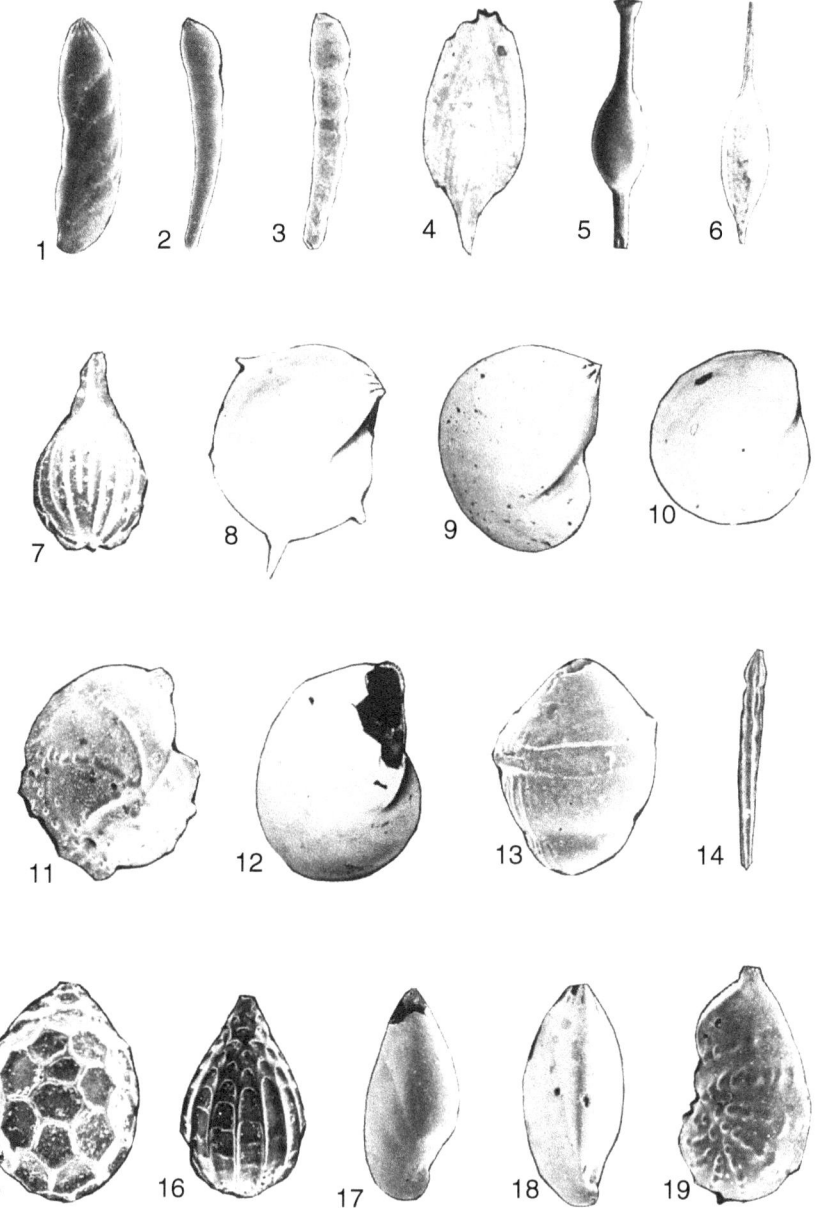

Pl. 68 **Some palaeoenvironmentally significant Foraminifera from the Cenozoic of the Gulf of Mexico, 2**. Nodosariida. 68.1 *Astacolus crepidulus*; 68.2 *Dentalina bradyensis*; 68.3 *Dentalina filiformis*; 68.4 *Frondicularia sagittula*; 68.5 *Grigelis monile*; 68.6 *Hyalinonetrion gracillima*; 68.7 *Lagena sulcata*; 68.8 *Lenticulina calcar*; 68.9 *Lenticulina gibba*; 68.10 *Lenticulina orbicularis*; 68.11 *Lenticulina submamilligera*; 68.12 *Lenticulina thalmanni*; 68.13 *Lingulina seminuda*; 68.14 *Nodosaria albatrossi*; 68.15 *Oolina hexagona*; 68.16 *Oolina squamosa*; 68.17 *Saracenaria altifrons*; 68.18 *Saracenaria latifrons*; 68.19 *Vaginulinopsis subaculeatus*.

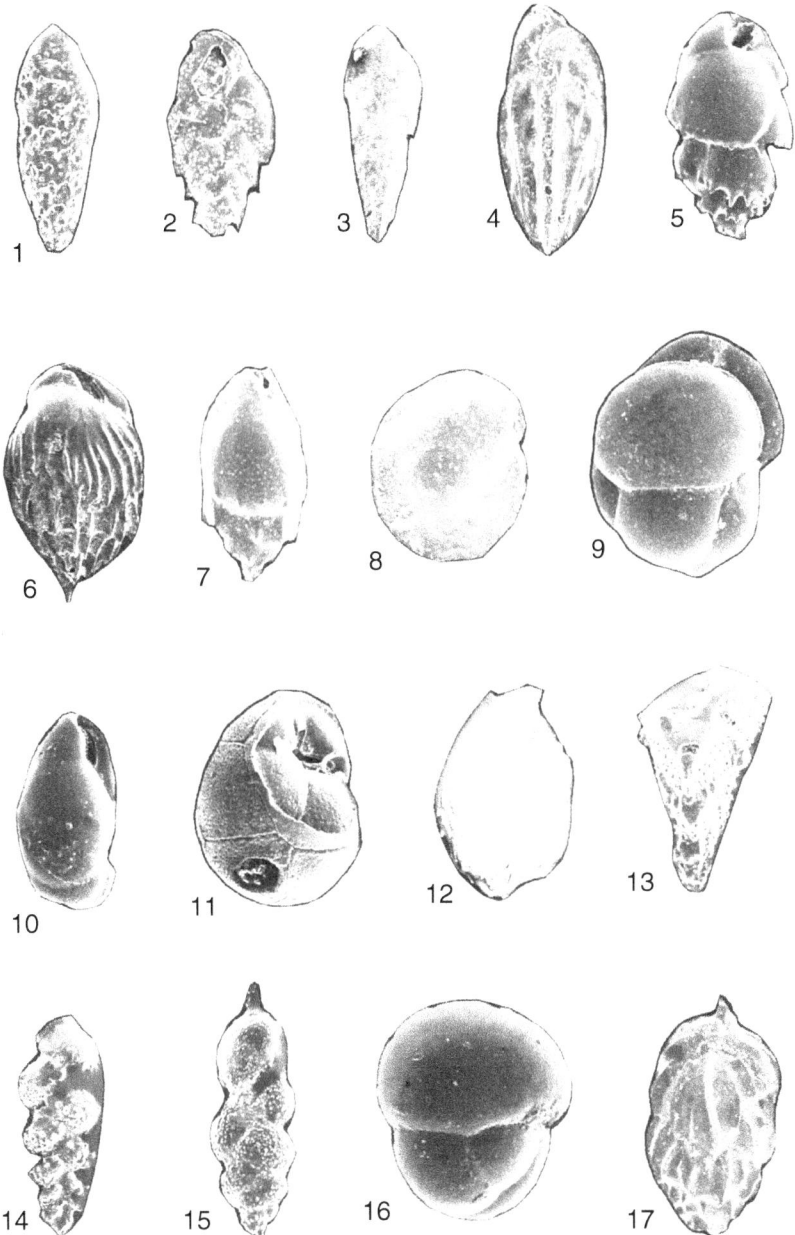

Pl. 69 **Some palaeoenvironmentally significant Foraminifera from the Cenozoic of the Gulf of Mexico, 3**. Buliminida. 69.1 *Bolivina albatrossi*; 69.2 *Brizalina alata*; 69.3 *Brizalina barbata*; 69.4 *Brizalina mexicana*; 69.5 *Bulimina marginata*; 69.6 *Bulimina mexicana*; 69.7 *Bulimina spicata*; 69.8 *Cassidulina australis*; 69.9 *Cassidulina obtusa*; 69.10 *Cassidulinoides bradyi*; 69.11 *Cassidulina* [*Globocassidulina*] *subglobosa*; 69.12 *Praeglobulimina spinescens*; 69.13 *Reussella atlantica*; 69.14 *Sagrinella lobata*; 69.15 *Siphouvigerina ampullacea*; 69.16 *Sphaeroidina bulloides*; 69.17 *Uvigerina peregrina*.

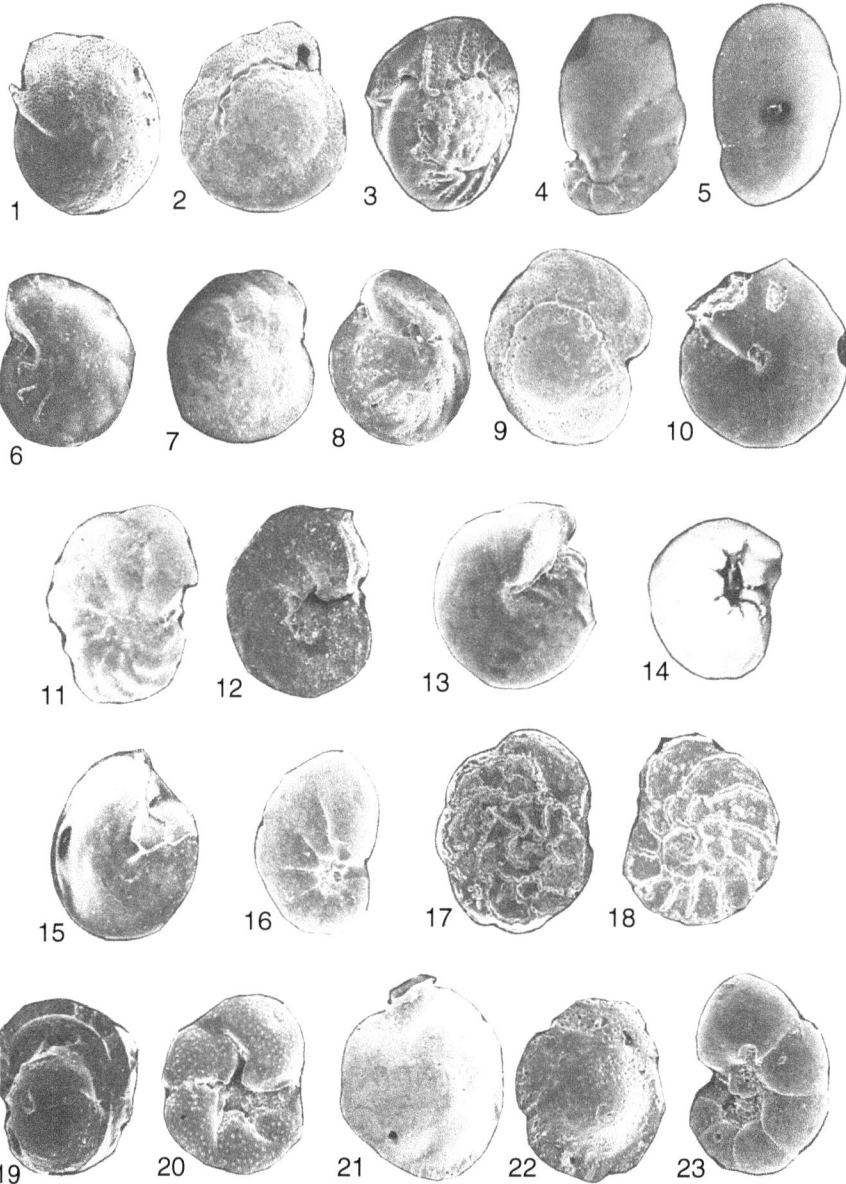

Pl. 70 **Some palaeoenvironmentally significant Foraminifera from the Cenozoic of the Gulf of Mexico, 4**. Rotaliida. 70.1–70.3 *Anomalinoides incrassatus*; 70.3 *Brizalina barbata*; 70.6 *Cibicidoides mollis*; 70.7–70.9 *Cibicidoides pachyderma*; 70.10 *Cibicidoides subhaidingerii*; 70.11–70.12 *Discorbinella bertheloti*; 70.13 *Gyroidina orbicularis*; 70.14 *Hansenisca soldanii/neosoldanii*; 70.15 H*eterolepa praecincta*; 70.16 *Nonionella atlantica*; 70.17–70.18 *Planulina foveolata*; 70.19 *Pullenia bulloides*; 70.20 *Rosalina floridana*; 70.21–70.22 *Siphonina bradyana*; 70.23 *Valvulineria minuta*.

*Planulina subtenuissimima*, *Cibicidoides alazanensis*, *Anomalinoides pseudogros-serugosus* and *Siphonina tenuicarinata* no younger than N9; *Bolivina byramensis* no younger than N8; *Planulina costata* no younger than N7; *Rectuvigerina stonei*, *Bolivina aliformis*, *Uvigerina spinulosa*, *Cibicidoides mexicanus*, *Uvigerina mexicana*, *Planulina ambigua*, *Bulimina impedens* and *Uvigerina havanensis* no younger than N5; and *Rectuvigerina nodifera*, *Bulimina glomarchallengeri* and *Cibicidoides grimsdalei* no younger than N4.

### *Palaeoenvironmental interpretation*

Benthic foraminifera are the most important group of fossils in palaeoenvironmental interpretation in the Gulf of Mexico (Curtis, 1955; Grimsdale & van Morkhoven, 1955; Clark & Bird, 1966; Tipsword *et al.*, 1966; Anisgard, 1970; Robinson, 1970; Leroy & Levinson, 1974; Robinson & Kohl, 1978; Katz & Miller, 1993; Schnitker, 1993; Jones *et al.*, in Repetski, 1996; Spencer, 1996; Castillo *et al.*, 1998; Breard *et al.*, 2000; Fang, in Olson & Leckie, 2003; Green *et al.*, in Bubik & Kaminski, 2004; Fluegeman, in Demchuk & Gary, 2009; Ingram *et al.*, 2010).

Interpretation is based on analogy with their living counterparts, and on associated fossils and sedimentary environments (Phleger & Parker, 1951; Phleger, 1955; Lankford, 1959; Phleger, 1960, 1963; Skinner, 1966; Krutak, 1975; Leroy & Hodgkinson, 1975; Bock, 1976; Pflum & Frerichs, 1976; Sen Gupta & Kilbourne, 1976; Brunner, 1979; Culver & Buzas, 1981; Poag, 1981; Poag & Tresslar, 1981; Sen Gupta *et al.*, 1981; Poag, 1984; Culver & Buzas, 1982b, 1983; Haman, in Verdenius *et al.*, 1983; Kohl *et al.*, in Bouma *et al.*, 1985; Hiltermann & Haman, 1986; Culver, 1988; Denne & Sen Gupta, 1988; Spencer, 1988; Corliss & Fois, 1990; Denne & Sen Gupta, 1991; Scott *et al.*, 1991; van der Zwaan & Jorissen, in Tyson & Pearson, 1991; Denne & Sen Gupta, 1993; Loubere *et al.*, 1993; Lagoe *et al.*, 1994; Sen Gupta & Aharon, 1994; Sen Gupta *et al.*, 1996; Osterman *et al.*, 2001; Denne & Sen Gupta, in Olson & Leckie, 2003; Osterman, 2003; Kohl *et al.*, in Anderson & Fillon, 2004; Osterman *et al.*, 2004; Brunner *et al.*, 2006; Buzas *et al.*, 2007; Bernhard *et al.*, 2008; Lobegeier & Sen Gupta, 2008; Fillon, in Demchuk & Gary, 2009; Osterman *et al.*, 2008, 2009; Sen Gupta *et al.*, in Tunnell *et al.*, 2009; Sen Gupta & Smith, 2010; Strauss *et al.*, 2012; Fig. 9.16).

Various quantitative techniques have been applied in the interpretation of palaeoenvironmental data (Robinson & Kohl, 1978; Jones *et al.*, in Repetski, 1996; Fang, in Olson & Leckie, 2003; Fluegeman, in Demchuk & Gary, 2009; see also Section 5.9).

*Palaeobathymetry* Benthic foraminifera are the principal group of use in palaeobathymetric interpretation in the region. The following bathymetric zones are generally distinguishable on the basis of their benthic foraminiferal assemblages:

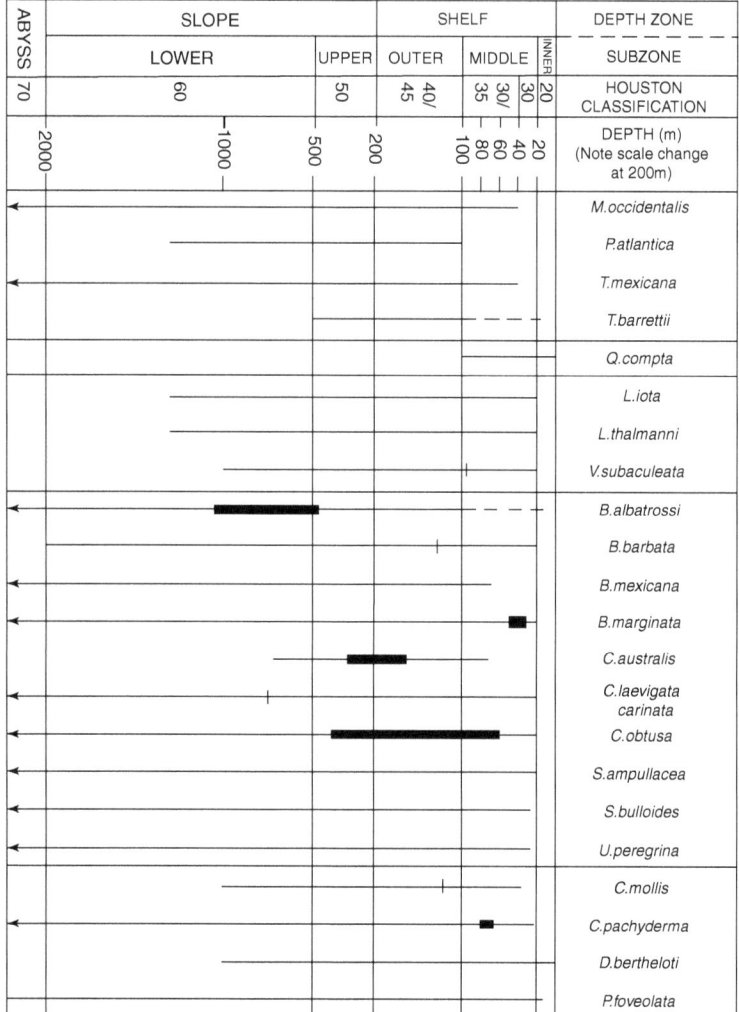

Fig. 9.16 **Bathymetric distribution of selected Foraminifera in the Gulf of Mexico**. *M. occidentalis* = *Martinottiella*; *P. atlantica* = *Pseudogaudryina*; *T. mexicana* = *Textularia*; *T. barretti* = *Textulariella*; *Q. compta* = *Quinqueloculina*; *L. iota, L. thalmanni* = *Lenticulina*; *V. subaculeata* = *Vaginulinopsis*; *B. albatrossi, B. barbata, B. mexicana* = *Bolivina* s.l.; *B. marginata* = *Bulimina*; *C. australis, C. laevigata carinata* = *Cassidulina*; *S. ampullacea* = *Siphouvigerina*; *S. bulloides* = *Sphaeroidina*; *U. peregrina* = *Uvigerina*; *C. mollis, C. pachyderma* = *Cibicidoides*; *D. bertheloti* = *Discorbinella*; *P. foveolata* = *Planulina*.

neritic (0–600 feet, typically associated with the surface mixed layer (SML) and Gulf/Loop Water (GLW)); upper bathyal (600–1500′, typically associated with Tropical Atlantic Central Water (TACW)); middle bathyal (1500–3000′, typically associated with Antarctic Intermediate Water (AIW)); lower bathyal (3000–6000′, typically associated with Caribbean Midwater (CMW)); and abyssal (> 6000′, typically associated with Gulf Basin Water (GBW)). The neritic zone is distinguished by *Eponides, Planulina, Pullenia, Sigmoilopsis, Melonis, Oridorsalis, Anomalinoides, Glomospira, Hoeglundina, Sphaeroidina, Uvigerina* etc.; the upper bathyal zone by *Bolivina, Bulimina, Chilostomella, Alabaminoides, Hansenisca* ['*Gyroidina*'], *Recurvoidella, Gavelinopsis, Valvulineria, Globobulimina, Lenticulina, Bathysiphon, Haplophragmoides, Osangulariella, Amphicoryna, Cassidulinoides, Ehrenbergina, Fursenkoina, Ammodiscoides, Ammodiscus, Cibicidoides, Veleroninoides, Eggerella, Karrerulina, Laticarinina, Reophax, Reticulophragmium, Tosaia* etc.; the middle bathyal zone by *Osangularia, Gaudryina, Recurvoides,* Oolininae, *Tolypammina, Martinottiella, Pleurostomella* etc.; the lower bathyal zone by *Allomorphina, Ioanella, Nonionella, Siphotextularia, Heronallenia, Pseudotrochammina, Pyrgo, Adelosina, Francesita* etc.; and the abyssal zone by *Francuscia, Bolivinita* etc. (Fig. 9.16).

*Sedimentary environments*   Benthic foraminifera are of particular use in the discrimination of sedimentary sub-environments in marginal to shallow marine, peri-deltaic, and in deep marine, submarine fan environments and associated reservoirs.

In the case of the modern and ancient Mississippi, delta, channel, levee and interdistributary bay sub-environments are distinguishable on the basis of benthic foraminiferal assemblages or cross-plots of foraminiferal morphogroups (see, for example, Haman, in Verdenius *et al.*, 1983; Hiltermann & Haman, 1986). Note that some species have modified (either elevated or depressed) upper depth limits in peri-deltaic sedimentary regimes (Pflum & Frerichs, 1976).

In the case of associated submarine fans, turbiditic sub-environments are distinguished either by allochthonous foraminiferal assemblages or by autochthonous ones dominated by DWAFs; and hemipelagic sub-environments by DWAFs and calcareous benthics; (Kohl *et al.*, in Bouma *et al.*, 1985; Kohl *et al.*, in Anderson & Fillon, 2004; Fillon, in Demchuk & Gary, 2009; see also Section 5.4.3). Pelagic sub-environments on the adjoining basin-plain are characterised by planktics as well as benthics.

### Integrated studies

A number of integrated bio- and sequence-stratigraphic schemes for the Gulf of Mexico have been published (Beard *et al.*, 1982; Armentrout, in Anon., 1987; Lamb *et al.*, in Ross & Haman, 1987; Baum & Vail, in Wilgus *et al.*, 1988; Greenlee & Moore, in Wilgus *et al.*, 1988; Scott *et al.*, in Wilgus *et al.*, 1988; Boyd *et al.*, 1989; Galloway, 1989b; Armentrout *et al.*, in Armentrout & Perkins, 1990; Hill & Rosen, in Armentrout & Perkins, 1990; Martin *et al.*, in Armentrout &

Perkins, 1990; Mitchum *et al.*, in Armentrout & Perkins, 1990; Pacht *et al.*, in Armentrout & Perkins, 1990; Riese *et al.*, in Armentrout & Perkins, 1990; Sangree *et al.*, in Armentrout & Perkins, 1990; Shaffer, 1990; Smith *et al.*, in Armentrout & Perkins, 1990; Taylor, in Armentrout & Perkins, 1990; Tobias, in Armentrout & Perkins, 1990; Vail & Wornardt, in Armentrout & Perkins, 1990; Wornardt & Vail, in Armentrout & Perkins, 1990; Rosen & Hill, 1990; Armentrout, in Weimer & Link, 1991; Mancini & Tew, 1991; Aubry, 1993; Pulham, in Weimer & Posamentier, 1993; Scott, in Simo *et al.*, 1993; Alexander & Flemings, 1995; Mancini & Tew, in Berggren *et al.*, 1995; Xue & Galloway, 1995; Armentrout, in Howell & Aitken, 1996; Mancini *et al.*, 1996; Morton & Suter, 1996; Pacht *et al.*, 1996; Reymond & Stampfli, 1996; Jiang, 1998; Villamil *et al.*, 1998; Galloway, 2001; Williams *et al.*, 2001; Armentrout, 2002; Henz & Hingliu Zeng, 2003; Brown *et al.*, 2005; Mancini *et al.*, 2008; Madof *et al.*, 2009). The ages of sequence boundaries are constrained by conventional biostratigraphy, supplemented by GC. The ages of MFSs are also constrained by conventional biostratigraphy. MFSs are characterised micropalaeontologically by maximum flooding, as indicated by foraminiferal palaeobathymetry, and also by maxima in abundance and diversity (Shaffer, in Anon, 1987; Loutit *et al.*, in Wilgus *et al.*, 1988; Armentrout *et al.*, in Armentrout & Perkins, 1990; Hill & Rosen, in Armentrout & Perkins, 1990; Pacht *et al.*, in Armentrout & Perkins, 1990; Vail & Wornardt, in Armentrout & Perkins, 1990; Wornardt & Vail, in Armentrout & Perkins, 1990; Rosen & Hill, 1990; Shaffer, 1990; Armentrout, in Weimer & Link, 1991; Pacht *et al.*, in Rhodes & Moslow, 1992). Associated – or non-associated – CSs are characterised lithologically, as a rule, in the case of shale-rich CSs (SRCSs), by gamma-log peaks, and as an exception, in the case of carbonate-rich CSs (CRCSs), by gamma-log troughs (Crews *et al.*, 2000). Systems tracts are characterised micropalaeontologically by palaeobathymetric trends: TSTs by upward deepening; HSTs by upward shallowing.

Palaeogeographic and facies maps of sequences have also been published (Bertagne, 1984; Riese *et al.*, in Armentrout & Perkins, 1990; Armentrout, in Weimer & Link, 1991; see also Fang, in Olson & Leckie, 2003). Source and/or reservoir distributions can be predicted from maps such as these.

## 9.4 Applications in reservoir exploitation

In reservoir exploitation, micropalaeontology assists specifically in the description and prediction of reservoir-rock distributions in time and in space, in a high-resolution sequence stratigraphic framework (Jones, 1996; Jones, in Jones & Simmons, 1999; Jones *et al.*, in Bubik & Kaminski, 2004; Jones, 2011a; see also the further reading list at the end of the chapter).

Note that micropalaeontological samples are acquired at a higher resolution in reservoir exploitation than would be the case in routine exploration. The fossil groups used will vary depending on the reservoir age and facies. Non-biostratigraphic technologies are applicable in non-fossiliferous reservoir ages and facies.

### 9.4.1 Integrated reservoir characterisation

In this, micropalaeontology plays a key role in the absolute chronostratigraphic age-dating and palaeoenvironmental interpretation of samples acquired during appraisal, development and production drilling. In the Middle East, foraminiferal and algal microfacies assists in the identification of reservoir 'sweet spots', there being in the Early Cretaceous Thamama group a link between depths of depositional environment of the order of 30–50 m, the occurrences of particular Foraminifera; and, importantly, of the Alga *Bacinella/Lithocodium*, characterised by a porous structure, and the optimal development of reservoir facies (Jones, 1996; Jones *et al.*, in Bubik & Kaminski, 2004; Jones, 2011a; see also Section 9.3.1 above, and the further reading list at the end of the chapter). In the North Sea, agglutinating foraminiferal microfacies further assist in the identification of stratigraphic heterogeneity and compartmentalisation, and of potential barriers and baffles to fluid flow, in deep marine, submarine fan reservoirs (Jones, in Jones & Simmons, 1999; see also Section 5.9 and 9.3.3, the further reading list at the end of the chapter, and Sections 9.5.4 and 9.7.3).

In the case of the Palaeogene reservoir in Fleming field in the North Sea, micropalaeontology has impacted not only integrated reservoir characterisation, but also the reservoir simulation model (see also Section 9.1.3, and the further reading list at the end of the chapter). Even more importantly, it has also enabled the refinement of the model so as to better match production history, and the revision of the production strategy and well planning programme.

## 9.5 Case studies of applications in reservoir exploitation

These case studies have been selected so as to include a reasonable representation of geological settings, reservoir ages and facies, and analytical techniques, with two examples from marginal to shallow marine, peri-deltaic, clastic reservoirs, and one each from shallow marine, platform carbonate, and deep marine, submarine fan clastic reservoirs (Jones, in Jones & Simmons, 1999; Jones *et al.*, in Jones & Simmons, 1999; Jones *et al.*, in Bubik & Kaminski, 2004; Jones, 2011a; see also the further reading list at the end of the chapter).

### 9.5.1 *Reservoir characterisation, marginal to shallow marine, peri-deltaic clastic reservoirs – Gullfaks, Snorre and Statfjord fields, Norwegian sector, and Ninian and Thistle fields, UK sector, North Sea*

Gullfaks, Snorre and Statfjord fields are located in the Norwegian sector of the North Sea, and Ninian and Thistle fields in the UK sector (Jones, 2011a; and additional references cited therein). They all produce from Jurassic Brent group marginal to shallow marine, peri-deltaic clastic reservoirs.

*Reservoir biostratigraphy.* Reservoir biostratigraphy is based on spore and pollen – and dinoflagellate – palynostratigraphy. The entire reservoir section appears to be of late Early?–Middle Jurassic age.

*Palaeobiology/palaeoenvironmental interpretation.* Palaeoenvironmental interpretation of the reservoir is based on integration of biofacies, that is, palynofacies and microfacies, and core, log and seismic facies. Palynofacies analysis has led to the recognition of a total of twelve reservoir facies in the Brent group reservoir in the Ninian and Thistle fields.

Microfacies analysis has led to the distinction of a range of peri-deltaic sub-environments in the Gullfaks, Snorre and Statfjord fields, and sub-regionally. Vegetated delta-top and to a lesser extent delta-front sub-environments have been distinguished on cross-plots of epiphytic, other epifaunal, and infaunal morphogroups of agglutinating Foraminifera on the basis of significant incidences of the epiphytic morphogroup, and prodelta sub-environments by significant incidences of the other epifaunal and infaunal morphogroups.

*Integration.* Integration of biostratigraphy and palaoenvironmental interpretation, with core, log and seismic data, has enabled the construction of time-slice palaeogeographic and facies maps that have proved of considerable use in sub-regional-scale mapping of peri-deltaic reservoir fairways.

### 9.5.2 *Reservoir characterisation, marginal to shallow marine, peri-deltaic clastic reservoir – Pedernales field, Venezuela, northern South America*

Pedernales field is located in the wetlands of the Orinoco delta some 100 km north-north-east of Maturin in the north-easternmost part of the eastern Venezuelan basin, sometimes referred to as the Maturin sub-basin (Jones *et al.*, in Jones & Simmons, 1999; Jones, 2011a; and additional references cited therein). It produces from a Pliocene La Pica formation, Pedernales member marginal to shallow marine, peri-deltaic clastic reservoir (and also from a Miocene Amacuro member deep marine, submarine fan clastic reservoir).

*Reservoir biostratigraphy.* The stratigraphic layering of the reservoir is based on integration of biostratigraphy, and core, log and seismic stratigraphy, in a work-station environment. Historically, the reservoir biostratigraphy was based mainly on qualitative micropalaeontology. More recently, it has come to be based more on integrated high-resolution quantitative micropalaeontology, nannopalaeontology, and spore and pollen, and to a lesser extent dinoflagellate, palynostratigraphy (although note that the palynostratigraphic data is of questionable usefulness on account of contamination of palynological samples by drilling fluid formulated from local river water and mud, and containing modern mangrove pollen and other palynomorphs effectively indistinguishable from their ancient counterparts in the subsurface). The formal reservoir biozonation has had to be based essentially on ecologically controlled benthic Foraminifera, with questionable regional chronostratigraphic and correlative significance, because of the rarity in the predominantly marginal marine facies of the reservoir of the open oceanic planktic foraminifera and calcareous nannoplankton that form the principal bases of global standard biozonations.

The zonation has been supplemented by an informal 'event stratigraphy' based on abundance and diversity peaks and troughs, biofacies and bathymetric trends etc. Abundance and diversity peaks have been used in the identification of sequence stratigraphically significant CSs in the same way that they have in Palo Seco field in south-western Trinidad.

*Palaeobiology/palaeoenvironmental interpretation.* Palaeoenvironmental interpretation of the reservoir is based on integration of biofacies, that is, microfacies and to a lesser extent palynofacies, and ichnofacies, and core, log and seismic facies. Microfacies analysis has led to the distinction of a range of delta-top, delta-front and prodelta environments with modern and/or ancient analogues elsewhere in the Orinoco system. Within the area of influence of the Orinoco, fluvially dominated delta-top environments have been distinguished on the basis of the *Centropyxis* microfacies; tidally influenced delta-top environments on the basis of the *Miliammina* and *Trochammina* microfacies; delta-front environments on the basis of the *Buliminella* microfacies, and *Skolithos* ichnofacies with *Ophiomorpha* and *Thalassinoides*; proximal prodelta environments on the basis of the *Eggerella* microfacies; and distal prodelta environments on the basis of the *Glomospira, Alveovalvulina/Silicosigmoilina* and *Cyclammina* microfacies.

*Integration.* Integration of reservoir bio- and sequence-stratigraphy and palaoenvironmental interpretation, with core, log and seismic data, has enabled the construction of time-slice palaeogeographic and facies maps that have proved of considerable use in field- to sub-regional-scale mapping of peri-deltaic reservoir fairways.

### 9.5.3 *Reservoir characterisation, shallow marine, peri-reefal carbonate reservoir – Al Huwaisah, Dhulaima, Lekhwair and Yibal fields, Oman, and Shaybah field, Saudi Arabia, Middle East*

Al Huwaisah, Dhulaima, Lekhwair and Yibal fields are located in Oman in the Middle East, and Shaybah field in Saudi Arabia, on the border with the United Arab Emirates (Jones *et al.*, in Bubik & Kaminski, 2004; Jones, 2011a; and additional references cited therein; see also the further reading list at the end of the chapter). They all produce from Early Cretaceous Thamama group, Shu'aiba formation shallow marine, peri-reefal carbonate reservoirs deposited on the margin of the 'Bab' intra-shelf basin (the reefs are constructed out of rudist bivalves, which are recognisable in cores, in computed tomography (CT) scans of cores, and on image logs).

*Al Huwaisah, Dhulaima, Lekhwair and Yibal fields    Reservoir biostratigraphy.* In the case of the Al Huwaisah field of central Oman, the reservoir can be dated to Early Aptian on the basis of the presence of *Mesorbitolina lotzei*, while in the cases of the Yibal, Dhulaima and Lekhwair field of northern Oman, the upper part of the reservoir can be dated to Late Aptian on the basis of the presence of *M. texana*.

   *Palaeobiology/palaeoenvironmental interpretation.* The observation that Shu'aiba formation reservoir deposition persisted into the Late Aptian in northern but not in central Oman is consistent with the emerging model of a relatively deep (Bab) intra-shelf basin in the northern area and emergence – through eustatic sea-level fall – of the formerly shallow-marine platform of the central area at this time.

   *Integration.* This is important from the petroleum geological point of view because the porosity and hence the quality of the Shu'aiba formation reservoir was enhanced – through subaerial exposure and leaching – only on the margin of the basin.

*Shaybah field*    Because the Shaybah field is located in the sand sea of the Rub' Al-Khali, romantically known as the 'Empty Quarter', it is extremely difficult to acquire seismic data over it. Reservoir characterisation is therefore more than usually reliant on other types of subsurface data such as core and log data. To assist with the process, some 17 000 feet (approximately 5200 m) of conventional core has been cut from over 50 wells, and semi-quantitative micro- and macro-palaeontological analysis has been undertaken on a substantial number of core samples.

*Reservoir biostratigraphy.* Reservoir biostratigraphy is based on calcareous nannoplankton, LBFs and rudist bivalves. The entire reservoir section appears to be of Aptian age.

*Palaeobiology/palaeoenvironmental interpretation.* Palaeoenvironmental interpretation of the reservoir is based on calcareous Algae, Foraminifera and rudist bivalves. A total of ten micropalaeontological biofacies have been recognised, and named after their dominant components. These are, more or less in order from distal to proximal: *Hedbergella* biofacies (ShBf-1), interpreted as basinal; *Hedbergella-Palorbitolina* biofacies (ShBf-2), interpreted as fore-reefal to basinal; *Praechrysalidina* biofacies (ShBf-3), interpreted as distal fore-reefal (and deep back-reef lagoonal); *Rotalia* biofacies (ShBf-4), interpreted as proximal fore-reefal; *Offneria* biofacies (ShBf-5), interpreted as peri-reefal; *Offneria-Oedomyophorus* biofacies (ShBf-6), interpreted as peri-reefal; *Glossomyophorus* biofacies (ShBf-7), interpreted as shallow back-reef lagoonal to peri-reefal; *Lithocodium* biofacies (ShBf-8), interpreted as shallow back-reef lagoonal; *Agriopleura* biofacies (ShBf-9), interpreted as deep back-reef lagoonal; and *Horiopleura* biofacies (ShBf-10), interpreted as fore-reefal.

*Integration.* Integration of core micro- and macro-palaeontological data and interpretation with core petrographic and sedimentological data and interpretations and wireline log petrophysical data has proved critical in developing reservoir layering and architecture schemes. Four stratigraphic subdivisions or layers of the reservoir have been recognised, namely, in ascending stratigraphic order: lower (Sh-1), interpreted as 'early'–'middle' Aptian; middle (Sh-2), interpreted as 'early'–'middle' Aptian; upper (Sh-3), interpreted as 'early'–'middle' Aptian; and, locally, onlapping the flanks of the field, and unrepresented on the crest, uppermost (Sh-4), interpreted as 'late' Aptian. Within each reservoir layer, a number of architectural elements have been recognised that appear to be related to the biofacies outlined above. Within the lower Shu'aiba reservoir layer (Sh-1), the interpreted distal *Hedbergella-Palorbitolina* biofacies (ShBf-2) and *Praechrysalidina* biofacies (ShBf-3) are developed over essentially the entire field; indicating transgression. Within middle Shu'aiba layer Sh-2, the interpreted peri-reefal *Offneria* biofacies (ShBf-5), *Offneria-Oedomyophorus* biofacies (ShBf-6) and *Glossomyophorus* biofacies (ShBf-7) are widely developed especially over the central part of the field; the interpreted shallow back-reef lagoonal *Lithocodium* biofacies (ShBf-8), only over the south-western part; and the interpreted fore-reefal to basinal *Hedbergella* biofacies (ShBf-1), *Hedbergella-Palorbitolina* biofacies (ShBf-2) and *Rotalia* biofacies (ShBf-4), only over the north-eastern part; indicating differentiation. Within upper Shu'aiba layer Sh-3, the interpreted deep back-reef lagoonal *Agriopleura* biofacies (ShBf-9) is developed over essentially the entire field; indicating regression.

### 9.5.4 *Reservoir characterisation, deep marine, submarine*
### *fan reservoir – Forties field, UK sector, North Sea*

Forties field is located in 104–128 m of water, some 180 km east-north-east of Aberdeen in licence blocks 22/10 (main part of field) and 22/6a (south-eastern extension) in the UK sector of the North Sea (Jones, in Jones & Simmons, 1999; Jones, 2011a; and additional references cited therein; see also the further reading list at the end of the chapter). It produces from a Palaeocene Montrose group, Forties formation deep marine, submarine fan reservoir (and also from an Eocene Rogaland deep marine, submarine fan group reservoir).

*Reservoir biostratigraphy.* The stratigraphic layering of the reservoir is based on integration of biostratigraphy, and core, log and seismic stratigraphy, in a work-station environment. Historically, the reservoir biostratigraphy was based mainly on semi-quantitative diatom and benthic foraminiferal micropalaeontology and dinoflagellate palynostratigraphy. More recently, it has come to be based more on high-resolution semi-quantitative to quantitative micropalaeontology and spore and pollen palynostratigraphy (see the further reading list at the end of the chapter).

The resolution of the reservoir biostratigraphy is now sufficiently high to demonstrate that historical layering models were wrong, and require righting as a matter of import, in view of the implications for the exploitation of the still substantial remaining reserves in the field. Notably, palynostratigraphy is now being applied at well-site for the first time, partly in order to direct drilling up through the reservoir to access previously bypassed attic oil from beneath; and partly in order to terminate drilling before penetrating the – incompetent – cap-rock (see the further reading list at the end of the chapter). Also notably, this palynos-tratigraphic work is being undertaken utilising green, non-acid well-site processing techniques.

*Palaeobiology/palaeoenvironmental interpretation.* Palaeoenvironmental inter-pretation of the reservoir is based on integration of biofacies, that is, microfacies and generally to a lesser extent palynofacies, and ichnofacies, and core, log and seismic facies. Palaeobathymetric interpretation based on the documented bathy-metric ranges of those species of agglutinating Foraminifera that occur in the reservoir and also occur in the Recent indicates that the reservoir was deposited in a deep marine, hence submarine fan, environment (Jones, 1994). Microfacies analysis has led to the distinction of a range of submarine fan sub-environments with outcrop analogues in the Austrian and Swiss Alps, Polish Carpathians and Spanish Pyrenees (Jones *et al.*, in Powell & Riding, 2005). Turbiditic channel axis sub-environments have been distinguished on cross-plots of morphogroups of epifaunal suspension-feeding, other epifaunal, and infaunal morphogroups of agglutinating Foraminifera on the basis of significant incidences of epifaunal suspension-feeding morphogroup A; inter-turbiditic channel off-axis sub-environments on the basis of significant incidences of other epifaunal morphogroup B; and hemipelagic channel levee and overbank to basin-plain sub-environments on the basis of significant incidences of infaunal morphogroup C (Jones & Charnock, 1985).

Interestingly, the palaeoenvironmental interpretation of the upper part of the reservoir, which is characterised by a restricted microfacies, is based as much if not more on palynofacies than on microfacies. The physics or chemistry of the benthic environment represented by the upper part of the reservoir evidently deteriorated initially to the point at which it was only habitable by opportunistic epifaunal Foraminifera, and subsequently to the point at which it was no longer habitable by Foraminifera. Concomitantly, the pelagic environment locally deteriorated to the point at which it was only habitable by opportunistic dinoflagellates.

*Integration.* Integration of reservoir bio- and sequence-stratigraphy and palaoenvironmental interpretation, with core, log and seismic data, has enabled the construction of time-slice palaeogeographic and facies maps that have proved of considerable use in field-scale mapping of the reservoir.

Time-lapse three-dimensional, or four-dimensional, seismic techniques are currently being used to track production through time, to identify unswept oil, and to gear the future infill injection and production drilling programmes to maximise sweep efficiency.

## 9.6 Applications in well-site operations

The most important application of micropalaeontology in well-site operations is in reservoir exploitation, in 'biosteering' (Jones *et al.*, 1999, 2002; Payne *et al.*, 2003; Jones *et al.*, in Koutsoukos, 2005; Jones, 2011a; and additional references cited therein; see also Sections 9.6.1 and 9.7.1–9.7.3, and the further reading list at the end of the chapter). Other applications, in petroleum exploration or reservoir exploitation, include monitoring of stratigraphic position while drilling, determination of casing, coring and terminal depths, and post-well analysis and auditing.

*Monitoring of stratigraphic position while drilling.* Monitoring while drilling exploration wells is seldom called for nowadays other than in frontier – 'wildcat' – areas.

*Determination of casing, coring and terminal depths.* Casing, coring and terminal depths are normally determined prior to drilling, with appropriate input from the micropalaeontologist. In many cases the depths are derived from two-way time to a seismic horizon, and may be inaccurate owing to uncertainty as to which is the most appropriate velocity function to use to convert time to depth. Well-site micropalaeontology provides an accurate determination, effectively in real time. It therefore enables decisions to be made also effectively in real time, thereby saving significant amounts of time and money. It can also have important safety implications, for example in the management of high-pressure, high-temperature or HPHT wells (see the further reading list at the end of the chapter).

*Post-well analysis and auditing. Post-well analysis* generally aims to provide a stratigraphic breakdown of the well within six weeks of the receipt of the final sample. Analytical data can be entered into software packages for the purposes of interpretation and integration. *Auditing* evaluates the technical and commercial success of the drilling. One of the most important parts involves the comparison of the prognosed well stratigraphy with the actual. This comparison is generally made by means of overlaying a synthetic seismogram of the well at the appropriate location, and transferring the actual well stratigraphy onto the seismic grid. This in turn helps to refine the predictive stratigraphic model.

### *9.6.1 Biosteering*

Biosteering involves real-time monitoring, through the well-site application of biostratigraphic technology, of the precise position relative to the reservoir in a – deviated – well, and, as necessary, for example when encountering a sub-seismic fault or a problem with seismic depth conversion or survey data, issuing instructions to redirect the well trajectory in order to ensure optimal well placement and reservoir penetration (Jones *et al.*, 1999, 2002; Payne *et al.*, 2003; Jones *et al.*, in Koutsoukos, 2005; Jones, 2011a; and additional references cited therein; see also the further reading list at the end of the chapter).

The high resolution of the technique (its 'window'), usually established by analyses of closely spaced samples from offset wells or a pilot hole, is often of the order of only a few feet or metres. The technique is, and will remain, critical to the exploitation of many petroleum reservoirs. The value added to date through the application of biosteering in BP runs into hundreds of millions of dollars. It is anticipated that this figure will further increase in the future, as the technology is transferred to fields in areas only now entering into production.

### 9.7 Case studies of applications in well-site operations

These case studies have been selected so as to include a reasonable representation of geological settings, reservoir ages and facies, and analytical techniques (Jones *et al.*, 1999, 2002; Payne *et al.*, 2003; Jones *et al.*, in Koutsoukos, 2005; Jones, 2011a; and additional references cited therein; see also the further reading list at the end of the chapter).

### 9.7.1 Biosteering, shallow marine, carbonate reservoir – Sajaa field, United Arab Emirates, Middle East

Sajaa field is located in the frontal thrust-sheets of the Oman Mountains, some 25 miles east of Sharjah town in Sharjah in the United Arab Emirates (Jones *et al.*, in Koutsoukos, 2005; Jones, 2011a; and additional references cited therein; see also the further reading list at the end of the chapter). It produces from Early Cretaceous, Barremian–Aptian Thamama group, Kharaib and Shu'aiba formation shallow marine, platformal to peri-reefal carbonate reservoirs. Production is through horizontal wells. Rates are of the order of 60 000 barrels per day.

A micropalaeontological biostratigraphic study enabled the reservoir interval to be divided into 21 zones of the order of a few tens of feet thick. Detailed palaeobathymetric interpretation of closely spaced samples has contributed significantly to the understanding of parasequence-scale reservoir facies and architecture, and to the identification of reservoir sweet spots. Subsequent biosteering – by means of thin-section micropalaeontology and microfacies – has kept the well bore within optimal parts of the reservoir over distances of several thousand feet (biosteering of multilaterals has also been possible). It has effectively replaced 'geosteering' that used coherency, which worked well in unfaulted but not in faulted sections, the approach taken on encountering a fault being to steer upwards to a known point in the stratigraphy and then back down again. This was not only time-consuming and expensive but potentially also dangerous, as it increased the risk of drilling out of the reservoir and into the seal, and causing the loss of control of the hole. Biosteering is currently being practised in conjunction with a combination of novel drilling technologies such as coiled tubing, slim-hole and under-balance (and also in conjunction with 'logging-while-drilling'). Even after the carbonate dissolution and sand-blasting effects associated with these drilling technologies, sufficient cuttings are still recoverable to allow the compositing of thin sections for micropalaeontological analysis. Note, though, that the drilling fluid has to be sufficiently buffered by the addition of alkaline substances to counter the acid dissolution (without damaging motors and seals).

Note that a similar biosteering methodology has been applied in the exploitation of the Late Permian Khuff formation reservoirs of various fields in Saudi Arabia (see the further reading list at the end of the chapter).

### 9.7.2 Biosteering, deep marine, carbonate reservoir – Valhall field, Norwegian sector, North Sea

Valhall field is located in the Norwegian sector of the North Sea (Jones *et al.*, in Koutsoukos, 2005; Jones, 2011a; and additional references cited therein; see also the further reading list at the end of the chapter). It produces from Late Cretaceous,

Maastrichtian Chalk group, Tor formation deep marine, autochthonous chalk (pelagic carbonates), allochthonous chalk, and chalky turbidite reservoirs. Production is through high-angle wells, of which some 50 have already been drilled. Rates are 105 000 barrels per day.

A micropalaeontological and nannopalaeontological biostratigraphic study enabled the reservoir interval to be divided into seven zones of the order of a few tens of feet thick. Subsequent biosteering – by means of micropalaeontology and nannopalaeontology – has targetted zones C and D – zones A and B possessing better reservoir properties in terms of porosity and permeability, but being unstable and prone to collapse under drawdown. One particularly successful well that was kept within zone D by biosteering produced 12 000 barrels per day. Other well-site applications of biostratigraphy include setting close to the base of the overburden without drilling overbalanced into the underpressured zone at the top of the reservoir, thereby causing formation damage; and terminating the well at the base of the reservoir, or 'biostopping'. Another application is identifying the origin of caved material and hence unstable zones in the tophole, impacting future well design. In terms of value added, 30% of the current production is attributed to optimal reservoir placement enabled by biosteering and associated technologies. Moreover, a cost saving of minimum of $1 000 000 per well – 7 days drilling, at $150 000 per day – is achieved by being able to set casing in the correct place by means of biostratigraphy.

Note that a similar biosteering methodology has been applied in the exploitation of the Chalk group reservoirs of other fields in the North Sea, for example Eldfisk and Arne (see the further reading list at the end of the chapter).

### 9.7.3 Biosteering, deep marine, clastic reservoir – Andrew field, UK sector, North Sea

Andrew field is located in the UK sector of the North Sea (Jones *et al.*, in Koutsoukos, 2005; Jones, 2011a; and additional references cited therein; see also the further reading list at the end of the chapter). It produces from Palaeocene, Montrose Chalk group, Andrew formation deep marine, submarine fan clastic reservoirs. The pay intervals are comparatively thin. Production is through horizontal wells. Rates are 64 000 barrels per day. Understanding of the reservoir facies and heterogeneities and consequences for fluid flow, and optimal placement of wells with respect to fluid contacts, are critical to maximisation of oil production prior to the inevitable early gas and/or water breakthough.

A micropalaeontological biostratigraphic study enabled the reservoir interval to be divided into seven zones of the order of a few tens of feet thick.

Microfacies, integrated with core sedimentology, was used to identify facies and heterogeneities with each zone, thereby establishing the spatio-temporal distribution of reservoir and non-reservoir units, and potential consequences for fluid flow. Mudstones A3 and A1 were interpreted as essentially hemipelagic on the basis of contained high abundance and diversity, low dominance assemblages of agglutinated Foraminifera further characterised by comparatively high incidences of complex infaunal morphogroup C of Jones & Charnock (1985). They were therefore also interpreted as potentially of regional or field-wide extent, and constituting barriers to fluid flow. This has subsequently been confirmed by pressure data. Mudstone A2 was interpreted as inter-turbiditic on the basis of contained low abundance and diversity, high dominance assemblages dominated by simple epifaunal morphogroups A and B. It was therefore also interpreted as of local extent, and constituting only a baffle to fluid flow. Subsequent biosteering – by means of agglutinated foraminiferal micropalaeontology and microfacies – has targetted reservoir unit B.

Fortuitously, the top of this unit is effectively coincident with the gas–oil contact over the crest of the field. Realisation of this fact has allowed the biosteered well-bore to be run more medially through the reservoir than in the initial well plan, with the overlying Mudstone A3 acting as a barrier to fluid flow and hence protecting it from gas invasion. It has been estimated that the optimal well placement and reservoir penetration enabled by the biosteering has added 10 million barrels of reserves to the books.

## 9.8 Unconventional petroleum geology

Unconventional petroleum geology is that in which the source-rock is also the reservoir (Jones, 2011a, b; and additional references cited therein; see also the further reading list at the end of the chapter). It covers coal-bed methane, or coal-seam gas, and shale gas. Gas is produced from coal-seam gas and shale gas reservoirs by first hydraulically fracturing them in order to improve their producibility. The 'fraccing' process can release gas into the subsurface, and raises important environmental concerns surrounding contamination of the subsurface water supply.

### 9.8.1 Shale gas

Shale is characterised by huge gas storage capacity (Jones, 2011a; and additional references cited therein; see also the further reading list at the end of the chapter). The storage capacity (porosity) of shale gas reservoirs appears to be in part controlled by the composition of the rock, and is highest in those rocks with the highest total organic carbon or TOC content, there being a positive correlation between TOC and gas sorption. The producibility (permeability) of shale gas reservoirs also appears to be in part controlled by the composition of the rock, and is highest in those rocks with the highest abiogenic or biogenic silica or calcite content, there being a positive correlation between silica or calcite content and brittle fracture propensity (note also that there is a positive correlation between TOC and biogenic silica and calcite content). The silica and calcite content of shale gas reservoirs is partly constituted of agglutinating Foraminifera in the case of the Middle Devonian Marcellus formation of the Appalachians of the north-eastern United States, and of calcareous benthic and planktic Foraminifera in the case of the 'Middle' Cretaceous Eagle Ford of Texas in the south-west (Jones, 2011b; see also the further reading list at the end of the chapter).

## 9.9 Micropalaeontology and Health, Safety and Environmental (HSE) issues in the petroleum industry

There are five principal areas in which micropalaeontology impacts on Health, Safety and Environmental (HSE) issues – or on which HSE issues impact on micropalaeontology – in the petroleum industry (Jones, 2006, 2011a; and additional references cited therein). These are: Environmental Impact Assessment (EIA); site investigation; pressure prediction; well-site operations; and environmental monitoring. Each is discussed in turn below.

### 9.9.1 Environmental Impact Assessment (EIA)

The living biota is of proven use in EIA in the petroleum industry, for example in baseline studies on the potential environmental impacts of drilling activities and infrastructure projects in and around British Petroleum's Prudhoe Bay and other oil fields on the North Slope of Alaska and Wytch Farm field under Poole Harbour in Dorset on the south coast of England, and in the Columbus basin off the east coast of Trinidad (see also Section 12.1).

### 9.9.2 Site investigation

Micropalaeontology has proved of use in site investigation in the Black Sea off Turkey, in the Caspian Sea off Azerbaijan and in the Mediterranean Sea off Egypt, specifically in helping to determine the safest sites for the locations of drilling rigs, pipelines and other facilities through the avoidance of geo-hazards (see also Section 11.1).

### *9.9.3 Pressure prediction*

Micropalaeontology provides critical inputs into pressure prediction from petroleum systems analysis and basin modelling (see also Section 9.1.1).

### *9.9.4 Well-site operations*

There are a number of areas in which micropalaeontology impacts HSE in the field of well-site operations, including not only making real-time casing and terminal depth – TD – calls, thus avoiding pressure kicks, but also providing an independent assessment of the condition and stability of the borehole, from cavings (see also Sections 9.1.4, 9.6 and 9.7).

### *9.9.5 Environmental monitoring*

The living biota is of proven use in environmental monitoring, for example in the monitoring of anthropogenically induced effects, including industrial pollution (see also Section 12.2).

### 9.10 Further reading

*Petroleum geology*

Goff *et al.*, in Al-Husseini, 1995; Ala, 1982; Altenbach *et al.*, 2012; Bkorlykke, 2010; Burdige, 2006; Carlsen & Ghori, 2005; Coburn & Yarus, 2000; Creaney & Passey, 1993; Grant *et al.*, in Dore & Sinding-Larsen, 1996; Emerson & Hedges, 2008; Gluyas & Swarbrick, 2004; Govindan, 2004; Gussow, 1954; Bohacs *et al.*, in Harris, 2005; Harris, 2005; Huc *et al.*, in Harris, 2005; Katz, in Harris, 2005; Tyson, in Harris, 2005; Tyson, in Hesselbo & Parkinson, 1996; Read & Horbury, in Horbury & Robinson, 1993; Copestake, in Jenkins, 1993; Jenkyns, 1980; Katz & Pratt, 1993; Kendall *et al.*, 2009; Klemme & Ulmishek, 1991; Klemme, in Magoon & Dow, 1999; Laudon, 1996; Martinez, 2003; McNeil *et al.*, 1996; Katz & Mello, in Mello & Katz, 2000; Millman & Farnsworth, 2011; Nummedal *et al.*, 2009; Passey *et al.*, 2010; Richardson & Rona, 1980; Bohacs *et al.*, in Roberts & Bally, 2012; Harris *et al.*, in Schlee, 1984; Schulz & Zabel, 2000; Selley, 1998; Summerhayes *et al.*, 1992; Sun, 1992; May *et al.*, in Vining & Pickering, 2010; Dunnington, in Weeks, 1958; Williams & Follows, 2011.

*Nummulite reservoirs*

Anketell & Mriheel, 2000; Anz & Ellouz, 1985; Beavington-Penney *et al.*, 2005; Carrillat *et al.*, 1999; Moody, in Hart, 1987; Jorry *et al.*, 2003; Macaulay *et al.*, 2001; Loucks *et al.*, in MacGregor *et al.*, 1998; Racey, 2001; Racey *et al.*, 2001; Bernasconi *et al.*, in Salem & Belaid, 1991; Vennin *et al.*, 2003.

*Applications in petroleum exploration*

Schenck, 1940; Fleisher & Lane, in Beaumont & Foster, 1999.

*Case studies of applications in petroleum exploration – Middle East*

Abdullah, 2001; Abdullah *et al.*, 1997; Abu Jaber *et al.*, 1989; Al-Jallal, in Al-Husseini, 1995; Hawas & Takezaki, in Al-Husseini, 1995; Al-Shamlan, 1980; Ala, 1982; Ala & Moss, 1979; Ala *et al.*, 1980; Aldabal & Alsharhan, 1989; Alkersan, 1975; Alsharhan, 1987, 1989; Alsharhan & Nairn, 1986, 1988, 1990, 1997; Alsharhan & Scott, 2000; Alsharhan *et al.*, 2001; Aqrawi *et al.*, 2010; ARAMCO, 1959; Ayres *et al.*, 1982; Best *et al.*, 1993; Beydoun, 1986, 1988, 1991, 1993; Bordenave & Burwood, 1990; Bordenave & Huc, 1995; BP, 1956; Burchette & Britton, in Brenchley & Williams, 1985; Stoneley, in Brooks, 1990; Stoneley, in Brooks & Fleet, 1987; Busnardo & Granier, 2011; Daniel, 1954; Droste, 1990; Dunnington, 1967; El-Naggar & El-Rifaiy, 1972, 1973; Forke *et al.*, 2010; Gaddo, 1971; Granier & Busnardo, 2013; Granier *et al.*, in Grosheny & Granier, 2011; Carmalt & St John, in Halbouty, 1984; Hughes Clarke, 1988; Ibrahim *et al.*, 2002; Immenhauser *et al.*, 1999; IPC, 1956; Jassim & Goff, 2006; Bordenave & Burwood, in Katz, 1994; Klemme & Ulmishek, 1991; Lehner *et al.*, 1985; Leturmy & Robin, 2010; Mitchell *et al.*, in Lomando & Harris, 1988; May, 1991; Metwalli *et al.*, 1974; Milner, 1998; Mitchell *et al.*, 1992; Murris, 1980a, b; Pitman *et al.*, 2004; Pittet *et al.*, 2002; Jordan *et al.*, in Roehl & Choquette, 1985; Sadooni, 1993; Harris *et al.*, in Schlee, 1984; Simmons *et al.*, 2007; Alsharhan *et al.*, in Simo *et al.*, 1993; Burchette, in Simo *et al.*, 1993; Sun, 1992; Terken, 1999; Vahren-kamp, 1996, 2010; Vahrenkamp *et al.*, 2012; van Buchem *et al.*, 1996, 2002a, b; Droste, in van Buchem *et al.*, 2010; Razin *et al.*, in van Buchem *et al.*, 2010; Sharp *et al.*, in van Buchem *et al.*, 2010; van Buchem *et al.*, in van Buchem *et al.*, 2010; Dunnington, in Weeks, 1958; Wilson, 1981, 1985.

*Indian subcontinent*

Ahmad & Ahmad, 2005; Ahsan *et al.*, 2012; Akhter *et al.*, 2007; Alam, 1989; Baillie *et al.*, 2006; Banerjee *et al.*, 2002; Bastia, 2006; Bastia & Nayak, 2006; Bastia *et al.*, 2007; Basu *et al.*, 1980; Biswas, 1982; Briggs 2003; Dolan, in Brooks, 1990; Brown & Dey, 1975; Copestake *et al.*, in Caughey *et al.*, 1986; Chatterjee *et al.*, 2013; Chowdhary, 2004; Daley & Alam, in Clift *et al.*, 2002; Smewing *et al.*, in Clift *et al.*, 2002; Compton, 2009; Curiale *et al.*, 2002; Dangwal *et al.*, 2008; Das *et al.*, 1999; Dhannawat & Mukherkjee, 1997; Dongil Kim *et al.*, 2012; Ellet *et al.*, 2005; Fazeelat *et al.*, 2010; Fitzsimmons *et al.*, 2005; Gakkhar *et al.*, 2012; Gombos *et al.*, 1995; Goswami *et al.*, 2007; Gupta, 2006; Rao & Talukdar, in Halbouty, 1980; Husain *et al.*, 2000; Imam, 2005; Imam & Hussain,

2002; Kassi *et al.*, 2009; Khan *et al.*, 1986; Biswas *et al.*, in Magoon & Dow, 1994; Mahanti *et al.*, 2006; Mahmud & Sheikh, 2012; Mohinuddin *et al.*, 1991; Mukherjee *et al.*, 2009; Nanda & Mathur, 2000; Negi *et al.*, 2006; Pande *et al.*, 2008; Pandey *et al.*, 2000; Quadri & Quadri, 1998; Quadri & Shuaib, 1986; Quallington, 2011; Raju, 1968, 2008; Raju *et al.*, 1999, 2005; Chandra *et al.*, in Raju *et al.*, 2005; Rao, 2001; Rao *et al.*, 1996; Rathore, 2007; Bhatnagar *et al.*, in Rathore, 2007; Raza *et al.*, 1990; Raza Khan *et al.*, 2000; Reddy *et al.*, 2005; Roychoudhury & Deshpande, 1982; Sheikh & Maseem, in Sahni, 1996; Sarma, 2012; Sastri *et al.*, 1973; Shuaib, 1982; Siddiqui, 2004; Singh *et al.*, 2008; Sinha *et al.*, 2012; Aswal *et al.*, in Swamy & Kapor, 2002; Gourshetty & Bhattacharya, in Swamy & Kapoor, 2002; Satyanarayana *et al.*, in Swamy & Kapoor, 2002; Singh *et al.*, in Swamy & Kapoor, 2002; Thomas *et al.*, in Swamy & Kapoor, 2002; Vyas *et al.*, in Swamy & Kapoor, 2002; Tozer *et al.*, 2008; Vaidyanadhan & Ramakrishnan, 2008; Wandrey, 2004; Wandrey *et al.*, 2004; Warwick *et al.*, 1998; Naini & Talwani, in Watkins, 1982; Watkinson *et al.*, 2007; Williams, 1958; Young Hoon Chung *et al.*, 2012; Zaigham & Mallick, 2000; Zutshi & Panwar, 1997.

### North Sea

Abbots, 1991; Johnson & Stewart, in Brenchley & Williams, 1985; Brooks & Glennie, 1987; Dore & Vining, 2005; Evans *et al.*, 2003; Glennie, 1998; Gluyas & Hichens, 2002; Sarg & Skjold, in Halbouty, 1982; Hardman, 1993; Hardman & Brooks, 1990; Pancost *et al.*, in Harris, 2005; Hart & Fitzpatrick, 1995; Tyson, in Hesselbo & Parkinson, 1996; Miller, in Huc, 1990; Illing & Hobson, 1981; Copestake, in Jenkins, 1993; Kjennerud & Gillmore, 2003; Parker, 1993; Pearce *et al.*, 2010; Rawson & Riley, 1982; Shanmugam *et al.*, 1995; Conort, in Spencer, 1986; Taylor, 1995; Tyson *et al.*, 1979; Underhill, 1998; Underhill & Partington, in Weimer & Posamentier, 1993; Vining & Pickering, 2010; Woodland, 1975; Ziegler *et al.*, 1997.

### South Atlantic

Guerra *et al.*, in Alsop *et al.*, 2012; Andresen *et al.*, 2011; Anka *et al.*, 2009, 2010; Burke *et al.*, in Arthur *et al.*, 2003; Hemsted, in Arthur *et al.*, 2003; Beglinger *et al.*, 2012; Brink, 1974; Brognon & Verrier, 1966; Mohriak *et al.*, in Brooks, 1990; Zimmermann *et al.*, in Brooks & Fleet, 1987; Burwood *et al.*, 1990, 1992; Cainelli & Mohriak, 1999; Burwood, in Cameron *et al.*, 1999; Coward *et al.*, in Cameron *et al.*, 1999; Schiefelbein *et al.*, in Cameron *et al.*, 1999; Cobbold *et al.*, 2001; Estrella *et al.*, in Demaison & Murris, 1984; Guardado *et al.*, in Edwards & Santogrossi, 1989; Fetter *et al.*, 2009; Cosmo *et al.*, in Foster & Beaumont, 1991; Clifford, in Halbouty, 1984; Hartwig *et al.*, 2012; Ho *et al.*, 2012; Hubbard, 1988;

Trindade *et al.*, in Katz, 1994; Mello *et al.*, in Katz & Pratt, 1993; Koutsoukos, 1982; Koutsoukos *et al.*, 1990; MacGregor, 2010; Magniez-Jannin & Jacquin, 1988; Mansurbeg *et al.*, 2012; Meisling *et al.*, 2001; Cole *et al.*, in Mello & Katz, 2000; Guardado *et al.*, in Mello & Katz, 2000; Katz & Mello, in Mello & Katz, 2000; Liro & Dawson, in Mello & Katz, 2000; Schoellkopf & Patterson, in Mello & Katz, 2000; Mello *et al.*, in Mohriak *et al.*, 2012; Moreira *et al.*, 2007; Mutti & Carminatti, 2012; Nairn & Stehli, 1973; Ojeda, 1982; Peres, 1993; Dickson *et al.*, in Post *et al.*, 2005; Reyre, 1966; Magnavita *et al.*, in Roberts & Bally, 2012; Mohriak & Fainstein, in Roberts & Bally, 2012; Schluter, 2006; Ala & Selley, in Selley, 1997; Seranne & Anka, 2005; Viana *et al.*, in Viana & Rebesco, 2007.

## Eastern Venezuelan basin

Bartolini *et al.*, 2003; Bermudez, 1962, 1966; Bermudez & Stainforth, 1975; Bingham *et al.*, 2012; Aymard, in Brooks, 1990; James, in Brooks, 1990; Cobos, 2005; de Romero & Galea-Alvarez, 1995; de Romero *et al.*, 2003; Donovan & Jackson, 1994; Audemard & Serrano, in Downey *et al.*, 2001; Erlich & Keens-Dumas, 2007; Escalona *et al.*, 2008; Eva *et al.*, 1989; Gonzalez de Juana *et al.*, 1980; Graham *et al.*, 1980; Hedberg, 1937; Higgins, 1996; Hoorn & Weselingh, 2010; Hoorn *et al.*, 1995; Hsu, 1999; Crespo de Cabrera *et al.*, in Huber *et al.*, 1999; James *et al.*, 2009; Kennan *et al.*, 2007; Lamb, 1964; Macellari & de Vries, 1987; di Croce *et al.*, in Mann, 1999; Mann & Escalona, 2010; Mann & Stein, 1997; Martinez, 1989; Martinez, 2003; Erlich & Barrett, in MacQueen & Leckie, 1992; Ministry of Energy and Energy Industries, 2009; Natural History Museum, Basel, 1996; Pindell & Perkins, 1992; Renz, 1962; Requejo *et al.*, 1994; Sydow *et al.*, in Roberts *et al.*, 2003; Rodrigues, 1993; Rohr, 1991; Ablewhite & Higgins, in Saunders, 1968; Lamb & Sulek, in Saunders, 1968; Metz, in Saunders, 1968; Stainforth, in Saunders, 1968; Stanley, 1960; Summa *et al.*, 2003; Gallango & Parnaud, in Tankard *et al.*, 1995; Kronman *et al.*, in Tankard *et al.*, 1995; Parnaud *et al.*, in Tankard *et al.*, 1995; Passalacqua *et al.*, in Tankard *et al.*, 1995; Tribovillard *et al.*, 1991; Wood, 2000; Woodside, 1981; Zapata *et al.*, 2003.

## Gulf of Mexico

Anderson & Fillon, 2004; Anon., 1989; Smith *et al.*, in Armentrout & Perkins, 1990; Taylor, in Armentrout & Perkins, 1990; Worrall & Snelson, in Bally & Palmer, 1989; Bartolini & Ramos, 2009; Bartolini *et al.*, 2003; Fails, in Brooks, 1990; Sassen, in Brooks, 1990; Carr *et al.*, 2009; Bowman *et al.*, in Demchuk & Gary, 2009; Fairbanks & Ruppel, 2012; Fillon *et al.*, 2001; Fritz *et al.*, 2000; Galloway, 1989b; Dolan, in Halbouty, 1984; Hentz, 1999; O'Neill *et al.*, in Jones &

Simmons, 1999; Mancini *et al.*, in Katz & Pratt, 1993; Kennan *et al.*, 2007; Madof *et al.*, 2009; Mancini & Puckett, 2005; Mancini *et al.*, 2012; McDonnell *et al.*, 2008; Miall, 2008; Moran-Zenteno, 1994; Murray *et al.*, 1985; Perkins & Martin, 1985; Pindell & Perkins, 1992; Post *et al.*, 2004, 2005; Pratson & Ryan, 1994; Reymond & Stampfli, 1996; Galloway, in Roberts & Bally, 2012; Miall & Blakey, in Roberts & Bally, 2012; Roberts *et al.*, 2003; Schofield *et al.*, 2008; Schumacher & Perkins, 1990; Steinhoff *et al.*, 2011; Weimer, 1990; Weimer & Link, 1991; Pulham, in Weimer & Posamentier, 1993; Weimer *et al.*, 1994; Scott *et al.*, in Wilgus *et al.*, 1988; Williams *et al.*, 2001; Zou *et al.*, 2012.

*Applications and case studies in reservoir exploitation*

Ainsworth *et al.*, 2000; Hughes, in Bubik & Kaminski, 2004; Cross *et al.*, 2010; Bowman *et al.*, in Demchuk & Gary, 2009; Dickinson *et al.*, 2001; Carter & Heale, in Gluyas & Hichens, 2003; Granier & Busnardo, 2013; Jolley *et al.*, 2010; Bergen & Sikora, in Jones & Simmons, 1999; Hill & Wood, 1980; Holmes, in Jones & Simmons, 1999; Jones, in Jones & Simmons, 1999; Mangerud *et al.*, in Jones & Simmons, 1999; Morris *et al.*, in Jones & Simmons, 1999; Payne *et al.*, in Jones & Simmons, 1999; Sikora *et al.*, in Jones & Simmons, 1999; Hughes, in Powell & Riding, 2005. Willliams *et al.*, in Powell & Riding, 2005; Racey *et al.*, 2001; Banner & Simmons, in Simmons, 1994; Slatt *et al.*, 2006; Wakefield *et al.*, 2001.

*Applications and case studies in well-site operations*

Shipp & Marshall, in Al-Husseini, 1995; Hart *et al.*, in Demchuk & Gary, 2009; Mears & Cullum, in Demchuk & Gary, 2009; Dickinson *et al.*, 2001; Hughes, 2010b; Holmes, in Jones & Simmons, 1999; Payne *et al.*, in Jones & Simmons, 1999; Shipp, in Jones & Simmons, 1999; Marshall, 2010; Marshall *et al.*, 2007; Yang-Logan *et al.*, 1996.

*Unconventionals*

Milliken, 2010; Ross & Bustin, 2009.

# 10

# Applications and case studies in mineral geology

This chapter deals with applications and case studies in mineral geology. It contains sections on applications and case studies in mineral exploration and in mineral exploitation. The section on mineral exploration includes a case study from the La Troya mine in Spain. The section on mineral exploitation includes case studies from the Pitstone quarry in Hertfordshire, and from the East Grimstead quarry in Wiltshire, both in the United Kingdom.

Foraminifera and other microfossils have proved of use in mineral exploration and exploitation (see, for example, Jones, 1996, 2006, 2011a; and additional references cited therein; see also the further reading list at the end of the chapter). Applications and case studies are given below. The two case studies of applications in mineral exploitation are both concerned with Late Cretaceous chalk, which is used in the cement and paper-whitening manufacturing industries. This reflects the peculiar problems faced in the detailed stratigraphic subdivision and surface and subsurface correlation of macroscopically essentially featureless chalky lithotypes, especially the identification of lithological and associated rock property changes associated with unconformities, faults and facies changes.

## 10.1 Application and case study in mineral exploration

### 10.1.1 La Troya mine, Spain

#### Material and methods

The study was undertaken in and around the Early Cretaceous, Barremian–Aptian rudist-reef-hosted lead–zinc ore body of the La Troya mine in the Basco–Cantabrian basin in the Basque country of Spain.

The primary objective was to collect samples from the site of the ore body itself and from – coeval – laterally equivalent sites at varying distances away, and to analyse them for their foraminiferal content. This was in order to test the hypothesis that foraminiferal assemblages from sites of ore-body deposition would be characterised by environmental stress, whereas those from laterally equivalent sites would not, enabling an 'environmental stress gradient', or, in three dimensions, a 'halo', to be defined, the likes of which might serve as pointers to the locations of hitherto undiscovered ore bodies elsewhere.

The secondary objective was to establish an ore-body depositional model (and to confirm the suspected syngenetic rather than epigenetic origin using palynological indications of thermal maturity above and below).

### Results and discussion

*'Environmental stress gradient'.* Unfortunately, in practice, it proved impossible to trace and sample the lateral equivalents of the ore body, owing in part to their extreme thinness (a function of the geochemically interpreted extremely short period of time over which the ore body formed), and in part to fault complications. The hypothesis of the environmental stress gradient thus remains effectively untrested either in the Basco–Cantabrian basin or elsewhere.

*Depositional model.* Foraminiferal analyses of the platform carbonates underlying the ore body indicated an Urgonian (late Barremian to early Aptian) age (and palynological analyses of the black pyritic shales ('margas negras') overlying it indicated a probable late Aptian age). Foraminiferal analyses of sediments adjacent, although, in view of the problems outlined above, not necessarily exactly laterally equivalent to, the orebody indicated a restricted marine environment, with only rare and stress-tolerant taxa present, including agglutinates and *Gyroidinoides*.

Palynological analyses yielded 'spore colour indices' in the range 5–7, indicating over-maturity. No conclusions could be drawn on the basis of this evidence with regard to the syngenetic or epigenetic origin of the ore body.

## 10.2  Applications and case studies in mineral exploitation

### 10.2.1  Pitstone quarry, Hertfordshire, United Kingdom

The Pitstone quarry in Hertfordshire extracts chalk for use in the cement manufacturing industry. A balance of high- and low-calcimetry chalks provides the optimum properties. At Pitstone, calcimetry is controlled primarily by stratigraphic horizon, being low in the lower part of the lower Chalk and *Plenus* marl and high in the upper part of the lower Chalk and the lower part of the middle Chalk. A micropalaeontological study was therefore undertaken with the aim of identifying the local extent of these horizons. Closely spaced samples from a series of boreholes were analysed and the sections subdivided and correlated using a high-resolution biostratigraphic zonation scheme. The data thus obtained were presented in the form of a series of profiles projecting calcimetry. These were utilised in formulating a pit design plan optimising the existing infrastructure and providing the most ergonomic and economic extraction of future supplies.

### 10.2.2  East Grimstead quarry, Wiltshire, United Kingdom

The East Grimstead quarry in Wiltshire extracts chalk for use in the paper-whitener manufacturing industry. 'Bright' chalks provide the optimum properties. At East Grimstead, 'brightness' is controlled by stratigraphic horizon, being highest in the *Offaster pilula* Sub-Zone of the upper Chalk. A micropalaeontological study was therefore undertaken with the aim of identifying the local extent of this horizon. Closely spaced samples from a series of boreholes were analysed and the sections subdivided and correlated using a high-resolution biostratigraphic zonation scheme. Results indicated that the zone of bright chalks only extended for a short distance to the south of the existing quarry works.

### 10.3  Further reading

Hart, in Jenkins, 1993.

# 11

# Applications and case studies in engineering geology

This chapter deals with applications and case studies in engineering geology. It contains sections on applications and case studies in site investigation and in seismic hazard assessment. The section on site investigation includes case studies from the Channel tunnel, the Thames barrier, and 'Project Orwell', all in the United Kingdom, and from Azerbaijan and Egypt. The section on seismic hazard assessment includes case histories from the Strait of Juan de Fuca, Vancouver Island, Canada, Iyo-nada Bay, Japan, and Hawke's Bay and Otago, New Zealand; and from Cabo de Gata Lagoon, Almeria, Spain.

Foraminifera and other microfossils have proved of use in engineering geology, in the fields of site investigation and of seismic hazard assessment (see, for example, Jones, 1996, 2006, 2011a; and additional references cited therein; see also the further reading list at the end of the chapter).

Applications and case studies are given below. Three of the four case studies of applications in mineral exploitation are concerned with Late Cretaceous chalk. As with the case studies in mineral exploitation (see Section 10.2), this reflects the peculiar problems faced in the detailed stratigraphic subdivision and correlation of macroscopically essentially featureless chalky lithotypes, especially the identification of lithological and engineering property changes associated with unconformities, faults and facies changes. Note in this context that calcareous nannofossil chalks have peculiar engineering properties, controlled primarily by their 'nannofabrics' (Bell, 2000). For example, some are peculiarly dense on account of the close packing of the subcubical calcareous nannofossil *Micula*.

## 11.1 Applications and case studies in site investigation

### 11.1.1 Channel tunnel, UK

The main application in this case was in the provision of precise and accurate stratigraphic control on the Albian Gault clay and Cenomanian lower Chalk

265

(Jones, 1996, 2006, 2011a; and additional references cited therein; see also the further reading list at the end of the chapter). Samples were taken with an average vertical spacing of as little of approximately 1 m from the appropriate stratigraphic section from a number of sites along a cross-channel transect. A high-resolution foraminiferal biostratigraphic zonation scheme was erected, the resolution and correlative value of which was enhanced by quantitative measures of percentages of planktic species and of planktic morphogroups (epipelagic *Hedbergella/Whiteinella*, mesopelagic *Dicarinella/Praeglobotruncana* and bathypelagic *Rotalipora*) (Carter & Hart, 1977; Hart *et al.*, in Jenkins & Murray, 1989). The resolution of the biozonation scheme was sufficient to allow recognition of local and regional onlap surfaces, as at the base of the lower Chalk, and erosional surfaces, as in the Middle Cenomanian part of the lower Chalk. Rock property changes associated with the east–west loss through onlap of the particularly clay-rich basal Chalk have important engineering implications.

### 11.1.2 Thames barrier, UK

A number of boreholes were drilled to investigate the rock properties of the upper Chalk in the subsurface in order to select a suitable founding level for the 3300-tonne moveable steel gates of the Thames barrier and their supporting concrete piers (Jones, 1996, 2006, 2011a; and additional references cited therein; see also the further reading list at the end of the chapter). Most of the variation in engineering properties as revealed by tests was in the vertical (stratigraphic) rather than horizontal sense. There was also some overprint associated with weathering, frost-shattering and solifluction. The principal objective of the parallel micropalaeontological study was therefore to provide a detailed correlation of different stratigraphic horizons and any offset by faults. A subsidiary aim was to distinguish solifluction from *in situ* chalk using the presence of exotic components (Tertiary sand grains etc.) introduced during movement. The sample spacing was again extremely close (of the order of 1 m). Again, a high-resolution foraminiferal biozonation scheme was erected (Bailey & Hart, in Wiedmann, 1979; Hart *et al.*, in Jenkins & Murray, 1989). The resolution was enhanced by separate plots, in the form of bar charts or kite diagrams, of selected superfamilies, and of the proportions of selected species or species groups within the superfamilies Cassidulinacea (*Globorotalites cushmani*, *Lingulogavelinella* aff. *vombensis*, other) and Globigerinacea (*Globotruncana bulloides/G. marginata*, *G. linneiana/G. pseudolinneiana*, *Hedbergella*). The resolution was sufficient to allow recognition not only of a widespread erosional unconformity of varying extent at the top of the Chalk, but also of cross-cutting faults. The engineering implications of rock property changes associated with faults at the foundation level were fully evaluated by engineering geologists prior to commencement of the construction operation.

### 11.1.3 Project Orwell, UK

As part of the recent 'Project Orwell', completed in 1999, a 5.5-km-long tunnel was constructed to reduce flooding in and around Ipswich in Suffolk (Jones, 2006, 2011a; and additional references cited therein; see also the further reading list at the end of the chapter). The tunnel had to be drilled between two resistant flint bands within the Campanian *Gonioteuthis quadrata* Zone of the Chalk, since encountering either would have constituted a significant hazard. The trajectory of the tunnel was successfully kept within this narrow 'window' by means of high-resolution micropalaeontology.

### 11.1.4 Site investigation in the petroleum industry – Azerbaijan and Egypt

In my working experience in the petroleum industry, micropalaeontology has proved of use in site investigation in the Black Sea off Turkey, in the Caspian Sea off Azerbaijan and in the Mediterranean Sea off Egypt, specifically in helping to determine the safest sites for the locations of drilling rigs, pipelines and other facilities through the avoidance of geo-hazards (Jones, 2011a).

#### Azerbaijan

In the Caspian Sea off Azerbaijan, high-resolution micro- and palyno-stratigraphy, essentially in the form of climatostratigraphy calibrated against the marine oxygen isotope record, has been used to date and correlate Quaternary sediments, and also to characterise potentially geo-hazardous sediments, in surface and shallow subsurface core samples (and high-resolution shallow seismic techniques have been used to map the potentially geo-hazardous sediments away from the areas of core control).

Micro- and palyno-facies has been used to characterise potentially geo-hazardous mud-volcano flows, on the basis of their non-contemporaneously reworked microfossil content.

#### Egypt

In the Mediterranean Sea off Egypt, high-resolution micro- and palyno-stratigraphy has been used, in conjunction with accelerator mass spectrometry, or AMS, and optically stimulated luminescence, or OSL, stratigraphy, to date and correlate Quaternary sediments, and also to characterise potentially geo-hazardous sediments, in surface and shallow subsurface core samples (and again high-resolution shallow seismic techniques have been used to map the potentially geo-hazardous sediments away from the areas of core control).

Micro- and palyno-facies stratigraphy has also been used to characterise potentially geo-hazardous land-slides on the basis of their contemporaneously transported microfossil content. It has also been used to demonstrate that land-slides significantly decreased in frequency between the 'early' and 'late' Holocene, and currently constitute a manageable geo-hazard. Palynofacies stratigraphy indicates that the 'early' Holocene was a time of humid climate and associated high sediment accumulation rate, and hence increased pore pressure, leading to increased land-slide frequency. In contrast, the 'late' Holocene was a time of arid climate, low sediment accumulation rate, and decreased pore pressure and land-slide frequency.

## 11.2 Applications and case studies in seismic hazard assessment

Foraminifera and other fossils have proved to be of use in seismic hazard assessment, specifically in palaeoseismology, that is, the study of the so-called 'ground effects' of past earthquakes, including vertical fault movements and tsunamis (Jones, 2011a; and additional references cited therein; see also the further reading list at the end of the chapter).

Case studies on fault movements in the Strait of Juan de Fuca, Vancouver Island, Canada, Iyo-nada Bay, Japan, and Hawke's Bay and Otago, New Zealand, and on a tsunami(te) in the Cabo de Gata Lagoon, Almeria, Spain are given below. Similar approaches have also been adopted elsewhere.

### 11.2.1 Strait of Juan de Fuca, Vancouver Island, Canada, Iyo-nada Bay, Japan, and Hawke's Bay and Otago, New Zealand

Rapid changes in bathymetry – of as little as 0.05–1.5 m – recorded by diatom and foraminiferal assemblages in dated marginal marine environments in the Strait of Juan de Fuca off Vancouver Island in Canada, in Iyo-nada Bay in Japan, and in Hawke's Bay and Otago in New Zealand, have been used to infer uplift associated with past earthquake activity (Jones, 2011a; and additional references cited therein; see also the further reading list at the end of the chapter). In the case of Hawke's Bay in New Zealand, they they have been used to infer uplift associated not only with the 1931 Napier earthquake (1.5 m), but also with previously unknown earlier earthquakes 600, 1600, 3000, 4200, 5800 and 7200 years before present, providing some measure of frequency probability, if not predictability.

### 11.2.2 Cabo de Gata Lagoon, Almeria, Spain

Admixtures of autochthonous marginal marine and allochthonous shallow to deep marine Foraminifera in dated marginal marine environments in the Cabo de Gata

Lagoon in Spain have been used to identify tsunamites and infer earthquake activity associated with the known 1522 Almeria earthquake (Jones, 2011a; and additional references cited therein; see also the further reading list at the end of the chapter).

## 11.3 Further reading

*Site investigation*

Hart *et al.*, in Demchuk & Gary, 2009; Hart, in Jenkins, 1993; Hart, in Martin, 2000; Wright, 2001.

*Seismic hazard assessment*

Atwater *et al.*, 2010; Goodman *et al.*, 2009a; Guidoboni & Ebel, 2009; Hayward *et al.*, 2006, 2008; Mamo *et al.*, 2009; Nixon *et al.*, 2009; Pilarczyk & Reinhardt, 2011; Reicherter & Becker-Heidmann, in Reicherter *et al.*, 2009 Reicherter *et al.*, 2009; Reinhardt *et al.*, 2006; Rothaus *et al.*, 2004; Sintubin *et al.*, 2010; Williams, 1999.

# 12

# Applications and case studies
# in environmental science

This chapter deals with applications and case studies in environmental science. It contains sections on applications and case studies in Environmental Impact Assessment (EIA), in environmental monitoring, and in anthropogenically mediated global change ('global warming'). The section on EIA includes a case study from the North Slope of Alaska. The section on environmental monitoring includes case studies on environmental monitoring of natural and anthropogenic effects on water quality, and on environmental monitoring of coral reef vitality. The section on anthropogenically mediated global change includes case studies on ocean acidification and on carbon dioxide sequestration.

Living Foraminifera and other micro- and macro-organisms, and dead Foraminifera and other micro- and macro-fossils, have proved of considerable use in environmental science, in EIA and in environmental monitoring, and in studies on anthropogenically mediated global change (see, for example, Jones, 1996, 2006, 2011a; and additional references cited therein; see also the further reading list at the end of the chapter).

Applications and case studies are given below.

## 12.1 Application and case study in Environmental Impact Assessment (EIA)

According to the Department of the Environment's definition, EIA is 'a process by which information about the environmental effects of a project is collected, both by the developer and other sources, and taken into account by the relevant decision-making body before a decision is given on whether the development should go ahead' (Jones, 2006, 2011a; and additional references cited therein; see also the further reading list at the end of the chapter).

Living Foraminifera and other micro- and macro-organisms have proved of use in EIA in the civil engineering industry, in baseline studies on the potential environmental impacts of industry projects, and the mitigation of these impacts, for example in the construction of the Rance barrage in France and the Severn barrage in Great Britain (see the further reading list at the end of the chapter).

They have also proved of use in EIA in the petroleum industry, in baseline studies on the potential environmental impacts of industry projects, and the mitigation of these impacts, for example on the North Slope of Alaska (see below), and in the Columbus basin off the east coast of Trinidad (author's unpublished observations).

### 12.1.1 Environmental Impact Assessment, North Slope, Alaska

Foraminifera have featured in baseline studies on the potential environmental impacts of industry projects on the 'Boulder Patch' kelp community in Stefansson Sound in the offshore area of the North Slope of Alaska, identified as highly sensitive to environmental disturbance (Dunton & Schonberg, in Truett & Johnson, 2000). *Cornuspira foliacea*, *C. involvens*, *C.* sp., *Elphidiella* sp., *Guttulina* sp., *Lagena* sp., *Miliolinella* sp., Nonionidae and *Triloculina* sp. have been recorded in dive samples from between and from under boulders.

## 12.2 Applications and case studies in environmental monitoring

### 12.2.1 Environmental monitoring of natural and anthropogenic effects on water quality

Living Foraminifera and other micro- and macro-organisms have proved of considerable use in environmental monitoring, in baseline studies on natural and artificial, anthropogenic effects on water quality, including domestic and industrial pollution and anthropogenically mediated eutrophication and anoxia, and also in remediation studies (Jones, 1996, 2006, 2011a; and additional references cited therein; see also the further reading list at the end of the chapter).

The effects of pollution on the foraminiferal microbiota vary considerably, but can include any or all of the following:

- modifications to the structure and composition of communities, including decreases in abundance and diversity, exclusion of certain species etc.;
- modifications to the reproductive cycles of individual species, including increases in incidences of asexually reproducing individuals;
- development of test deformities; and
- changes in shell chemistry, including increased absorption of metals, sulphur etc., and pyritisation;
- (all of which appear to become progressively more pronounced as the pollution continues)
- and, at least locally, death.

As far as can be ascertained from the available data, the effects appear to be more or less localised around the source of the pollution. Note, though, that the eutrophication associated with nitrogen-rich runoff from sugar-cane plantations, shrimp farms and sewage outlets in eastern South America is experienced at least as far afield as Trinidad in the eastern Caribbean, owing to the effect of the Guyana current (Wilson, 2006).

The effects appear to disappear more or less as soon as the source of the pollution disappears (although this would probably not be so in the case of radioactive pollution). In a recent well-documented case involving the disposal of oil-contaminated cuttings samples from oilfield operations offshore Congo, concentric zones of comparatively highly, moderately and lowly stressed environments were observed around the disposal site over a horizontal distance initially of 750 m, and four years subsequently of 250 m (Mojtahid *et al.*, 2006; see also Jorissen *et al.*, 2009; Bicchi *et al.*, 2010; Hess *et al.*, 2010). The highly stressed environments, observed initially between 0 and 70 m from the disposal site, are characterised by extremely impoverished benthic foraminiferal assemblages; the moderately stressed environments, observed initially 70–250 m from the disposal site, by locally enriched benthic foraminiferal assemblages containing high proportions (> 50%) of opportunistic – infaunal – species tolerant of eutrophication and/or dysoxia, including *Bolivina* spp., *Bulimina aculeata*, *B. marginata*, *Textularia sagittula* and *Trifarina bradyi*; and the lowly stressed environments, observed initially 250–750 m from the disposal site, by impoverished assemblages containing lower proportions (<50%) of opportunistic species. The proportion of opportunistic species appears to be a useful measure of the extent of anthropogenic as well as natural eutrophication.

### *European Water Framework Directive and European Marine Strategy Framework Directive*

Such studies will surely increase in importance in the future, not least in view of the recently issued European Water Framework Directive or EWFD, due to be implemented by 2015, and European Marine Strategy Framework Directive or EMSFD, due to be implemented by 2020 (see, for example, Barras *et al.*, 2010; Jorissen, 2010; Schonfeld, 2012; Schonfeld *et al.*, 2012). Preparatory work is already well underway on a range of FOraminiferal BIo-MOnitoring or FOBIMO initiatives, aimed at standardising foraminiferal sampling and preparation methods and taxonomy; on identifying 'sensitive', 'indifferent', 'tolerant', 'second-order opportunist' and 'first-order opportunist' indicator species; and on identifying 'high', 'good', 'average', 'poor' and 'bad' environmental quality categories on the basis of these indicator species (Schonfeld *et al.*, 2012). Potentially useful indices, actually already used in macrofaunal studies, include the AZTI Marine

Biotic Index or AMBI, the Benthic Quality Index or BQI, and the Infaunal Trophic Index or ITI (van Hoey *et al.*, 2010; Labune *et al.*, 2012).

### 12.2.2 Environmental monitoring of coral reef vitality

Living Foraminifera have also proved of use in studies on coral reef vitality, for example, in Florida and on Hawaii in the United States, on the Great Barrier Reef in Australia, and on the Abrolhos reef in Brazil (Jones, 2006, 2011a; and additional references cited therein; see also the further reading list at the end of the chapter). An index based on foraminiferal assemblages, the so-called FORaminifera in Assessment and Monitoring, or FORAM, Index, or FI, has been used to provide an indication of water quality and its ability to support healthy coral reefs (Hallock *et al.*, 2003). The FI is given by the formula: $FI = (10 \times Ps) + (Po) + (2 \times Ph)$; where Ps is the proportion of symbiont-bearing Larger Benthic Foraminifera or LBFs, such as are characteristic of natural, oligotrophic environments; Po is the proportion of opportunistic Foraminifera, such as are characteristic of naturally or artificially eutrophic environments, for example, those affected by nutrient-rich run-off associated with agricultural activities; and Ph is the proportion of other Foraminifera. The results of a recent study on the Low Isles Reef on the Great Barrier Reef in Australia revealed low FIs, and naturally or artificially eutrophic environments, unfavourable for coral reef vitality, only over a relatively small area of the reef, adjacent to a mangrove swamp providing a local, natural source of nutrient input in the form of leaf litter and bird guano from thousands of Torres Strait pigeons (Scheuth & Frank, 2008). Incidentally, this area had been essentially cleared of reef-forming corals by a particularly destructive cyclone in 1950, and subsequently recolonised by Algae and soft-bodied corals. Having visited the Low Isles Reef following Cyclone Joy in 1989, I can attest to the damage done to certain reef-forming corals, such as branching morphotypes of *Acropora*, by such natural events.

## 12.3 Applications and case studies in anthropogenically mediated global change ('global warming')

An Inter-governmental Panel on Climate Change (IPCC) concluded in 1996 that anthropogenic greenhouse gas emissions were responsible for the observed present, or rather recent, global warming and sea-level rise, as had long been suspected, and that action was required to limit such emission, the case and timetable for which was set out in the 'Kyoto Protocol' in 1997 (Jones, 2011a; and additional references cited therein; see also the further reading list at the end of the chapter). (Interestingly, anthropogenic greenhouse gas emissions and associated effects on the climate system have recently been shown to have begun in the pre-industrial rather than in the industrial era, with the clearance of forests for agriculture. Decreased anthropogenic greenhouse gas emissions associated with decreases in human populations during periods of pandemic in the pre-industrial era resulted in global cooling.)

Foraminifera and other micro- and macro-fossils provide a long-term record of global environmental and climatic change, including natural as well as anthropogenic global warming; and, importantly, they also provide pointers as to rates of change. For example, they indicate that the past natural global warming associated with events such as the Palaeocene–Eocene Thermal Maximum or PETM was relatively rapid, implying that the present anthropogenic global warming could be, too, and with similarly widespread consequences (Aubry *et al.*, 1998; Wing *et al.*, 2003; Sluijs *et al.*, in Williams *et al.*, 2007; Dunkley Jones *et al.*, 2010; Whidden & Jones, 2012; and additional references cited therein).

### *12.3.1  Ocean acidification*

Living Foraminifera and other calcareous marine micro- and macro-organisms have proved of use in studies on increasing ocean acidification in response to increasing, anthropogenically mediated atmospheric carbon dioxide concentration, the results of which suggest that their calcification, growth and even survival could be adversely affected if the oceans continue to acidify (Jones, 2011a; and additional references cited therein; see also the further reading list at the end of the chapter).

### *12.3.2  Carbon dioxide sequestration*

One of the proposed counters to recent global warming and associated adverse effects (see above) is carbon dioxide sequestration (Jones, 2011a; and additional references cited therein; see also the further reading list at the end of the chapter).

Living Foraminifera have proved of use in a recent study on carbon dioxide sequestration in a deep-sea reservoir in Monterey Bay in California (Ricketts *et al.*, 2009). They have also proved of use in laboratory studies, the results of which suggest that calcareous benthic species 'will face considerable challenges to maintain ... populations' in environments characterised by conditions of elevated carbon dioxide concentrations (Bernhard *et al.*, 2009).

### 12.4  Further reading

#### *General*

Barbieri *et al.*, 2006; Bennett & Doyle, 1997; Botkin & Keller, 2010; Crawford & Crawford, 1996; Ernst, 2000; Evans, 1997; Harris, 2004; Harrison, 2001; Martin,

2000; Masters & Ela, 2008; Pickering & Owen, 1994; Scott & Lipps, 1995; Spitz & Trudinger, 2009; Truett & Johnson, 2000.

### Environmental Impact Assessment

Morris & Therivel, 2001; Rouvillois, 1967, 1972a, b; Wood, 2003.

### Environmental monitoring

Akimoto *et al.*, 2004; Alve, 1991a, b; Alve & Dolven, 2010; Antony, 1980; Armynot du Chatelet & Debenay, 2010; Bandy *et al.*, 1964a, b, 1965a, b; Banerji, 1973, 1989; Barbosa *et al.*, 2010; Bates & Spencer, 1979; Bhalla & Nigam, 1986; Bhalla *et al.*, 2007; Boltovskoy & Boltovskoy, 1968; Bouchet *et al.*, 2009, 2010; Buckley *et al.*, 1974; Carnahan *et al.*, 2009; Vilela *et al.*, in Carvalho *et al.*, 2007; Dabbous & Scott, 2012; de Freitas Prazeres *et al.*, 2012; Debenay & Bui Thi Luan, 2006; Denoyelle *et al.*, 2010; Dermitzakis & Alafonsou, 1987; Dijkstra *et al.*, 2010; Ellison *et al.*, 1986; Ernst *et al.*, 2006; Frezza *et al.*, 2005; Frontalini & Coccioni, 2011; Gooday *et al.*, 2010c; Gooday *et al.*, 2009b; Hayward *et al.*, 2004b; Hess et al., 2010b; Hornung & Kress, 1991; Hornung *et al.*, 1989; Kameswara Rao & Satyanarayana Rao, 1979; le Furgey & St Jean, 1976; Magno *et al.*, 2012; Alve, in Martin, 2000; Bresler & Yanko-Hombach, in Martin, 2000; Coccioni, in Martin, 2000; Debenay *et al.*, in Martin, 2000; Ebrahim, in Martin, 2000; Geslin *et al.*, in Martin, 2000; Hallock, in Martin, 2000; Hippensteel & Martin, in Martin, 2000; Ishman, in Martin, 2000; van der Zwaan, in Martin, 2000; Mateu, 1974; Mojtahid *et al.*, 2008; Nagy & Alve, 1987; Nigam, 2005; Cato *et al.*, in Olausson & Cato, 1980; Oliveira-Silva *et al.*, 2012; Osterman *et al.*, 2005; Resig, in Pearson, 1960; Reinhardt *et al.*, 2001; Naidu, in Samanta, 1985; Schafer, 1970, 1973, 1982; Schafer & Cole, 1974; Schafer *et al.*, 1975, 1991; Scott *et al.*, 2001, 2005; Seibold & Seibold, 1981; Seiglie, 1968, 1971, 1974, 1975; Yanko *et al.*, in Sen Gupta, 1999; Setty, 1976, 1982, 1984; Setty & Nigam, 1984; Sharifi *et al.*, 1991; Tsujimoto *et al.*, 2008; Venec-Peyre, 1981; Venec-Peyre *et al.*, 2010; Vilela *et al.*, 2011; Watkins, 1961; Wright, 1968; Yanko & Flexer, 1991; Yanko & Kronfeld, 1992, 1993; Yanko *et al.*, 1992, 1994; Zhang *et al.*, 2010.

### Anthropogenically mediated global change

Archer & Rahmstorf, 2009; Battarbee *et al.*, 2003; Cornell *et al.*, 2012; Cowie, 2013; Dessler, 2012; Dessler & Parson, 2010; Edenhofer *et al.*, 2012; Eggleton, 2012; Field *et al.*, 2012; Gerhard *et al.*, 2001; Gore, 2006; Grubb, 1999; Houghton, 2009; Kaplan *et al.*, 2009; Kennedy, 2004; Launder & Thomas, 2009; Lovell, 2006, 2009; Neelin, 2010; Nevle & Bird, 2008; Ruddiman, 2003, 2005; Walker & King, 2008; Xuefeng Yu *et al.*, 2010.

*Ocean acidification*

Bordelon & Schneider, 2010; Dias *et al.*, 2010; Doney *et al.*, 2009; Foster, 2008; Andersson *et al.*, in Gattuso & Hansson, 2011; Greene *et al.*, 2012; Hall & Chan, 2004; Hart *et al.*, 2010; Hikami *et al.*, 2010, 2011; Khanna *et al.*, 2010; Kuroya-nagi *et al.*, 2009; Lear & Rosenthal, 2006; Morse *et al.*, 2006; Orr *et al.*, 2005; Riebesell *et al.*, 2000; Sanyal *et al.*, 1996; Yu & Elderfield, 2007.

*Carbon dioxide sequestration*

Launder & Thompson, 2009; Lovell, 2009.

# 13

# Applications and case studies in archaeology

This chapter deals with applications and case studies in archaeology. It contains sections on applications – in archaeostratigraphy and in environmental archaeology – and on case studies of applications. The section on case studies includes ones on the palaeoenvironmental interpretation of the Pleistocene–Holocene of the British Isles, using proxy Recent benthic foraminiferal distribution data; on the Early Palaeolithic of Boxgrove and Valdoe, West Sussex (500 000 years BP); and on the Medieval of the City of London (AD 1200?–1350).

## 13.1 Applications

In archaeology, Foraminifera and other micro- and macro-fossils have been used to provide information on the stratigraphic and, in particular, palaeoenvironmental – including palaeoclimatic – context of human evolution, dispersal and activity (including settlement and land use) (see Whittaker *et al.*, 2003; Jones, 2006; Jones & Whittaker, in Whittaker & Hart, 2010; Jones, 2011a; and additional references cited therein; see also the further reading list at the end of the chapter, and Sections 13.1.1 and 13.1.2).

They have also been used to provide information as to the provenance of clay and other materials used in the manufacture of pottery, mosaics and earthworks (Friedrich *et al.*, in McGuire *et al.*, 2000; Quinn & Day, 2007; Tasker *et al.*, 2009; Wilkinson *et al.*, 2010; White *et al.*, 2012; see also below). Also to provide information as to the provenance of building and decorative stone, and hence as to the appropriate materials to be used in the restoration on ancient buildings and monuments. And of flint, amber and other trade goods, and ship's ballast, and hence as to ancient trade links.

For example, Foraminifera have been used to pinpoint the precise source of the clay used in the manufacture of the Minoan pottery discovered in the recent excavation of Akrotiri on the island of Thera in the Aegean, which turned out to be most likely somewhere around Cape Loumaravi, only some 5 km to the north-west (Friedrich *et al.*, in McGuire *et al.*, 2000). (Thera was part of the much larger volcanic island of Santorini until the explosive volcanic eruption of 1640 BC, which, incidentally, is interpreted as having been partly responsible for the decline of the Minoan civilisation on Crete.)

They have also been used, alongside calcareous nannofossils, to identify the source of the chalk used in the manufacture of the individual tesserae for the Roman mosaics of Calleva Atrebatum or modern Silchester in Hampshire, which turned out to be in Dorset, some 100 km to the south-west (Tasker *et al.*, 2009). And similarly to identify the source of the chalk used in the mosaics of Fishbourne on the Isle of Wight (White *et al.*, 2012).

Finally, Foraminifera have also been used to pinpoint the precise source of the stone used in the construction of the English Civil War defensive earthworks at Wallingford Castle in Oxfordshire, which turned out to be an outcrop of the Glauconitic Marl member of the West Melbury Marly Chalk formation local to the castle site (Wilkinson *et al.*, 2010) In this particular case the use of a local source of stone may have been born of necessity, the castle having been besieged by plucky Parliamentarian forces for the 65 days leading up to the Royalist surrender on July 27, 1646! Whether or not this was so, the stone, when com-pacted, turned out to be eminently suitable to the construction of defensive earthworks and even gun emplacements.

### 13.1.1 Archaeostratigraphy

A number of Quaternary dating methods are available for use in archaeostratigra-phy, including not only conventional marine micropalaeontological (planktic for-aminiferal) and nannopalaeontological (calcareous nannoplankton), and non-marine palynological (spore and pollen) and macropalaeontological (mollusc and verte-brate) biostratigraphy; but also Marine or Oxygen Isotope Stage (MIS or OIS) stratigraphy, accelerator mass spectometry (AMS), amino-acid racemisation, cos-mogenic chlorine-36 rock exposure, dendrochronology, electron spin resonance (ESR), magnetic susceptibility, molecular stratigraphy, optically stimulated lumi-nescence (OSL), radio-carbon dating, thermoluminescence (TL), and uranium-series dating (Jones, 2006, 2011a; and additional references cited therein; see also Section 5.7, and the further reading list at the end of the chapter).

### 13.1.2 Environmental archaeology

The exacting ecological requirements and tolerances of many species of Forami-nifera and other micro- and macro-fossils, and their rapid response to changing environmental and climatic conditions, render them useful in empirical or in transfer-function-based palaeoenvironmental and palaeoclimatic interpretation, and hence in environmental archaeology (Whittaker *et al.*, 2003; Jones, 2006; Jones & Whittaker, in Whittaker & Hart, 2010; Jones, 2011a; and additional references cited therein; see also the further reading list at the end of the chapter).

Selected case studies, all of them incidentally from the British Isles, are given in Sections 13.2.1–13.2.3. Other case studies, from further afield (many of them from the Near East), include those on the ancient harbour of Herod the Great in Caesarea Maritima, Israel, destroyed by a tsunami in AD 115 (Reinhardt *et al.*, 1994, Reinhardt *et al.*, 1998a, Raban *et al.*, 1999, Reinhardt, 1999, Reinhardt & Patterson, 1999, Reinhardt & Raban, 1999; Reinhardt *et al.*, 2006); the Amerindian Montague Harbour in British Columbia in Canada (Reinhardt *et al.*, 1996); the Bronze Age to Byzantine harbour of Tyre in Phoenicia (Mariner *et al.*, 2005); the ancient harbour and coastline of Liman Tepe in Turkey (Goodman *et al.*, 2009b); the coastline of Marennes–Oleron Bay on the Atlantic coast of France (Poirier *et al.*, 2010); a Pharaonic harbour on the Red Sea (Hein *et al.*, 2011); and the Roman harbour of Luna in Italy (Bini *et al.*, 2012).

## 13.2 Case studies of applications

### *13.2.1 Palaeoenvironmental interpretation of the Pleistocene–Holocene of the British Isles, using proxy Recent benthic foraminiferal distribution data*

#### *Introduction*

The palaeoenvironments of the Pleistocene to sub-Recent Holocene of the British Isles have been interpreted using proxy data on the environmental, and in particular the biogeographic, distributions of benthic Foraminifera from the Recent (Jones, 2006; Jones & Whittaker, in Whittaker & Hart, 2010; Jones, 2011a; and additional references cited therein). The approach adopted has been qualitative and uniformitarian, that is to say, to assume that the empirically observed modern distributions are directly applicable to the interpretation of ancient environments.

The biogeographic provinces referred to are as follows (see also Fig. 13.1):

- the Arctic Province (which includes the Norwegian Sea), the southern boundary of which is approximately coincident with the 0 °C winter/5 °C summer isotherm;

- the Subarctic Province (which includes the northern North Sea, Scandinavia and the Baltic), the southern boundary of which is approximately coincident with the 5 °C winter/10 °C summer isotherm;

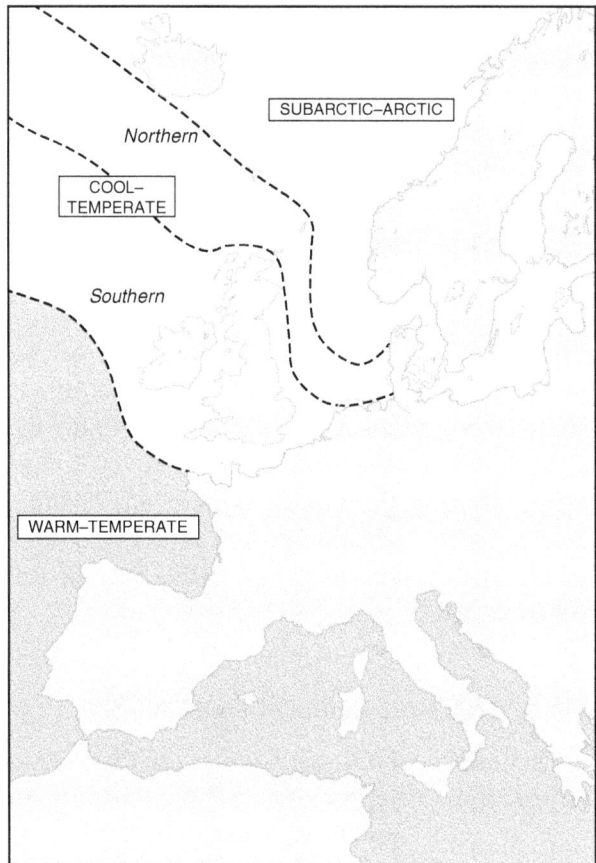

Fig. 13.1  **Biogeographic distribution of benthic Foraminifera in the north-east Atlantic**.

- the (Atlantic) Cool–Temperate Province (which includes the west of Scotland, the west of Ireland, the Irish Sea, the English Channel and the southern North Sea), the southern boundary of which is approximately coincident with the 10 °C winter/15 °C summer isotherm;
- the (Atlantic) Warm–Temperate Province (which includes the Bay of Biscay, Iberia and the Mediterranean), the southern boundary of which is approximately coincident with the 15 °C winter/20 °C summer isotherm.

In the context of the palaeoclimatic interpretation of the Pleistocene to sub-Recent Holocene of the British Isles, cold-water conditions have been inferred on the basis of the occurrence of the Subarctic–Arctic benthic Foraminifera *Astacolus hyalacrulus, Astrononion gallowayi, Buccella tenerrima, Cassidulina norcrossi, C. reniformis s.l., C. teretis s.l., Cibicides grossus, Dentalina baggi, D. frobisherensis, D. ittai, Elphidiella arctica, E. groenlandica, E. nitida, Elphidium albiumbilicatum,*

*E. bartletti, E. clavatum* (common-abundant), *E. hallandense, Esosyrinx curta, Fissurina serrata, Gordiospira arctica, Guttulina glacialis, Lagena parri, Laryngosigma hyalascidia, Miliolinella chukchiensis, Nonion orbicularis, Nonionellina labradorica, Oolina scalariformissulcata, Pseudopolymorphina novangliae, Pyrgo williamsoni, Quinqueloculina agglutinata, Q. arctica, Q. stalkeri, Stainforthia feylingi, S. loeblichi, Trichohyalus bartletti, Trifarina fluens* and *Triloculina trihedra*, the southern limits of distribution of which at the present time are north of the British Isles on the warmer, western margin, or in the North Sea on the cooler, eastern margin (Fig. 13.1). Key species are illustrated on Pl. 71.

Comparatively warm-water conditions have been inferred on the basis of the occurrence of the southern Cool–Temperate species *Aubignyna perlucida, Bulimina elongata, Lagena perlucida, Laryngosigma harrisi, Nonionella* sp. A and *Siphonina georgiana*, the northern limits of distribution of which are in the southern half of the British Isles at the present time. Comparatively warm-water conditions could also be inferred on the basis of the occurrence of the essentially southern Cool–Temperate species *Ammonia aberdoveyensis, A. falsobeccarii, A. flevensis, Bolivina variabilis, Cancris auricula, Cornuspira selseyensis, Elphidium crispum, E. incertum, Gaudryina rudis, Ophthalmidium balkwilli, Quinqueloculina bicornis, Q. cliarensis, Q. intricata, Q. lata, Q. oblonga, Rosalina anomala, R. milletti* and *Rotaliella chasteri*, the northern limits of distribution of which are in the northern half of the British Isles on the warmer, western margin, but in the southern half on the cooler, eastern margin (Fig. 13.1). Key species are illustrated on Pl. 72.

Conditions warmer than those obtaining at the present time have been inferred on the occurrence of the Warm–Temperate species *Ammonia parkinsoniana, Elphidium* cf. *advenum* and *Elphidium fichtellianum*, the northern limits of distribution of which are south of the British Isles at the present time. Conditions warmer than those obtaining at the present time have also been inferred on the occurrence of the southern Cool–Temperate species *Aubigyna perlucida* in the Saalian Equivalent, OIS9 of Norfolk and Lincolnshire and *Bulimina elongata* in the Saalian Equivalent, OIS7 of Co. Durham, as these species do not range this far north on the eastern margin of the British Isles at the present time.

### Results and discussion

*Early Pleistocene, Pastonian.* The Pastonian Paston member of the Cromer Forest-Bed formation of Paston in Norfolk (the stratotype locality) is developed in non-marine to marginal marine facies, and, to the best of our knowledge, contains no benthic Foraminifera.

The slightly older Pastonian or pre-Pastonian Sidestrand member of the Norwich Crag Formation of Weybourne Hope in Norfolk (formerly known as

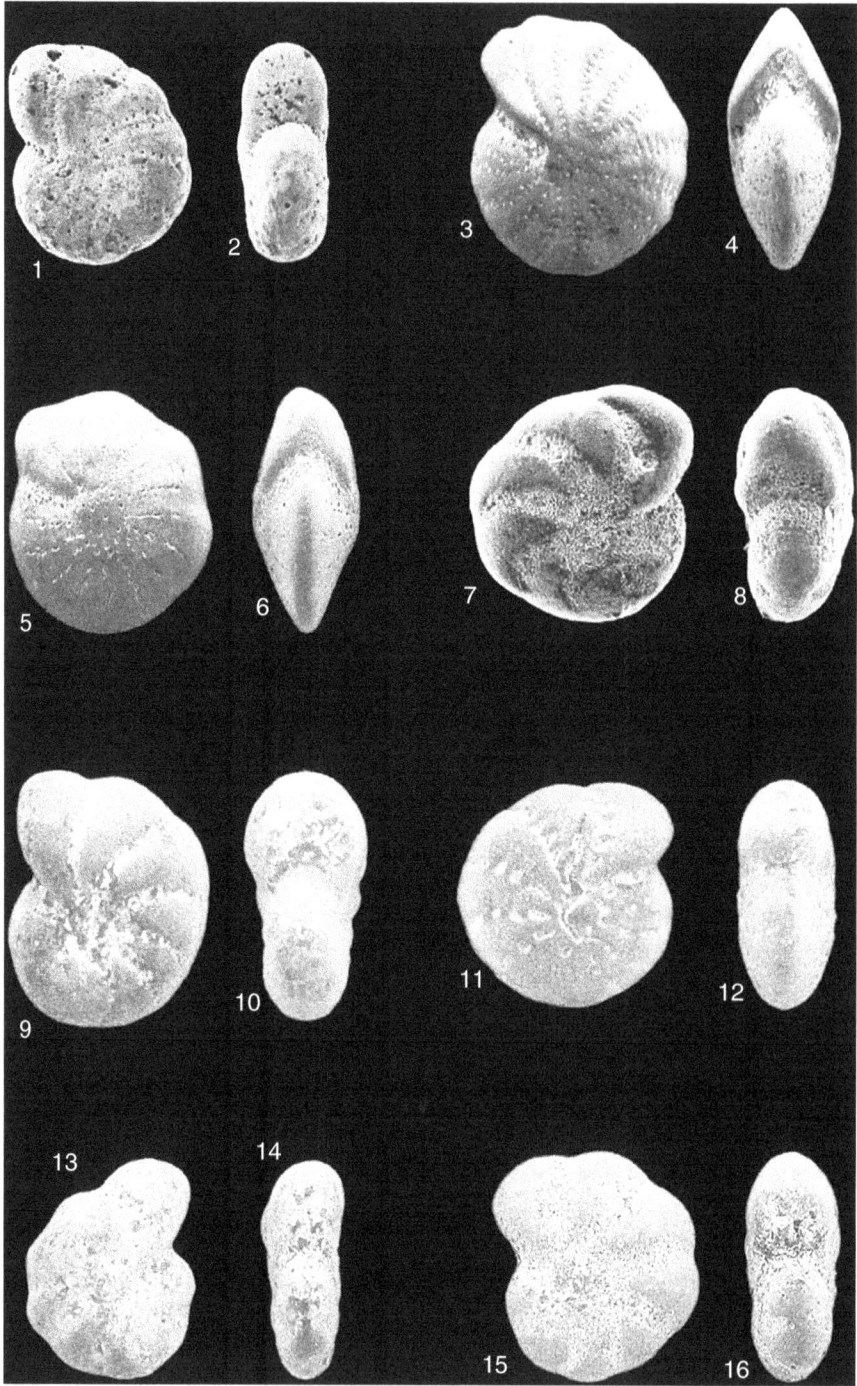

Pl. 71  **Some palaeoenvironmentally significant Foraminifera from the Pleistocene–Holocene of the British Isles, 1**. Cool-water Elphidiidae. 71.1–71.2 *Elphidiella arctica*; 71.3–71.4 *Elphidiella groenlandica*; 71.5–71.6 *Elphidiella nitida*; 71.7–71.8 *Elphidium albiumbilicatum*; 71.9–71.10 *Elphidium bartletti*; 71.11–71.12 *Elphidium clavatum*; 71.13–71.14 *Elphidium frigidum*; 71.15–71.16 *Elphidium subarcticum*.

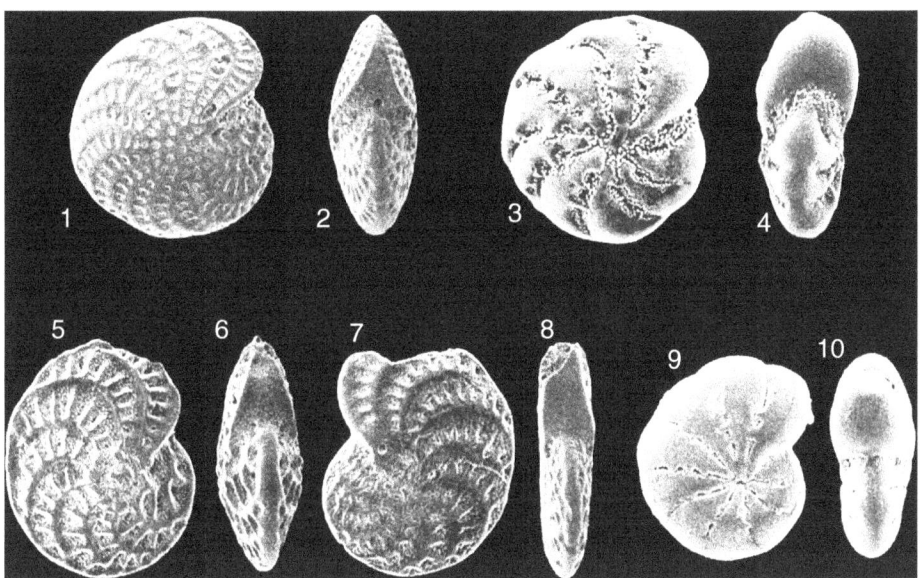

Pl. 72 **Some palaeoenvironmentally significant Foraminifera from the Pleistocene–Holocene of the British Isles, 2**. Warm-water Elphidiidae. 72.1–72.2 *Elphidium crispum*; 72.3–72.4 *Elphidium earlandi sensu* Murray, 1971; 72.5–72.8 *Elphidium fichtellianum/complanatum*; 72.9–72.10 *Elphidium incertum*.

the 'Weybourne Crag') is characterised by the benthic Foraminifer *Elphidium bartletti*, indicating Subarctic–Arctic climatic conditions.

*Middle Pleistocene, Beestonian.* The Beestonian Runton member of the Cromer Forest-Bed formation of West Runton in Norfolk (the stratotype locality) is developed in non-marine to marginal marine facies, and, to the best of our knowledge, contains no benthic Foraminifera.

*Cromerian Complex, OIS 17.* The Cromerian, Oxygen Isotope Stage 17 (OIS17) Interglacial appears on benthic foraminiferal evidence to have been characterised, at least on the east coast, by southern Cool–Temperate to Warm–Temperate conditions, warmer than those obtaining at the present time.

The Cromerian Cromer Forest-Bed formation of Pakefield in Suffolk is characterised by *Elphidium incertum*, indicating marginal marine, southern Cool–Temperate climatic conditions. The marine Subarctic–Arctic species *Elphidium bartletti* is also present, but rare, and interpreted as probably reworked, possibly from the equivalent of the pre-Pastonian Sidestrand member of the Norwich Crag formation. It is worthy of note that the Pakefield site has yielded one of the earliest records of human activity in northern Europe, in the form of 32 worked flint artefacts (the nearby Happisburgh site has recently yielded *the* earliest record, dating to OIS21 or OIS25, again in the form of worked flint artefacts).

The 'Cromer Forest-Bed formation' of Norton Subcourse Quarry near Great Yarmouth in Norfolk, which, although not as yet positively dated, is possibly at least in part correlative with the Cromer Forest-Bed formation of Pakefield, is characterised by the Mediterranean species *Elphidium fichtellianum*, indicating Warm–Temperate conditions, warmer than those obtaining at the present time. The Subarctic–Arctic species *Elphidium bartletti* is also present, and interpreted as probably reworked. Note, incidentally, that the Norton Subcourse Quarry section also contains reworked Foraminifera of mixed, including probable Jurassic, and Lower Cretaceous, rather than predominantly Upper Cretaceous ('Chalk with Flints') provenance, suggesting some sort of correlation with the comparatively flint-poor Kesgrave (or Bytham) Sands and Gravels.

*Cromerian Complex, OIS13.* The Cromerian, OIS13 Interglacial appears to have been characterised on the south coast by southern Cool–Temperate to Warm–Temperate conditions, warmer than those obtaining at the present time; and in the North Sea by Subarctic–Arctic conditions. The preferred interpretation is that this was because warm waters were unable to flow eastwards into the North Sea from the English Channel, in turn because the Strait of Dover was closed by a land bridge, as has been proposed by numerous authors. Another interpretation is that it was because cool waters were able to flow westwards into the North Sea from the White Sea, because the 'Karelian Portal' (in Finland and Russia) was open, irrespective of whether the Strait of Dover was open or closed, as is thought to have been the case in the Ipswichian. Yet another is that it was because cool waters were able to flow southwards into the North Sea from the Norwegian Sea, again irrespective of whether the Strait of Dover was open or closed, as at the present time.

The Cromerian, OIS13 Steyne Wood member of the Solent formation of Bembridge on the Isle of Wight is characterised by *Aubignyna perlucida* and *Lagena perlucida*, indicating southern Cool–Temperate conditions.

The Cromerian, OIS13 Slindon Sand member of the West Sussex Coast formation of Boxgrove, Trumley Copse and Valdoe on the Goodwood–Slindon Raised Beach in West Sussex is characterised by *Ammonia parkinsoniana* and *Elphidium fichtellianum*, indicating Warm–Temperate conditions, warmer than those obtaining at the present time (see also Section 13.2.2). The overlying Slindon Silt member of Boxgrove has yielded the oldest actual hominid remains in Britain, in the form of teeth and a leg-bone.

The interglacial identified in Foraminiferal Zone 34K in the 81/34 Borehole in the Devil's Hole area in the North Sea, underlain and overlain by units yielding isoleucine epimerisation ratios of 0.365 and 0.343, respectively, and interpreted as Cromerian, OIS13, is characterised by *Cassidulina reniformis s.l.*, *C. teretis*, *Elphidium albiumbilicatum*, *E. bartletti*, *E. clavatum*, *E. hallandense*, *Nonion orbicularis* and *Trifarina fluens*, indicating Subarctic–Arctic conditions.

*Anglian, OIS12*. The Anglian, OIS12 Glacial appears to have been characterised on the east coast by Subarctic–Arctic conditions.

The Anglian Corton member of the North Sea Drift formation of Corton in East Anglia is characterised by reworked Jurassic, Cretaceous, Palaeogene and Neogene Foraminifera, and by the possibly also reworked Quaternary Foraminifera *Elphidiella arctica* and *Nonion orbicularis*, indicating Subarctic–Arctic conditions. The Anglian Woolpit Beds of Woolpit near Bury St Edmunds are characterised by *Cassidulina reniformis s.l.*, *C. teretis*, *Elphidium clavatum*, *Elphidiella nitida*, *Elphidium hallandense* and *Nonion orbicularis*, again indicating Subarctic–Arctic conditions.

The Anglian Leet Hill member of the North Sea Drift formation of Leet Hill is barren of *in situ* taxa, and contains only reworked Jurassic, Cretaceous and Quaternary taxa.

*Hoxnian, OIS11*. The Hoxnian, OIS11 Interglacial appears to have been characterised on the south coast by Temperate conditions, and in the North Sea by Subarctic conditions.

The Hoxnian of Earnley, Bracklesham Bay, in West Sussex is characterised by *Ammonia 'beccarii'* [?*A. batavus*], *A. 'beccarii'* var., *Elphidium margaritaceum* and *E. williamsoni*, indicating either Temperate to Subarctic conditions, or Temperate conditions if *A. 'beccarii'* var. is a species other than *A. batavus*.

The interglacial identified in Foraminiferal Zone 34H in the 81/34 Borehole in the Devil's Hole area in the North Sea, marked by isoleucine epimerisation ratios ranging from 0.27 to 0.328 and averaging 0.285, and thus interpreted as Hoxnian, OIS11, is characterised by *Cassidulina norcrossi*, *C. reniformis s.l.*, *Elphidium albiumbilicatum*, *E. clavatum* and *Nonion orbicularis*, indicating Arctic–Subarctic conditions, and by *Spiroplectinella wrightii*, indicating Temperate–Subarctic conditions.

*Saalian Equivalent, OISs10–6*. The 'Saalian Equivalent' as herein defined equates to OISs10–6. It includes the Wolstonian *s.l.* of East Anglia, but not the Wolstonian s.s. of Wolston in the East Midlands, which is actually of Anglian, OIS12 age.

*OIS9 ('Purfleet' Interglacial)*. The Saalian Equivalent, OIS9 Interglacial appears to have been characterised, at least on the east coast, by southern Cool–Temperate conditions, slightly warmer than those obtaining at the present time.

The Saalian Equivalent, OIS9 Nar Valley Clay of the Nar member of the Nar Valley formation of the East Winch No. 1 Borehole and of a now backfilled trench near Tottenhill, both close to the stratotype locality elsewhere in the Nar Valley in Norfolk, is characterised by *Aubignyna perlucida*, indicating southern Cool–Temperate conditions, slightly warmer than those obtaining at the present time (this species does not occur this far north at the present time). The Nar member of the actual stratotype locality has been dated as Saalian Equivalent, OIS9 on uranium-series evidence.

The Saalian Equivalent, OIS9 Kirmington formation of Kirmington in Lincoln-shire is dominated by *Aubignyna perlucida*, again indicating southern Cool–Temperate conditions, slightly warmer than those obtaining at the present time.

*OIS7 ('Aveley' Interglacial).* The Saalian, OIS7 Interglacial appears to have been characterised generally on the south and east coasts by southern Cool–Temperate conditions, slightly warmer than those obtaining at the present time; and in the North Sea by Subarctic–Arctic conditions. However, it also appears to have been characterised locally on the south coast, on the Brighton-Norton Raised Beach, by Subarctic–Arctic conditions.

The OIS7 Lifeboat Station member of the Lifeboat Station Channel, Selsey, West Sussex is characterised by *Lagena perlucida*, also indicating southern Cool–Temperate conditions. The 'early' – OIS7 Aldingbourne member of the West Sussex Coast formation of Aldingbourne and Tangmere on the Aldingbourne Raised Beach is characterised by *Aubignyna perlucida*, *Bulimina elongata* and *Nonionella* sp. A, indicating southern Cool–Temperate conditions. In contrast, the 'late' OIS7 Black Rock and Norton members of the West Sussex Coast formation of numerous localities on the Brighton-Norton Raised Beach are characterised by *Cassidulina reniformis s.l.*, *Elphidium albiumbilicatum*, *E. clavatum*, *E. hallan-dense* and *Oolina scalariformissulcata*, indicating Subarctic–Arctic conditions.

The OIS7 Easington formation of Easington in Co. Durham is characterised by *Bulimina elongata*, indicating southern Cool–Temperate conditions, slightly warmer than those obtaining at the present time (this species does not occur this far north at the present time).

The interglacial identified in Foraminiferal Zone E in the 81/34 Borehole in the Devil's Hole area in the North Sea, marked by an isoleucine epimerisation ratio of 0.189 and thus interpreted as Saalian Equivalent, OIS7, is characterised by *Cassidulina norcrossi*, *C. reniformis s.l.*, *C. teretis*, *Elphidium albiumbilicatum*, *E. bartletti*, *E. clavatum*, *Nonion orbicularis* and *Trifarina fluens*, indicating Subarctic–Arctic conditions.

*Late Pleistocene, Ipswichian, OIS5e.* The Ipswichian, OIS5e Interglacial appears to have been characterised generally on the south and east coasts by southern Cool–Temperate to Warm–Temperate conditions, warmer than those obtaining at the present time; and in the North Sea by Temperate–Subarctic conditions. However, it also appears to have been characterised locally on the south coast, on the Pagham Raised Beach, by Subarctic–Arctic conditions.

The Ipswichian of Cardigan Bay is characterised by the Mediterranean species *Elphidium fichtellianum*, indicating Warm–Temperate conditions, warmer than those obtaining at the present time.

The Ipswichian of Goldcliff on the Gwent Levels is characterised by *Rosalina milletti*, indicating southern Cool–Temperate conditions.

The Ipswichian of the Middlezoy member of the Burtle formation of Greylake on the Somerset Levels is characterised by the Mediterranean species *Elphidium* cf. *advenum*, indicating Warm–Temperate conditions, warmer than those obtaining at the present time.

The Ipswichian 'blown sand' of the Thatcher Stone member of the Torbay formation of Hope's Nose and North Thatcher in Devon is characterised by the Mediterranean species *Elphidium fichtellianum*, indicating Warm–Temperate conditions, warmer than those obtaining at the present time.

The Ipswichian Pagham member of the West Sussex Coast formation of numerous localities on the Pagham Raised Beach is characterised by *Cassidulina reniformis s.l.* and *Elphidium albiumbilicatum*, indicating Subarctic–Arctic conditions.

The Ipswichian Equivalent Eemian 'Troll Interglacial' of Core 5.1/5.2 in the Troll area of the Norwegian sector of the North Sea is characterised by *Spiroplectinella wrightii*, indicating Temperate–Subarctic conditions.

*Devensian, OISs5d-2*. The Devensian, OISs5d-2 Glacial appears to have been characterised generally by Subarctic–Arctic conditions. However, it also appears to have been characterised locally, in the Windermere Stadial, by Temperate influence.

*OIS3*. The Devensian, OIS3 Sistrakeel formation of Sistrakeel in Ireland is characterised by dominant (90%) *Elphidium clavatum*, indicating Subarctic–Arctic conditions.

*OIS2, Late Glacial, including Last Glacial Maximum*. The radio-carbon dated Late Glacial of Glenulra in Ireland is characterised by *Buccella tenerrima*, *Cassidulina reniformis s.l.*, *C. teretis*, *Dentalina baggi*, *Elphidiella groenlandica*, *Elphidium bartletti*, *E. clavatum*, *E. hallandense*, *Miliolinella chukchiensis*, *Nonion orbicularis*, *Nonionellina labradorica* and *Quinqueloculina* arctica, indicating Subarctic–Arctic conditions. The Glenulra succession has been interpreted as representing the locally developed Glenavy Stadial. The radio-carbon dated Belderg formation of Belderg is characterised by *Cassidulina reniformis s.l.*, *Elphidium bartletti*, *E. hallandense*, *Miliolinella chukchiensis*, *Nonion orbicularis*, *Nonionellina labradorica*, *Pyrgo williamsoni*, *Quinqueloculina agglutinata* and *Q. arctica*, indicating Subarctic–Arctic conditions. The Belderg formation represents the locally developed Belderg Stadial. The acceletator mass spectroscopy (AMS)-dated Devensian Cooley Farm member of the Louth formation of the margins of Dundalk Bay, Co. Louth and Derryogue member of the Mourne formation of the Mourne region are dominated by *Elphidium clavatum*, indicating Subarctic–Arctic conditions. The Cooley Farm and Derryogue members represent the locally developed Cooley Point Interstadial.

The undifferentiated Late Glacial Dog Mills member of the Orrisdale formation of Dog Mills on the Isle of Man is characterised by *Elphidiella hannai* (possibly reworked from an older glacial) and *Elphidium clavatum*, indicating Subarctic–Arctic conditions.

The undifferentiated Late Glacial of the Western Irish Sea formation of the Irish Sea is characterised by *Cassidulina reniformis s.l.*, *Elphidiella nitida*, *Elphidium bartletti*, *E. clavatum*, *E. hallandense*, *Nonion orbicularis*, *Nonionellina labradorica*, *Quinqueloculina arctica* and *Q. stalkeri*, indicating Subarctic–Arctic conditions.

The undifferentiated Late Glacial of Cardigan Bay is characterised by *Buccella tenerrima*, *Cassidulina reniformis s.l.*, *Elphidium albiumbilicatum*, *E. clavatum* and *Nonion orbicularis*, indicating Subarctic–Arctic conditions.

The undifferentiated Late Glacial of Broughton Bay in Glamorgan is characterised by *Astrononion gallowayi*, *Cassidulina reniformis s.l.*, *Elphidium albiumbilicatum* and *E. hallandense*, indicating Subarctic–Arctic conditions.

The Last Glacial Maximum, Dimlington Stadial (or Late Glacial Stadial) Bridlington member of the Holderness formation of Dimlington near Bridlington in Yorkshire is characterised by *Astacolus hyalacrulus*, *Astrononion gallowayi*, *Buccella tenerrima*, *Cassidulina norcrossi*, *C. reniformis s.l.*, *C. teretis*, *Cibicides grossa*, *Dentalina baggi*, *D. frobisherensis*, *Elphidiella arctica*, *E. groenlandica*, *E. nitida*, *Elphidium bartletti*, *E. clavatum*, *E. hallandense*, *Guttulina glacialis*, *Nonion orbicularis*, *Nonionellina labradorica*, *Oolina scalariformissulcata*, *Pyrgo williamsoni*, *Quinqueloculina agglutinata*, *Q. stalkeri*, *Trichohyalus bartletti* and *Trifarina fluens*, indicating Subarctic–Arctic conditions. Reworking from various stratigraphic horizons – including Jurassic, Cretaceous, Tertiary and possibly also older Quaternary – is a conspicuous feature of the Bridlington Crag. Essentially tropical to subtropical Tertiary larger benthic species (*Operculina* sp.) and planktic species (*Hantkenina alabamensis*, *Morozovella* sp.) are present in some collections, and perhaps best interpreted as contaminants.

*Late Glacial [Oldest* Dryas *Equivalent] Stadial*. The radio-carbon dated Oldest *Dryas* Equivalent Stadial of the Hebridean Shelf is characterised by *Cassidulina reniformis s.l.*, *Elphidium albiumbilicatum*, *E. bartletti*, *E. clavatum*, *Nonion orbicularis*, *Nonionellina labradorica* and *Quinqueloculina stalkeri*, indicating Subarctic–Arctic conditions.

The interpreted Oldest *Dryas* Equivalent Stadial Killellan member of the Ardyne formation of Ardyne in Argyll and a biostratigraphically equivalent unit in the '100-foot Beach Clay' of Benderloch are characterised by *Astacolus hyalacrulus*, *Buccella tenerrima*, *Elphidium bartletti*, *E. clavatum*, *E. hallandense*, *Laryngosigma hyalascidia*, *Nonion orbicularis*, *Pyrgo williamsoni* and *Quinqueloculina agglutinata*, indicating Subarctic–Arctic conditions.

The radio-carbon dated Late Glacial Stadial of Corvish in Co. Donegal in Ireland is characterised by *Cassidulina reniformis s.l.*, *Elphidium clavatum*, *Pseudopolymorphina novangliae*, *Pyrgo williamsoni* and *Quinqueloculina stalkeri*, indicating Subarctic–Arctic conditions.

The AMS dated Devensian Killard Point formation of the drumlin fields around Killard Point in Co. Down is dominated by *Elphidium clavatum*, indicating Subarctic–Arctic conditions. The Killard Point formation represents the Late Glacial Killard Point Stadial (Heinrich Event 1).

*Windermere Interstadial = Bolling–Allerod or Greenland Interstadial 1 Equivalent.* The radio-carbon dated Bolling and Allerod Equivalent Interstadials of the Hebridean Shelf are characterised by mixed, interpreted time-averaged assemblages including the essentially Temperate species *Ammonia batavus* (?including the southern? Cool–Temperate species *Ammonia falsobeccarii*), and the Subarctic–Arctic species *Cassidulina reniformis s.l.*, *Elphidium albiumbilicatum*, *E. bartletti*, *E. clavatum*, *Nonion orbicularis*, *Nonionellina labradorica* and *Quinqueloculina stalkeri*. The intervening Older *Dryas* Equivalent Stadial of the Hebridean Shelf (Wester Ross Re-advance) is characterised by *Cassidulina reniformis s.l.*, *Elphidium albiumbilicatum*, *E. bartletti*, *E. clavatum*, *Nonion orbicularis*, *Nonionellina labradorica* and *Quinqueloculina stalkeri*, indicating Subarctic–Arctic conditions.

The radio-carbon dated Toward member of the Ardyne formation of Ardyne in Argyll and a biostratigraphically equivalent unit in the '100-foot Beach Clay' of Benderloch are characterised by mixed, interpreted time-averaged assemblages of including the southern? Cool–Temperate species *Ammonia falsobeccarii*, and the Subarctic–Arctic species *Astacolus hyalacrulus*, *Astrononion gallowayi*, *Dentalina frobisherensis*, *D. ittai*, *Elphidiella arctica*, *Elphidium albiumbilicatum*, *E. bartletti*, *E. clavatum*, *E. hallandense*, *Gordiospira arctica*, *Lagena parri*, *Laryngosigma hyalascidia*, *Nonion orbicularis*, *Nonionelllina labradorica*, *Pseudopolymorphina novangliae*, *Pyrgo williamsoni*, *Quinqueloculina agglutinata*, *Q. stalkeri*, *Trifarina fluens* and *Triloculina trihedra*.

*Loch Lomond [Younger Dryas Equivalent] Stadial.* The radio-carbon dated Younger *Dryas* Stadial of the Hebridean Shelf is characterised by *Cassidulina reniformis s.l.*, *Elphidium albiumbilicatum*, *E. bartletti*, *E. clavatum*, *Nonion orbicularis*, *Nonionellina labradorica* and *Quinqueloculina stalkeri*, indicating Subarctic–Arctic conditions.

The radio-carbon dated Ardyne Point member of the Ardyne formation of Ardyne in Argyll and a biostratigraphically equivalent unit in the '100-foot Beach Clay' of Benderloch are characterised by *Astacolus hyalacrulus*, *Buccella tenerrima*, *Dentalina ittai*, *Elphidiella arctica*, *Elphidium albiumbilicatum*, *E. bartletti*, *E. clavatum*, *E. hallandense*, *Esosyrinx curta*, *Gordiospira arctica*, *Guttulina glacialis, Laryngosigma hyalascidia, Nonion orbicularis, Pseudopolymorphina novangliae, Pyrgo williamsoni, Quinqueloculina agglutinata, Trifarina fluens* and *Triloculina trihedra*, indicating Subarctic–Arctic conditions.

*Holocene, Flandrian, OIS1*. The Flandrian appears to have been characterised generally by Temperate conditions. However, it also appears to have been characterised locally, in the Pre-Boreal and Boreal stages, by Subarctic–Arctic influence.

The radio-carbon dated Flandrian of the Hebridean Shelf is characterised by the Subarctic–Arctic species *Cassidulina reniformis s.l., Elphidium albiumbilicatum, E. bartletti, E. clavatum, Nonion orbicularis, Nonionellina labradorica* and *Quinqueloculina stalkeri*, and the essentially Temperate species *Ammonia batavus*. Subarctic–Arctic species are commonest in the Pre-Boreal and Boreal stages, and Temperate species in the Atlantic Stage.

The Flandrian of the Dovey Marshes and of Cardigan Bay is characterised by the Subarctic–Arctic species *Cassidulina reniformis s.l., Elphidium albiumbilicatum, E. bartletti, E. clavatum, Nonion orbicularis, Nonionellina labradorica, Quinqueloculina agglutinata* and *Triloculina trihedra*, and the southern Cool–Temperate species *Aubignyna perlucida, Bulimina elongata, Lagena perlucida, Ophthalmidium balkwilli* and *Rosalina milletti*.

The Flandrian of Start Bay in Devon is characterised by *Lagena perlucida*, indicating southern Cool–Temperate conditions.


### 13.2.2  *Early Palaeolithic, Boxgrove and Valdoe, West Sussex (500 000 years BP)*

The raised beach at Boxgrove on the West Sussex coastal plain has yielded evidence of archaic human occupation of Britain dating back to the early Middle Pleistocene, or Early Palaeolithic (Whittaker *et al.*, 2003; Jones, 2006; Jones & Whittaker, in Whittaker & Hart, 2010; Jones, 2011a; and additional references cited therein; see also the further reading list at the end of the chapter). The evidence is in the form of *Homo heidelbergensis* remains – a tibia and a tooth – and associated Acheulian flint hand-axes and other artefacts. The human remains and associated artefacts are concentrated on an Early Palaeolithic foreshore occupation surface in the Slindon Silt member of the Slindon formation of the West Sussex Coast group.

Foraminifera and other marine organisms have been found in the Slindon Sand at Boxgrove, immediately underlying the occupation surface in the Slindon Silt; and also in in two equivalent sections at the associated adjacent site of Valdoe, one of them an important, complete, 10-m section resting on Chalk bedrock in Valdoe field. The Foraminifera found at Boxgrove and Valdoe include *Ammonia parkinsoniana* and *Elphidium fichtellianum*, which are of Warm–Temperate aspect (see Section 13.2.1), *Elphidium fichtellianum* being a Mediterranean species, and

*Ammonia parkinsoniana* occurring both in the Mediterranean and on the eastern seaboard of the United States from Long Island to North Carolina), indicating warmer temperatures than those of today. They also include *Ammonia falsobeccarii, Aubignyna perlucida, Bolivina variabilis, Elphidium crispum* and *E. incertum*, which are of southern Cool–Temperate aspect (see Section 13.2.1), the northern limits of their distributions being in the southern half of the British Isles (or in the northern half on the warmer, western margin). The Warm–Temperate species, indicating warmer temperatures than those of today, are found throughout the lower part of the Slindon Sand in the sections at Valdoe. However, only southern Cool–Temperate species are found in samples from the upper part of the sections. This could indicate a slight cooling. However, considering the fact that there are no diagnostic Warm–Temperate indicators in marginal as against shallow marine environments, it could equally be due to a shallowing and/or freshening. Note in this context that a number of shallow marine species not known to be tolerant of reduced depth and/or salinity, including *Ammonia falsobeccarii, Cassidulina obtusa, Elphidium crispum, E. fichtellianum* and *E. macellum*, become less common towards the tops of the sections at Valdoe; and marginal to shallow marine species that are known to be tolerant of such conditions, including *Ammonia batavus, Aubignyna perlucida, Cibicides lobatulus, Elphidium excavatum, E. gerthi, E. incertum, E. williamsoni* and *Nonion depressulus*, become more common. (Some shallow marine species not known to be tolerant of reduced depth and/or salinity, including *Bolivina difformis, B. pseudoplicata, B. spathulata, B. variabilis, Buccella frigida, Lagena perlucida, Patellina corrugata, Planorbulina distoma* and *Trifarina angulosa*, remain common in the upper parts of the sections, although, significantly, these tend to be those that are either small or otherwise hydrodynamically unstable, or phytal in habit, and thus disproportionately prone to transportation.)

### 13.2.3 Medieval, City of London (AD 1200?–1350)

Archaeological excavation at a site in Tudor Street in the City of London in 1978, preparatory to site redevelopment, has yielded interesting evidence of domestic and industrial activity in the Middle Ages (Jones, 2006, 2011a, 2012; and additional references cited therein; see also the further reading list at the end of the chapter). The evidence is essentially in the form of the record of ancient organismal remains in the medieval deposits of the River Fleet, a tributary of the Thames, and of the influence on it of human activity. Incidentally, the word 'fleet' is Anglo-Saxon in origin, and refers to the ability of the river to float boats on the rising tide. The post-Anglo-Saxon history of the River Fleet has been described as 'a decline from a river to a brook, from a brook to a ditch, and from a ditch to a drain'.

The medieval deposits of the Fleet are approximately 1 m thick, between −0.4 and + 0.6 m OD, and unconformably overlie a bedrock of London clay and a layer of interpreted pre-Roman occupation deposits, and underlie a layer of post-medieval industrial waste, post-dating the period of land reclamation of AD 1350–1360. They have been provisionally dated to the thirteenth and fourteenth centuries on the basis of pot-sherds.

They contain over 140 species of organisms, including the Foraminifera *Ammonia flevensis*, *Elphidium williamsoni*, *Fissurina lucida*, *Miliammina fusca*, *Nonion germanicum* and *Trochammina inflata*. The principal elements of the biota are essentially fresh-water to only slightly brackish, and it is thought that the remaining minor, marine elements, including the Foraminifera, are allochthonous. It is also thought that the salinity of the medieval Fleet may have been artificially low on account of the effects of the old London bridge, built by Peter Colechurch in 1176 and not removed until as recently as 1831.

They also contain abundant industrial and domestic refuse such as wood chips (waste from wood-working), horn cores (waste from horn-working), hide (waste from butchery or tannery), leather shoes, meat- and domestic-animal bones, and animal and human dung. The refuse and organic waste dumped into the river evidently resulted initially in eutrophication, and the proliferation of some, opportunistic, species, including fresh-water molluscs and the macrophytes *Zannicellia* and *Groenlandia* (in Bed X). However, it also resulted untimately in hypertrophication, and the effective elimination of all species (in Bed Y), as recorded in archive records for 1343. Incidentally, the pollution became so bad that it was ordered in 1357 that 'no man shall take … any manner of rubbish … or dung, from out of his stable or elsewhere, to throw … the same into the rivers of Thames and Fleet …. And if any one should be found doing the contrary thereof, let him have the prison for his body, and other heavy punishment as well, at the discretion of the Mayor and the Aldermen'!

## 13.3 Further reading

*Applications – general*

Anderson *et al.*, 2007; Bell & Walker, 1992; Dincauze, 2000; Haslett, 2002; Horton & Edwards, 2006; Lowe & Walker, 1997; Shennan & Andrews, 2000; Walker, 2005; Wilkinson & Stevens, 2003; Williams *et al.*, 2003; Wilson *et al.*, 2000.

*Archaeostratigraphy*

Gibbard & van Kolfschoten, in Gradstein *et al.*, 2004; Grossman, in Gradstein *et al.*, 2012; Saltzman & Thomas, in Gradstein *et al.*, 2012; Walker, 2005.

*Environmental archaeology*

Bates *et al.*, 2010; Horton & Edwards, 2000, 2006; Horton *et al.*, 1999; Massey *et al.*, 2006; Whittaker, in Roberts & Parfitt, 1999; Sejrup *et al.*, 2004; Horton *et al.*, in Shennan & Andrews, 2000.

*Case Studies – Palaeolithic, Boxgrove and Valdoe*

Gamble, 1999; Hunter & Ralston, 2009; Pettitt & White, 2012; Pitts & Roberts, 1997; Pryor, 2003; Roberts & Parfitt, 1999; Whittaker, in Roberts & Parfitt, 1999; Stringer, 2006;.

*Medieval City of London*

Barton, 1962; Boyd, in Neale & Brasier, 1981.

# References

Abawi, T.S., 1989. Foraminifera, stratigraphy and sedimentary environment of the Euphates formation, Lower Miocene, Sinjar area, northwestern Iraq. *Newsletters on Stratigraphy*, **21**(1): 15–24.

Abawi, T.S. & Maroof, R.A., 1992. Planktonic foraminiferal biostratigraphy of the Serikagni formation (Oligocene/Miocene), Sinjar area, NW Iraq. *Neues Jahrbuch fur Geologie und Palaontologie, Monatshefte*, **1992**(12): 709–20.

Abbots, I.L. (ed.), 1991. *United Kingdom Oil and Gas Fields: 25 Years Commemorative Volume*. London; The Geological Society (Special Memoir, No. 14).

Abdelghany, O., 2003. Late Campanian–Maastrichtian Foraminifera from the Simsima formation on the western side of the northern Oman mountains. *Cretaceous Research*, **24**: 391–405.

Abdelghany, O., 2006. Early Maastrichtian larger Foraminifera from the Qahlah formation, United Arab Emirates and Sultanate of Oman border region. *Cretaceous Research*, **27**: 898–906.

Abdullah, F.H.A., 2001. A preliminary evaluation of Jurassic source rock potential in Kuwait. *Journal of Petroleum Geology*, **24**(3): 361–78.

Abdullah, F.H.A., Nederlof, P.J.R., Ormerod, M.P. & Kinghorn, R.R.F., 1997. Thermal history of the Lower and Middle Cretaceous source rocks in Kuwait. *GeoArabia*, **2**(2): 151–64.

Abdulsamad, E.O. & Barbieri, R., 1999. Foraminiferal distribution and palaeoecological interpretation of the Eocene–Miocene carbonates at Al Jabal al Akhdar (Northeast Libya). *Journal of Micropalaeontology*, **18**: 45–65.

Abramovich, S., Keller, G., Stuben, D. & Berner, Z., 2003. Characterisation of late Campanian and Maastrichtian planktonic foraminiferal depth habitats and vital activities based on stable isotopes. *Palaeogeography, Palaeoclimatology, Palaeoecology*, **202**(1–2): 1–29.

Abu Jaber, N.S., Kimberley, M.M. & Cavaroc, V.V., 1989. Mesozoic–Palaeogene basin development within the eastern Mediterranean borderland. *Journal of Petroleumn Geology*, **12**(4): 419–35.

Abu-Zied, R.H., Rohling, E.J., Jorissen, F.J. *et al.*, 2008. Benthic foraminiferal response to changes in bottom-water oxygenation and organic carbon flux in the eastern Mediterranean during LGM to Recent times. *Marine Micropaleontology*, **67**(1–2): 46–68.

Adams, C.G., 1960. A note on two important collections of Foraminifera in the British Museum (Natural History). *Micropalaeontology*, **6**: 417–418.

Adams, C.G., 1970. A reconsideration of the East Indian letter classification of the Tertiary. *Bulletin of the British Museum (Natural History), Geology*, **19**(3): 87–137.

Adams, C.G., 1976. Larger Foraminifera and the late Cenozoic history of the Mediterranean region. *Palaeogeography, Palaeoclimatology, Palaeoecology*, **20**: 47–66.

Adams, C.G., 1980. Larger Foraminifera and the late Cenozoic history of the Mediterranean region. *Annales des Mines et de la Geologie*, **III**: 535–6.

Adams, C.G. & Ager, D.V. (eds.), 1967. *Aspects of Tethyan Biogeography*. London; Systematics Association (Publication, No. 7).

Adams, C.G., Harrison, C.A. & Hodgkinson, R.L., 1980. Some primary type specimens of Foraminifera in the British Museum

(Natural History). *Bulletin of the British Museum (Natural History), Geology*, **32**: 3–17.

Adams, C.G., Lee, D.E. & Rosen, B.R., 1990. Conflicting isotopic and biotic evidence for tropical sea-surface temperatures during the Tertiary. *Palaeogeography, Palaeoclimatology, Palaeoecology*, **77**: 289–313.

Adegoke, O.S., Omatsala, N.E. & Salami, N.B., 1976. Benthonic foraminiferal biofacies off the Niger Delta. *Maritime Sediments Special Publication*, **1**: 279–92.

Adl, S.M. Simpson, A.G., Farmer, M.A. *et al.*, 2005. The new higher level classification of eukaryotes with emphasis on the taxonomy of protists. *Journal of Eukaryotic Microbiology*, **52**(5): 399–451.

Afzal, J., Khan, F.R., Khan, S.N. *et al.*, 2005. Foraminiferal biostratigraphy and paleoenvironments of the Paleocene Lockhart limestone from Kotal Pass, Kohat, northern Pakistan. *Pakistan Journal of Hydrocarbon Research*, **15**: 9–23.

Afzal, J., Williams, M. & Aldridge, R.J., 2009. Revised stratigraphy of the lower Cenozoic succession of the greater Indus basin in Pakistan. *Journal of Micropalaeontology*, **28**: 7–23.

Agterberg, F. & Gradstein, F.M., 1988. Recent developments in quantitative stratigraphy. *Earth-Science Reviews*, **25**: 1–73.

Agterberg, F.P. & Gradstein, F.M., 1999. The RASC method for ranking and scaling of biostratigraphic events. *Earth-Science Reviews*, **46**: 1–25.

Agusti, J., Rook, L. & Andrews, P. (eds.), 1999. *Hominoid Evolution and Climatic Change in Europe, Volume 1: The Evolution of Neogene Terrestrial Ecosystems in Europe.* Cambridge; Cambridge University Press.

Ahmad, A. & Ahmad, N., 2005. Paleocene petroleum system and its significance for exploration in the southwest Lower Indus basin and nearby offshore of Pakistan. *SPE/PAPG Annual Technical Conference, Islamabad*: 1–22.

Ahmad, N., Fink, P., Sturrock, S. *et al.*, 2012. Sequence strtaigraphy as predictive tool in Lower Goru fairway, lower and middle Indus platform, Pakistan. *Search and Discovery Article*, 10404.

Ahsan, S.A., Khan, R., Naveed, Y. & Saqab, M.M., 2012. Physico-chemical controls on source rock in offshore Indus – comparative study of some major Tertiary deltas of the world. *Search and Discovery Article*, 50560.

Aigner, T., 1983. Facies and origin of nummulitic build-ups: an example from the Giza Pyramids plateau (Middle Eocene, Egypt). *Neues Jahrbuch für Geologie und Palaontologie, Abhandlungen*, **166**(3): 347–68.

Ainsworth, N.R., Riley, L.A. & Gallagher, L.T., 2000. An Early Cretaceous lithostratigraphic and biostratigraphic framework for the Britannia field reservoir (Late Barremian–Late Aptian), UK North Sea. *Petroleum Geoscience*, **6**: 345–67.

Akhter, S., Bennett, J., Carmichael, S. *et al.*, 2007. Structure and stratigraphy of the offshore Indus basin, Pakistan. *Abstracts, 'Emerging Plays in Australasia' Conference*,. London; The Geological Society

Akimoto, K., Hattori, M., Uematsu, K. & Kato, C., 2001. The deepest living Foraminifera, Challenger Deep, Mariana Trench. *Marine Micropaleontology*, **42**(1–2): 95–7.

Akimoto, K., Nakahara, K., Kondo, H. *et al.*, 2004. Environmental reconsruction based on heavy metals, diatoms and benthic foraminifers in the Isahaya reclamation area, Nagasaki, Japan. *Journal of Environmental Micropaleontology, Microbiology, Meiobenthology*, **1**: 83–104.

Ala, M.A., 1982. Chronology of trap formation and migration of hydrocarbons in Zagros sector of southwest Iran. *American Association of Petroleum Geologists Bulletin*, **66**(10): 1535–41.

Ala, M.A., Kinghorn, R.R.F. & Rahman, M., 1980. Organic geochemistry and source rock characteristics of the Zagros petroleum province, southwest Iran. *Journal of Petroleum Geology*, **3**(1): 61–89.

Ala, M.A. & Moss, B.J., 1979. Comparative petroleum geology of southeast Turkey and northeast Syria. *Journal of Petroleum Geology*, **4**(4): 3–27.

Alam, M., 1989. Geology and depositional history of Cenozoic sediments of the Bengal basin in Bangladesh. *Palaeogeography, Palaeoclimatology, Palaeoecology*, **69**: 125–39.

Al-Aswad, A.A., 1997. Stratigraphy, sedimentary environment and depositional evolution of the Khuff formation in south-central Saudi Arabia. *Journal of Petroleum Geology*, **20**(3): 307–26.

Aldabal, M.A. & Alsharhan, A.S., 1989. Geological model and reservoir evaluation of the Lower Cretaceous Bab member in the Zakum field, Abu Dhabi, U.A.E. *Society of Petroleum Engineers, Paper*, **18007**: 797–810.

Alexander, L.L. & Flemings, P.B., 1995. Geologic evolution of a Pliocene–Pleistocene salt-withdrawal mini-basin: Eugene Island Block 330, offshore Louisiana. *American Association of Petroleum Geologists Bulletin*, **79**(12): 1737–56.

Al-Fares, A.A., Bouman, M. & Jeans, P., 1998. A new look at the Middle to Lower Cretaceous stratigraphy, offshore Kuwait, *GeoArabia*, **3**(4): 543–60.

Al-Hashimi, H.A.J., 1980. Biostratigraphy of Eocene–Lower Oligocene of Western Desert, Iraq. *Annales des Mines et de la Geologie*, **28**(3): 209–29.

Al-Hashimi, H.A.J. & Amer, R.M., 1985. *Tertiary Microfacies of Iraq*. Baghdad; Directorate General for Geological Survey and Mineral Investigation.

Al-Hashimi, W.S. & Sadooni, F.N. (eds.), 1985. *Proceedings of the Second Geological Congress on the Middle East (GEOCOME-II)*. Baghdad; Arab Geologists Association.

Al-Husseini, M.I. (ed.), 1995. *Proceedings of the Middle East Geosciences Conference, Bahrain, 1994*. Manama, Bahrain; Gulf PetroLink.

Al-Husseini, M.I., 1997. Jurassic sequence stratigraphy of the western and southern Arabian Gulf. *GeoArabia*, **2**(4): 361–82.

Al-Husseini, M.I. (ed.), 2004. *Carboniferous, Permian and Early Triassic Arabian Stratigraphy*. Manama, Bahrain; Gulf PetroLink (GeoArabia Special Publication, No. 3).

Al-Husseini, M.I., 2007. Stratigraphic note: revised ages (Ma) and accuracy of Arabian Plate maximum flooding surfaces. *GeoArabia*, **12**(4): 167–70.

xAli, W., Paul, W. & Young-On, V. (eds.), 1998. *Transactions of the 3rd Geological Conference of the Geological Society of Trinidad and Tobago and the 14th Caribbean Geological Conference*. The Geological Society of Trinidad and Tobago.

Alkersan, H.F., 1975. Depositional environments and geologic history of the Mishrif formation in southern Iraq. *Proceedings of the Ninth Arab Petroleum Congress, Paper* **121**(B-3).

Allison, P.A., Hesselbo, S.P. & Brett, C.E., 2008. Methane seeps on an Early Jurassic dysoxic seafloor. *Palaeogeography, Palaeoclimatology, Palaeoecology*, **270**: 230–8.

Al-Qayim, B.A.J., Khaiwka, M.H. & Ronca, L.B., 1980. Depositional environment and diagenesis of the Oligocene reef cycles, Kirkuk oil field, northern Iraq. *Modern Geology*, **7**: 177–90.

Al-Saad, H., 2008. Stratigraphic distribution of Middle Jurassic Foraminifera in the Middle East. *Revue de Paleobiologie*, **27**(1): 1–13.

Al-Shamlan, A.A., 1980. Microfacies and petrographic analysis of the Mauddud formation in Kuwait and nearby regions. *Journal of the University of Kuwait (Science)*, **7**: 187–203.

Alsharhan, A.A., 1987. Geology and reservoir characteristics of carbonate buildups in giant Bu Hasa oil field, Abu Dhabi, United Arab Emirates. *American Association of Petroleum Geologists Bulletin*, **71**(10): 1304–18.

Alsharhan, A.S., 1989. Petroleum geology of the United Arab Emirates. *Journal of Petroleum Geology*, **12**(3): 253–88.

Alsharhan, A.S. & Kendall, C.G.St.C., 1991. Cretaceous chronostratigraphy, unconformities and eustatic sea-level changes in the sediments of Abu Dhabi, United Arab Emirates. *Cretaceous Research*, **12**: 379–401.

Alsharhan, A.A. & Nairn, A.E.M., 1986. A review of the Cretaceous formations in the Arabian Peninsula and Gulf, part I: Lower Cretaceous (Thamama group) stratigraphy and paleogeography. *Journal of Petroleum Geology*, **9**(4): 365–92.

Alsharhan, A.A. & Nairn, A.E.M., 1988. A review of the Cretaceous formations in the Arabian Peninsula and Gulf, part II: mid Cretaceous (Wasia group) stratigraphy and paleogeography. *Journal of Petroleum Geology*, **11**(1): 89–112.

Alsharhan, A.A. & Nairn, A.E.M., 1990. A review of the Cretaceous formations in the Arabian Peninsula and Gulf, part III: Lower Cretaceous (Aruma group) stratigraphy and paleogeography. *Journal of Petroleum Geology*, **13**(3): 247–66.

Alsharhan, A.S. & Nairn, A.E.M., 1997. *Sedimentary Basins and Petroleum Geology of the Middle East*. Amsterdam; Elsevier.

Alsharhan, A.S., Rizk, Z.A., Nairn, A.E.M. *et al.*, 2001. *Hydrogeology of an Arid Region: The Arabian Gulf and Adjoining Areas*. Amsterdam; Elsevier.

Alsharhan, A.S. & Scott, R.W. (eds.), 2000. *Middle East Models of Jurassic/Cretaceous Carbonate Systems*. Tulsa, OK; Society of Economic Paleontologists and Mineralogists (Special Publication, No. 69).

Alsharhan, A.S. & Whittle, G.L., 1995. Sedimentary–diagenetic interpretation and reservoir characteristics of the Middle Jurassic (Areaj formation) in the southern Arabian Gulf. *Marine and Petroleum Geology*, **12**(6): 615–28.

Al-Shdidi, S., Thomas, G. & Delfaud, J., 1995. Sedimentology, diagenesis and oil habitat of Lower Cretaceius Qamchuga group, northern Iraq. *American Association of Petroleum Geologists Bulletin*, **79**(5): 763–70.

Al-Silwadi, M.S., Kirkham, A., Simmons, M.D. & Twombley, B.N., 1996. New insights into regional correlation and sedimentology,

Arab formation (Upper Jurassic), offshore Abu Dhabi. *GeoArabia*, **1**(1): 6–27.

Alsop, G.I., Archer, S.G., Hartley, A.J. *et al.* (eds.), 2012. *Salt Tectonics, Sediments and Prospectivity*. London; The Geological Society (Special Publication, No. 363).

Altenbach, A.V., Bernhard, Joan M. & Seckbach, Joseph (eds.), 2012. *Anoxia: Evidence for Eukaryote Survival and Paleontological Strategies*. Hamburg; Springer.

Altenbach, A.V., Pflaumann, U., Schiebel, R. *et al.*, 1999. Scaling percentages and distributional patterns of benthic Foraminifera with flux rates of organic carbon. *Journal of Foraminiferal Research*, **29**(3): 173–85.

Altenbach, A.V. & Struck, U., 2001. On the coherence of organic carbon flux and benthic foraminiferal biomass. *Journal of Foraminiferal Research*, **31**(2): 79–85.

Alvaro, J.J., Aretz, M., Boulvain, F. *et al.*, 2007. *Palaeozoic Reefs and Bioaccumulations: Climatic and Evolutionary Controls*. London; The Geological Society (Special Publication, No. 275).

Alve, E., 1991a. Benthic Foraminifera in sediment cores reflecting heavy metal pollution in Sorfjord, western Norway. *Journal of Foraminiferal Research*, **21**(1): 1–19.

Alve, E., 1991b. Foraminifera, climatic change and pollution: a study of late Holocene sediments in Drammensfjord, southeast Norway. *The Holocene*, **1**(3): 243–61.

Alve, E., 1995. Benthic foraminiferal distribution and recolonization of formerly anoxic environments in Drammensfjord, southern Norway. *Marine Micropaleontology*, **25**: 169–86.

Alve, E., 1999. Colonization of new habitats by benthic Foraminifera: a review. *Earth-Science Reviews*, **46**: 167–85.

Alve, E., 2010a. Benthic foraminiferal responses to absence of fresh phytodetritus: a two-year experiment. *Marine Micropaleontology*, **76**: 67–75.

Alve, E., 2010b. Defining 'reference' conditions: monitoring inner Oslofjord, Norway. *Abstracts, Forams2010 (International Symposium on Foraminifera), Bonn*: 46.

Alve, E. & Bernhard, J.M., 1995. Vertical migratory response of benthic Foraminifera to controlled oxygen concentrations in an experimental mesocosm. *Marine Ecology Progress Series*, **116**: 137–51.

Alve, E. & Dolven, 2010. Defining 'reference' conditions: monitoring inner Oslofjord, Norway. *Abstracts, Forams2010 International Symposium on Foraminifera, Bonn*: 46.

Amirshahkarami, M., Vaziri-Moghaddam, H. & Taheri, A., 2007. Sedimentary facies and sequence stratigraphy of the Asmari formation at Chaman-Bolbol, Zagros basin, Iran. *Journal of Asian Earth Sciences*, **29**: 947–59.

Anderson, D.E., Goudie, A.S. & Parker, A.G., 2007. *Global Environments through the Quaternary*. Oxford; Oxford University Press.

Anderson, J.B., 1975. Ecology and distribution of Foraminifera in the Weddell Sea, Antarctica. *Micropaleontology*, **21**(1): 69–96.

Anderson, J.B. & Fillon, R.H. (eds.), 2004. *Late Quaternary Stratigraphic Evolution of the northern Gulf of Mexico*. Tulsa, OK; Society of Economic Paleontologists and Mineralogists (Special Publication, No. 79).

Ando, A., Huber, B.T. & MacLeod, K.G., 2010. Depth–habitat reorganization of planktonic Foraminifera across the Albian/Cenomanian boundary. *Paleobiology*, **36**(3): 357–73.

Andresen, K.J., Huuse, M., Schodt, N.H. *et al.*, 2011. Hydrocarbon plumbing systems of salt minibasins offshore Angola revealed by three-dimensional seismic analysis. *American Association of Petroleum Geologists Bulletin*, **95**(6): 1039–65.

Angell, R.W., 1967. The process of chamber formation in the foraminifer *Rosalina floridana*. *Journal of Protozoology*, **14**: 566–74.

Angiolini, L., Nicora, A., Bucher, H. *et al.*, 1998. Evidence of a Guadalupian age for the Khuff formation of southeastern Oman: preliminary report. *Rivista Italiana di Paleontologia e Stratigrafia*, **104**(3): 329–40.

Anglada, R., Jouval, J. & M'Boro, R., 1981a. Evolution paleogeographique du plateau continental du Gulfe de Guinee au Tertiaire superieur. *Cahiers de Micropaleontologie*, **2**: 7–21.

Anglada, R., Jouval, J. & M'Boro, R., 1981b. Foraminiferes Tertiaires dans les basins oust-Africains: queqlques representants de la famille des Bolivinitidae Cushman (Foraminiferida). *Cahiers de Micropaleontologie*, **2**: 23–9.

Anglada, R. & Radrianasolo, A., 1985. Utilisation des foraminiferes planctoniques dans la paleooceanographie de la Tethys au Cretace. *Bulletin de la Societe Geologique de France, Ser. 8*, **I**(5): 747–55.

Anisgard, H.W., 1970. Causes of dominantly arenaceous foraminiferal assemblages in downdip Wilcox of Louisiana. *Transactions of the Gulf Coast Association of Geological Societies*, **20**: 210–7.

Anka, Z., Seranne, M. & di Primio, R., 2010. Evidence of large Upper Cretaceous depocentre across the Continent–Ocean Boundary of the Congo–Angola basin. Implications for palaeo-drainage and potential ultra-deep source rocks. *Marine and Petroleum Geology*, **27**: 601–11.

Anka, Z., Seranne, M., Lopez, M. *et al.*, 2009. The long-term evolution of the Congo deepsea fan: a basin-wide view of the interaction between a giant submarine fan and a mature passive margin. *Tectonophysics*, **470**: 42–56.

Anketell, J.M. & Mriheel, I.Y., 2000. Depositional environment and diagenesis of the Eocene Jdeir formation, Gabes-Tripoli basin, western offshore Libya. *Journal of Petroleum Geology*, **23**(4): 425–47.

Anon., 1987. *Innovative Biostratigraphic Approaches to Sequence Analysis: New Exploration Opportunities*. Gulf Coast Section – Society of Economic Paleontologists and Mineralogists (GCS-SEPM) Foundation (Proceedings of the 8th Annual Research Conference).

Anon., 1989. *Gulf of Mexico Salt Tectonics, Associated Processes and Exploration Potential*. Gulf Coast Section – Society of Economic Paleontologists and Mineralogists (GCS-SEPM) Foundation (Proceedings of the 10th Annual Research Conference).

Anschutz, P., Jorissen, F.J., Chaillou, G. *et al.*, 2002. Recent turbidite deposition in the eastern Atlantic: early diagenesis and biotic recovery. *Journal of Marine Research*, **60**: 835–54.

Antony, A., 1980. Foraminifera of the Vembanad estuary. *Bulletin of the Department of Marine Science, University of Cochin*, **11**: 25–63.

Anz, J.H. & Ellouz, M., 1985. Development and operation of the El Gueria reservoir, Ashtart field, offshore Tunisia. *Journal of Petroleum Technology, Tunis*, **37**: 481–7.

Aqrawi, A.A.M., Horbury, A.D., Goff, J.C. & Sadooni, F.N., 2010. *The Petroleum Geology of Iraq*. Beaconsfield, Bucks.; Scientific Press.

ARAMCO, 1959. Ghawar oil field, Saudi Arabia. *American Association of Petroleum Geologists Bulletin*, **43**: 434–54.

Archer, D. & Rahmstorf, S., 2009. *The Climate Crisis: An Introductory Guide to Climate Change*. Cambridge; Cambridge University Press.

Archibald, J.M., Longet, D., Pawlowski, J. & Keeling, P.J. 2003. A novel polyubiquitin structure in Cercozoa and Foraminifera: evidence for a new eukaryotic supergroup. *Molecular Biology and Evolution*, **20**(1): 62–6.

Armentrout, J. (ed.), 2002. *Sequence Stratigraphic Models for Exploration and Production: Evolving Methodology, Emerging Models and Application Histories*. Gulf Coast Section – Society of Economic Paleontologists and Mineralogists (GCS-SEPM) Foundation (Proceedings of the 22nd Annual Research Conference).

Armentrout, J.M. & Perkins, B.F. (eds.), 1990. *Sequence Stratigraphy as an Exploration Tool: Concepts and Practices in the Gulf Coast*. Gulf Coast Section – Society of Economic Paleontologists and Mineralogists (GCS-SEPM) Foundation (Proceedings of the 11th Annual Research Conference).

Armentrout, J. & Rosen, N. (eds.), 2002. *Sequence Stratigraphic Models for Exploration and Production*. Gulf Coast Section – Society of Economic Paleontologists and Mineralogists (GCS-SEPM) Foundation (Proceedings of the 22nd Annual Research Conference)

Armynot de Chatelet, E., Bout-Roumazeilles, V., Riboulleau, A. & Trentesaux, A., 2009. Sediment (grain size and clay mineralogy) and organic matter quality control on living benthic Foraminifera. *Revue de Micropaleontologie*, **52**: 75–84.

Armynot du Chatelet, E. & Debenay, J.-P., 2010. The anthropogenic impact on the western French coasts as revealed by Foraminifera: a review. *Revue de Micropaleopntiologie*, **53**: 129–37.

Arnold, Z.M., 1954. Culture methods in the study of living Foraminifera. *Journal of Paleontology*, **28**: 404–16.

Arnold, Z.M., 1966. A laboratory system for maintaining small-volume cultures of Foraminifera and other organisms. *Micropaleontology*, **12**: 109–18.

Arthur, T.J., MacGregor, D.S. & Cameron, N.R., 2003. *Petroleum Geology of Africa: New Themes and Developing Technologies*. London; The Geological Society (Special Publication, No. 207).

Ashckenazi-Polivoda, S., Edelman-Furstenberg, Y., Almogi-Labin, A. & Benjamini, C., 2010. Characterisation of lowest oxygen concentrations within ancient upwelling environments. *Palaeogeography, Palaeoclimatology, Palaeoecology*, **289**: 134–44.

Athersuch, J., Banner, F.T. & Simmons, M.D., 1992. On *Trochamijiella gollesstanehi* gen. nov. et sp.nov. (Foraminiferida, Loftusiacea), an index for the Middle Eastern marine late Bathonian. *Journal of Micropalaeontology*, **11**(1): 7–12.

Atwater B.F., ten Brink U.S., Buckley M., *et al.*, 2010. Geomorphic and stratigraphic evidence for an unusual tsunami or storm a few centuries ago at Anegada, British Virgin Islands. *Natural Hazards* (doi 10.1007/s11069–010–9622–6).

Aubry, M.-P., 1993. Neogene allostratigraphy and depositional history of the De Soto Canyon area, western Gulf of Mexico. *Micropaleontology*, **39**(4): 327–66.

Aubry, M.-P., Lucas, S.G. & Berggren, W.A. (eds.), 1998. *Late Paleocene-Early Eocene Climatic and Biotic Events in the Marine and Terrestrial Records*. New York, NY; Columbia University Press.

Aurahs, R., Goker, M., Grimm, G.W. *et al.*, 2009. Using the multiple analysis approach to reconstruct phylogenetic relationships among planktonic Foraminifera from highly divergent and length-polymorphic SSU rDNA sequences. *Bioinformatics and Biology Insights*, **2009**(3): 155–77.

Aurahs, R., Treis, Y., Darling, K. & Kucera, M., 2011. A revised taxonomic and phylogenetic concept for the planktonc foraminifer species *Globigerinoides ruber* based on molecular and morphometric evidence. *Marine Micropaleontology*, **79**: 1–14.

Austin, W.E.N. & James, R.H. (eds.), 2008. *Biogeochemical Controls on Palaeoceanographic Environmental Proxies*. London; The Geological Society (Special Publication, No. 303).

Ayres, M.G., Bilal, M., Jones, R.W. *et al.*, 1982. Hydrocarbon habitat in main producing areas, Saudi Arabia. *American Association of Petroleum Geologists Bulletin*, **66**(1): 1–9.

Aziz, S.K. & Abd El-Sattar, M.M., 1997. Sequence stratigraphic modelling of the lower Thamama group, east onshore Abu Dhabi, United Arab Emirates. *GeoArabia*, **2**(2): 179–202.

Bailey, H.W., Dungworth, G., Hardy, M. *et al.*, 1989. A fresh approach to the Metlaoui. *Memoire, Enterprise Tunisienne d'Activites Petrolieres (Actes Deuxieme Journee de Geologie Tunisienne appliqué a la Recherche des Hydrocarbures)*, **3**: 281–308.

Baillie, P., Barber, P.M., Deighton, I. *et al.*, 2006. Petroleum systems of the deepwater Mannar basin, offshore Sri Lanka. *Proceedings, IPA-AAPG Deepwater and Frontier Exploration in Asia & Australasia Symposium, 2004*, DFE04-PO-015.

Baldi, K. & Hohenegger, J., 2008. Paleoecology of benthic Foraminifera of Baden-Sooss section (Badenian, Middle Miocene, Vienna basin, Austria). *Geologia Carpathica*, **59**: 411–24.

Bally, A.W. & Palmer, A.R. (eds.), 1989. *The Geology of North America – An Overview*. Boulder, CO; Geological Society of America.

Bandy, O.L., 1970. Upper Cretaceous–Cenozoic paleobathymetric cycles, eastern Panama and northern Colombia. *Transactions of the Gulf Coast Association of Geological Societies*, **20**: 181–93.

Bandy, O.L. & Arnal, R.E., 1960. Concepts of foraminiferal paleoecology. *Bulletin of the American Association of Petroleum Geologists*, **44**(12): 1921–32.

Bandy, O.L., Ingle, J.C. & Resig, J.M., 1964a. Foraminiferal trends, Laguna Beach outfall area, California. *Limnology & Oceanography*, **9**: 112–23.

Bandy, O.L., Ingle, J.C. & Resig, J.M., 1964b. Foraminifera: Los Angeles County outfall area, California. *Limnology & Oceanography*, **9**: 124–37.

Bandy, O.L., Ingle, J.C. & Resig, J.M., 1965a. Foraminiferal trends, Hyperion outfall, California. *Limnology & Oceanography*, **10**: 314–22.

Bandy, O.L., Ingle, J.C. & Resig, J.M., 1965b. Modifications of Foraminifera distributions by the Orange County outfall, California. *Marine Technology Society Transactions*, **1965**: 54–76.

Banerjee, A., Pahari, S., Jha, M. *et al.*, 2002. The effective source rocks in the Cambay basin. *American Association of Petroleum Geologists Bulletin*, **86**(3): 433–56.

Banerji, R.K., 1973. Benthic Foraminifera as an aid to recognise polluted environments. *Proceedings of the 60th Session of the Indian Science Congress*: 116.

Banerji, R.K., 1989. Foraminifera and discrimination of polluted environments along the Bombay coast. *Proceedings of the Indian Colloquium on Micropalaeontology and Stratigraphy, Delhi*, **33** 98–117.

Bang, I., 1990. Biostratigraphy of the North Sea Danian based on planktonic Foraminifera. *Tertiary Research*, **11**(2–4): 191–3.

Banner, F.T., Copestake, P. & White, M.R., 1993. Barremian–Aptian Praehedbergellidae of the North Sea area: a reconnaissance. *Bulletins of the British Mueum (Natural History), Geology*, **49**(1): 1–30.

Banner, F.T. & Desai, D., 1988. A review and revision of the Jurassic–Early Cretaceous Globigerinina, with especial reference to the Aptian assemblages of Speeton (North

Yorkshire, England). *Journal of Micropalaeontology*, **7**(2): 143–85.

Banner, F.T. & Highton, J., 1990. On *Everticyclammina* Redmond (Foraminifera), especially *E. kelleri* (Henson). *Journal of Micropalaeontology*, **9**(1): 1–14.

Banner, F.T. & Lowry, F.M.D., 1985. The stratigraphical record of planktonic Foraminifera and its evolutionary implications. *Special Papers in Paleontology*, **33**: 117–30.

Banner, F.T., Simmons, M.D. & Whittaker, J.E. 1991. The Mesozoic Chrysalidinidae (Foraminifera, Textulariacea) of the Middle East: the Redmond (Aramco) taxa and their relatives. *Bulletin of the British Museum (Natural History), (Geology Series)*, **47**: 101–52.

Banner, F.T. & Whittaker, J.E. 1991. Redmond's 'New lituolid foraminifera' from the Mesozoic of Saudi Arabia. *Micropaleontology*, **37**: 41–59.

Barbieri, R., Hohenegger, J. & Pugliese, N. (eds.), 2006. Foraminifera and environmental micropaleontology. *Marine Micropaleontology*, **61**: 1–154.

Barbosa, C.F., Scott, D.B., Seoane, J.C.S. & Turcq, B.J., 2005. Foraminiferal zonations as base lines for Quaternary sea-level fluctuations in south-southeastern Brazilian mangroves and marshes. *Journal of Foraminiferal Research*, **35**(1): 22–43.

Barbosa, C.F., Seoane, J.C.S., Ferreira, B.P. *et al.*, 2010. Taxonomy and ecology of *Amphistegina* spp. in coral reef health status. *Abstracts, Forams2010 (International Symposium on Foraminifera), Bonn*: 52–3.

Barras, C., Geslin, E., Jorissen, F. *et al.*, 2010. Benthic Foraminifera as bio-indicators of coastal water quality in the Mediterranean Sea in relation to the implementation of the Water Framework Directive. *Abstracts, Forams2010 (International Symposium on Foraminifera), Bonn*: 53.

Bartenstein, H., 1985. Stratigraphic pattern of index Foraminifera in the Lower Cretaceous of Trinidad. *Newsleters on Stratigraphy*, **14**(2): 110–7.

Bartenstein, H., 1987. Micropaleontological synthesis of the Lower Cretaceous in Trinidad, West Indies: remarks on the Aptian/Albian boundary. *Newsletters on Stratigraphy*, **17**(3): 143–54.

Bartolini, C., Buffler, R.T. & Blickwede, J.F. (eds.), 2003. *The Circum-Gulf of Mexico and the Caribbean: Hydrocarbon Habitats, Basin Formation and Plate Tectonics*. Tulsa, OK; American Association of Petroleum Geologists (Memoir, No.79).

Bartolini, C. & Ramos, J.R.R. (eds.), 2009. *Petroleum Systems in the Southern Gulf of Mexico*. Tulsa, OK; American Association of Petroleum Geologists (Memoir, No. 90).

Barton, N., 1962. *The Lost Rivers of London*. Leicester; Leicester University Press.

Bassoullet, J.-P., Fourcade, E. & Peybernes, B., 1985. Paleobiogeographie des grands foraminiferes benthiques des marges neo-tethysiennes au Jurassique et au Cretace inferieur. *Bulletin de la Societe Geologique de France, Ser. 8*, **I**(5): 699–713.

Bastia, R., 2006. An overview of Indian sedimentary basins with special focus on emerging east coast deep water frontiers. *The Leading Edge*, July 2006: 818–29.

Bastia, R. & Nayak, P.K., 2006. Tectonostratigraphy and depositional patterns in Krishna offshore basin, Bay of Bengal. *The Leading Edge*, July 2006: 839–45.

Bastia, R., Nayak, P. & Singh, P., 2007. Shelf delta to deepwater basin: a depositional model for Krishna-Godavari basin. *Search and Discovery Article*, 40231.

Basu, D.N., Banerjee, A. & Tamhane, D.M., 1980. Source areas and migration trends of oil and gas in Bombay offshore basin, India. *American Association of Petroleum Geologists Bulletin*, **64**(2): 209–20.

Bates, J.M. & Spencer, R.S., 1979. Modifications of foraminiferal trends by the Chesapeake–Elisabeth sewage outfall, Virginia Beach, Virginia. *Journal of Foraminiferal Research*, **9**(2): 125–40.

Bates, M., Horne, D.J., Jones, R.W. & Whittaker, J.E., 2010. Marine microfossils from the West Sussex coastal plain: problems in the interpretation of non-analogue faunas. *Abstracts, INQUA 'Past and Present Land–Ocean Interactions in the Geological Record' Field Meeting, Arundel*.

Battarbee, R., MacKay, A., Birks, J. & Oldfield, F., 2003. *Global Change in the Holocene*. London; Arnold Hodder.

Beaumont, E.A. & Foster, N.H. (eds.), 1999. *Treatise of Petroleum Geology: Exploring for Oil and Gas Traps*. Tulsa, OK; American Association of Petroleum Geologists.

Beard, J.H., Sangree, J.B. & Smith, L.A., 1982. Quaternary chronology, paleoclimate, depositional sequences, and eustatic cycles. *American Association of Petroleum Geologists Bulletin*, **66**(2): 158–69.

Beavington-Penney, S.J., 2004. Analysis of the effects of abrasion on the test of *Palaeonummulites venosus*: implications for the origin of nummulithoclastic sediments. *Palaois*, **19**(2): 143–55.

Beavington-Penney, S.J. & Racey, A., 2004. Ecology of extant nummulitids and other larger benthic Foraminifera: applications in palaeoenvironmental analysis. *Earth-Science Reviews*, **67**: 219–65.

Beavington-Penney, S.J., Wright, V.P. & Racey, A., 2005. Sediment production and dispersal on Foraminifera-dominated early Tertiary ramps: the Eocene El Garia formation, Tunisia. *Sedimentology*, **52**: 537–69.

Beavington-Penney, S.J., Wright, V.P. & Racey, A., 2006. The Middle Eocene Seeb formation of Oman: an investigation of acyclicity, stratigraphic completeness and accumulation rates in shallow marine carbonate settings. *Journal of Sedimentary Research*, **76**(10): 1137–61.

Beglinger, S.E., Doust, H. & Cloetingh, S., 2012. Relating petroleum system and play development to basin evolution: west African South Atlantic basins. *Marine and Petroleum Geology*, **30**: 1–25.

Belanger, C.L., 2011. Evaluating taphonomic bias of paleoecological data in fossil benthic foraminiferal assemblages. *Palaios*, **26**: 767–78.

Belasky, P., 1996. Biogeography of Indo-Pacific larger Foraminifera and scleractinian corals: a probabilistic approach to estimating taxonomic diversity, faunal similarity, and sampling bias. *Palaeogeography, Palaeoclimatology, Palaeoecology*, **122**: 119–41.

Bell, F.G., 2000. *Engineering Properties of Soils and Rocks. Fourth Edition*. Oxford; Blackwell Science.

Bell, M. & Walker, M.J.C., 1992. *Late Quaternary Environmental Change: Physical & Human Perspectives*. London; Longman.

Belopolsky, A., Boyd, K., Carmichael, S. *et al.*, 2006. Controls on deepwater sandstone deposition in the Campos and Espirito Santo basins of Brazil: implications for reservoir prediction. *Abstracts, American Association of Petroleum Geologists International Conference and Exhibition, Perth, Australia.*

Belopolsky, A., Boyd, K., Carmichael, S. *et al.*, 2007. Controls on deepwater sandstone deposition in the Campos and Espirito Santo basins of Brazil: implications for reservoir prediction. *Abstracts, Geological Society 'South Atlantic Petroleum Systems' Conference, London.*

Bennett, M.R. & Doyle, P., 1997. *Environmental Geology: Geology and the Human Environment*. New York; John Wiley.

Benton, M.J. (ed.), 1993. *The Fossil Record*. London; Chapman & Hall.

Berger, A.L., Imbrie, J., Hays, G. *et al.* (eds.), 1984. *Milankovitch and Climate*. Dordrecht; D. Riedel.

Berger, W.H., 1969. Ecological patterns of living planktonic Foraminifera. *Deep-Sea Research*, **16**: 1–24.

Berger, W.H. & Diester-Haass, L., 1988. Paleoproductivity: the benthic/planktonic ratio in Foraminifera as a productivity index. *Marine Geology*, **81**: 15–25.

Berger, W.H., Smetacek, V.S. & Wefer, G. (eds.), 1989. *Productivity in the Ocean: Present and Past*. New York; John Wiley.

Berggren, W.A., 1969. Cenozoic stratigraphic planktonic foraminiferal zonation and the radiometric time scale. *Nature*, **224**: 1072–5.

Berggren, W.A. & Miller, K.G., 1988. Paleogene tropical planktonic foraminiferal biostratigraphy and magnetochronology. *Micropaleontology*, **34**(4): 362–80.

Berggren, W.A., Kent, D.V., Aubry, M.-P. & Hardenbol, J., 1995. *Geochronology, Time Scales and Global Stratigraphic Correlation*. Tulsa, OK; Society of Economic Paleontologists and Mineralogists/Society for Sedimentary Geology (Special Publication, No. 54).

Berkeley, A., Perry, C.T. & Smithers, S.G., 2009. Taphonomic signatures and patterns of test degradation on tropical, intertidal benthic Foraminifera. *Marine Micropaleontology*, **73**: 148–63.

Bermudez, P.J., 1962. Foraminiferos de los lutitos de Punta Tolete, Territorio Delta Amacuro (Venezuela). *Geos, Caracas*, **8**: 35–8.

Bermudez, P.J., 1966. Consideraciones sobre os sedimentos del Mioceno Medio al Reciente de los costas central y oriental de Venezuela, primera parte. *Boletin de Geologia*, **14**: 333–412.

Bermudez, P.J. & Stainforth, R.M., 1975. Aplicaciones de foraminiferos planctonicos e la bioestratigrafia del Terciario en Venezuela. *Revista Espanola de Micropaleontologia*, **7**(3): 373–89.

Berney, C. & Pawlowski, J., 2003. Revised small subunit rRNA analysis provides further evidence that Foraminifera are related to Cercozoa. *Journal of Molecular Evolution*, **57**: S120–7.

Bernhard, J., 1986. Characteristic assemblages and morphologies of benthic Foraminifera from anoxic, organic-rich deposits. *Journal of Foraminiferal Research*, **16**: 207–15.

Bernhard, J., 1988. *Post mortem* vital staining in benthic Foraminifera: duration and importance in population and distribution studies. *Journal of Foraminiferal Research*, **18**: 143–6.

Bernhard, J., 1989. The distribution of benthic Foraminifera with respect to oxygen concentration and organic carbon levels in shallow-water Antarctic sediments. *Limnology and Oceanography*, **34**(6): 1131–41.

Bernhard, J.M., 1993. Experimental and field evidence of Antarctic foraminiferal tolerance to anoxia and hydrogen sulphide. *Marine Micropaleontology*, **20**(3–4): 203–13.

Bernhard, J.M., 1996. Microaerophilic and facultative anaerobic benthic Foraminifera: a review of experimental and ultrastructural evidence. *Revue de Paleobiologie*, **15**: 261–75.

Bernhard, J., 2000. Distinguishing live from dead Foraminifera: methods review and proper applications. *Micropaleontology*, **46**(1): 38–46.

Bernhard, J., 2003. Possible symbionts in bathyal Foraminifera. *Science*, **200**: 861.

Bernhard, J. & Alve, E., 1996. Survival, ATP pool, and ultrastructural characterization of benthic Foraminifera from Drammensfjord (Norway): response to anoxia. *Marine Micropaleontology*, **28**(1): 5–18.

Bernhard, J.M., Barry, J.P., Buck, K.R. & Starczak, V.R., 2009. Impact of intentionally injected carbon dioxide hydrate on deep-sea benthic foraminiferal survivial. *Global Change Biology*, 2009 (doi: 10.1111/j.1365–2486.2008.01822.x).

Bernhard, J.M. & Bowser, S.S., 1999. Benthic Foraminifera of dysoxic sediments: chloroplast sequestration and functional morphology. *Earth-Science Reviews*, **46**(1–4): 149–65.

Bernhard, J.M., Casciotti, K., McIlvin, M.R. *et al.*, 2012. Potential importance of physiologically diverse benthic Foraminifera in sedimentary nitrate storage and respiration. *Journal of Geophysical Research*, **117**: G03002 (doi: 10.1029/2012J9001949).

Bernhard, J.M., Ostermann, D.R., Williams, D.S. & Blanks, J.K., 2006. Comparison of two methods to identify live benthic Foraminifera: a test between Rose Bengal and Cell-Tracker Green with implications for stable isotope paleoreconstruction. *Paleoceanography*, **21**(4): PA210.

Bernhard, J.M., Sen Gupta, B.K. & Baguley, J.G., 2008. Benthic Foraminifera living in Gulf of Mexico bathyal and abyssal sediments: community analysis and comparison to metazoan meiofaunal biomass and density. *Deep Sea Research Part II: Topical Studies in Oceanography*, **55**(24–26): 2617–26.

Bernhard, J.M., Sen Gupta, B.K. & Borne, P.F., 1997. Benthic foraminiferal proxy to estimate dysoxic bottom-water oxygen concentrations: Santa Barbara basin, US Pacific continental margin. *Journal of Foraminiferal Research*, **27**: 301–10.

Bernhard, J.M., Visscher, P.T. & Bowser, S.S., 2003. Submillimeter life positions of bacteria, protists and metazoans in laminated sediments of the Santa Barbara basin. *Limnology and Oceanography*, **48**(2): 813–28.

Bernstein, B.B., Hessler, R., Smith, R. & Jumars, P., 1978. Spatial dispersion of benthic Foraminifera in the abyssal central North Pacific. *Limnology and Oceanography*, **23**(3): 401–16.

Bertagne, A.J., 1984. Seismic stratigraphy of Veracruz tongue, deep southwestern Gulf of Mexico. American *Association of Petroleum Geologists Bulletin*, **68**: 1894–907.

Best, J.A., Barazangi, M., Al-Saad, D. *et al.*, 1993. Continental margin evolution of the northern Arabian platform in Syria. *American Association of Petroleum Geologists Bulletin*, **77**(2): 173–93.

Beydoun, Z.R., 1986. The petroleum resources of the Middle East: a review. *Journal of Petroleum Geology*, **9**(1): 5–28.

Beydoun, Z.R., 1988. *The Middle East: Regional Geology and Petroleum Resources*. Beaconsfield; Scientific Press.

Beydoun, Z.R., 1991. *Arabian Plate Hydrocarbon Geology and Potential: A Plate Tectonic Approach*. Tulsa, OK; American Association of Petroleum Geologists (Studies in Geology, No. 33).

Beydoun, Z.R., 1993. Evolution of the northeastern Arabian plate margin and shelf: hydrocarbon habitat and conceptual future potential. *Revue de l'Institut Francais du Petrole*, **48**(4): 311–45.

Bhalla, S.N., 1969. Foraminifera from the type Raghavapuram Shales, east coast Gondwanas, India. *Micropaleontology*, **15**(1): 61–84.

Bhalla, S.N., Khare, N., Shaunkukha, D.H. & Henriques, P.J., 2007. Foraminiferal studies in nearshore regions of western coast of India and Laccadives Islands: a review. *Indian Journal of Marine Sciences*, **36**(4): 272–87.

Bhalla, S.N. & Nigam, R., 1986. Recent Foraminifera from polluted marine environments of Velsao Beach, south Goa, India. *Revue de Paleobiologie*, **5**: 43–6.

Bhalla, S.N. & Talib, A., 1991. Callovian–Oxfordian Foraminifera from Jhurio Hill, Kutch, western India. *Revue de Paleobiologie*, **10**(1): 85–114.

Bhaumik, A.K. & Sen Gupta, A.K., 2005. Deep sea benthic Foraminifera from gas hydrate rich zone, Blake Ridge, North Atlantic (ODP Hole 997A). *Current Science*, **88**(12): 1969–73.

Bicchi, E., Barras, C., Denoyelle, M. *et al.*, 2010. The impact of offshore drilling activities monitored by recent and subfossil assemblages of benthic Foraminifera. *Abstracts, Forams2010 (International Symposium on Foraminifera), Bonn*: 57.

Bicchi, E., Ferrero, E. & Gonera, M., 2003. Palaeoclimatic interpretation based on Middle Miocene planktonic Foraminifera: the Silesia basin (Paratethys) and Monferrato (Tethys) records. *Palaeogeography, Palaeoclimatology, Palaeoecology*, **196**: 265–303.

Bignot, C., 1985. *Elements of Micropalaeontology*. London; Graham & Trotman.

Billman, H., Hottinger, L. & Oesterle, H., 1980. Neogene to Recent rotaliid Foraminifera from the Indo-Pacific Ocean: their canal system, their classification and their stratigraphic use. *Schweizerische Palaontologische Abhandlungen*, **101**: 71–113.

Bingham, L., Zurita-Milla, R. & Escalona, A., 2012. Geographic Information System-based fuzzy-logic analysis for petroleum exploration with a case study of northern South America. *American Association of Petroleum Geologists Bulletin*, **96**(11): 2121–42.

Bini, M., Bruckner, H., Chelli, A. *et al.*, 2012. Palaeogeographies of the Magra Valley coastal plain to constrain the location of the Roman harbour of Luna (NW Italy). *Palaeogeography, Palaeoclimatology, Palaeoecology*, **337**–8: 37–51.

Biswas, S.K., 1982. Rift basins in western margin of India and their hydrocarbon prospects with special reference to Kutch basin. *American Association of Petroleum Geologists Bulletin*, **66**(10): 1497–513.

Bkorlykke, K., 2010. *Petroleum Geosciences: From Sedimentary Environments to Rock Physics*. Berlin; Springer.

Blow, W.H., 1979. *The Cainozoic Globigerinida*. Leiden; E.J. Brill.

Bock, W.D., 1976. Distribution and significance of Foraminifera in the MAFLA area. *Maritime Sediments Special Publication*, **1**: 221–38.

Boersma, A. & Premoli Silva, I., 1983. Paleocene planktonic foraminiferal biogeography and paleoceanography of the Atlantic Ocean. *Micropaleontology*, **29**(4): 355–81.

Boersma, A. & Premoli Silva, I., 1991. Distribution of Paleogene planktonic Foraminifera – analogies with the Recent? *Paleoceanography, Palaeoclimatology, Palaeoecology*, **83**: 29–48.

Bolli, H.M., 1969. Zonacion de sedimentos marinos del Cretaceo hasa el Plioceno basada en foraminiferos planctonicos. *Instituto Mexicano del Petroleo*, **69**(AE/047): 1–36.

Bolli, H.M., Beckmann, J.-P. & Saunders, J.B., 1994. *Benthic Foraminiferal Biostratigraphy of the South Caribbean Region*. Cambridge; Cambridge University Press.

Bolli, H.M., Saunders, J.B. & Perch-Nielsen, K. (eds.), 1985. *Plankton Stratigraphy*. Cambridge; Cambridge University Press.

Boltovskoy, E., 1963. The littoral foraminiferal biocoenoses of Puerto Deseado (Patagonia, Argentina). *Contributions, Cushman Foundation for Foraminiferal Research*, **14**: 58–70.

Boltovskoy, E., 1984. Foraminifera of mangrove swamps. *Physis, Buenos Aires, Secc. A*, **42**: 1–9.

Boltovskoy, E. & Boltovskoy, A., 1968. Foraminiferos y tecamebas de la parte inferior del Rio Quequen Grande, Provincia de Buenos Aires, Argentina. *Revista del Museo Argentino de Ciencias Naturales 'Bernardino Ruvadavia', Hidrobiologia*, **2**: 127–64.

Boltovskoy, E., Scott, D.B. & Medioli, F.S., 1991. Morphological variations of benthic foraminiferal tests in response to changes in ecological parameters: a review. *Journal of Paleontology*, **65**(2): 175–85.

Boltovskoy, E. & Wright, R.A., 1976. *Recent Foraminifera*. The Hague; Dr W. Junk.

Bond, D.P.G. & Wignall, P.B., 2009. Latitudinal selectivity of foraminiferal extinctions during the late Guadalupian crisis. *Paleobiology*, **35**(4): 465–83.

Borcic, A., Bogner, D. & Popadic, S., 2010. Using Artificial Neural Networks (ANN) and Self Organising Maps (SOM) for better understanding of Recent benthic Foraminifera distribution. *Abstracts, Forams2010 (International Symposium on Foraminifera), Bonn*: 60.

Bordelon, L. & Schneider, B., 2010. How are Foraminifera affected by ocean acidification? *Abstracts, Forams2010 (International Symposium on Foraminifera), Bonn*: 60–1.

Bordenave, M.L. & Burwood, R., 1990. Source rock distribution and maturation in the Zagros orogenic belt: provenance of the Asmari and Bangestan reservoir oil accumulation. *Organic Geochemistry*, **16**: 369–87.

Bordenave, M.L. & Huc, A.-Y., 1995. The Cretaceous source rocks of the Zagros foothills of Iran. *Revue de l'Institut Francais du Petrole*, **50**(6): 727–52.

Bornmalm, L., 1997. *Taxonomy and Palaeoecology of Late Neogene Benthic Foraminifera from the Caribbean Sea and Eastern Equatorial Pacific Ocean*. Oslo; Scandinavian University Press (Fossils and Strata, No. 41).

Bosak, T., Lahr, D.J.G., Pruss, S.B. *et al.*, 2012. Possible early Foraminifera in post-Sturtian

(716–635Ma) cap carbonates. *Geology*, **40**(1): 67–70.

Bosence, D.W.J. & Allison, P.A., 1995. *Marine Palaeoenvironmental Analysis from Fossils*. London; The Geological Society (Special Publication, No. 83).

Botkin, D.B. & Keller, E.A., 2010. *Environmental Science. Seventh Edition*. New York; John Wiley.

Bouchet, V.M.P., Alve, E., Rygg, B. *et al.*, 2010. Determining ecological quality status of coastal waters using benthic foraminifera: calibration with benthic macrofauna. *Abstracts, Forams2010 (International Symposium on Foraminifera), Bonn*: 61.

Bouchet, V.M.P., Sauriau, P.-G., Debenay, J.-P. *et al.*, 2009. Influence of the mode of macrofauna-mediated bioturbation on the vertical distribution of living benthic Foraminifera: first insight from axial tomodensitometry. *Journal of Experimental Marine Biology and Ecology*, **371**(1): 20–33.

BouDagher-Fadel, M.K., 2002. The stratigraphical relationship between planktonic and larger benthic Foraminifera in Middle Miocene to Lower Pliocene carbonate facies of Sulawesi, Indonesia. *Micropaleontology*, **48**: 153–76.

BouDagher-Fadel, M.K., 2008. *Evolution and Geological Significance of Larger Benthic Foraminifera*. Oxford; Elsevier.

BouDagher-Fadel, M.K., 2012. *Evolution and Geological Significance of Planktonic Foraminifera*. Oxford; Elsevier.

BouDagher-Fadel, M.K. & Banner, F.T., 1999. Revision of the stratigraphic significance of the Oligocene–Miocene 'Letter Stages'. *Revue de Micropaleontologie*, **42**: 93–7.

BouDagher-Fadel, M.K., Banner, F.T. & Whittaker, J.E., 1997. *The Early Evolutionary History of Planktonic Foraminifera*. London; Chapman & Hall.

BouDagher-Fadel, M.K. & Price, G.D., 2010. Evolution and palaeogeographic distribution of the lepidocyclinids. *Journal of Foraminiferal Research*, **40**(1): 79–108.

Boudreau, R.E.A., Patterson, R.T., Dalby, A.P. & McKillop, W.B., 2001. Non-marine occurrence of the foraminifer *Cribroelphidium gunteri* in northern Lake Winnipegosis, Manitoba, Canada. *Journal of Foraminiferal Research*, **31**(2): 108–19.

Bouma, A.H., Normark, W.R. & Barnes, N.E. (eds.), 1985. *Submarine Fans and Related Turbidite Systems*. Berlin; Springer-Verlag.

Bowden, A.J., Gregory, F.J. & Henderson, A.S. (eds.), in press. *History of Foraminiferal Micropalaeontology*. London; The Geological Society (for The Micropalaeontological Society).

Boyd, R., Suter, J. & Penland, S., 1989. Sequence stratigraphy of the Mississippi delta. *Transactions of the Gulf Coast Association of Geological Societies*, **39**: 331–40.

Bozorgnia, F., 1964. *Microfacies and Microorganisms of Paleozoic through Tertiary Sediments of Some Parts of Iran*. Tehran; National Iranian Oil Company.

Bozorgnia, F., 1973. *Paleozoic Foraminiferal Biostratigraphy of Central and East Alborz Mountains, Iran*. Tehran; National Iranian Oil Company (Geological Laboratories Publication No. 4).

Bozorgnia, F. & Kalantari, A., 1965. *Nummulites of Parts of Central and East Iran*. Tehran; National Iranian Oil Company.

BP, 1956. Oil and gas in southwest Iran. *Proceedings, 20th International Geological Congress, Mexico*, **II**: 33–72.

Brady, H.B., 1876. *Monograph on Carboniferous and Permian Foraminifera (the genus Fusulina excepted)*. London; The Palaeontographical Society.

Brady, H.B., 1884. Report on the Foraminifera dredged by H.M.S. *Challenger* during the years 1873–1876. *Reports of the Scientific Results of the Voyage of H.M.S.* Challenger, *Zoology*, 9.

Brady, H.B., 1888. Note on the so-called 'soapstone' of Fiji. *Quarterly Journal of the Geological Society*, **44**(1–4): 1–10.

Brasier, M.D., 1975. An outline history of seagrass communities. *Palaeontology*, **18**(4): 681–702.

Brasier, M.D., 1980. *Microfossils*. London; George Allen & Unwin.

Brasier, M.D. & Donahue, J., 1985. Barbuda – an emerging reef and lagoon complex on the edge of the Lesser Antilles island arc. *Journal of the Geological Society*, **142**: 1101–17.

Brasier, M.D. & Green, O.R., 1993. Winners and losers: stable isotopes and microhabitats of living Archaiasidae and Eocene Nummulitidae (larger Foraminifera). *Marine Micropaleontology*, **20**(3/4): 267–76.

Breard, S.Q., Callender, A.D., Denne, R.A. & Nault, M.J., 2000. Taxonomic uniformitarianism in Gulf of Mexico basin Cenozoic foraminiferal paleoecology: is the present always the key to the past? *Gulf Coast Association of Geological Societies Transactions*, **50**: 725–36.

Brenchley, P. (ed.), 1984. *Fossils and Climate*. New York; John Wiley

Brenchley, P.J. & Williams, B.P.J. (eds.), 1985. *Sedimentology: Recent Developments and Applied Aspects.* Oxford; Blackwell.

Brenckle, P.L., 2005. *A Compendium of Upper Devonian-Carboniferous Type Foraminifers from the Former Soviet Union.* Cushman Foundation for Foraminiferal Research (Special Publication, No. 38).

Briggs, J.C., 2003. The tectonic and biogeographic history of India. *Journal of Biogeography*, **30**: 381–8.

Briggs, D.E.G. & Crowther, P.R. (eds.), 2001. *Palaeobiology. Second Edition.* Oxford; Blackwell.

Briguglio, A. & Hohenegger, J., 2009. Nummulitids hydrodynamics: an example using *Nummulites globulus* Leymerie, 1846. *Bollettino della Societa Geologica Italiana*, **48**(2): 105–11.

Briguglio, A. & Hohenegger, J., 2010. Quantifying forams growth with high-resolution X-ray computed tomography: ontogeny, phylogeny and paleoceanographic applications. *Abstracts, Forams2010 (International Symposium on Foraminifera), Bonn*: 62–3.

Briguglio, A. & Hohenegger, J., 2011. How to react to shallow water hydrodynamics: the larger benthic Foraminifera solution. *Marine Micropaleontology*, **81**: 63–76.

Brink, A.H., 1974. Petroleum geology of Gabon basin. *American Association of Petroleum Geologists Bulletin*, **58**: 216–35.

Brognon, G.P. & Verrier, G.R., 1966. Oil and geology in Cuanza basin of Angola. *American Association of Petroleum Geologists Bulletin*, **50**: 108–58.

Bromley, R.G. & Nordmann, E., 1971. Maastrichtian adherent Foraminifera encircling clionid pores. *Bulletin of the Geological Society of Denmark*, **20**: 362–8.

Brönnimann, P., 1978. Recent benthonic Foraminifera from Brazil. Morphology and ecology. Part III. Notes on *Asterotrochammina* Bermudez & Seiglie. *Notes du Laboratoire Paleontologie, Universite de Geneve*, **3**: 1–8.

Brönnimann, P. & Maisonneuve, E., 1980. Revision of the trochamminid genus *Remaneica* Rhumbler, 1938 (Foraminiferida). *Notes du Laboratoire Paleontologie, Universite de Geneve*, **6**: 1–14.

Brönnimann, P. & Renz, H.H. (eds.), 1969. *Proceedings of the First International Conference on Planktonic Microfossils.* Leiden; E.J. Brill.

Brönnimann, P. & Whittaker, J.E., 1983. *Deuterammina* (*Lepidodeuterammina*) subgen. nov., and a redescription of *Rotalina ochracea* Williamson (Protozoa: Foraminiferida).

*Bulletin of the British Museum (Natural History), Zoology*, **45**(5): 233–8.

Brönnimann, P. & Whittaker, J.E., 1993. Taxonomic revision of some Recent agglutinated Foraminifera from the Malay Archipelago in the Millett collection, British Museum (Natural History). *Bulletin of the British Museum (Natural History)*, **59**(2): 107–24.

Brönnimann P., Whitttaker, J.E. & Zaninetti, L., 1978. *Shanita*, a new pillared miliolacean foraminifer from the Late Pemian of Burma and Thailand. *Rivista Italiana di Paleontologia*, **84**(1): 63–92

Brönnimann P., Whittaker, J.E. & Zaninetti, L., 1992. Brackish water Foraminifera from mangrove sediments of southwestern Viti Levu, Fiji Islands, southwest Pacific. *Revue de Paleobiologie*, **11**(1): 13–65.

Brooks, J. (ed.), 1990. *Classic Petroleum Provinces.* London; The Geological Society (Special Publication, No. 50).

Brooks, J. & Fleet, A.J. (eds.), 1987. *Marine Petroleum Source Rocks.* London; The Geological Society (Special Publication, No. 26).

Brooks, J. & Glennie, K. (eds.), 1987. *Petroleum Geology of North-West Europe.* London; Graham & Trotman.

Brouwer, J., 1965. Agglutinated foraminiferal faunas from turbiditic sequences, I & II. *Proceedings, Koninklijke Nederlandse Akademie van Wetenschappen, Section B*, **68**(5): 309–34.

Brown, J.C. & Dey, A.K., 1975. *The Mineral Fields of the Indian Subcontinent and Burma.* Oxford; Oxford University Press.

Brown, J.F., 1995. *Macroevolution.* Chicago; University of Chicago Press.

Brown, L.F., Jr., Loucks, R.G. & Trevino, R.H., 2005. Site-specific sequence-stratigraphic section benchmark charts are key to regional chronostratigraphic systems tract analysis in growth-faulted basins. *American Association of Petroleum Geologists Bulletin*, **89**(6): 715–24.

Brown, L.F., Jr., Benson, J.M., Brink, G.J. *et al.*, 1995. *Sequence Stratigraphy in Offshore South African Divergent Basins.* Tulsa, OK; American Association of Petroleum Geologists (Studies in Geology, No. 41).

Brown, R.E., Anderson, L.D., Thomas, E. & Zachos, J.C., 2011. A core-top calibration of B/Ca in the benthic foraminifers *Nuttallides umbonifera* and *Oridorsalis umbonatus*: a proxy for Cenozoic bottom water carbonate saturation. *Earth and Planetary Science Letters*, **310**: 360–8.

Bruckner, S. & Mackensen, A., 2008. Organic matter rain rates, oxygen availability, and vital effects from benthic Foraminifera del13C in the historic Skagerrak, North Sea. *Marine Micropaleontology*, **66**(3–4): 192–207.

Brunner, C.A., 1979. Distribution of planktonic Foraminifera in surface sediments of the Gulf of Mexico. *Micropaleontology*, **25**(3): 325–35.

Brunner, C.A., Beall, J.M., Bentley, S.J. & Furukawa, Y., 2006. Hypoxia hotspots in the Mississippi Bight. *Journal of Foraminiferal Research*, **36**(2): 95–107.

Brunner, C.A. & Ledbetter, M.T., 1987. Sedimentological and micropaleontological detection of turbidite muds in hemipelagic sequences: an example from the Late Pleistocene levee of Monterey fan, central California continental margin. *Marine Micropaleontology*, **12**: 223–39.

Brunner, C.A. & Normark, W.R., 1985. Biostratigraphic implications for turbidite depositional processes on the Monterey deep-sea fan, central California. *Journal of Sedimentary Petrology*, **55**(4): 495–505.

Bubik, M. & Kaminski, M.A. (eds.), 2004. *Proceedings of the Sixth International Workshop on Agglutinated Foraminifera*. Grzybowski Foundation (Special Publication, No. 8).

Buchanan, J.B. & Hedley, R.H., 1960. A contribution to the biology of *Astrorhiza limicola*. *Journal of the Marine Biological Association of the United Kingdom*, **39**: 549–60.

Buckley, D.E., Owens, E.H., Schafer, C.T. *et al.*, 1974. Canso Strait and Chedabucto Bay: a multidisciplinary study of the impact of man on the marine environment. *Geological Survey of Canada Paper*, **74**–30(1): 133–60.

Bujak, J. & Mudge, D., 1994. A high-resolution North Sea Eocene dinocyst zonation. *Journal of the Geological Society*, **151**(3): 449–62.

Burdige, D.J., 2006. *Geochemistry of Marine Sediments*. Princeton, NJ; Princeton University Press.

Burollet, R.F., Boichard, R., Lambert, B. & Villain, J.M., 1986. Sedimentation and ecology of the Paternoster Platform, East Kalimantan. *Proceedings of the Indonesian Petroleum Association 15th Annual Convention*: 155–70.

Burwood, R., Cornet, P.J., Jacobs, L. & Paulet, J., 1990. Organofacies variation control on hydrocarbon generation: a lower Congo coastal basin (Angola) case history. *Organic Geochemistry*, **16**(1–3): 325–38.

Burwood, R., Leplat, P., Mycke, B. & Paulet, J., 1992. Rifted margin source rock deposition: a carbon isotope and biomarker study of a west African Lower Cretaceous 'lacustrine'

section. *Organic Geochemistry*, **19**(1–3): 41–52.

Busnardo, R. & Granier, B., 2011. Aptian ammonites of Abu Dhabi (United Arab Emirates). *Carnets de Geologie*, **2011**(04): 117–35.

Butt, A., 1981. *Depositional Environments of the Upper Cretaceous Rocks in the Northern Part of the Eastern Alps*. Cushman Foundation for Foraminiferal Research (Special Publication, No. 20).

Buxton, M.W.N. & Pedley, H.M., 1989. A standardized model for Tethyan Tertiary carbonate ramps. *Journal of the Geological Society*, **146**(5): 746–8.

Buzas, M.A., 1968. On the spatial distribution of Foraminifera. *Contributions from the Cushman Foundation for Foraminiferal Research*, **19**: 1–11.

Buzas, M.A., 1974. Vertical distribution of *Ammobaculites* in Rhode River, Maryland. *Journal of Foraminiferal Research*, **4**(3): 144–7.

Buzas, M.A., Culver, S.J. & Jorissen, F.J., 1993. A statistical evaluation of the microhabitats of living (stained) infaunal benthic Foraminifera. *Marine Micropaleontology*, **20**: 311–20.

Buzas, M.A., Hayek, L.-A.C. & Culver, S.J., 2007. Community structure of benthic Foraminifera in the Gulf of Mexico. *Marine Micropaleontology*, **65**: 43–53.

Buzas, M.A. & Sen Gupta, B.K. (eds.), 1982. *Foraminifera: Notes for a Short Course*. Knoxville, TN; University of Tennessee (Studies in Geology, No. 6).

Cahuzac, B. & Poignant, A., 1997. Essai de biozonation dans les basins europeens a l'aide des grands foraminiferes neritiques. *Bulletin de la Societe Geologique de France*, **168**(2): 155–69.

Cainelli, C. & Mohriak, W., 1999. Some remarks on the evolution of sedimentary basins along the eastern Brazilian continental margin. *Episodes*, **22**(3): 206–16.

Callard, S.L., Gehrels, W.R., Morrison, B.V. & Grenfell, H.R., 2011. Suitability of saltmarsh Foraminifera as proxy indicators of sea level in Tasmania. *Marine Micropaleontology*, **79**: 121–31.

Cameron, N.R., Bate, R.H. & Clure, V.S. (eds.), 1999. *The Oil and Gas Habitats of the South Atlantic*. London; The Geological Society (Special Publication, No. 153).

Cann, J.H., Belperio, A.P. & Murray-Wallace, C.V., 2000. Late Quaternary paleosealevels and paleoenvironments inferred from Foraminifera, northern Spencer Gulf, South Australia.

*Journal of Foraminiferal Research*, **30**(1): 29–53.

Cannariato, K.J., Kennett, J.P. & Behl, R.J., 1999. Biotic response to late Quaternary rapid climate switches in Santa Barbara basin: ecological and evolutionary implications. *Geology*, **27**: 63–6.

Carbonel, P. & Moyes, J., 1987. Late Quaternary palaeoenvironments of the Mahakam delta. *Palaeogeography, Palaeoclimatology, Palaeoecology*, **61**: 265–84.

Cariou, E. & Hantzpergue, P. (co-ords.), 1997. *Biostratigraphie du Jurassique ouest-Eureopeen et Mediterraneen*. Bulletin des Centres de Recherches Exploration–Production Elf-Aquitaine. (Memoire, No. 17).

Carlsen, G.M. & Ghori, K.A.R., 2005. Canning Basin and global Palaeozoic petroleum systems. *APPEA Journal*, **2005**: 349–63.

Carmichael, S.M., Akhter, S., Bennett, J.K. *et al.*, 2009. Geology and hydrocarbon potential of the offshore Indus basin, Pakistan. *Petroleum Geoscience*, **15**: 117–30.

Carnahan, E.A., Hoare, A.M., Hallock, P. *et al.*, 2009. Foraminiferal assemblages in Biscayne Bay, Florida, USA: responses to urban and agricultural pollution in a subtropical estuary. *Marine Ecology*, **59**: 221–33.

Caron, M. & Homewood, P., 1983. Evolution of early planktonic foraminifers. *Marine Micropaleontology*, **7**: 453–62.

Carozzi, A.V., Reyes, M.V. & Ocango, V.P., 1976. *Microfacies and Microfossils of the Miocene Reef Carbonates of the Philippines*. Manila, Philippines; Philippines Oil Development Company (Special Publication).

Carr, T., d'Agostino, T., Amrose, W. *et al.* (eds.), 2009. *Unconventional Energy Resources: Making the Unconventional Conventional*. Gulf Coast Section – Society of Economic Paleontologists and Mineralogists (GCS-SEPM) Foundation (Proceedings of the 29th Annual Research Conference).

Carr-Brown, B., 1972. The Holocene/Pleistocene contact in the offshore area east of Galeota Point, Trinidad, West Indies. *VI Conferencia Geologica del Caribe, Margarita, Venezuela, Memorias*: 381–97.

Carrillat, A., Caline, B. & Davaud, E., 1999. Prediction of poroperm properties from rock fabric in a nummulite carbonate reservoir (El Garia fm., offshore Libya). *Journal of Conference Abstracts*, **4**(2): 907.

Carter, D.J. & Hart, M.B., 1977. Aspects of mid-Cretaceous stratigraphical micropalaeontology. *Bulletin of the British Museum (Natural History), Geology*, **29**: 1–135.

Carvalho, I.S., Cassab, R.C.T. & Schwanke, C. (eds.), 2007. *Paleontologia: Cenarios de Vida*. Rio de Janeiro; Editora Interciencia.

Castillo, M.-L.M., Sen Gupta, B.K. & Alcala-Herrera, J.A., 1998. Late Quaternary change in deep bathyal and abyssal waters of the Gulf of Mexico: preservation record of the foraminifer Biloculinella iregularis. *Journal of Foraminiferal Research*, **28**(2): 95–101.

Catuneanu, O., 2006. *Principles of Sequence Stratigraphy*. Amsterdam; Elsevier.

Catuneanu, O., Abreu, V., Bhattacharya, J. *et al.*, 2009. Towards the standardisation of sequence stratigraphy. *Earth-Science Reviews*, **92**: 1–33.

Catuneanu, O., Galloway, W.E., Kendall, G.St.C. *et al.*, 2011. Sequence stratigraphy: methodology and nomenclature. *Newsletters on Stratigraphy*, **44**(3): 173–245.

Caudri, C.M.B., 1996. The larger Foraminifera of Trinidad (West Indies). *Eclogae Geologicae Helvetiae*, **89**(3): 1137–309.

Caughey, C.A., Carter, D.C. & Clure, J. (eds.), 1996. *Proceedings of the International Symposium on Sequence Stratigraphy in SE Asia, May 1995*. Indonesian Petroleum Association.

Caus, E., Bernaus, J.M., Calonge, E. & Martin-Chivelet, J., 2009. Mid-Cenomanian separation of Atlantic and Tethyan domains in Iberia by a land-bridge: the origin of larger Foraminifera provinces? *Palaeogeography, Palaeoclimatology, Palaeoecology*, **283**: 172–81.

Caus, E., Gomez-Garrido, A. & Rodes, D., 1988. Reevaluation of the *Lepidorbitoides* evolution as a function of the age relations between species established with nannofossil biostratigraphy. *Revue de Paleobiologie, Volume Special*, **2**: 421–8.

Cavalier Smith, T., 1993. Kingdom Protozoa and its 18 phyla. *Microbiological Reviews*, **57**: 953–94.

Cavalier-Smith, T., 2003. Protist phylogeny and the high-level classification of Protozoa. *European Journal of Protistology*, **39**: 338–48.

Cavalier-Smith, T., 2004. Only six kingdoms of life. *Proceedings of the Royal Society of London, B*, **271**: 1251–62.

Cedhagen, T., Goldstein, S.T. & Gooday, A.J., (eds.), 2002. Biology and biodiversity of allogromiid foraminifera. *Journal of Foraminiferal Research*, **32**(4).

Cetean, C.G., Balc, R., Kaminski, M.A. & Filipescu, S. 2011. Integrated biostratigraphy and palaeoenvironments of an upper

Santonian–upper Campanian succession from the southern part of the Eastern Carpathians, Romania. *Cretaceous Research*, **32**(5): 575–90.

Cetean, C.G. & Kaminski, M.A., 2011. New deep-water agglutinated Foraminifera from the Upper Oligocene of Angola. *Micropaleontology*, **57**(3): 255–62.

Chamney, T.P., 1976. Foraminiferal morphogroup symbol for paleoenvironmental interpretation of drill cuttings samples: Arctic America. *Maritime Sediments Special Publication*, **1**: 585–624.

Chaproniere, G.C.H., 1975. Palaeoecology of Oligocene larger Foraminiferida, Australia. *Alcheringa*, **1**: 37–58.

Chaproniere, G.C.H., 1980. Oligocene and Miocene larger Foraminiferida from Australia and New Zealand. *Bureau of Mineral Resources Bulletin*, **188**: 1–98.

Chaproniere, G.C.H., 1981. Australasian mid-Tertiary larger foraminferal associations and their bearing on the East Indian Letter Classification. *Bureau of Mineral Resources Journal of Australian Geology and Geophysics*, **6**: 145–51.

Charnock, M.A. & Jones, R.W., 1997. North Sea lituolid Foraminifera with complex inner structures: taxonomy, stratigraphy and evolutionary relationships. *Annales Societatis Geologorum Poloniae*, **67**: 183–96.

Chatterjee, S., Goswami, A. & Scotese, C.R., 2013. The longest voyage: tectonic, magmatic and paleoclimatic evolution of the Indian plate during its northward flight from Gondwana to Asia. *Gondwana Research*, **23**: 238–67.

Cherchi, A. & Schroder, R., 2013. The *Praeorbitolina/Palorbitolinoides* association: an Aptian biostratigraphic key-interval at the southern margin of the Neo-Tethys. *Cretaceous Research*, **39**: 70–7.

Cherif, O.H., A-Rifaiy, I.A. & El-Deeb, W.Z.M., 1992. 'Post-nappes' early Tertiary foraminiferal paleoecology of the northern Hafit area, south of Al-Ain city (United Arab Emirates). *Micropaleontology*, **38**(1): 37–56.

Chowdhary, L.R., 2004. *Petroleum Geology of the Cambay Basin, Gujurat, India*. Dahra Dun; Indian Petroleum Publishers.

Christiansen, B., 1958. The foraminifer fauna in the Drobak Sound in the Oslo Fjord (Norway). *Nytt Magasin for Zoologi*, **6**: 1–91.

Christiansen, O., 1971. Notes on the biology of Foraminifera. *3me Symposium Europeen de Biologie Marine, Supplement, Vie et Milieu*, **22**: 465–78.

Cifelli, R. & Richardson, S., 1990. *The History of the Classification of Foraminifera 1826–1933*. Cushman Foundation for Foraminiferal Research (Special Publication, No. 27).

Cifelli, R. & Scott, G., 1986. Stratigraphic record of the Neogene globorotaliid radiation (planktonic Foraminiferida). *Smithsonian Contributions to Paleobiology*, **58**: 1–101.

Clark, D.L. & Bird, K.J., 1966. Foraminifera and paleoecology of the upper Austin and lower Taylor (Cretaceous) strata in north Texas. *Journal of Paleontology*, **40**(2): 315–27.

Clift, P.D., Kroon, D., Gaedicke, C. & Craig, J. (eds.), 2002. *The Tectonic and Climatic Evolution of the Arabian Sea Region*. London; The Geological Society (Special Publication, No. 195).

CLIMAP Project Members, 1976. The surface of the ice-age Earth. *Science*, **191**: 1131–7.

CLIMAP Project Members, 1981. *Seasonal Reconstruction of the Earth's Surface at the Last Glacial Maximum*. Boulder, CO; Geological Society of America (Map and Chart Series, No. 36).

Cloetingh, S., 1986. Tectonics of passive margins: implications for the stratigraphic record. *Geologie en Mijnbouw*, **65**: 103–17.

Cobbold, P.R., Meisling, K.E. & Mount, V.S., 2001. Reactivation of an obliquely rifted margin, Campos and Santos basins, southeastern Brazil. *American Association of Petroleum Geologists Bulletin*, **85**(11): 1925–44.

Cobos, S., 2005. Structural interpretation of the Monagas foreland thrust belt. *Search and Discovery Article*, 30031.

Coburn, T.C. & Yarus, J.M. (eds.), 2000. *Geographic Information Systems in Petroleum Exploration and Development*. Tulsa, OK; American Association of Petroleum Geologists (Applications in Geology, No.4).

Coccioni, R. & Galeotti, S., 1993. Orbitally induced cycles in benthonic foraminiferal morphogroups and trophic structure distribution patterns from the Late Albian 'Amadeus Segment' (central Italy). *Journal of Micropalaeontology*, **12**(2): 227–39.

Coccioni, R., Luciani, V. & Marsili, A., 2006. Cretaceous Oceanic Anoxic Events and radially elongated chambered planktonic Foraminifera: paleoecological and paleoceanographic implications. *Palaeogeography, Palaeoclimatology, Palaeoecology*, **235**: 66–92.

Coe, A.L. (ed.), 2003. *The Sedimentary Record of Sea-Level Change*. Milton Keynes; The Open University.

Coe, A.L., 2010. *Geological Field Techniques.* Oxford; Wiley-Blackwell.

Coimbra, J.C., Carreno, A.L. & de Santana dos Anjos-Zerfass, G., 2009. Biostratigraphy and palaeoceanographical significance of the Neogene planktonic Foraminifera from the Pelotas basin, southernmost Brazil. *Revue de Micropaleontologie*, **52**: 1–14.

Collinson, P., 1980. Vertical distribution of Foraminifera off the coast of Northumberland, England. *Journal of Foraminiferal Research*, **10**(1): 75–8.

Compton, P.M., 2009. The geology of the Barmer basin, Rajasthan, India, and the origins of its major oil reservoir, the Fatehgarh formation. *Petroleum Geoscience*, **15**: 107–16.

Cooper, R.A., Crampton, J.S., Raine, I. *et al.*, 2001. Quantitative biostratigraphy of the Taranaki basin, New Zealand: a deterministic and probabilistic approach. *American Association of Petroleum Geologists Bulletin*, **85**(8): 1469–98.

Corliss, B.H., 1985. Microhabitats of benthic Foraminifera within deep-sea sediments. *Nature*, **314**: 435–8.

Corliss, B.H., 1991. Morphology and microhabitat preferences of benthic Foraminifera from the northwest Atlantic Ocean. *Marine Micropaleontology*, **17**: 195–236.

Corliss, B.H. & Chen, C., 1988. Morphotype patterns of Norwegian Sea deep-sea benthic Foraminifera and ecological implications. *Geology*, **16**: 716–9.

Corliss, B.H. & Fois, E., 1990. Morphotype analysis of deep-sea benthic Foraminifera from the northwest Gulf of Mexico. *Palaios*, **5**: 589–605.

Cornell, S.E., Prentice, I.C., House, J.I. & Downy, C.J. (eds.), 2012. *Understanding the Earth System: Global Change Science for Application.* Cambridge; Cambridge University Press.

Cowie, J., 2013. *Climate Change: Biological and Human Aspects.* Cambridge; Cambridge University Press.

Cox, P.T., 1937. The genus *Loftusia* in south western Iran. *Eclogae Geologicae Helvetiae*, **30**(2): 431–50.

Coxall, H.K., Wilson, P.A., Pearson, P.N. & Sexton, P.F., 2007. Iterative evolution of digitate planktonic Foraminifera. *Paleobiology*, **33**(4): 495–516.

Crame, J.A. & Owen, A.W. (eds.), 2002. *Palaeobiogeography and Biodiversity Change: the Ordovician and Mesozoic-Cenozoic Radiations.* London; The Geological Society (Special Publication, No. 194).

Crawford, R.L. & Crawford, D.L. (eds.), 1996. *Bioremediation: Principles and Applications.* Cambridge; Cambridge University Press.

Creaney, S. & Passey, Q.R., 1993. Recurring patterns of total organic carbon and source rock quality within a sequence stratigraphic framework. *American Association of Petroleum Geologists Bulletin*, **77**(3): 386–401.

Crespo de Cabrera, S. & di Gianni Canudas, N., 1994. Biostratigraphy and paleogeography of the eastern Venezualan basin during the Oligo/Miocene. *V Simposio Bolivariano (Exploracion Perolera en las Cuenas Subandinas)*: 231–3.

Crews, J.R., Weimer, P., Pulham, A.J. & Waterman, A.S., 2000. Integrated approach to condensed section identification in intraslope basins, Pliocene–Pleistocene, northern Gulf of Mexico. *American Association of Petroleum Geologists Bulletin*, **84**(10): 1519–36.

Crittenden, S., 1986. Planktonic Foraminifera and biostratigraphy of the early Tertiary strata of the North Sea basin: a bried discussion. *Newsletters on Stratigraphy*, **15**: 163–71.

Cronblad, H.G. & Malmgren, B.A., 1981. Climatically controlled variation of Sr and Mg in Quaternary planktonic Foraminifera. *Nature*, **291**: 61–4.

Cross, N., Goodall, I., Hollis, C. *et al.*, 2010. Reservoir description of a mid-Cretaceous siliciclastic–carbonate ramp reservoir: Mauddud formation in the Raudhatain and Sabiriyah fields, north Kuwait. *GeoArabia*, **15**(2): 17–50.

Crux, J.A., Gard, I.G., Griggs, P.H. *et al.*, 2005. Fossil age group plots: a rapid interpretation technique for complex structural areas. *Search and Discovery Article*, 40161.

Cubaynes, R., Rey, J., Ruget, C. *et al.*, 1990. Relations between systems tract and micropaleontological assemblages on a Toarcian carbonate shelf (Quercy, southwest France). *Bulletin de la Societe Geologique de France*, **6**(6): 989–93.

Cubitt, J.M. & Reyment, R.A. (eds.), 1982. *Quantitative Stratigraphic Correlation.* Chichester; John Wiley.

Culver, S.J., 1988. New foraminiferal depth zonation of the northwestern Gulf of Mexico. *Palaios*, **3**: 69–85.

Culver, S.J., 1990. Benthic Foraminifera of Puerto Rican mangrove-lagoon systems: potential for paleoenvironmental interpretation. *Palaios*, **5**: 34–51.

Culver, S.J. & Buzas, M.A., 1981. *Distribution of Recent Benthic Foraminifera in the Gulf of*

*Mexico*. Washington, DC; Smithsonian University Press (Smithsonian Contributions to the Marine Sciences, No. 8).

Culver, S.J. & Buzas, M.A., 1982a. *Distribution of Recent Benthic Foraminifera in the Caribbean Region*. Washington, DC; Smithsonian University Press (Smithsonian Contributions to the Marine Sciences, No. 14).

Culver, S.J. & Buzas, M.A., 1982b. *Distribution of Recent Benthic Foraminifera in the Gulf of Mexico. Volume II*. Washington, DC; Smithsonian Institution Press (Smithsonian Contributions to Marine Science, No. 8).

Culver, S.J. & Buzas, M.A., 1983. Recent benthic foraminiferal provinces in the Gulf of Mexico. *Journal of Foraminiferal Research*, **13**(1): 21–31.

Culver, S.J. & Rawson, P.F. (eds.), 2000. *Biotic Response to Global Change: The Last 145 Million Years*. Cambridge; Cambridge University Press.

Curiale, J.A., Covington, G.H., Shamsuddin, A.H.M. *et al.*, 2002. Origin of petroleum in Bangladesh. *American Association of Petroleum Geologists Bulletin*, **86**(4): 625–52.

Curtis, N.M., Jr., 1955. Paleoecology of the Viesca member of the Weches formation at Smithville, Texas. *Journal of Paleontology*, **29**(2): 263–82.

Cushman, J.A., 1922. Shallow water Foraminifera of the Tortugas region. *Papers, Tortugas Laboratory*, **17**: 1–85.

Dabbous, S.A. & Scott, D.B., 2012. Short-term monitoring of Halifax Harbour (Nova Scotia, Canada) pollution remediation using benthonic Foraminifera as proxies. *Journal of Foraminiferal Research*, **42**(3): 187–205.

Dalby, A.P., Patterson, R.T. & Haggart, J.W., 2009. Distribution of Albian–Cenomanian Foraminifera from Queen Charlotte Islands, British Columbia, Canada: constraints on the timing of the northward migration of the Wrangella Terrane. *Journal of Foraminiferal Research*, **39**(3): 231–45.

Daniel, E.J., 1954. Fractured oil reservoirs of the Middle East. *American Association of Petroleum Geologists Bulletin*, **38**: 774–815.

Dangwal, V., Sengupta, S. & Desai, A.G., 2008. Speculated petroleum systems in deep offshore Mahanadi basin in Bay of Bengal, India. *7th International Conference & Exposition on Petroleum Geophysics, Hyderabad*, P-01.

Darling, K.F., Kroon, D., Wade, C.M., & Leigh Brown, A.J., 1996. Molecular phylogeny of the planktonic Foraminifera. *Journal of Foraminiferal Research*, **26**(4): 324–30.

Darling, K.F., Kucera, M., Kroon, D. & Wade, C.M., 2006. A resolution for the coiling direction paradox in *Neogloboquadrina pachyderma*. *Paleoceanography*, **21**: PA2011.

Darling, K.F., Kucera, M., Pudsey, C.J. & Wade, C.M., 2004. Molecular evidence links cryptic diversification in polar plankton to Quaternary climate dynamics. *Proceedings of the National Academy of Sciences, USA*, **101**: 7657–62.

Darling, K.F., Kucera, M. & Wade, C.M., 2007. Global molecular phylogeography reveals persistent Arctic circumpolar isolation in a marine planktonic protist. *Proceedings of the National Academy of Sciences, USA*, **104**: 5002–7.

Darling, K.F., Kucera, M., Wade, C.M. *et al.*, 2003. Seasonal distribution of genetic types of planktonic foraminifer morphospecies in the Santa Barbara Channel and its paleoceanographic implications. *Paleoceanography*, **18**(2): 1032.

Darling, F. & Wade, C.M., 2008. The genetic diversity of planktonic Foraminifera and the global distribution of ribosomal RNA genotypes. *Marine Micropaleontology*, **67**: 216–38.

Darling, K.F., Wade, C.M., Kroon, D. & Leigh Brown, A. J., 1997. Planktonic foraminiferal molecular evolution and their polyphyletic origins from benthic taxa. *Marine Micropaleontology*, **30**: 251–66.

Darling, K.F., Wade, C.M., Kroon, D. *et al.*, 1999. The diversity and distribution of modern planktonic foraminiferal small subunit ribosomal RNA genotypes and their potential as tracers of present and past ocean circulation. *Paleoceanography*, **14**: 3–12.

Darling, K.F., Wade, C.M., Steward, I.A. *et al.*, 2000. Molecular evidence for genetic mixing of Arctic and Antarctic subpolar populations of planktonic foraminifers. *Nature*, **405**: 43–7.

da Silva, A.A., Pawlowski, J. & Gooday, A.J., 2006. High diversity of deep-sea *Gromia* from the Arabian Sea revealed by small subunit rDNA sequence analysis. *Marine Biology*, **148**(4): 769–77.

Das, A.K., Gupta, R.P., Hussam, R. & Maurya, S.N., 1999. Cenozoic sequence stratigraphy of Krishna mouth area and beyond in eastern offshore, India, with special reference to depositional setting and hydrocarbon habitat in Miocene sediments. *Geohorizons*, **4**(2): 1–10.

Davies, L.M. & Pinfold, E.S., 1937. *The Eocene Beds of the Punjab Salt Range*. Calcutta; The

Geological Survey of India (Palaeontologia Indica, XXIV (Memoir, No. 1)).

Davies, R.B., Casey, D.M., Horbury, A.D. *et al.*, 2002. Early to mid-Cretaceous mixed carbonate–clastic shelfal systems: examples, issues and models from the Arabian Plate. *GeoArabia*, **7**(3): 541–98.

Davydov, V.I., 1988a. About a phylogenetic criterion of weighing specific features in foraminifer systematics (exemplified by fusulinids). *Revue de Paleobiologie, Volume Special*, **2**: 47–55.

Davydov, V.I., 1988b. Archaediscidae in the Upper Carboniferous and Lower Permian. *Revue de Paleobiologie, Volume Special*, **2**: 39–46.

Dawson, O., 1993. Fusuline foraminiferal biostratigraphy and carbonate facies of the Permian Ratburi limestone, Saraburi, central Thailand. *Journal of Micropalaeontology*, **12**(1): 9–34.

Dawson, O. & Racey, A., 1993. Fusuline–calcareous algal biofacies of the Permian Ratburi limestone, Saraburi, central Thailand. *Journal of Southeast Asian Earth Sciences*, **8** (1–4): 49–65.

Dawson, O., Racey, A. & Whittaker, J.E., 1993. The palaeoecological and palaeobiogeographic significance of Shanita (Foraminifera) and associated Foraminifera/Algae from the Permian of peninsular Thailand. *Proceedings of the International Symposium on the Biostratigraphy of Mainland South-East Asia (Facies and Paleontology)*: 282–95.

De, S. & Gupta, A.K., 2010. Deep-sea faunal provinces and their inferred environments in the Indian Ocean based on distribution of Recent benthic Foraminifera. *Palaeogeography, Palaeoclimatology, Palaeoecology*, **291**: 429–42.

de Cizancourt, M., 1951. *Grands Foraminiferes du Paleocene, de l'Eocene Inferieur et de l'Eocene Moyen du Venezuela*. Paris; Societe Geologique de France (Memoire, No. 64).

de Freitas Prazeres, M., Martins, S.E. & Bianchini, A., 2012. Assessment of water quality in coastal waters of Fernando de Noronha, Brazil: biomarker analyses in *Amphistegina lessonii. Journal of Foraminiferal Research*, **42**(1): 56–65.

de Graciansky, P.-C., Hardenbol, J., Jacquin, Th. & Vail, P.R. (eds.). 1998. *Mesozoic and Cenozoic Sequence Stratigraphy of European Basins*. Tulsa, OK; Society of Economic Paleontologists and Mineralogists/ Society for Sedimentary Geology (Special Publication, No. 60).

de Klasz, I. & Gageonnet, R., 1965. Biostratigraphie du basin Gabonais. *Memoires du Bureau de Recherches Geologiques et Minieres*, **32**: 277–303.

de Klasz, I. & Rerat, D., 1961. Quelques nouveaux foraminiferes du Cretace et du Tertiaire du Gabon (Afrique equatoriale). *Revue de Micropaleontologie*, **4**(4): 175–89.

de Klasz, I. & Rerat, D., 1963. The stratigraphic range of the foraminiferal genus *Gabonella* in the Upper Cretaceous of Gabon (equatorial Africa). *Micropaleontology*, **9**(3): 325–6.

de Klasz, I. & van Hinte, J.E., 1977. Remarcques sue le genre *Gabonella* (foraminiferes) et description de deux nouvelles especes du Cretace Superieur du Gabon. *Annales des Mines et de la Geologie*, **28**: 481–97.

de Man, E. & van Simaeys, S., 2004. Late Oligocene warming event in the southern North Sea basin: benthic Foraminifera as paleotemperature proxies. *Geologie en Mijnbouw*, **83**: 227–39.

de Matos, J.E., Hulstrand, R. & Walkden, G.M., 1994. The biostratigraphy of the Lower Jurassic of the UAE: outcrop and subsurface compared. *Abu Dhabi Society of Petroleum Engineers, Paper*, 51.

de Mello e Sousa, S., Fairchild, T.R. & Tibana, P., 2003. Cenozoic biostratigraphy of larger Foraminifera from the Foz do Amazonas basin, Brazil. *Micropaleontology*, **49**(3): 253–66.

de Mello e Sousa, S., Passos, R.F., Fukumoto, M. *et al.*, 2006. Mid–lower bathyal benthic Foraminifera of the Campos basin, southeastern Brazilian margin: biotopes and controlling ecological factors. *Marine Micropaleontology*, **61**: 40–57.

de Mello e Sousa, S., Rossetti, D., Fairchild, T.R. *et al.*, 2009. Microfacies and sequence stratigraphy of the Amapa formation, Late Paleocene to Early Eocene, Foz do Amazonas basin, Brazil. *Palaeogeography, Palaeoclimatology, Palaeoecology*, **280**: 440–55.

de Moura, J. Costa, de Moraes Rios-Netto, A., Wanderley, M.D. & de Sousa, F.P., 1999. Using acids to extract calcareous microfossils from carbonate rocks. *Micropaleontology*, **45**(4): 429–36.

de Romero, L.M. & Galea-Alvarez, F.A., 1995. Campanian Bolivinoides and microfacies from the La Luna formation, western Venezuela. *Marine Micropaleontology*, **26**: 383–404.

de Romero, L.M., Truskowski, I.E., Bralower, T.J. *et al.*, 2003. An integrated calcareous microfossil biostratigraphic and carbon-isotope stratigraphic framework for the La Luna formation, western Venezuela. *Palaios*, **18**: 349–66.

de Vargas, C., Bonzon, M., Rees, N.W. *et al.*, 2002. A molecular approach to diversity and biogeography in the planktonic foraminifer *Globigerinella siphonifera* (d'Orbigny). *Marine Micropaleontology*, **45**(2): 101–16.

de Vargas, C., Norris, R., Zaninetti, L. *et al.*, 1999. Molecular evidence of cryptic speciation in planktonic foraminifers and their relation to oceanic provinces. *Proceedings of the National Academy of Science, USA*, **96**: 2864–8.

de Vargas, C. & Pawlowski, J., 1998. Molecular versus taxonomic rates of evolution in planktonic Foraminifera. *Molecular Phylogenetics and Evolution*, **9**(3): 463–9.

de Vargas, C., Renaud, S., Hilbrecht, H. & Pawlowski, J., 2001. Pleistocene adaptive radiation in *Globorotalia truncatulinoides*: genetic, morphologic, and environmental evidence. *Paleobiology*, **27**(1): 104–25.

de Vargas, C., Zaninetti, L., Hilbrecht, H. & Pawlowski, J., 1997. Phylogeny and rates of molecular evolution of planktonic Foraminifera: SSU rDNA sequences compared to the fossil record. *Journal of Molecular Evolution*, **45**(3), 285–94.

Dean, W.E. & Arthur, H.A. (eds.), 1998. *Stratigraphy and Paleoenvironments of the Cretaceous Western Interior Seaway, USA*. Tulsa, OK; Society of Economic Paleontologists and Mineralogists (Concepts in Sedimentology and Paleontology, No. 6).

Debenay, J.P., 1988. Foraminifera larger than 0.5 mm in the southwestern lagoon of New Caledonia: distribution related to abiotic properties. *Journal of Foraminiferal Research*, **18**(2): 158–75.

Debenay, J.-P., Arfi, R. & Konate, S., 1987. Foraminiferes recents des milieu paraliques des cotes d'Afrique de l'ouest. *Geologie Mediterraneene*, **XIV**(1): 5–13.

Debenay, J.-P. & Basov, I., 1993. Distribution of Recent benthic Foraminifera on the west African shelf and slope: a synthesis. *Revue de Paleobiologie*, **12**(1): 265–300.

Debenay, J.-P., Beck-Eichler, B., Fernandez-Gonzalez, M. *et al.*, 1996. Les foraminiferes paraliques des cotes d'Afrique et d'Amerique du Sud de part et d'autre de l'Atlantique: comparaison – discussion. *Actes Colloques Angers*, **1994**: 463–71.

Debenay, J.-P. & Bui Thi Luan, 2006. Foraminiferal assemblages and the Confinement Index

as tools for assessment of saline intrusion and human impact in the Mekong delta and neighbouring areas (Vietnam). *Revue de Micropaleontologie*, **49**(2): 74–85.

Debenay, J.-P., Geslin, E., Eichler, B.B. *et al.*, 2001. Foraminiferal assemblages in a hypersaline lagoon, Araruama (R.J.) Brazil. *Journal of Foraminiferal Research*, **31**(2): 133–51.

Debenay, J.-P. & Guiral, D., 2006. Mangrove swamp Foraminifera, indicators of sea level or paleoclimate. *Revue de Paleobiologie*, **25**(2): 567–74.

Debenay, J.-P., Guiral, D. & Parra, M., 2004. Behaviour and taphonomic loss in foraminiferal assemblages of French Guiana. *Marine Geology*, **208**(2–4): 295–314.

Debenay, J.-P., Sigura, A. & Justine, J.-L., 2011. Foraminifera in the diet of coral reef fish from the lagoon of New Caledonia: predation, digestion, dispersion. *Revue de Micropalaeontologie*, **54**: 87–103.

Decrouez, D. & Lanterno, E., 1979. Les 'bancs a nummulites' de l'Eocene et leurs implications. *Archives des Sciences, Geneve*, **32**(1): 67–94.

DeLaca, T.E., 1982. Use of dissolved amino acids by the foraminifer *Notodendrodes antarctikos*. *American Zoologist*, **2**: 683–90.

DeLaca, T.E., Karl, D.M. & Lipps, J.H. 1981. Direct use of dissolved organic carbon by agglutinated Foraminifera. *Nature*, **289**: 287–9.

DeLaca, T.E., Lipps, J.H. & Hessler, R.R., 1980. The morphology and ecology of a large agglutinated Antarctic foraminifer (Textulariina: Notodendrodidae *nov.*). *Zoological Journal of the Linnean Society*, **69**: 205–24.

Delaney, M.L. & Boyle, E.A., 1987. Cd/Ca in Late Miocene benthic Foraminifera and changes in the global organic-carbon budget. *Nature*, **330**: 156–9.

Demaison, G. & Murris, R.J. (eds.), 1984. *Petroleum Geochemistry and Basin Evalution*. Tulsa, OK; American Association of Petroleum Geologists (Memoir, No. 35).

Demchuk, T.D. & Gary, A.C. (eds.), 2009. *Geologic Problem Solving with Microfossils: a Volume in Honour of Garry D. Jones*. Tulsa, OK; Society of Economic Paleontologists and Mineralogists (Special Publication, No. 93).

den Dulk, M., Reichart, G.J., Memon, G.M. *et al.*, 1998. Benthic foraminiferal response to variations in surface water productivity and oxygenation in the northern Arabian Sea. *Marine Micropaleontology*, **35**(1–2): 43–66.

den Dulk, M., Reichart, G.J., van Heyst, S. *et al.*, 2000. Benthic Foraminifera as proxies of organic matter flux and bottom water oxygenation? A case history from the Arabian Sea. *Palaeogeography, Palaeoclimatology, Palaeoecology*, **161**(3): 337–59.

Denne, R.A. & Sen Gupta, B.K., 1988. Abundance variations of dominant benthic Foraminifera on the northwestern Gulf of Mexico slope: relationship to bathymetry and water mass boundaries. *Bulletin de l'Institut Geologique du Bassin d'Aquitaine*, **44**: 33–43.

Denne, R.A. & Sen Gupta, B.K., 1991. Association of bathyal Foraminifera with water masses in the northwestern Gulf of Mexico. *Marine Micropaleontology*, **17**(3/4): 173–93.

Denne, R.A. & Sen Gupta, B.K., 1993. Matching of benthic foraminiferal depth limits and water-mass boundaries in the northwestern Gulf of Mexico: an investigation of species occurrences. *Journal of Foraminiferal Research*, **23**(2): 108–17.

Denoyelle, M., Geslin, E., Galgani, F. *et al.*, 2010. Innovative use of Foraminifera in ecotoxicology: testing potential toxicity. *Abstracts, Forams2010 (International Symposium on Foraminifera), Bonn*: 75.

Derin, B. & Reiss, Z., 1966. *Jurassic Microfacies of Israel*. Tel Aviv; Israel Institute of Petroleum.

Dermitzakis, M.D. & Alafonsou, P., 1987. Geological framework and the observed oil seeps of Zakynthos Island: their possible influence on the pollution of the marine environment. *Thalassographica*, **10**(2): 7–22.

Dessler, A., 2012. *Introduction to Modern Climate Change*. Cambridge; Cambridge University Press.

Dessler, A. & Parson, E.A., 2010. *The Science and Politics of Global Climate Change: A Guide to the debate*. Cambridge; Cambridge University Press.

Dettmering, C., Rottger, R., Hohenegger, J. & Schmaljohann, R., 1998. The trimorphic life cycle in foraminifera: observations from cultures allow new evaluation. *European Journal of Protistology*, **34**(4): 363–8.

Dhannawat, B.S. & Mukherkjee, M.K., 1997. Source rock studies in Jaisalmer basin, India. *Indian Journal of Petroleum Geology*, **6**(1): 25–42.

d'Hondt, S. & Zachos, J.C., 1998. Cretaceous Foraminifera and the history of planktonic photosymbiosis. *Paleobiology*, **24**(4): 512–23.

di Bari, D., 1999. *Globigerina*-like Foraminifera (Oberhauserellidae) from the Triassic (Ladinian–Carnian) of thc San Cassiano formation (Dolomites, Italy). *Abhandlungen der Geologischen Bundesanstalt*, **56**(2): 49–67.

Dias, B.B., Hart, M.B., Smart, C.W. & Hall-Spencer, J.M., 2010. Modern seawater acidification: the response of foraminifers to high-$CO_2$ conditions in the Mediterranean Sea. *Journal of the Geological Society*, **167**: 843–6.

Dickinson, B., Waterhouse, M., Goodall, J. & Holmes, N., 2001. Blenheim field: the appraisal of a small oil field with a horizontal well. *Petroleum Geoscience*, **7**: 81–95.

Dijkstra, N., Juntilla, J., Carroll, J. *et al.*, 2010. Benthic foraminiferal assemblages as indicators of anthropogenic activity in the SW Barents Sea. *Abstracts, Forams2010 (International Symposium on Foraminifera), Bonn*: 77.

Dincauze, D.F., 2000. *Environmental Archaeology: Principles and Practice*. Cambridge; Cambridge University Press.

Dixon, J., Dietrich, J., McNeil, D.H. *et al.* (eds.), 1985. *Geology, Biostratigraphy and Organic Geochemistry of Jurassic to Pleistocene Strata, Beaufort–Mackenzie Area, Northwest Canada: Course Notes*. Calgary, Alberta; Canadian Society for Petroleum Geology.

Dixon, J.E. & Robertson, A.H.F. (eds.), 1984. *The Geological Evolution of the Eastern Mediterranean*. Oxford; Blackwell.

Diz, P., Barras, C., Geslin, E. *et al.*, 2012. Incorporation of Mg and Sr and oxygen and carbon stable isotope fractionation in cultured *Ammonia tepida*. *Marine Micropaleontology*, **92**–93: 16–28.

Dobson, M. & Haynes, J., 1973. Association of Foraminifera with hydroids on the deep shelf. *Micropaleontology*, **19**(1): 78–90.

Doney, S.C., Fabry, V.J., Feely, R.A. & Kleypas, J.A., 2009. Ocean acidification: the other $CO_2$ problem. *Annual Review of Marine Science*, **1**: 169–92.

Dongil Kim, Su-Yeong Yang & Jaewoo Kim, 2012. Geological modeling with seismic inversion for deepwater turbidite fields offshore northwestern Myanmar. *Search and Discovery Article*, 40877.

Donovan, S.K. & Jackson, T.A. (eds.), 1994. *Caribbean Geology: An Introduction*. Kingston, Jamaica; University of the West Indies Publishers' Association.

Dore, A.G. & Sinding-Larsen, R. (eds.), 1996. *Quantification and Prediction of Petroleum Resources*. Amsterdam; Elsevier (Norwegian Petroleum Society Special Publication, No. 6).

Dore, A.G. & Vining, B.A., 2005. *Petroleum Geology: North-West Europe and Global Perspectives (Proceedings of the Sixth Petroleum Geology Conference, 2003)*. London; The Geological Society.

Douglas, J.A., 1936. A Permo-Carboniferous fauna from south-west Persia (Iran). *Palaeontologia Indica, N.S.*, **22**(6).

Douglas, J.A., 1950. The Carboniferous and Permian faunas of south Iran and Iranian Baluchistan. *Palaeontologia Indica, N.S.*, **22**(7).

Downey, M.W., Threet, J.C. & Morgan, W.A. (eds.), 2001. *Petroleum Provinces of the Twenty-First Century*. Tulsa, OK; American Association of Petroleum Geologists (Memoir, No. 74).

Doyle, W.L. & Doyle, M.M., 1940. The structure of Zooxanthellae. *Papers, Tortugas Laboratory*, **32**: 127–42.

Drinia, H., Antonarakou, A., Tsaparas, N. & Dermitzakis, M.D., 2007. Foraminiferal stratigraphy and palaeoecological implications in turbidite-like deposits from the early Tortonian (Late Miocene) of Greece. *Journal of Micropalaeontology*, **26**: 145–58.

Drobne, K., 1988. Structural elements and stratigraphic distribution of larger miliolids in the foraminferal family Fabulariidae. *Revue de Paleobiologie, Volume Special*, **2**: 643–61.

Drooger, C.W., 1979. Marine connections of the Neogene Mediterranean, deduced from the evolution and distribution of larger Foraminifera. *Annales Geologiques des Pays Hellenique, Tome hors serie*, **1979**(1): 361–9.

Drooger, C.W. & Kaasschieter, J.P.H., 1958. Foraminifera of the Orinoco–Trinidad–Paria Shelf. *Verhandelingen der Koninklijke Nederlandsche Akademie van Wetenschappen, Afdeeling Natuurkunde, Eerste Reeks*, **XXII**: 1–108.

Droste, H., 1990. Depositional cycles and source-rock development in an epiric intra-platform basin: the Hanifa formation of the Arabian peninsula. *Sedimentary Geology*, **69**: 281–96.

Ducassou, E., Migeon, S., Capotondi, L. & Mascle, J., 2013. Run-out distance and erosion of debris-flows in the Nile deep-sea fan system: evidence from lithofacies and micropalaeontological analyses. *Marine and Petroleum Geology*, **39**: 102–23.

Duchemin, G., Jorissen, F.J., Redois, F. & Debenay, J.-P., 2005. Foraminiferal microhabitats in a high marsh: consequences for reconstructing past sea levels. *Palaeogeography, Palaeoclimatology, Palaeoecology*, **226**: 167–85.

Duguay, L.E., 1983. Comparative laboratory and field studies on calcification and carbon fixation in foraminiferal–algal associations. *Journal of Foraminiferal Research*, **13**: 252–61.

Duleba, W. & Debenay, J.-P., 2003. Hydrodynamic circulation in the estuaries of Estaçâo Ecologica Jureia-Itatins, Brazil, inferred from Foraminifera and thecamoebian assemblages. *Journal of Foraminiferal Research*, **33**(1): 62–93.

Dullo, W.-C. Hotzl, H. & Jado, R.A., 1983. New stratigraphical results from the Tertiary sequence of the Midyan area, NW Saudi Arabia. *Newsletters on Stratigraphy*, **12**(2): 75–83.

Dunkley Jones, T., Lunt, D.J., Schmidt, D.N. *et al.*, 2010. A review of the Paleocene–Eocene Thermal Maximum temperature anomaly. *Abstracts, International Palaeontological Congress, London*: 153.

Dunnington, H.V., 1955. Close zonation of Upper Cretaceous globigerinal sediments by abundance ratios of *Globotruncana* species groups. *Micropaleontology*, **1**(3): 207–19.

Dunnington, H.V., 1967. Stratigraphical distribution of oilfields in the Iran–Iraq–Arabia basin. *Journal of the Institute of Petroleum*, **53**(520): 129–61.

Duros, P., Fontanier, C., de Stigter, H.C. *et al.*, 2012. Live and dead foraminiferal faunas from Whittard Canyon (NE Atlantic): focus on taphonomic processes and paleoenvironmental applications. *Marine Micropaleontology*, **94**–95: 25–44.

Eames, F.E., 1952. A contribution to the study of the Eocene in western Pakistan and western India. *Quarterly Journal of the Geological Society*, **CVII**: 159–200.

Eames, F.E. & Smout, A.H., 1955. Complanate alveolinids and associated Foraminifera from the Upper Cretaceous of the Middle East. *Annals and Magazine of Natural History, Ser. 12*, **8**: 505–12.

Edenhofer, O., Pichs-Madruga, R., Sokona, Y. *et al.* (eds.), 2012. *Renewable Energy Sources and Climate Change Mitigation: Special Report of the Intergovernmental Panel on Climate Change*. Cambridge; Cambridge University Press.

Edwards, J.D. & Santogrossi, P.A. (eds.), 1989. *Divergent/Passive Margin Basins*. Tulsa, OK; American Association of Petroleum Geologists (Memoir, No. 48).

Edwards, L.E., 1984. Insights on why graphic correlation (Shaw's method) works. *Journal of Geology*, **92**: 583–97.

Edwards, R.J., van de Plassche, O., Gehrels, W.R. & Wright, A.J., 2004. Assessing sea-level data from Connecticut, USA, using a foraminiferal transfer function for tide level. *Marine Micropaleontology*, **51**(3–4): 239–55.

Eggleton, T., 2012. *A Short Introduction to Climate Change*. Cambridge; Cambridge University Press.

Einsele, G. (ed.), 1991. *Cycles and Events in Stratigraphy*. Berlin; Springer Verlag.

El-Bishlawy, S.H., 1985. Geology and hydrocarbon occurrence of the Khuff formation in Abu Dhabi. *Society of Petroleum Engineers, Paper*, 13678.

El-Khayal, A.E.-M., 1974. Foraminiferal biostratigraphy of the Umm Er Radhuma formation (Paleocene–Lower Eocene) of eastern Saudi Arabia. *Bulletin of the Faculty of Science, Riyadh University*, **6**: 195–213.

Elkhazri, A., Abdallah, H., Razgallah, S. *et al.*, 2013. Carbon-isotope and microfaunal stratigraphy bounding the lower Aptian Oceanic Anoxic Event 1a in northeastern Tunisia. *Cretaceous Research*, **39**: 133–48.

Ellet, P., Heaton, R. & Watts, M., 2005. The Barmer basin, Rajasthan, India – the ingredients which led to success. *Abstracts, American Association of Petroleum Geologists Annual Convention*.

Ellison, R.L., 1972. *Ammobaculites*, foraminiferal proprietor of Chesapeake Bay estuaries. *Geological Society of America Memoir*, **133**: 247–62.

Ellison, R.L., Broome, R. & Oglivie, R., 1986. Foraminiferal response to trace metal contamination in the Patapso River and Baltimore Harbor, Maryland. *Marine Pollution Bulletin*, **17**: 419–23.

Ellison, R.L. & Peck, G.E., 1983. Foraminiferal recolonization on the continental shelf. *Journal of Foraminiferal Research*, **13**(4): 231–41.

El-Naggar, Z.R. & El-Rifaiy, I.A., 1972. Stratigraphy and microfacies of the type Magwa formation of Kuwait, Arabia, part 1: Rumaila Limestone member. *American Association of Petroleum Geologists Bulletin*, **56**(8): 1464–93.

El-Naggar, Z.R. & El-Rifaiy, I.A., 1973. Stratigraphy and microfacies of the type Magwa formation of Kuwait, Arabia, part 2: Mishrif Limestone member. *American Association of Petroleum Geologists Bulletin*, **57**(11): 2263–79.

El-Naggar, Z.R. & El-Nakhal, H.A., 1986. *Stratigraphy and Planktonic Foraminifera of the Wasia Group (Cretaceous) in Kuwait*. Riyadh; College of Science, King Saud University.

El-Naggar, Z.R. & Kamel, S., 1988. Biostratigraphic analysis of the Late Cretaceous–early Paleogene succession in the Turayf area, NW Saudi Arabia. *Revista Espanola de Micropaleontologia*, **20**(3): 329–53.

El-Nakhal, H.A., 1990. Surdud group, a new lithostratigraphic unit of Jurassic age in the Yemen Arab Republic. *Journal of King Saud University, Science*, **2**: 125–43.

Embry, A.F., 2009. *Practical Sequence Stratigraphy*. Canadian Society of Petroleum Geologists (available online at www.cspg.org).

Emerson, S.R. & Hedges, J.I., 2008. *Chemical Oceanography and the Marine Carbon Cycle*. Cambridge; Cambridge University Press.

Emery, D. & Myers, K. (eds.), 1996. *Sequence Stratigraphy*. Oxford; Blackwell.

Emiliani, C., 1955. Pleistocene temperatures. *Journal of Geology*, **63**: 538–78.

Emiliani, C. (ed.), 1981. *The Oceanic Lithosphere: The Sea, Volume 7*. Cambridge, MA; Harvard University Press.

Epstein, S., Buchsbaum, R., Lowenstam, H.A. & Urey, H.C., 1953. Revised carbonate–water isotopic temperature scale. *Geological Society of America Bulletin*, **64**: 1315–25.

Erbacher, J., Hemleben, C., Huber, B.T. & Markey, M., 1999. Correlating environmental changes during early Albian Oceanic Anoxic Event 1B using benthic foraminiferal paleoecology. *Marine Micropaleontology*, **38**(1): 7–28.

Erbacher, J. & Nelskamp, S., 2006. Comparison of benthic Foraminifera inside and outside a sulphur-oxidising bacterial mat from the present Oxygen-Minimum Zone off Pakistan (NE Arabian Sea). *Deep-Sea Research II*, **53**: 751–75.

Erlich, R.N. & Keens-Dumas, J., 2007. Late Cretaceous palaeogeography of north-eastern South America: implications for source and reservoir development. *Transactions of the Fourth Geological Conference of the Geological Society of Trinidad and Tobago, Port-of-Spain*: 1–34.

Ernst, S., Bours, R., Duijnstee, I. & van der Zwaan, B., 2005. Experimental effects of an organic matter pulse and oxygen depletion on a benthic foraminiferal shelf community. *Journal of Foraminiferal Research*, **35**(3): 177–97.

Ernst, S., Duijnstee, I. & van der Zwaan, B., 2002. The dynamics of the benthic foraminiferal microhabitats: recovery after experimental disturbance. *Marine Micropaleontology*, **46** (3–4): 343–62.

Ernst, S.R., Morvan, J., Geslin, E. *et al.*, 2006. Benthic foraminiferal response to experimentally induced Erika oil pollution. *Marine Micropaleontology*, **61**: 76–93.

Ernst, W.G. (ed.), 2000. *Earth Systems: Processes and Issues*. Cambridge; Cambridge University Press.

Ertan, K., Hemleben, V. & Hemleben, C., 2004. Molecular evolution of some selected benthic Foraminifera as inferred from sequences of the small subunit ribosomal DNA. *Marine Micropaleontology*, **53**: 367–88.

Erwin, D.H., 1998. The end and the beginning: biotic recoveries from mass extinction. *Trends in Ecology and Evolution*, **13**: 344–9.

Escalona, A., Mann, P. & Bingham, L., 2008. Hydrocarbon exploration plays in the great Caribbean region and neighboring provinces. *Search and Discovery Article*, 10047.

Eva, A.N., 1976. The paleoecology and sedimentology of Middle Eocene larger Foraminifera in Jamaica. *Maritime Sediments Special Publication*, **1B**: 467–76.

Eva, A.N., 1980. Pre-Miocene seagrass communities in the Caribbean. *Palaeontology*, **23** (1): 231–6.

Eva, A.N., Burke, K., Mann, P. & Wadge, G., 1989. Four-phase tectonostratigraphic development of the southern Caribbean. *Marine and Petroleum Geology*, **6**: 9–21.

Eva, A.N., Casey, D.M., Daly, M.C. *et al.*, 1990. Tectonostratigraphic evolution of Arabia. *Abstracts, Middle East Geosciences Conference, Kuwait, 1990*.

Evans, A.M., 1997. *An Introduction to Economic Geology and its Environmental Impact*. Oxford; Blackwell Science.

Evans, D., Graham, C., Armour, A. & Bathurst, P. (eds. and co-ords.), 2003. *The Millennium Atlas: Petroleum Geology of the Central and Northern North Sea*. London; Geological Society/Norwegian Petroleum Society/Geological Survey of Denmark and Greenland.

Faber, W.W., Anderson, O.R., Lindsey, J.L., & Caron, D.A., 1985. Algal–foraminiferal symbiosis in the planktonic foraminifer *Globigerinella aequilateralis*, I: occurrence and stability of two mutually exclusive chrysophyte endosymbionts and their ultrastructure. *Journal of Foraminiferal Research*, **18**: 334–43.

Fairbanks, M.D. & Ruppel, S.C., 2012. High-resolution stratigraphy and facies architecture of the Upper Cretaceous (Cenomanian–Turonian) Eagle Ford formation, central Texas. *Search and Discovery Article*, 10408.

Fatela, F., Moreno, J., Moreno, F. *et al.*, 2009. Environmental constraints of foraminiferal assemblages distribution across a brackish tidal marsh (Caminha, NW Portugal). *Marine Micropaleontology*, **70**: 80–88.

Fazeelat, T., Jalees, M.I. & Bianchi, T.S., 2010. Source-rock potential of Eocene, Paleocene and Jurassic deposits on the subsurface of the Otwar basin, northern Pakistan. *Journal of Petroleum Geology*, **33**(1): 87–96.

Fermont, W.J.J., 1982. Discocyclinidae from Ein Avedat (Israel). *Utrecht Micropaleontological Bulletin*, **27**: 1–173.

Fetter, M., de Ros, L.F. & Bruhn, C.H.L., 2009. Petrographic and seismic evidence for the depositional setting of giant turbidite reservoirs and the paleogeographic evolution of Campos basin, offshore Brazil. *Marine and Petroleum Geology*, **26**: 824–53.

Field, C.B., Barros, V., Stocker, T.F. & Qin Dahe (eds.), 2012. *Managing the Risks of Extreme Events and Disasters to Advance Climate Change Adaptation: Special Report of the Intergovernmental Panel on Climate Change*. Cambridge; Cambridge University Press.

Figueira, B.O., Grenfell, H.R., Hayward, B.W., & Alfaro, A.C., 2012. Comparison of Rose Bengal and CellTracker Green staining for identification of live salt-marsh Foraminifera. *Journal of Foraminiferal Research*, **42**(3): 206–15.

Filipsson, H.L., 2008. Culturing of benthic Foraminifera for improved paleoceanographic reconstruction. *Palaios*, **23**(1): 1–2.

Filipsson, H.L., Bernhard, J.M., Lincoln, S.A. & McCorkle, D.C., 2010. A culture-based calibration of benthic foraminiferal paleotemperature proxies: del18O and Mg/Ca results. *Biogeosciences*, **7**: 1335–47.

Fillon, R.H., Rosen, N.C. & Weimer, P. (eds.), 2001. *Petroleum Systems of Deep-Water Basins: Global and Gulf of Mexico Experience*. Gulf Coast Section – Society of Economic Paleontologists and Mineralogists (GCS-SEPM) Foundation (Proceedings of the Annual Research Conference).

Finger, K.L., 1990. *Atlas of California Neogene Foraminifera*. Cushman Foundation for Foraminiferal Research (Special Publication, No. 28).

Finger, K.L., 1992. *Biostratigraphic Atlas of Miocene Foraminifera from the Monterey and Modeno Formations, central and southern California*. Cushman Foundation for Foraminiferal Research (Special Publication, No. 29).

Fiorini, F., 2010. Benthic foraminiferal distribution from the Colombian Caribbean continental slope and shelf. *Abstracts, Forams2010*

(*International Symposium on Foraminifera*), *Bonn*: 85.

Fiorini, F., Scott, D.B. & Wach, G.D., 2010. Characterization of paralic paleoenvironments using benthic Foraminifera from Lower Cretaceous deposits (Scotian shelf, Canada). *Marine Micropaleontology*, **76**: 11–22.

Fisher, C.G., Sageman, B.B., Asure, S.E. *et al.*, 2003. Planktonic foraminiferal porosity analysis as a tool for paleoceanographic reconstruction, mid-Cretaceous Western Interior Sea. *Palaios*, **18**: 34–46.

Fitzsimmons, R., Buchanan, J. & Izatt, C., 2005. The role of outcrop geology in predicting reservoir presence in the Cretaceous and Paleocene successions of the Sulaiman range, Pakistan. *American Association of Petroleum Geologists Bulletin*, **89**(2): 231–54.

Flakowski, J., Bolivar, I., Fahrni, J. & Pawlowski, J., 2005. Actin phylogeny of Foraminifera. *Journal of Foraminiferal Research*, **35**(2): 93–102.

Flecker, R. & Ellam, R.M., 1999. Distinguishing climatic and tectonic signals of marginal basins using Sr isotopes: an example from the Messinian salinity crisis, eastern Mediterranean. *Journal of the Geological Society, London*, **156**: 847–54.

Flugelman, R.H., Jr., Berggren, W.A. & Briskin, M., 1990. Paleocene benthonic foraminiferal biostratigraphy of the eastern Gulf coastal plain. *Micropaleontology*, **36**(1): 56–64.

Fontanier, C., Jorissen, F.J., Chaillou, G. *et al.*, 2005. Live foraminferal faunas from a 2800 m deep lower canyon station from the Bay of Biscay: faunal response to focusing of refractory organic matter, *Deep-Sea Research 1*, **52**: 1189–227.

Fontanier, C., Jorissen, F.J., Geslin, E. *et al.*, 2008. Live and dead foraminiferal faunas from Saint-Tropez canyon (Bay of Frejus): observations based on *in situ* and incubated cores. *Journal of Foraminiferal Research*, **38** (2): 137–56

Forbes, C.L., 1960. Carboniferous and Permian Fusulinidae from Spitsbergen. *Paleontology*, **2**(2): 210–25.

Forke, H.C., Bendias, D., Walz, L. *et al.*, 2010. Larger benthic foraminiferal biostratigraphy of a Lower Khuff outcrop equivalent: Saiq formation, Al Jabal al-Akhdar, Oman Mountains. *Abstracts, Forams2010 (International Symposium on Foraminifera), Bonn*: 85–6.

Foster, G.L., 2008. Seawater pH, $pCO_2$ and $[CO2–3]$ variations in the Caribbean Sea over the last 130 kyr: a boron isotope and B/Ca study of planktonic Foraminifera. *Earth and Planetary Science Letters*, **271** (1–4): 254–66.

Foster, N.H. & Beaumont, E.A. (eds.), 1991. *Stratigraphic Traps II*. Tulsa, OK; American Association of Petroleum Geologists.

Frankel, L., 1972. Subsurface reproduction in Foraminifera. *Journal of Paleontology*, **46**: 62–5.

Frankel, L., 1974. Observations and speculations on the habitat and habits of *Trochammina ochracea* (Williamson) in subsurface sediments. *Journal of Paleontology*, **48**(1): 143–8.

Frankel, L., 1975a. Pseudopodia of surface and subsurface dwelling *Miliammina fusca* (Brady). *Journal of Foraminiferal Research*, **5**(3): 211–7.

Frankel, L., 1975b. Subsurface feeding in Foraminifera. *Journal of Paleontology*, **49**: 563–5.

Fraser, A., Goff, J., Jones, R.W. *et al.*, 2006. A regional overview of the exploration potential of the Middle East. *Abstracts, American Association of Petroleum Geologists International Conference and Exhibition, Perth, Australia.*

Fraser, A., Goff, J., Jones, R.W. *et al.*, 2007. A regional overview of the exploration potential of the Middle East. *Abstracts, European Association of Geoscientists and Engineers Conference, London.*

Fraser, A., Goff, J., Simpson, I. *et al.*, 2003. A regional overview of the petroleum systems of the Middle East. *Abstracts, Petroleum Geology of the Middle East Conference, The Geological Society, London.*

Frerichs, W.E., 1970. Paleobathymetry, paleotemperature, and tectonism. *Geological Society of America Bulletin*, **81**: 3445–52.

Frezza, V., Bergamin, L. & di Bella, L., 2005. Opportunistic benthic Foraminifera as indicators of eutrophicated environments. Actualistic study and comparison with the Santernian middle Tiber valley (central Italy). *Bollettino della Societa Paleontologica Italiana*, **44**(3): 193–201.

Friedrich, O., 2010. Benthic Foraminifera and their role to decipher paleoenvironment during mid-Cretaceous Oceanic Anoxic Events – the 'anoxic benthic Foraminifera' paradox. *Revue de Micropaleontologie*, **53**: 175–92.

Fricdrich, O., Reichelt, K., Herrle, O. *et al.*, 2003. Formation of the Late Aptian Niveau Fallot black shales in the Vocontian Basin (SE

France): evidence from foraminifera, palynomorphs and stable isotopes. *Marine Micropaleontology*, **49**(1–2): 65–86.

Friedrich, O., Voigt, S., Kuhnt, T. *et al.*, 2011. Repeated bottom-water oxygenation during OAE2: timing and duration of short-lived benthic foraminiferal repopulation events (Wunstorf, northern Germany). *Journal of Micropalaeontology*, **30**: 119–28.

Fritsen, A., Bailey, H., Gallagher, L., *et al.*, 1999. *A Joint Chalk Stratigraphic Framework [Joint Chalk Research Phase V]*. Stavanger; Norwegian Petroleum Directorate, Stavanger.

Fritz, D.A., Belsher, T.W., Medlin, J.M. *et al.*, 2000. New exploration concepts for the Edwards and Sligo margins, Cretaceous of onshore Texas. *American Association of Petroleum Geologists Bulletin*, **84**(7): 905–22.

Frontalini, F. & Coccioni, R., 2011. Benthic Foraminifera as indicators of pollution: a review of Italian research over the last three decades. *Revue de Micropalaontologie*, **54**: 115–27.

Funnell, B.M. & Riedel, W.R. (eds.), 1971. *The Micropaleontology of Oceans*. Cambridge; Cambridge University Press.

Gaddo, J.Z.H., 1971. The Mishrif formation palaeoenvironment in the Rumaila/Tuba/Zubair region of S. Iraq. *Journal of the Geological Society of Iraq*, **4**: 1–12.

Gaillot, J. & Vachard, D., 2007. The Khuff formation (Middle East) and time-equivalents in Turkey and south China: biostratigraphy from Capitanian to Changhsingian times (Permian), new foraminiferal taxa, and palaeogeographical implications. *Coloquios de Paleontologia*, **57**: 37–223.

Gallagher, S.J., 1998. Controls on the distribution of calcareous Foraminifera in the Lower Carboniferous of Ireland. *Marine Micropaleontology*, **34**: 187–211.

Gakkhar, R.A., Bechtel, A. & Gratzer, R., 2012. Source-rock potential and origin of hydrocarbons in the Cretaceous and Jurassic sediments of the Punjab platform (Indus basin), Pakistan. *Search and Discovery Article*, 50572.

Galea-Alvarez, F.A. & Moreno-Vasquez, J., 1994. Foraminiferal biofacies of the Carapita formation and their relationship with tectonic events. *V Simposio Bolivariano (Exploracion Perolera en las Cuenas Subandinas)*: 225–7.

Galeron, J., Sibuet, M., Vanreusel, A., *et al.*, 2001. Temporal patterns among meiofauna and macrofauna taxa related to changes in sediment geochemistry at an abyssal NE Atlantic site. *Progress in Oceanography*, **50** (1/4): 303–24.

Galloway, W.E., 1989a. Genetic stratigraphic sequences in basin analysis I: architecture and genesis of flooding-surface bounded depositional units. *American Association of Petroleum Geologists Bulletin*, **73**(2): 125–42.

Galloway, W.E., 1989b. Genetic stratigraphic sequences in basin analysis II: application to northwest Gulf of Mexico Cenozoic basin. *American Association of Petroleum Geologists Bulletin*, **73**(2): 143–64.

Galloway, W.E., 2001. Cenozoic evolution of sediment accumulation in deltaic and shorezone depositional systems, northern Gulf of Mexico basin. *Marine and Petroleum Geology*, **18**: 1031–40.

Galtsoff, P.S., Lutz, F.E., Welch, P.S. & Needham, J.G. (eds.), 1937. *Culture Methods for Invertebrate Animals*. New York, NY; Comstock Publishing Co.

Gamble, C., 1999. *The Palaeolithic Societies of Europe*. Cambridge; Cambridge University Press.

Garcia-Mondejar, J. & Fernandez-Mendiola, P.A., 1993. Sequence stratigraphy and systems tracts of a mixed carbonate and siliciclastic platform-basin setting: the Albian of Lunada and Soba, northern Spain. *American Association of Petroleum Geologists Bulletin*, **77**(2): 245–75.

Gargouri, S., 1988. Paleoecologic distribution of the Cenomanian Foraminifera in central Tunisia. *Revue de Paleobiologie, Volume Special*, **2**: 431–6.

Gattuso, J.-P. & Hansson, L. (eds.), 2011. *Ocean Acidification*. Oxford; Oxford University Press.

Gaucher, Sial, A.N., Halverson, G.P. & Frimmel, H.E. (eds.), 2010. *Neoproterozoic–Cambrian Tectonics, Global Change and Evolution*. Amsterdam; Elsevier (Developments in Precambrian Geology Series, No. 16).

Gaucher, C. & Sprechman, P., 1999. Upper Vendian skeletal fauna of the Arroyo de Soldado group, Uruguay. *Beringeria*, **23**: 55–91.

Gaur, K.N. & Talib, A., 2009. Middle–Upper Jurassic Foraminifera from Jumara hills, Kutch, India. *Revue de Micropalaeontologie*, **52**: 227–48.

Gebhardt, H., 2006. Resolving the calibration problem in Cretaceous benthic Foraminifera paleoecological interpretation: Cenomanian to Coniacian assemblages from the Benue Trough analyzed by conventional methods and correspondence analysis. *Micropaleontology*, **52**(2): 151–76.

Gebhardt, H., Kuhnt, W. & Holbourn, A., 2004. Foraminiferal response to sea-level change, organic flux and oxygen deficiency in the Cenomanian of the Tarfaya Basin, southern Morocco. *Marine Micropaleontology*, **53** (1–2): 133–58.

Geel, T., 2000. Recognition of stratigraphic sequences in carbonate platform and slope deposits: empirical models based on microfacies analysis of Palaeogene deposits in southeastern Spain. *Palaeogeography, Palaeoclimatology, Palaeoecology*, **155**: 211–38.

Gerhard, L.C., Harrison, W.E. & Hanson, B.M. (eds.), 2001. *Geological Perspectives on Global Climate Change*. Tulsa, OK; American Association of Petroleum Geologists (Studies in Geology, No. 47).

Ghabeishavi, A., Vaziri-Moghaddam, H., Taheri, A. & Taati, F., 2010. Microfacies and depositional environment of the Cenomanian of the Bangestan anticline, SW Iran. *Journal of Asian Earth Sciences*, **37**: 275–85.

Ghose, B.K., 1977. Paleoecology of the Cenozoic reefal foraminifers and algae – a brief review. *Palaeogeography, Palaeoclimatology, Palaeoecology*, **22**: 231–56.

Giere, O., 2009. *Meiobenthology. Second Edition*. Berlin; Springer.

Gieskes, J., Rathburn, A.E., Martin, J.B. *et al.*, 2011. Cold seeps in Monterey Bay, California: geochemistry of pore waters and relationship to benthic foraminiferal calcite. *Applied Geochemistry*, **26**: 738–46.

Ghosh, A., Saha, S., Saraswati, P.K. *et al.*, 2009. Intertidal Foraminifera in the macro-tidal estuaries of the Gulf of Cambay: implications for interpreting sea-level change in palaeo-estuaries. *Marine and Petroleum Geology*, **26**: 1592–9.

Ginsburg, R.N., 1997. Perspectives on the blind test. *Marine Micropaleontology*, **29**(2): 101–3.

Glennie, K.G. (ed.), 1998. *Introduction to the Petroleum Geology of the North Sea. Fourth Edition*. Oxford; Blackwell.

Gluyas, J. & Hichens, H.M. (eds.), 2003. *United Kingdom Oil and Gas Fields: Millennium Commemorative Volume*. London; The Geological Society (Memoir, No. 20).

Gluyas, J. & Swarbrick, R., 2004. *Petroleum Geoscience*. Oxford; Blackwell.

Goff, J., Jones, R.W. & Lehmann, C., 2005. Origin and potential of unconventional Jurassic oil reservoirs on the northern Arabian plate. *SPE*, 93505.

Goff, J., Shamshiri, M., Jahani, S. *et al.*, 2004. Discovery and geology of a giant fossil Jurassic oil field in the Zagros mountains, southwest Iran. *GeoArabia*, **9**(1): 60–61 (abstract).

Goineau, A., Fontanier, C., Jorissen, F.J. *et al.*, 2011. Live (stained) benthic Foraminifera from the Rhone prodelta (Gulf of Lion, NW Mediterranean): environmental controls on a river-dominated shelf. *Journal of Sea Research*, **65**: 58–75.

Goldstein, S.T., 1988. On the life cycle of *Saccammina alba* Hedley, 1962. *Journal of Foraminiferal Research*, **18**: 311–25.

Goldstein, S.T. & Bernhard, J.M. (eds.), 1997. Biology of Foraminiferida: applications in paleoceanography, paleobiology, and environmental sciences. *Journal of Foraminiferal Research*, **27**(4).

Goldstein, S.T. & Corliss, B.H., 1994. Deposit feeding in selected deep-sea and shallow-water benthic Foraminifera. *Deep-Sea Research*, **41**: 229–41.

Goldstein, S.T. & Moodley, L., 1993. Gametogenesis and life cycle of the foraminifer *Ammonia beccarii* (Linné) forma *tepida* (Cushman). *Journal of Foraminiferal Research*, **23**(4): 213–20.

Gollesstaneh, A., 1974. The biostratigraphy of the 'Khami group' and the Jurassic–Cretaceous boundary in Fars province (southern Iran). *Bulletin du Bureau de Recherches Geologiques et Minieres, Deuxieme Serie*, **IV**(3): 165–97.

Gombos, A.M., Jr., Powell, W.G. & Norton, I.O., *et al.*, 1995. The tectonic evolution of western India and its impact on hydrocarbon occurrences: an overview. *Sedimentary Geology*, **96**: 119–29.

Gooday, A.J., 1983a. Primitive Foraminifera and Xenophyophorea in IOS epibenthic sledge samples from the northeast Atlantic. *Institute of Oceanographic Sciences Report*, **156**: 1–33.

Gooday, A.J., 1983b. *Bathysiphn rusticus* de Folin, 1886 and *Bathysiphon folini* n.sp.: two large agglutinated Foraminifera abundant in abyssal N.E. Atlantic epibenthic sledge samples. *Journal of Foraminiferal Research*, **13**(4): 262–76.

Gooday, A.J., 1988. A response by benthic Foraminifera to the deposition of phytodetritus in the deep sea. *Nature*, **332**(6159): 70–3.

Gooday, A.J., 1993. Deep-sea benthic foraminiferal species which exploit phytodetritus: characteristic features and controls on distribution. *Marine Micropaleontology*, **22**: 187–205.

Gooday, A.J., 1994. The biology of deep-sea Foraminifera: a review of some advances and their applications in paleoceanography. *Palaios*, **9**(1): 14–31.

Gooday, A.J., 2002a. Biological responses to seasonally varying fluxes of organic matter to the ocean floor: a review. *Journal of Oceanography*, **58**(2): 305–32.

Gooday, A.J., 2002b. Organic-walled allogromiids: aspects of their occurrence, diversity and ecology in marine habitats. *Journal of Foraminiferal Research*, **32**(4): 384–99.

Gooday, A.J., 2003. Benthic Foraminifera (Protista) as tools in deep-water palaeoceanography: environmental influences on faunal characteristics. *Advances in Marine Biology*, **46**: 1–90.

Gooday, A.J., Anikeeva, O.V. & Pawlowski, J., 2011. New genera and species of monothalamous Foraminifera from Balaclava and Kazach'ya Bays (Crimean Peninsula, Black Sea). *Marine Biodiversity*, **41**(4): 481–94.

Gooday, A.J., Anikeeva, O.V. & Sergeeva, N.G., 2006. *Tinogullmia lukyanovae sp. nov. – a* monothalamous, organic-walled foraminiferan from the coastal Black Sea. *Journal of the Marine Biological Association of the United Kingdom*, **86**(1): 43–9.

Gooday, A.J. & Aranda da Silva, A., 2009. Large organic-walled Protista (*Gromia*) in the Arabian Sea: Density, diversity, distribution and ecology. *Deep-Sea Research II*, **56**(6–7): 422–33.

Gooday, A.J., Bernhard, J.M., Levin, L.A. & Suhr, S.B., 2000a. Foraminifera in the Arabian Sea oxygen minimum zone and other oxygen-deficient settings: taxonomic composition, diversity, and relation to metazoan faunas. *Deep-Sea Research II*, **47**(1/2): 25–54.

Gooday, A.J. & Bowser, S.S., 2005. The second species of *Gromia* (Protista) from the deep sea: its natural history and association with the Pakistan Margin Oxygen Minimum Zone. *Protist*, **156**(1): 113–26.

Gooday, A.J., Bett, B.J., Escobar-Briones, E. *et al.*, 2010c. Habitat heterogeneity and its influence on benthic biodiversity in oxygen minimum zones. *Marine Ecology*, **31**(1): 125–47.

Gooday, A.J., Bowser, S.S., Bett, B.J. & Smith, C.R., 2000b. A large testate protist, *Gromia sphaerica sp.nov.* (Order Filosea), from the bathyal Arabian Sea. *Deep-Sea Research II*, **47**(1/2): 55–73.

Gooday, A.J., Bowser, S.S., Cedhagen, T. *et al.*, 2005. Monothalamous foraminiferans and gromiids (Protista) from western Svalbard: a preliminary survey. *Marine Biology Research*, **1**(4): 290–312.

Gooday, A.J. & Haynes, J.R., 1983. Abyssal Foraminifera, including two new genera, encrusting the interior of *Bathysiphon rusticus* tubes. *Deep-Sea Research*, **30**(6A): 591–614.

Gooday, A.J., Holzmann, M., Cornelius, N. & Pawlowski, J., 2004a. A new monothalamous foraminiferan from 1000–6300 m water depth in the Weddell Sea: morphological and molecular characterisation. *Deep-Sea Research II*, **51**: (14–16): 1603–16.

Gooday, A.J., Hori, S., Todo, Y. *et al.*, 2004b. Soft-walled, monothalamous benthic foraminiferans in the Pacific, Indian and Atlantic Oceans: aspects of biodiversity and biogeography. *Deep Sea Research I*, **51**(1): 33–53.

Gooday, A.J. & Hughes, J.A., 2002. Foraminifera associated with phytodetritus deposits at a bathyal site in the northern Rockall Trough (NE Atlantic): seasonal contrasts and a comparison of stained and dead assemblages. *Marine Micropaleontology*, **46**(1–2): 83–110.

Gooday, A.J., Hughes, J.A. & Levin, L.A., 2001a. The foraminiferan macrofauna from three North Carolina (USA) slope sites with contrasting carbon flux: a comparison with the metazoan macrofauna. *Deep-Sea Research I*, **48**(7): 1709–39.

Gooday, A.J. & Jorissen, F.J., 2012. Benthic foraminiferal biogeography: controls on global distribution patterns in deep-water settings. *Annual Review of Marine Science*, **4**(1): 237–62.

Gooday, A.J., Jorissen, F., Levin, L.A. *et al.*, 2009b. Historical records of coastal eutrophication-induced hypoxia. *Biogeosciences*, **6**(8): 1707–45.

Gooday, A.J., Kamenskaya, O.E. & Kitazato, H., 2008a. The enigmatic, deep-sea, organic-walled genera *Chitinosiphon, Nodellum* and *Resigella* (Protista, Foraminifera): a taxonomic re-evaluation. *Systematics and Biodiversity*, **6**(3): 385–404.

Gooday, A.J., Kamenskaya, O.E. & Soltwedel, T., 2010b. The organic-walled genera *Resigella* and *Conicotheca* (Protista, Foraminifera) at two Arctic deep-sea sites (North Pole and Barents Sea), including the description of a new species of *Resigella*. *Marine Biodiversity*, **40**(1): 33–44.

Gooday, A.J., Kitazato, H., Hori, S. & Toyofuku, T., 2001b. Monothalamous soft-shelled Foraminifera at an abyssal site in the North Pacific: a preliminary report. *Journal of Oceanography*, **57**(3): 377–84.

Gooday, A.J. & Lambshead, P.J.D., 1989. Influence of seasonally deposited phytodetritus on benthic foraminiferal populations in the bathyal northeast Atlantic: the species response. *Marine Ecology Progress Series*, **58**: 53–67.

Gooday, A.J., Levin, L.A., Aranda da Silva, A. et al., 2009c. Faunal responses to oxygen gradients on the Pakistan margin: a comparison of foraminiferans, macrofauna and megafauna. *Deep-Sea Research II*, **56**(6–7): 488–502.

Gooday, A.J., Malzone, M.G., Bett, B.J. & Lamont, P.A., 2010a. Decadal-scale changes in shallow-infaunal foraminiferal assemblages at the Porcupine Abyssal Plain, NE Atlantic. *Deep Sea Research II*, **57**(15): 1362–82.

Gooday, A.J. & Pawlowski, J., 2004. *Conqueria laevis* gen. and sp. nov., a new soft-walled, monothalamous foraminiferan from the deep Weddell Sea. *Journal of the Marine Biological Association of the United Kingdom*, **84**(5): 919–24.

Gooday, A.J., Rothe, N., Cedhagen, T. et al., 2009a. Three new species of deep-sea *Gromia* (Protista, Rhizaria) from the bathyal and abyssal Weddell Sea, Antarctica. *Zoological Journal of the Linnean Society*, **157**(3): 451–69.

Gooday, A.J., Todo, Y., Uematsu, K. & Kitazato, H., 2008b. New organic-walled Foraminifera (Protista) from the ocean's deepest point, the Challenger Deep (western Pacific Ocean). *Zoological Journal of the Linnean Society*, **153**(3): 399–423.

Gooday, A.J. & Turley, C.M., 1990. Responses by benthic organisms to inputs of organic material to the sea floor: a review. *Philosophical Transactions of the Royal Society, London, A*, **331**: 119–38.

Goodman, B., Dey, H., Reinhardt, E.G. et al., 2009a. Tsunami waves generated by the Santorini eruption reached eastern Mediterranean shores. *Geology*, **37**(10): 943–6.

Goodman, B., Reinhardt, E., Boyce, J. et al., 2009b. A geoarchaeological multi-proxy study unveils the ancient paleocoastline and harbors of Liman Tepe, Turkey. *Terra Nova*, **21**(2): 97–104.

Gonzalez de Juana, C., Iturralde, J.M. & Picard, X., 1980. *Geologia de Venezuela y de sus Cuencas Petroliferas*. Caracas; Ediciones Foninves.

Gore, A., 2006. *An Inconvenient Truth: The Planetary Emergency of Global Warming and What We Can Do About It*. London; Bloomsbury

Gorog, A., Szinger, B., Toth, E. & Viszkok, J., 2012. Methodology of the micro-computer tomography on foraminifera. *Palaeontologia Electronica*, **15**(1).

Goswami, B.G., Singh, H., Bhatnagar, A.K. et al., 2007. Petroleum systems of the Mumbai offshore basin, India. *Search and Discovery Article*, 10154.

Govindan, A., 2004. Miocene deep water agglutinated Foraminifera from offshore Krishna-Godavari basin, India. *Micropaleontology*, **50**(3): 213–52.

Gradstein, F.M. & Berggren, W.A., 1981. Flysch-type agglutinated Foraminifera and the Maastrichtian to Paleogene history of the Labrador and North seas. *Marine Micropaleontology*, **6**: 211–68.

Gradstein, F.M., Bowman, A., Lugowski, A. & Hammer, O., 2008. Increasing resolution in exploration biostratigraphy – part I. *Natural Resources Research*, **17**(3) (doi: 10.1007/s11053–008–9070–0).

Gradstein, F.M. & Backstrom, S., 1996. Cainozoic biostratigraphy and paleobathymetry, northern North Sea and Haltenbanken. *Norsk Geologisk Tidsskrift*, **76**: 3–32.

Gradstein, F.M. & Kaminski, M.A., 1989. Taxonomy and biostratigraphy of new and emended species of Cenozoic deep-water agglutinated Foraminifera from the Labrador and North seas. *Micropaleontology*, **35**(1): 72–92.

Gradstein, F.M. & Kaminski, M.A., 1997. New species of Paleogene deep-water agglutinated Foraminifera from the North Sea and Norwegian Sea. *Annnales Societatis Geologorum Poloniae*, **67**: 217–29.

Gradstein, F.M., Kaminski, M.A. & Agterberg, F.P., 1999. Biostratigraphy and paleoceanography of the Cretaceous seaway between Norway and Greenland. *Earth-Science Reviews*, **46**: 27–98.

Gradstein, F.M., Kaminski, M.A., Berggren, W.A. et al., 1994. Cenozoic biostratigraphy of the North Sea and Labrador shelf. *Micropaleontology*, **40** (Supplement).

Gradstein, F.M., Kristiansen, I.L., Loemo, L. & Kaminski, M.A., 1992. Cenozoic foraminiferal and dinoflagellate cyst biostratigraphy of the central North Sea. *Micropaleontology*, **38**(2): 101–37.

Gradstein, F.M., Ogg, J.G., Schmitz, M.D. & Ogg, G.M., 2012. *A Geologic Time Scale 2012*. Oxford; Elsevier.

Gradstein, F., Ogg, J. & Smith, A. (eds.), 2004. *A Geologic Time Scale 2004*. Cambridge; Cambridge University Press.

Grafe, K.-U., 1999. Sedimentary cycles, burial history and foraminiferal indicators for systems tracts and sequence boundaries in the Cretaceous of the Basco-Cantabrian basin (northern Spain). *Neues Jahrbuch für Geologie und Palaontologie, Abhandlungen*, **212**(1–3): 85–130.

Graham, J.J., de Lasz, I. & Rerat, D., 1966. Quelques importants foraminiferes du Tertiaire du Gabon (Afrique equatorial). *Revue de Micropaleontologie*, **82**(2): 71–84.

Graham, J.J., Iturralde, J.M. & Picard, X., 1980. *Geologia de Venezuela y de sus Cuencas Petroliferas*. Ediciones Foninves.

Granier, B.R.C., 2000. Lower Cretaceous stratigraphy of Abu Dhabi and the United Arab Emirates – a reappraisal. *9th Abu Dhabi International Petroleum Exhibition and Conference*, 0918.

Granier, B., Al-Suwaidi, A.S., Busnardo, R. *et al.*, 2003. New insight on the stratigraphy of the 'Upper Thamama' in offshore Abu Dhabi (U.A.E.). *Carnets de Geologie*, **2003**(05): 1–17.

Granier, B. & Busnardo, R., 2013. New stratigraphic data on the Aptian of the Persian Gulf. *Cretaceous Research*, **39**: 170–82.

Grant, S.F., Stewart, I.J. & Jones, R.W., 2000. *Abstracts, GeoLuanda 2000 International Conference (14th African Colloquium on Micropalaeontology/Fourth Colloquium on the Stratigraphy and Palaeogeography of the South Atlantic), Luanda, Angola*: 76.

Greene, S.E., Stewart, I.J. & Jones, R.W., 2000. Lower Congo basin chronostratigraphy. *Abstracts, GeoLuanda 2000 International Conference (14th African Colloquium on Micropalaeontology/Fourth Colloquium on the Stratigraphy and Palaeogeography of the South Atlantic), Luanda, Angola*: 76.

Green, O.R., 2001. *A Manual of Practical Laboratory and Field Techniques in Palaeobiology*. Dordrecht; Kluwer Academic Publishers.

Greene, S.E., Martindale, R.C., Ritterbush, K.A. *et al.*, 2012. Recognising ocean acidification in deep time: an evaluation of the evidence for acidification across the Triassic–Jurassic boundary. *Earth-Science Reviews*, **113**: 72–93.

Gregory, F.J., Armstrong, H.A., Boomer, I. *et al.*, 2006. Celebrating 25 years of advances in micropalaeontology: a review. *Journal of Micropalaeontology*, **25**: 97–112.

Grell, K.G., 1954. Der Generationswechsel der Polythalmen Foraminifera *Rotaliella heterocaryotica*. *Archiv der Protistenkunde*, **100**, 211–35.

Grimsdale, T.F., 1952. Cretaceous and Tertiary Foraminifera from the Middle East. *Bulletin of the British Museum (Natural History)*, **1**(8): 221–48.

Grimsdale, T.F. & van Morkhoven, F.P.C.M., 1955. The ratio between pelagic and benthonic Foraminifera as a means of estimating depth of deposition of sedimentary rocks. *Procedings of the Fourth World Petroleum Congress*, **1/D**: 473–91.

Grocke, D.R. & Wortman, U.G. (eds.), 2008. Investigating climates, environments and biology using stable isotopes. *Palaeogeography, Palaeoclimatology, Palaeoecology*, 266.

Grosheny, D. & Granier, B. (eds.), 2011. Platform to basin correlations in Cretaceous times. *Boletin del Instituto de Fisiografia y Geologia*, 79–81.

Grosheny, D. & Tronchetti, G., 1988. Response of benthic Foraminifera to variations in the environment: an example from the Santonian of La Cadiere d'Azur (S.E. France). *Revue de Paleobiologie, Volume Special*, **2**: 537–45.

Groves, J.R. & Altiner, D., 2005. Survival and recovery of calcareous Foraminifera pursuant to the end-Permian mass extinction. *Comptes Rendus Palevol*, **4**: 419–32.

Groves, J.R., Altiner, D. & Rettori, R., 2003. Origin and evolutionary radiation of the Order Lagenida (Foraminifera). *Journal of Paleontology*, **77**(5): 831–43.

Groves, J.R. & Brenckle, P.L., 1997. Graphic correlation in frontier petroleum provinces: application to Upper Paleozoic sections in the Tarim basin, western China. *American Association of Petroleum Geologists Bulletin*, **81**(8): 1259–66.

Groves, J.R. & Lee, A., 2008. Accelerated rates of foraminiferal origination and extinction during the Late Paleozoic ice age. *Journal of Foraminiferal Research*, **38**(1): 74–84.

Groves, J.R., Rettori, R. & Altiner, D., 2004. Wall structures in selected Paleozoic lagenide foraminifera. *Journal of Paleontology*, **78**(2): 245–56.

Groves, J.R. & Wang Yue, 2009. Foraminiferal diversification during the late Paleozoic ice age. *Paleobiology*, **35**(3): 367–92.

Grubb, M., 1999. *The Kyoto Protocol: A Guide and Assessment*. London; The Royal Institute of International Affairs.

Grun, W., Lauer, G., Niedermayr, G. & Schnabel, W., 1964. Die Kreide-Tetrtiar-Grenze im Wienerwaldflysch bei Hochstrass (Niederosterreich). *Sonderabdruck aus den Verhandlungen der Geologischen Bundesanstalt*, **1964**(2): 226–83.

Gu Songzhu, Feng Qinglai & He Weihyong, 2007. The Late Permian deep-water fauna:

latest Changhsingian small foraminifers from southwestern Guangxi, south China. *Micropaleontology*, **53**(4): 311–30.

Guidoboni, E. & Ebel, J.E., 2009. *Earthquakes and Tsunamis in the Past: A Guide to Techniques in Historical Seismology*. Cambridge; Cambridge University Press.

Guillaume, H. A., Bolli, H. M. & Beckmann, J. P., 1972. Estratigrafía del Cretáceo inferior en la Serranía del Interior, Oriente de Venezuela. Boletin de Geologia, Ministerio de Minas e Hidrocarburos, Venezuela, III(5): 1619–55.

Gupta, S.K., 2006. Basin architecture and petroleum system of Krishna Godavari basin, east coast of India. *The Leading Edge*, July 2006: 830–7.

Gusic, I., Jelaska, V. & Velic, I., 1988. Foraminiferal assemblages, facies and environments in the Upper Cretaceous of the Island of Brac, Yugoslavia. *Revue de Paleobiologie, Volume Special*, **2**: 447–56.

Gussow, W.C., 1954. Differential entrapment of oil and gas: a fundamental principle. *American Association of Petroleum Geologists Bulletin*, **38**: 816–53.

Haak, R. & Postuma, J.A., 1975. The relation between the tropical planktonic foraminiferal zonation and the Tertiary Far East letter classification. *Geologie en Mijnbouw*, **54**: 195–8.

Haas, H.C. & Kaminski, M.A. (eds.), 1997. *Contribution to the Micropaleontology and Paleoceanography of the northern North Atlantic*. Grzybowski Foundation (Special Publication, No. 5).

Habura, A., Goldstein, S.T., Broderick, S. & Bowser S.S., 2008. A bush, not a tree: the extraordinary diversity of cold-water basal foraminiferans extends to warm water environments. *Limnology & Oceanography*, **53** (4): 1339–51.

Habura, A., Goldstein, S.T., Parfrey, L.W. & Bowser, S.S., 2006. Phylogeny and ultrastructure of *Miliammina fusca*: evidence for secondary loss of calcification in a miliolid foraminifer. *Journal of Eurykaryote Microbiology*, **53**(3): 204–10.

Haig, D.W., 1979. Global distribution patterns for mid-Cretaceous foraminiferids. *Journal of Foraminiferal Research*, **9**(1): 29–40.

Haig, D.W., 2003. Palaeobathymetric zonation of Foraminifera from Lower Permian shale deposits of a high-latitude southern interior sea. *Marine Micropaleontology*, **49**(4): 317–34.

Haig, D.W., 2005. Foraminiferal evidence for inner neritic deposition of Lower Cretaceous

(Upper Aptian) radiolarian-rich black shales on the Western Australian margin. *Journal of Micropalaeontology*, **24**(1): 55–76.

Haig, D.W., 2012. Palaeobathymetric gradients across Timor during 5.7–3.3 Ma (latest Miocene–Pliocene) and implications for collision uplift. *Palaeogeography, Palaeoclimatology, Palaeoecology*, **331**–332: 50–9.

Halbouty, M.T. (ed.), 1980. *Giant Oil and Gas Fields of the Decade 1968–1978*. Tulsa, OK; American Association of Petroleum Geologists (Memoir, No. 30).

Halbouty, M.T. (ed.), 1982. *The Deliberate Search for the Subtle Trap*. Tulsa, OK; American Association of Petroleum Geologists (Memoir, No. 32).

Halbouty, M.T. (ed.), 1984. *Future Petroleum Provinces of the World*. Tulsa, OK; American Association of Petroleum Geologists (Memoir, No. 40).

Hall, J., Harrison, D.E. & Stammer, D. (eds.), 2010. *Proceedings of OceanObs'09: Sustained Ocean Observations and Information for Society, Vol. 2*. Noordwijk; European Space Agency.

Hall, J.M. & Chan, L.H., 2004. Li/Ca in multiple species of benthic and planktonic Foraminifera: thermocline, latitudinal and glacial–interglacial variation. *Geochimica et Cosmochimica Acta*, **68**(3): 529–45.

Hallam, A. (ed.), 1967. Depth indicators in marine sedimentary environments. *Marine Geology Special Issue*, **5**(5/6): 329–567.

Hallam, A.(ed.), 1973. *Atlas of Palaeobiogeography*. Amsterdam; Elsevier.

Hallock, P., 1988. Diversification of algal-symbiont bearing Foraminifera: a response to oligotrophy. *Revue de Paleobiologie, Volume Special*, **2**: 789–97.

Hallock, P. & Glenn, E.C., 1985. Numerical analysis of foraminiferal assemblages: a tool for recognizing depositional facies in Lower Miocene reef complexes. *Journal of Paleontology*, **59**(6): 1382–94.

Hallock, P. & Glenn, E.C., 1986. Larger Foraminifera: a tool for palaeoenvironmental analysis of Cenozoic carbonate depositional facies. *Palaios*, **1**(1): 55–64.

Hallock, P., Lidz, B.H., Cockey-Burkhard, E.M. & Donnelly, K., 2003. Foraminifera as bioindicators in coral reef assessment and monitoring: the FORAM Index. *Environmental Monitoring and Assessment*, **81**(1–3): 221–38.

Hallock, P. & Peebles, M.W., 1993. Foraminifera with chlorophyte endosymbionts: habitats of six species in the Florida Keys. *Marine Micropaleontology*, **20**(3/4): 277–92.

Hallock, P. & Pomar, L., 2008. Cenozoic photic reef and carbonate ramp habitats: a new look using paleoceanographic evidence. *Proceedings of the 11th International Coral Reef Symposium, Fort Lauderdale*.

Hallock, P., Premoli Silva, I. & Boersma, A., 1991. Similarities between planktonic and larger foraminiferal evolutionary ternds through Paleogene paleoceanographic changes. *Palaeogeography, Palaeoclimatology, Palaeoecology*, **83**: 49–64.

Hallock Muller, P., 1976. Sediment production by shallow-water benthic Foraminifera at selected sites on Oahu, Hawaii. *Maritime Sediments Special Publication*, **1A**: 263–6.

Hamaoui, M., 1976. The use of benthonic Foraminifera as paleoecological markers. *Maritime Sediments Special Publication*, **1A**: 477–80.

Hanse, A., 1965. Les microfaunes en Angola. *Memoires du Bureau de Recherches Geologiques et Minieres*, **32**: 327–34.

Haq, B.U. & Boersma, A. (eds.), 1978. *Introduction to Marine Micropalaeontology*. Amsterdam; Elsevier.

Haq, B.U. & Boersma, A. (eds.), 1998. *Introduction to Marine Micropalaeontology. Second Edition*. Amsterdam; Elsevier.

Haq, B.U. & Al-Qahtani, A.M., 2005. Phanerozoic cycles of sea-level change on the Arabian Platform. *GeoArabia*, **10**(2): 127–60.

Haq, B.U., Hardenbol, J. & Vail, P.R., 1987. Chronology of fluctuating sea levels since the Triassic. *Science*, **235**: 1156–67.

Haque, A.F.M.M., 1956. The Foraminifera of the Ranikot and the Laki of the Nammal Gorge, Salt Range. *Paleontologia Pakistanica*, **1**: 1–300

Hardman, R.F.P. (ed.), 1993 *Exploration Britain: Geological Insights for the Next Decade*. London; The Geological Society (Special Publication, No. 67).

Hardman, R.F.P. & Brooks, J. (eds.), 1990. *Tectonic Events Responsible for Britain's Oil and Gas Reserves*. London; The Geological Society (Special Publication, No. 67).

Harland, W.B., Armstrong, R.L., Cox, A.V. *et al.*, 1990. *A Geologic Time Scale 1989*. Cambridge; Cambridge University Press.

Harland, W.B., Cox, A.V., Llewellyn, P.G. *et al.*, 1982. *A Geologic Time Scale*. Cambridge; Cambridge University Press.

Harney, J.N., Hallock, P. & Talge, H.K., 1998. Observations on a trimorphic life cycle in *Amphistegina gibbosa* populations from the Florida Keys. *Journal of Foraminiferal Research*, **28**(2): 141–7.

Harper, C.W., Jr. & Crowley, K.D., 1985. Insights on why graphic correlation (Shaw's method) works: a discussion. *Journal of Geology*, **93**: 503–6.

Harries, P.J. (ed.), 2003. *Approaches to High-Resolution Stratigraphic Paleontology*. Dordrecht; Kluwer Academic Publishers.

Harris, F. (ed.), 2004. *Global Environmental Issues*. Cambridge; John Wiley.

Harris, N.B. (ed.), 2005. *The Deposition of Organic-Carbon-Rich Sediments: Models, Mechanisms and Consequences*. Tulsa, OK; Society of Economic Paleontologists and Mineralogists (Special Publication, No.82).

Harris, P.M. (ed.), 1983. *Carbonate Buildups: a Core Workshop*. Tulsa, OK; Society of Economic Paleontologists and Mineralogists.

Harrison, R.M. (ed.), 2001. *Pollution: Causes, Effects and Control. Fourth Edition*. Cambridge; Royal Society of Chemistry.

Hart, M.B., 1980a. The recognition of mid-Cretaceous sea-level changes by means of Foraminifera. *Cretaceous Research*, **1**: 189–97.

Hart, M.B., 1980b. A water depth model for the evolution of the planktonic Foraminifera. *Nature*, **286**: 252–4.

Hart, M.B. (ed.), 1987. *Micropalaeontology of Carbonate Environments*. Chichester; Ellis Horwood.

Hart, M.B., 2000. Foraminifera, sequence stratigraphy and regional correlation: an example from the uppermost Albian of southern England. *Revue de Micropaleontologie*, **43**(1–2): 27–45.

Hart, M.B., Dias, B.B., Smart, C.W. & Hall-Spencer, J.M., 2010. Modern seawater acidification: the response of Foraminifera to high $CO_2$ conditions in the Mediterranean Sea. *Abstracts, Forams2010 (International Symposium on Foraminifera), Bonn*: 101–2.

Hart, M.B. & Fitzpatrick, M.E.J., 1995. Kimmeridgian palaeoenvironments: a micropalaeontological perspective. *Proceedings of the Ussher Society*, **8**: 433–6.

Hart, M.B., Hudson, W., Smart, C.W. & Tyszka, J., 2012. A reassessment of '*Globigerina bathoniana*' Pazdrowa, 1969 and the palaeoceanographic significance of Jurassic planktonic Foraminifera from southern Poland. *Journal of Micropalaeontology*, **31**: 97–109.

Hart, M.B., Hylton, M.D., Oxford, M.J. *et al.*, 2003. The search for the origin of the planktonic Foraminifera. *Journal of the Geological Society, London*, **160**: 341–3.

Hart, M.B., Kaminski, M.A. & Smart, C.W. (eds.), 2000. *Proceedings of the Fifth*

*International Workshop on Agglutinated Foraminifera.* Grzybowski Foundation (Special Publication, No. 7).

Hartman, S.E., Lampitt, R.S., Larkin, K.E. *et al.*, 2012. The Porcupine Abyssal Plain fixed-point Sustained Observatory (PAP-SO): variations and trends from the Northeast Atlantic fixed-point time-series. *ICES Journal of Marine Science*, **69**(5), 776–83.

Hartwig, A., di Primio, R., Anka, Z. & Horsfield, B., 2012. Source rock characteristics and compositional kinetic models of Cretaceous organic rich black shales offshore southwestern Africa. *Organic Geochemistry*, **51**: 17–34.

Haslett, S.K. (ed.), 2002. *Quaternary Environmental Archaeology.* London; Arnold.

Hass, C. & Kaminski, M.A. (eds.), 1997. *Contributions to the Micropaleontology and Paleoceanography of the northern North Atlantic.* Grzybowski Foundation (Special Publication, No. 5).

Hasson, P.F., 1985. New observations on the biostratigraphy of the Saudi Arabian Umm Er Radhuma formation (Paleogene) and its correlation with neighboring regions. *Micropaleontology*, **31**(4): 335–64.

Haward, N.J.B. & Haynes, J.R., 1976. *Chlamys opercularis* (Linnaeus) as a mobile substrate for Foraminifera. *Journal of Foraminiferal Research*, **6**(1): 30–8.

Hay, W.W. & Floegel, S., 2012. New thoughts about the Cretaceous climate and oceans. *Earth-Science Reviews*, **115**: 262–72.

Haynes, J.R., 1965. Symbiosis, wall structure and habitat in Foraminifera. *Contributions from the Cushman Foundation for Foraminiferal Research*, **16**: 40–3.

Haynes, J.R., 1981. *Foraminifera.* Cambridge; Macmillan.

Haynes, J.R., 1990. The classification of the Foraminifera: a review of historical and philosophical perspectives. *Palaeontology*, **33**(3): 503–28.

Haynes, J.R. & Dobson, M.R., 1969. Physiography, Foraminifera and sedimentation in the Dovey Estuary (Wales). *Geological Journal*, **6**(2): 217–56.

Hayward, B.W., 1986. A guide to palaeoenvironmental assessment using New Zealand Cenozoic foraminiferal faunas. *New Zealand Geological Survey Special Report (Palaeontology)*, **109**: 1–31.

Hayward, B.W., 1990. Use of foraminiferal data in analysis of Taranaki basin, New Zealand. *Journal of Foraminiferal Research*, **20**(1): 71–83.

Hayward, B.W., 2001. Global deep-sea extinctions during the Pleistocene ice ages. *Geology*, **29**(7): 599–602.

Hayward, B.W., 2002. Late Pliocene to Middle Pleistocene extinctions of deep-sea benthic Foraminifera ('*Stilostomella* extinction') in the southwest Pacific. *Journal of Foraminiferal Research*, **32**(3): 274–307.

Hayward, B.W., Gregory, M.R. & Kennett, J.P., 2011. An extinct Foraminifera endemic to hydrocarbon seeps? *Geology*, **39**(6): 603–5.

Hayward, B.W., Grenfell, H.R., Nicholson, K. *et al.*, 2004b. Foraminiferal record of human impact on intertidal estuarine environments in New Zealand's largest city. *Marine Micropaleontology*, **53**: 37–66.

Hayward, B.W., Grenfell, H.R., Sabaa, A.T. *et al.*, 2006. Micropaleontological evidence of large earthquakes in the past 7200 years in southern Hawke's Bay, New Zealand. *Quaternary Science Reviews*, **25**: 1186–207.

Hayward, B.W., Grenfell, H.R., Sabaa, A.T. *et al.*, 2008. Foraminiferal evidence of Holocene subsidence and fault displacements, coastal South Otago, New Zealand. *Journal of Foraminiferal Research*, **37**(4): 344–59.

Hayward, B.W., Holzmann, M., Grenfell, H.R. *et al.*, 2004a. Morphological distinction of molecular types in *Ammonia*: towards a taxonomic revision of the world's most commonly misidentified foraminifera. *Marine Micropaleontology*, **50**(3–4): 237–71.

Hayward, B.W., Johnson, K., Sabaa, A.T. *et al.*, 2010. Cenozoic record of elongate, cylindrical deep-sea benthic Foraminifera in the North Atlantic and Equatorial Pacific Oceans. *Marine Micropaleontology*, **74**: 75–95.

Hayward, B.W. & Kawagata, S., 2005. Extinct Foraminifera figured in Brady's *Challenger* Report. *Journal of Micropalaeontology*, **24**: 171–5.

Hayward, B.W. & Triggs, C.M., 1994. Computer analysis of benthic foraminiferal associations in a New Zealand tidal inlet. *Journal of Micropalaeontology*, **13**(2): 93–101.

Haywick, D.W. & Henderson, R.A., 1991. Foraminiferal paleobathymetry of Plio-Pleistocene cyclothemic sequences, Petane group, New Zealand. *Palaios*, **6**: 586–99.

Hazel, J.E., Edwards, L.E. & Bybell, L.M. Significant unconformities and the hiatuses represented by them in the Paleogene of the Atlantic and Gulf coastal province.

Hedberg, H.D., 1937. Foraminifera of the middle Tertiary Carapita formation of northeastern Venezuela. *Journal of Paleontology*, **11**(8): 661–97.

Hedgpeth, J.W. (eds.), 1957. *Treatise on Marine Ecology and Paleoecology*. Boulder, CO; Geological Society of America (Memoir, No. 1).

Hedges, S.B. & Kumar, S., 2009. *The Timetree of Life*. Oxford; Oxford University Press.

Hedley, R.H., 1958. A contribution to the biology and cytology of *Haliphysema* (Foraminifera). *Proceedings of the Zoological Society of London*, **130**(4): 569–76.

Hedley, R.H., 1962. The significance of an 'inner chitinous lining' in saccamminid organisation, with special reference to a new species of *Saccammina* (Foraminifera) from New Zealand. *New Zealand Journal of Science*, **5**(3): 375–89.

Hedley, R.H., 1964. The biology of Foraminifera. *International Review of General and Experimental Zoology*, **1**: 1–44.

Hedley, R.H. & Adams, C.G (eds.), 1974. *Foraminifera, Volume 1*. London; Academic Press.

Hedley, R.H. & Adams, C.G. (eds.), 1978. *Foraminifera, Volume 3*. London; Academic Press.

Hedley, R.H. & Wakefield, J.St.J., 1967. A collagen-like sheath in the arenaceous foraminifer *Haliphysema* (Protozoa). *Journal of the Royal Microscopical Society*, **87** (3–4): 475–81.

Hein, C.J., FitzGerald, D.M., Milne, G.A. *et al.*, 2011. Evolution of a Pharaonic harbour on the Red Sea: implications for coastal response to changes in sea level and climate. *Geology*, **39**(7): 687–90.

Heinz, P., Kitazato, H., Schmiedl, G. & Hemleben, C., 2001. Response of deep-sea benthic Foraminifera from the Mediterranean Sea to simulated phytoplankton pulses under laboratory conditions. *Journal of Foraminiferal Research*, **31**(3): 210–27.

Hemleben, C., Kaminski, M.A., Kuhnt, W. & Scott, D.B. (eds.), 1990. *Paleoecology, Biostratigraphy, Paleoceanography and Taxonomy of Agglutinated Foraminifera [Proceedings of the Third International Workshop on Agglutinating Foraminifera]*. Dordrecht; Kluwer Academic Publishers.

Hemleben, C., Spindler M. & Anderson, O.R., 1989. *Modern Planktonic Foraminifera*. Berlin; Springer Verlag.

Henderson, A.S. & Jones, R.W., 2006. The Challenger Foraminifera Online. *Abstracts, The Micropalaeontological Society Foraminifera and Nannofossil Groups Joint Meeting, Liverpool*.

Henson, F.R.S., 1948. *Larger Imperforate Foraminifera of South-Western Asia*. London; British Museum (Natural History).

Henson, F.R.S., 1950a. Cretaceous and Tertiary reef formations and associated sediments in the Middle East. *American Association of Petroleum Geologists Bulletin*, **34**(2): 215–38.

Henson, F.R.S., 1950b. *Middle Eastern Tertiary Peneroplidae (Foraminifera), with Remarks on the Phylogeny and Taxonomy of the Family*. Wakefield; West Yorkshire Printing Co. Ltd.

Hentz, T.F. (ed.), 1999. *Advanced Reservoir Characterization for the 21st Century*. Gulf Coast Section – Society of Economic Paleontologists and Mineralogists (GCS-SEPM) Foundation (Proceedings of the 19th Annual Research Conference).

Henz, T.F. & Hingliu Zeng, 2003. High-frequency Miocene sequence stratigraphy, offshore Louisiana: cycle framework and influence on production distribution in a mature shelf province. *American Association of Petroleum Geologists Bulletin*, **87**(2): 197–230.

Herguera, J.C. & Berger, W.H., 1991. Paleoproductivity from benthic Foraminifera abundance: glacial to postglacial change in the west equatorial Pacific. *Geology*, **19**: 1173–6.

Heron-Allen, E. & Earland, A., 1913. On some Foraminifera from the North Sea etc. dredged by the Fisheries Cruiser 'Goldseeker' (International North Sea Investigations – Scotland). Part 2 – On the distribution of *Saccammina sphaerica* (M. Sars) and *Psammosphaera fusca*(Schulze) in the North Sea, particularly with reference to the suggested identity of the two species. *Journal of the Royal Microscopical Society*: 1–26.

Hess, S., Alve, E., Rygg, B. & Telford, R.J., 2010b. Monitoring benthic community recovery in the Oslofjord: responses to capping and re-oxygenation. *Abstracts, Forams2010 (International Symposium on Foraminifera), Bonn*: 106–7.

Hess, S., Alve, E. & Trannum, H.C., 2010a. Effects of water-based drill cuttings vs. physical burial on benthic Foraminifera and macrofauna: a mesocosm experiment. *Abstracts, Forams2010 (International Symposium on Foraminifera), Bonn*: 106.

Hess, S., Jorissen, F.J., Venet, V. & Abu Zied, R., 2005. Benthic foraminiferal recovery after recent turbidite deposition in Cap Breton Canyon, Bay of Biscay. *Journal of Foraminiferal Research*, **35**(2): 114–29.

Hess, S. & Kuhnt, W., 1996. Deep-sea benthic foraminiferal recolonization of the 1991 Mt

Pinatubo ash layer in the South China Sea. *Marine Micropaleontology*, **28**: 171–97.

Hess, S., Kuhnt, W., Hill, S., *et al.*, 2001. Monitoring the recolonization of the Mt Pinatubo 1991 ash layer by benthic foraminifera. *Marine Micropaleontology*, **43**(1–2): 119–42.

Hesselbo, S.P. & Parkinson, D.N. (eds.), 1996. *Sequence Stratigraphy in British Geology*. London; The Geological Society (Special Publication, No.103).

Hewaidy, A.G.A. & Al-Saad, H.A., 2000. Foraminiferal biostratigraphy of the Lower–Middle Jurassic sequences in eastern Arabia. *GeoResearch Forum*, **6**: 95–104.

Higgins, G.E., 1996. *A History of Trinidad Oil*. Trinidad Express Newspapers Ltd.

Highton, P.J.C., Racey, A., Wakefield, M.I. *et al.*, 1997. Quantitative biostratigraphy: an example from the Neogene of the Gulf of Thailand. *The International Conference on Stratigraphy and Tectonic Evolution of Southeast Asia and the South Pacific, Bangkok, Thailand*: 563–85.

Hikami, M., Fujita, K., Suzuki, A. *et al.*, 2010. Effects of ocean acidification on calcification of symbiont-bearing reef foraminfers. *Abstracts, Forams2010 (International Symposium on Foraminifera), Bonn*: 107.

Hikami, M., Ushie, H., Irie, T. *et al.*, 2011. Contrasting calcification responses to ocean acidification between two reef foraminifers harboring different algal symbionts. *Geophysical Research Letters*, **38**: L19601.

Hill, P.J. & Wood, G.V., 1980. Geology of the Forties field, UK Continental Shelf (North Sea). *American Association of Petroleum Geologists Bulletin*, **64**(8): 81–93.

Hill, T.M., Kennett, J.P. & Spero, H.J., 2003. Formation as indicators of methane-rich environments: a study of modern methane seeps in Santa Barbara Channel, California. *Marine Micropaleontology*, **49**(1–2): 123–38.

Hillaire-Marcel, C. & de Vernal, A. (eds.), 2007. *Proxies in Late Cenozoic Paleoceanography*. Amsterdam; Elsevier.

Hillebrandt, A. von & Urlichs, M., 2008. Foraminifera and Ostracoda from the northern Calcareous Alps and the end-Triassic biotic crisis. *Berichte der Geologischen Bundesanstalt*, **76**: 30–8.

Hilterman, H., 1968. Neuere palaontologischen Daten zum Flysch-Problem. *Erdoel-Erdgas-Zeitschrift*, **84**: 151–7.

Hiltermann, H. & Haman, D., 1986. Sociology and synecology of brackish-water Foraminifera and thecamoebinids of the Balize delta, Louisiana. *Facies*, **13**: 287–94.

Hippensteel, S.P., Martin, R.E., Nikitina, D. & Pizzuto, J.E., 2000. The formation of Holocene marsh foraminiferal assemblages, middle Atlantic coast, USA: implications for Holocene sea-level change. *Journal of Foraminiferal Research*, **30**(4): 272–93.

Ho, S., Cartwright, J.A. & Imbert, P., 2012. Vertical evolution of fluid venting structures in relation to gas flux in the Neogene–Qauternary of the Lower Congo basin, offshore Angola. *Marine Geology*, **332**–4: 40–55.

Hofker, J., 1976. Further studies on Caribbean Foraminifera. *Studies on the Fauna of Curacao and Other Caribbean Islands*, **162**.

Hohenegger, J., 2005. Estimation of environmental paleogradient values based on presence/absence data: a case study using benthic Foraminifera for paleodepth estimation. *Palaeogeography, Palaeoclimatology, Palaeoecology*, **217**: 115–30.

Hohenegger, J., Andersen, N., Baldi, K. *et al.*, 2008. Paleoenvironment of the early Badenian (Middle Miocene) in the southern Vienna basin (Austria) – multivariate analysis of the Baden-Sooss section. *Geologia Carpathica*, **59**: 461–87.

Holbourn, A., Kuhnt, W. & Erbacher, J., 2001. Benthic Foraminifera from Lower Albian black shales (Site 1049, ODP Leg 171): evidence for a non-uniformitarian record. *Journal of Foraminiferal Research*, **31**(1): 60–74.

Holcova, K., 2003. Foraminiferal assemblages in acid residues from the 'Cisarska Rokle' gorge at Srbsko (the Lower/Middle Devonian boundary interval, Barrandian area) and their paleoenvironmental significance. *Bulletin of Geosciences*, **78**(4): 393–403.

Holcova, K., 2004. Foraminifers from the Lower/Middle Devonian boundary beds of the Barrandian area, Czech Republic, and their paleoecology. *Journal of Foraminiferal Research*, **34**(3): 214–31.

Holcova, K., 2010. The Early Devonian morphogroups of agglutinated foraminifers from the Barrandian area (Czech Republic). *Abstracts, Forams2010 (International Symposium on Foraminifera), Bonn*: 109.

Holland, S.M., 1993. Sequence stratigraphy of a carbonate-clastic ramp: the Cincinnatian Series (Upper Ordovician) in its type area. *Geological Society of America Bulletin*, **105**: 306–22.

Holzmann, M., 2000. Species concept in foraminifera: *Ammonia* as a case study. *Micropaleontology*, **46** (suppl., 1): 21–37.

Holzmann, M., Hohenegger, J., Hallock, P. *et al.*, 2001. Molecular phylogeny of large miliolid

Foraminifera (Soritacea Ehrenberg 1839). *Marine Micropaleontology*, **43**(1–2): 57–74.

Holzmann, M., Hohenegger, J. & Pawlowski, J., 2003. Molecular data reveal parallel evolution in nummulitid foraminifera. *Journal of Foraminiferal Research*, **33**(4): 277–84.

Holzmann, M. & Pawlowski, J., 2000. Taxonomic relationships in the genus *Ammonia* (Foraminifera) based on ribosomal DNA sequences. *Journal of Micropalaeontology*, **19**: 85–95.

Holzmann, M. & Pawlowski, J., 2002. Freshwater foraminiferans from Lake Geneva: past and present. *Journal of Foraminiferal Research*, **2**: 344–50.

Holzmann, M., Piller, W.E., Zaninetti, L. et al., 1998. Molecular versus morphology variability in *Ammonia* spp. (Foraminifera, Protozoa) from the Lagoon of Venice, Italy. *Revue de Micropaleontologie*, **41**(1): 59–69.

Honisch, B., Bijma, J., Russell, A.D. et al., 2003. The influence of symbiont photosynthesis on the boron isotope composition of Foraminifera shells. *Marine Micropaleontology*, **49** (1–2): 87–96.

Hoorn, C., Guerrero, J., Sarmiento, G.A. & Lorente, M.A., 1995. Andean tectonics as a cause for changing drainage patterns in Miocene northern South America. *Geology*, **23** (3): 237–40.

Hoorn, C. & Wesselingh, F. (eds.), 2010. *Amazonia: Landscape and Species Evolution*. Oxford; Wiley-Blackwell.

Horbury, A.D. & Robinson, A.G. (eds.), 1993. *Diagenesis and Basin Development*. Tulsa, OK; American Association of Petroleum Geologists (Studies in Geology, No. 36).

Hornibrook, N. de B., 1968. *A Handbook of New Zealand Microfossils (Foraminifera and Ostracods)*. New Zealand Geological Survey.

Hornung, H. & Kress, N., 1991. Monitoring of heavy metals in sediments, benthic fauna and fishes along the Mediterranean coast of Israel. *Israel Oceanographic and Limnological Research Report*, H15/91.

Hornung, H., Krom, M.D. & Cohen, Y., 1989. Trace metal distribution in sediments and benthic fauna of Haifa Bay. *Estuarine, Coastal and Shelf Sciences*, **29**: 43–56.

Horton, B.P., Culver, S.J., Hardbattle, M.I.J. et al., 2007. Reconstructing Holocene sea-level change for the central Great Barrier Reef (Australia) using subtidal foraminifera. *Journal of Foraminiferal Research*, **37**(4): 327–43.

Horton, B.P. & Edwards, R.J., 2000. Quantitative palaeoenvironmental reconstruction techniques in sea-level studies. *Archaeology in the Severn Estuary*, **11**: 105–19.

Horton, B.P. & Edwards, R.J., 2006. *Quantifying Holocene Sea Level Change using Intertidal Foraminifera: Lessons from the British Isles*. Cushman Foundation for Foraminiferal Research (Special Publication, No. 40).

Horton, B.P., Edwards, R.J. & Lloyd, J.M., 1999. A foraminifer-based transfer function: implications for sea-level studies. *Journal of Foraminiferal Research*, **29**(2): 117–29.

Horton, B.P., Whittaker, J.E., Thomson, K.H. et al., 2005. The development of a modern foraminiferal data set for sea-level reconstructions, Wakatobi Marine National Park, southeast Sulawesi, Indonesia. *Journal of Foraminiferal Research*, **35**(1): 1–14.

Hosseini, S.A. & Conrad, M.A., 2008. Calcareous algae, Foraminifera and sequence stratigraphy of the Fahliyan formation at Kuh-e-Surmeh (Zagros basin, SW of Iran). *Geologia Croatica*, **61**(2–3): 215–37.

Hottinger, L., 1962. Recherches sur les alveolines du Paleocene et de l'Eocene. *Abhandlungen der Schweizerischen Palaontologischen Gesellschaft*, **75**/76: 1–243.

Hottinger, L., 1977. Foraminifers operculiniformes. *Memoires du Museum National d'Histoire Naturelle, Serie C (Sciences de la Terre)*, **40**: 1–159.

Hottinger, L., 2000. Functional morphology of benthic foraminiferal shells, envelopes of cells beyond measure. *Micropaleontology*, **46** (Suppl., 1): 57–86.

Hottinger, L. & Drobne, K., 1988. Tertiary alveolinids: problems linked to the conception of species. *Revue de Paleobiologie, Volume Special*, **2**: 665–81.

Houghton, J., 2009. *Global Warming: The Complete Briefing. Fourth Edition*. Cambridge; Cambridge University Press.

Houston, R.M. & Huber, B.T., 1998. Evidence of symbiosis in fossil taxa? Ontogenetic stable isotope trends in some Late Cretaceous planktonic foraminifera. *Marine Micropaleontology*, **34**(1/2): 29–46.

Howell, J.A. & Aitken, J.F. (eds.), 1996. *High-Resolution Sequence Stratigraphy: Innovations and Applications*. London; The Geological Society (Special Publication, No. 104).

Hsu, K.J., 1999. *Caribbean Basins*. Amsterdam; Elsevier (Sedimentary Basins of the World Series, No. 4).

Huang, B., Jian, Z., Cheng, X. & Wang, P., 2003. Foraminiferal responses to upwelling variations in the South China Sea over the last 220,000 years. *Marine Micropaleontology*, **47**(1–2): 1–16.

Hubbard, R.J., 1988. Age and significance of sequence boundaries on Jurassic and Early Cretaceous rifted continental margins. *American Association of Petroleum Geologists Bulletin*, **72**(1): 49–72.

Huber, B.T., Bijma, J. & Darling, K., 1997. Cryptic speciation in the living planktonic Foraminifera *Globigerinella siphonifera* (d'Orbigny). *Paleobiology*, **23**(1): 33–62.

Huber, B.T., Bralower, T.J. & Leckie, R.M. (eds.), 1999. Paleoecological and geochemical signatures of Cretaceous anoxic events: a tribute to William V. Sliter. *Journal of Foraminiferal Research*, **29**(4).

Huc, A.-Y. (ed.), 1990. *Deposition of Organic Facies*. Tulsa, OK; American Association of Petroleum Geologists (Studies in Geology, No. 30).

Huc, A.-Y., 1995. *Paleogeography, Paleoclimate and Source Rocks*. Tulsa, OK; American Association of Petroleum Geologists (Studies in Geology, No. 40).

Hudson, R.G.S. & Chatton, M., 1959. The Musandam limestone (Jurassic to Lower Cretaceous) of Oman, Arabia. *Notes et Memoires sur le Moyen-Orient*, **VII**: 69–93

Hudson, W., Hart, M.B. & Smart, C.W., 2009. Palaeobiogeography of early planktonic Foraminifera. *Bulletin de la Societe Geologique de France*, **180**(1): 27–38.

Hughes, G.W., 1996. A new bioevent stratigraphy of Late Jurassic Arab-D carbonates of Saudi Arabia. *GeoArabia*, **1**(3): 417–34.

Hughes, G.W., 2000. Bioecostratigraphy of the Shu'aiba formation, Shaybah field, Saudi Arabia. *GeoArabia*, **5**(4): 545–78.

Hughes, G.W., 2002. The stratigraphic significance of microbiofacies to selected Saudi Arabian reservoirs. *GeoArabia*, **7**: 248–50.

Hughes, G.W., 2004a. Middle to Late Jurassic biofacies of Saudi Arabia. *Rivista Italiana di Paleontologia e Stratigrafia*, **110**(1): 173–9.

Hughes, G.W., 2004b. Middle to Upper Jurassic Saudi Arabian carbonate petroleum reservoirs: biostratigraphy, micropalaeontology and palaeoenvironments. *GeoArabia*, **9**(3): 79–114.

Hughes, G.W., 2009. Biofacies and palaeoenvironments of the Jurassic Shaqra group of Saudi Arabia. *Volumina Jurassica*, **VI**: 33–45.

Hughes, G.W., 2010a. Palaeoenvironments of Upper Permian Foraminifera of Saudi Arabia. *Abstracts, Forams2010 (International Symposium on Foraminifera), Bonn*: 111–2.

Hughes, G.W., 2010b. Using Foraminifera to biosteer the Upper Permian Khuff reservoirs in Saudi Arabia. *Abstracts, Forams2010*

*(International Symposium on Foraminifera), Bonn*: 112.

Hughes, G.W., Al-Khalid, M. & Varol, O., 2009. Oxfordian biofacies and palaoeenvironments of Saudi Arabia. *Volumina Jurassica*, **VI**: 47–60.

Hughes, G.W., & Beydoun, Z.R., 1992. The Red Sea-Gulf of Aden: biostratigraphy, lithostratigraphy and palaeoenvironments. *Journal of Petroleum Geology*, **15**(2): 135–56.

Hughes, G.W. & Johnson, R.S., 2005. Lithostratigraphy of the Red Sea region. *GeoArabia*, **10**(3): 49–126.

Hughes, G.W. & Nani, N., 2009. Sedimentological and micropalaeontological evidence to elucidate post-evaporitic carbonate palaeoenvironments of the Saudi Arabian latest Jurassic. *Volumina Jurassica*, **VI**: 61–73.

Hughes, G.W., Perincek, D., Grainger, D.J. *et al.*, 1999. Lithostratigraphy and depositional history of part of the Midyan region of Saudi Arabia. *GeoArabia*, **4**(4): 503–42.

Hughes, G.W., Varol, O. & Al-Khalid, M., 2008. Late Oxfordian micropalaeontology, nannopalaeontology and palaeoenvironments of Saudi Arabia. *GeoArabia*, **13**(2): 15–46.

Hughes, G.W., Varol, O. & Beydoun, Z., 1991. Evidence for Middle Oligocene rifting in the Gulf of Aden and Late Oligocene rifting in the southern Red Sea. *Marine and Petroleum Geology*, **8**: 354–8.

Hughes Clarke, M.W., 1988. Stratigraphy and rock unit nomenclature in the oil-prodcing area of interior Oman. *Journal of Petroleum Geology*, **11**(1): 5–60.

Hunter, J. & Ralston, I., 2009. *The Archaeology of Britain*. New York; Routledge.

Hunter, V.F., 1976. Benthonic microfaunal shelfal assemblages and Neogene depositional pattern from northern Venezuela. *Maritime Sediments Special Publication*, **1**: 459–66.

Hunter, V.F., 1978. Foraminiferal correlation of Tertiary mollusc horizons of the southern Caribbean area. *Geologie en Mijnbouw*, **57**: 193–203.

Hurst, A., Johnson, H.D., Burley, S.D. *et al.* (eds.), 1996. *Geology of the Humber Group: Central Graben and Moray Firth, UKCS*. London; The Geological Society (Special Publication, No. 114).

Husain, R., Gupta, R.P. & Lal, N.K., 2000. Tectono-stratigraphic evolution and petroleum systems of Krishna-Godavari basin, India. *Proceedings, Fifth International Conference & Exhibition, Petroleum Geochemistry & Exploration in the Afro-Asian Region, New Dehli*: 443–58.

Husum, K. & Hald, M., 2012. Arctic planktonic foraminiferal assemblages: implications for subsurface temperature reconstructions. *Marine Micropaleontology*, **96–97**: 38–47.

Husum, K., Hald, M. & Burhol, A., 2010. Fine tuning high latitude sea surface temperature reconstructions: new modern training sets of planktonic Foraminifera. *Abstracts, Forams2010 (International Symposium on Foraminifera), Bonn*: 112–2.

Ibrahim, M.I.A., Al-Saad, H. & Kholeif, S.E., 2002. Chronostratigraphy, palynofacies, source-rock potential and organic thermal maturity of Jurassic rocks from Qatar. *GeoArabia*, **7**(4): 675–96.

Ikebe, N. & Tsuchi, R. (eds.), 1984. *Neogene Pacific Datum Planes*. Tokyo; University of Tokyo Press.

Illing, L.V. & Hobson, D.G. (eds.), 1981.*Petroleum Geology of the Continental Shelf of North-West Europe*. London; Heyden.

Imam, B., 2005. *Energy Resurces of Bangladesh*. Dhaka; University Grants Commission of Bangladesh.

Imam, B. & Husain, M., 2002. A review of hydrocarbon habitats in Bangladesh. *Journal of Petroleum Geology*, **25**(1): 31–52.

Imbrie, J., van Donk, J. & Kipp, N.G., 1973. Paleoclimatic investigation of a late Pleistocene Caribbean deep-sea core: comparison of isotopic and faunal methods. *Quaternary Research*, **3**: 10–38.

Immenhauser, A.A., Schlager, W., Burns, S.J. *et al.*, 1999. Late Aptian to late Albian sea-level fluctuations constrained by geochemical and biological evidence (Nahr UMr formation, Oman). *Journal of Sedimentary Research*, **69**: 434–46.

Ingram, W.C., Meyers, S.R., Brunner, C.A. & Martens, C.S., 2010. Late Pleistocene–Holocene sedimentation surrounding an active seafloor gas-hydrate and cold-seep field on the northern Gulf of Mexico slope. *Marine Geology*, **278**: 43–53.

Insalaco, E., Virgone, A., Courme, B. *et al.*, 2006. Upper Dalan member and Kangan formation between the Zagros mountains and offshore Fars, Iran: depositional system, biostratigraphy and stratigraphic archirecture. *GeoArabia*, **11**(2): 75–176.

IPC, 1956. Geological occurrence of oil and gas in Iraq. *Proceedings, 20th International Geological Congress, Mexico*, **II**: 73–101.

Ishizaki, K. & Saito, T. (eds.), 1992. *Centenary of Japanese Micropaleontology*. Tokyo; Terra Scientific Publishing Company.

Ishitani, Y., Ishikawa, S.A., Inagaki, Y. *et al.*, 2011. Multigene phylogenetic analyses including diverse radiolarian species support the 'Retaria' hypothesis: the sister relationship of Radiolaria and Foraminifera. *Marine Micropaleontology*, **81**: 32–42.

Isik, U. & Hakyemez, A., 2011. Integrated Oligocene–Lower Miocene larger and planktonic foraminferal biostratigraphy of the Kahramanmaras basin (southern Anatolia, Turkey). *Turkish Journal of Earth Sciences*, **20**: 185–202.

Izart, A., Vaslet, D., Briand, C. *et al.*, 1998. Stratigraphic correlations between the continental and marine Tethyan and Peri-Tethyan basins during the Late Carboniferous and the Early Permian. *Geodiversitas*, **20**(4): 521–93.

Jablonski, D. & Raup, D.M., 1995. Selectivity of end-Cretaceous marine bivalve extinctions. *Science*, **268**: 389–91.

Jacquin, T., Arnaud-Vanneau, A., Arnaud, H. *et al.*, 1991. Systems tracts and depositional sequences in a carbonate setting: a study of continuous outcrops from platform to basin at the scale of seismic lines. *Marine and Petroleum Geology*, **8**: 122–39.

James, G.A. & Wynd, J.G., 1965. Stratigraphic nomenclature or Iranian oil consortium agreement area. *Bulletin of the American Association of Petroleum Geologists*, **49** (12): 2182–245.

James, K.H., Lorente, M.A. & Pindell, J.L. (eds.), 2009. *The Origin and Evolution of the Caribbean Plate*. London; The Geological Society (Special Publication, No. 328).

Jannink, N.T., Zachariasse, W.J. & van der Zwaan, G.J., 1998. Living (Rose Bengal stained) benthic Foraminifera from the Pakistan continental margin (northern Arabian Sea). *Deep Sea Research I*, **45**: 1483–1513.

Jarvis, I., Carson, G.A., Cooper, M.K.E. *et al.*, 1988. Microfossil assemblages and the Cenomanian–Turonian (Late Cretaceous) Oceanic Anoxic Event. *Cretaceous Research*, **9**: 3–103.

Jassim, S.Z. & Goff, J.C., 2006. *Geology of Iraq*. Brno; Moravian Museum and Dolin.

Jenkins, D.G. (ed), 1993. *Applied Micropalaeontology*. Dordrecht; Kluwer Academic Publishers.

Jenkins, D.G. & Murray, J.W. (eds.), 1989. *Stratigraphical Atlas of Fossil Foraminifera. Second Edition*. Chichester; Ellis Horwood.

Jenkyns, H., 1980. Cretaceous anoxic events: from continents to oceans. *Journal of the Geological Society*, **137**(2): 171–88.

Jepps, M.W., 1942. Studies on *Polystomella Lamarck* (Foraminifera). *Journal of the Marine Biological Association of the United Kingdom*, **25**: 607–66.

Jepps, M.W., 1953. Nucleii of *Cycloclypeus carpenteri* Brady. *Nature*, **171**: 1114–5.

Jiang, M.M., 1998. Middle Eocene through basal Miocene sequence bio-stratigraphy of the western Gulf Coast region. *Marine and Petroleum Geology*, **14**(7/8): 855–66.

Iin Yugan, Wardlaw, B.R., Glenister, B.F. & Kotlyar, G.V., 1997. Permian chronostratigraphic subdivisions. *Episodes*, **20**(1): 10–5.

Johnson, B., 1976. Ecological ranges of selected Toarcian and Domerian (Jurassic) foraminiferal species from Wales. *Maritime Sediments Special Publication*, **1B**: 545–56.

Johnson, K.A., Culver, S.J. & Kamola, D.L., 2005. Marginal marine Foraminifera from the Blackhawk formation (Late Cretaceous, Utah). *Journal of Foraminiferal Research*, **35**(1): 50–64.

Jolley, S.J., Fisher, Q.J., Ainsworth, R.B. *et al.*, 2010. *Reservoir Compartmentalization*. London; The Geological Society (Special Publication, No. 197).

Jones, Rina W. & Milton, N.J., 1994. Sequence development during uplift: Palaeogene stratigraphy and relative sea-level history of the outer Moray Firth, UK North Sea. *Marine and Petroleum Geology*, **11**(2): 157–65.

Jones, R.W., 1983. *Late Quaternary Benthonic Foraminifera from Deep-Water Sites in the N.E. Atlantic and Arctic*. University College of Wales, Aberystwyth; unpublished doctoral thesis.

Jones, R.W., 1984. A revised classification of the unilocular Nodosariida and Buliminida (Foraminifera). *Revista Espanola de Micropaleontologia*, **XVI**: 91–160.

Jones, R.W., 1986. Distribution of 'morphogroups' of agglutinating Foraminifera in the Rockall Trough: a synopsis. *Proceedings of the Royal Society of Edinburgh*, **88B**: 55–8.

Jones, R.W., 1990. The *Challenger* expedition, Henry Bowman Brady (1835–1891) and the *Challenger* Foraminifera. *Bulletin, British Museum (Natural History), Historical Series*, **18**(2): 115–43.

Jones, R.W., 1992. Benthonic Foraminifera associated with natural petroleum seeps. *European Journal of Protistology*, **28**(3): 344–5 (abstract).

Jones, R.W., 1993. Sequence stratigraphic models for the detached platform carbonates of south-east Asia. *Abstracts, Carbonate Petroleum Reservoirs Conference, Geological Society, London*.

Jones, R.W., 1994. *The Challenger Foraminifera*. Oxford; Oxford University Press.

Jones, R.W., 1996. *Micropalaeontology in Petroleum Exploration*. Oxford; Oxford University Press.

Jones, R.W., 1997. Aspects of the Cenozoic stratigraphy of the northern Sulaiman ranges, Pakistan. *Journal of Micropalaeontology*, **16**: 51–8.

Jones, R.W., 2000. Proterozoic to Palaeozoic sequence stratigraphy of south-west Iran. *Abstracts, MEGSTRAT1 Workshop, Dubai*.

Jones, R.W., 2001. Biostratigraphic characterisation of submarine fan sub-environments, deep-water offshore Angola. *Abstracts, Sixth International Workshop on Agglutinated Foraminfera, Prague*.

Jones, R.W., 2003a. Micropalaeontological Characterisation of Submarine Fan/Channel Sub-Environments, Deep-Water Angola. *Abstracts, William Smith Conference ('Wrestling with Mud'), The Geological Society, London*.

Jones, R.W., 2003b. Micropalaeontological Characterisation of Mudrock Seal Capacity. *Abstracts, William Smith Conference ('Wrestling with Mud'), The Geological Society, London*.

Jones, R.W., 2006. *Applied Palaeontology*. Cambridge; Cambridge University Press.

Jones, R.W., 2007. Henry Bowman Brady, hero of foraminiferology. The Man, the Scientist and the Scientific Lagacy. *Abstracts, The Micropalaeontological Society Annual General Meeting Presentations*.

Jones, R.W., 2009. Stratigraphy, palaeoenvironmental interpretation and uplift history of Barbados based on foraminiferal and other palaeontological evidence. *Journal of Micropalaeontology*, **28**: 37–44.

Jones, R.W., 2010. Functional morphology of agglutinating Foraminifera: 25 years of progress. *Abstracts, Forams 2010, Bonn, Germany*: 116.

Jones, R.W., 2011a. *Applications of Palaeontology: Techniques and Case Studies*. Cambridge; Cambridge University Press.

Jones, R.W., 2011b. The importance of biogenic silica in the exploration for and exploitation of unconventional 'shale gas' reservoirs. *Abstracts, Geobiology and Environments of Silica Biomineralizers' Conference, Lille*: 18.

Jones, R.W., 2012. *The Lost City of London: Before the Great Fire of 1666*. Stroud; Amberley.

Jones, R.W. & Charnock, M.A., 1985. 'Morphogroups' of agglutinating Foraminifera, their life positions and feeding habits, and potential applicability in (paleo)ecological studies. *Revue de Micropalaeontologie*, **4**(2): 311–20.

Jones, R.W., Lowe, S., Milner, P. *et al.*, 1999. Biosteering in BP-AMOCO – more bang for your bug! *Geological Society of America Annual Meeting & Exposition, Denver, Colorado, 1999, Abstracts:* A-356.

Jones, R.W., Lowe, S., Milner, P. *et al.*, 2002. The role and value of 'biosteering' in hydrocarbon reservoir exploitation. *Abstracts, AAPG International Conference and Exhibition, Cairo.*

Jones, R.W., Pickering, K., BouDagher-Fadel, M. & Matthews, S., 2003. Micropalaeontological characterisation of submarine fan/channel sub-environments, Ainsa system (Middle Eocene), south-central Pyrenees, Spain. *Extended Abstract, International Conference on Deep-Water Processes in Ancient and Modern Environments, Barcelona and Ainsa.*

Jones, R.W. & Pudsey, C.A., 1994. Recent benthonic Foraminfera from the western Antarctic Ocean. *Journal of Micropalaeontology*, **13**(1): 17–23.

Jones, R.W. & Simmons, M.D., 1999. *Biostratigraphy in Production and Development Geology.* London; The Geological Society (Special Publication, No. 152).

Jones, R.W., Simmons, M.D. & Whittaker, J.E., 2006. On the stratigraphical and palaeobiogeographical significance of *Borelis melo melo* (Fichtel & Moll, 1798) and *B. melo curdica* (Reichel, 1937) (Foraminifera, Miliolida, Alveolinidae). *Journal of Micropalaeontology*, **25**(2): 175–85.

Jones, R.W. & Wonders, A.A.H., 1992. Benthic foraminifers and paleobathymetry of Barrow group (Berriasian–Valanginian) deltaic sequences, sites 762 and 763, northwest shelf, Australia. *Proceedings of the Ocean Drilling Program, Scientific Results*, **122**: 557–68.

Jordt, H., Faleide, J.I., Bjorlykke, K. & Ibrahim, M.T., 1995. Cenozoic sequence stratigraphy of the central and northern North Sea basin: tectonic development, sequence distribution and provenance areas. *Marine and Petroleum Geology*, **12**(8): 845–79.

Jorissen, F., 1999. Benthic foraminiferal successions across Late Quaternary Mediterranean sapropels. *Marine Geology*, **153**: 91–101.

Jorissen, F., 2010. Foraminifera and pollution monitoring: room for improvement. *Abstracts, Forams2010 (International Symposium on Foraminifera), Bonn*: 116–7.

Jorissen, F.J., Bicchi, E., Duchemin, G. *et al.*, 2009. Impact of oil-based drill mud disposal on benthic foraminiferal assemblages on the continental margin off Angola. *Deep-Sea Research II*, **56**: 2270–91.

Jorissen, F.J. & Rohling, E.J. (eds.), 2000. Foraminiferal proxies of paleoproductivity. *Marine Micropaleontology*, **40**(3).

Jorry, S., Davaud, E. & Caline, B., 2003. Controls on the distribution of nummulite facies: a case study from the late Ypresian El Garia formation (Kesra plateau, central Tunisia). *Journal of Petroleum Geology*, **26**(3): 283–306.

Jukes-Browne, A.J. & Harrison, J.B., 1892. The geology of Barbados, part II. The Oceanic Deposits. *Quarterly Journal of the Geological Society*, **48**: 170–226.

Kaddouri, N. & Al-Shaibani, S.K., 1993. *Palorbitolina lenticularis* (Blumenbach), an Upper Barremian–Lower Aptian biozone in Iraq. *Iraqi Geological Journal*, **26**(2): 98–106.

Kaiho, K., 1991. Global changes of Paleogene aerobic/anaerobic benthic Foraminifera and deep-sea circulation. *Palaeogeography, Palaeoclimatology, Palaeoecology*, **83**: 65–85.

Kaiho, K., 1994. Benthic foraminiferal dissolved-oxygen index and dissolved-oxygen levels in the modern ocean. *Geology*, **22**(8): 719–22.

Kaiho, K., 1999. Effect of organic carbon flux and dissolved oxygen on the Benthic Foraminiferal Oxygen Index (BFOI). *Marine Micropaleontology*, **38**(1): 67–76.

Kalantari, A., 1969. *Foraminifera from the Middle Jurassic–Cretaceous Successions of Koppet Dagh Region (N.E. Iran).* Tehran; National Iranian Oil Company (Geological Laboratories Publication No. 3).

Kalantari, A., 1976. *Microbiostratigraphy of the Sarvestan Area, Southwestern Iran.* Tehran; National Iranian Oil Company (Geological Laboratories Publication No. 4).

Kalantari, A., 1982. Microbiostratigraphy of Paleozoic through Jurassic sediments of Ahmadi anticline, southwestern Iran. *Revista Espanola de Micropaleontologia*, **XIV**: 263–90.

Kalantari, A., 1986. *Microfacies of Carbonate Rocks of Iran.* Tehran; National Iranian Oil Company (Geological Laboratories Publication No. 11).

Kale, A.S., 2011. Comments on 'Sequence surfaces and paleobathymetric trends in Albian to Maastrichtian sediments of Arilayur area, Cauvery basin, India'. *Marine and Petroleum Geology*, **28**(6): 1252–9.

Kalogeropoulou, V., Bett, B.J., Gooday, A.J. et al., 2010. Temporal changes (1989–1999) in deep-sea metazoan meiofaunal assemblages on the Porcupine Abyssal Plain, NE Atlantic. *Deep Sea Research II*, **57**(15): 1383–95.

Kameswara Rao, K. & Satyanarayana Rao, T.S., 1979. Studies on pollution ecology of Foraminifera of the Trivandrum coast. *Indian Journal of Marine Science*, **8**(1): 31–5.

Kaminski, M.A., 1985. Evidence for control of abyssal agglutinated foraminiferal community structure by substrate disturbance: results from the HEBBLE area. *Marine Geology*, **66**: 113–31.

Kaminski, M.A. & Coccioni, R. (eds.), 2008. *Proceedings of the Seventh International Workshop on Agglutinated Foraminifera, Krakow, Poland*. Grzybowski Foundation (Special Publication, No. 13).

Kaminski, M.A. & Crespo de Cabrera, S., 1999. A new species of primitive *Reticulophragmium* (Foraminifera) from the Paleocene Vidono formation of northeastern Venezuela. *Annales Societatis Geologorum Poloniae*, **69**: 189–93.

Kaminski, M.A. & Filipescu, S. (eds.), 2011. *Proceedings of the Eighth International Workshop on Agglutinated Foraminifera, Krakow, Poland*. Grzybowski Foundation (Special Publication, No. 16).

Kaminski, M.A. & Geroch, S., 1987. Two new species of *Phenacophragma* from the Paleogene of Trinidad and Poland. *Micropaleontology*, **33**: 185–8.

Kaminski, M.A., Geroch, S. & Gasinski, M.A. (eds.), 1995. *Proceedings of the Fourth International Workshop on Agglutinated Foraminifera, Krakow, Poland*. Grzybowski Foundation (Special Publication, No. 3).

Kaminski, M.A., Geroch, S. & Kaminski, D., 1993. *The Origins of Applied Micropaleontology: The School of Josef Grzybowski*. The Grzybowski Foundation (Special Publication, No. 1).

Kaminski, M.A. & Gradstein, F.M., 1987. Palaeobiogeography of Paleogene flysch-type foraminiferal assemblages in the north Atlantic. Gulf Coast Section – Society of Economic Paleontologists and Mineralogists (GCS-SEPM) Foundation (Proceedings of the 8th Annual Research Conference).

Kaminski, M.A. & Gradstein, F.M., 2005. *Atlas of Paleogene Cosmopolitan Deep-Water Agglutinated Foraminifera*. The Grzybowski Foundation (Special Publication, No. 10).

Kaminski, M.A. & Schroder, C., 1987. Environmental analysis of deep-sea agglutinated Foraminifera: can we distinguish tranquil from disturbed environments? Gulf Coast Section – Society of Economic Paleontologists and Mineralogists (GCS-SEPM) Foundation (Proceedings of the 8th Annual Research Conference).

Kaminski, M.A., Setoyama, E. & Cetean, C.G., 2010. The Phanerozoic diversity of agglutinated foraminifera: origination and extinction rates. *Acta Palaeontologica Polonica*, **55**(3): 529–39.

Kanmera, K., Ishi, K. & Toriyama, R., 1976. The evolution and extinction patterns of Permian fusulinaceans. *Geology and Palaeontology of South-East Asia*, **17**: 129–54.

Kaplan, J.O., Krumhardt, K.M. & Zimmermann, N., 2009. The prehistoric and preindustrial deforestation of Europe. *Quaternary Science Reviews*, **28**: 3016–34.

Kassab, I.I., 1978. Biostratigraphy of Upper Cretaceous–lower Tertiary of north Iraq. *Annales des Mines et dela Geologie*, **28**(2): 277–325.

Kassi, A.M., Kelling, G., Kassi, A.K. et al., 2009. Cretaceous–Palaeocene lithostratigraphic successions across the Bibai thrust, western Sulaiman fold–thrust belt, Pakistan: their significance in deciphering the early collisional history of the NW Indian plate margin. *Journal of Asian Earth Sciences*, **35**: 435–44.

Katz, B.J. (ed.), 1994. *Petroleum Source Rocks*. Berlin; Springer.

Katz, B.J. & Pratt, L.M., 1993. *Source Rocks in a Sequence Stratigraphic Framework*. Tulsa, OK; American Association of Petroleum Geologists (Studies in Geology, No. 37).

Katz, M.E., Cramer, B.S., Franzese, A. et al., 2010. Traditional and emerging geochemical proxies in Foraminifera. *Journal of Foraminiferal Research*, **40**(2): 165–92.

Katz, M.E. & Miller, K.G., 1993. Latest Oligocene to earliest Pliocene benthic foraminiferal facies of the northeastern Gulf of Mexico. *Micropaleontology*, **39**(4): 367–403.

Katz, M.E. & Thunell, R.C., 1984. Benthic foraminiferal biofacies associated with Middle Miocene to Early Pliocene oxygen-deficient conditions in the eastern Mediterranean. *Journal of Foraminiferal Research*, **14**(3): 187–202.

Kauffman, E.G. & Hazel, J.E. (eds.), 1977. *Concepts and Methods in Biostratigraphy*. Stroudsburg, PA: Dowden, Hutchinson & Ross.

Kawagata, S., Hayward, B.W., Grenfell, H.R. & Sabaa, A.T., 2005. Mid-Pleistocene extinction of deep-sea Foraminifera in the North

Atlantic Gateway (ODP sites 980 and 982). *Palaeogeography, Palaeoclimatology, Palaeoecology*, **221**: 267–91.

Kavary, E. & Frizzell, D.L., 1963. Upper Cretaceous and lower Cenozoic Foraminifera from west-central Iran. *Bulletin of the School of Mines and Metallurgy, University of Missouri*, **102**: 1–89.

Kazmierczak, J., 1973. *Tolypammina vagans* (Foraminiferida) as inhabitant to the Oxfordian siliceous sponges. *Acta Palaeontologica Polonica*, **18**: 95–115.

Keeling, P.J., 2001. Foraminifera and Cercozoa are related in actin phylogeny: two orphans find a home? *Molecular Biology and Evolution*, **18**(8): 1551–7.

Keller, G., Khosla, S.C., Sharma, R. *et al.*, 2009. Early Danian planktonic Foraminifera from Cretaceous/Tertiary Intertrappean beds at Jhilmili, Chhindwara district, Madhya Pradesh, India. *Journal of Foraminiferal Research*, **39**(1): 40–55.

Kemp, A.C., Horton, B.P. & Culver, S.J., 2009. Distribution of modern salt-marsh Foraminifera in the Albemarle–Pamlico estuarine system of North Carolina, USA: implications for sea-level research. *Marine Micropaleontology*, **72**: 222–38.

Kendall, G.St.C., Chiarenzelli, J. & Hassan, H.S., *et al.*, 2009. World source rock potential through geological time: a function of basin restriction, nutrient level, sedimentation rate, and sea-level rise. *Search and Discovery Article*, 40472.

Kender, S., Kaminski, M.A. & Jones, R.W., 2006. Four new species of deep-water agglutinated Foraminifera from the Oligocene–Miocene of the Congo fan (offshore Angola). *Micropaleontology*, **52**: 465–70.

Kender, S., Kaminski, M.A. & Jones, R.W., 2007a. Benthic foraminiferal response to reduced oceanic acidification and global warming during the Miocene Monterey Excursion, Congo fan, southeastern Atlantic. *Abstracts, American Geophysical Union Fall Meeting*: OS11A-1083.

Kender, S., Kaminski, M.A. & Jones, R.W., 2007b. Responses of benthic foraminiferal assemblages to isotope shifts and sedimentological changes during the Early–Middle Miocene in the deep-sea Congo fan, eastern South Atlantic. *Abstracts, 6th Polish Micropalaeontological Workshop, Gdansk*: 35.

Kender, S., Kaminski, M.A. & Jones, R.W., 2008. Early to Middle Miocene Foraminifera from the deep sea Congo fan, offshore Angola. *Micropaleontology*, **54**: 477–568.

Kender, S., Peck, V.L., Jones, R.W. & Kaminski, M.A., 2009. Middle Miocene Oxygen Minimum Zone expansion offshore West Africa: evidence for global cooling precursor events. *Geology*, **37**(8): 699–702.

Kennan, L., Pindell, J. & Rosen, N.C. (eds.), 2007.*The Paleogene of the Gulf of Mexico and Caribbean Basins: Processes, Events and Petroleum Systems*. Gulf Coast Section – Society of Economic Paleontologists and Mineralogists (GCS-SEPM) Foundation (Proceedings of the 27th Annual Research Conference).

Kennedy, R.F., Jr., 2004. *Crimes Against Nature: Standing up to Bush and the Kyoto Killers: How George W. Bush and His Corporate Pals Are Plundering the Country*. New York, NY; Harper Collins.

Kennedy, W.J. & Simmons, M.D., 1991. Mid-Cretaceous ammonites and associated microfossils from the central Oman mountains. *Newsletters on Stratigraphy*, **25**(3): 127–54.

Kennett, J.P. (ed.), 1985. *The Miocene Ocean: Paleoceanography and Biogeography*. Boulder, CO; Geological Society of America (Memoir, No. 163).

Kennett, J.P., & Srinivasan, M.S., 1983. *Neogene Planktonic Foraminifer: A Phylogenetic Atlas*. Stroudsburg, PA; Hutchinson Ross Publishing Company.

Khan, M.A., Ahmed, R., Raza, H.A. & Kemal, A., 1986. Geology of petroleum in Kohat-Potwar depression, Pakistan. *American Association of Petroleum Geologists Bulletin*, **70**(4): 396–414.

Khanna, N., Austin, W.E.N. & Paterson, D.M., 2010. Biological response of Foraminifera to ocean acidification. *Abstracts, Forams2010 (International Symposium on Foraminifera), Bonn*: 120.

Kiel, S. (ed.), 2010. *The Vent and Seep Biota: Aspects from Microbes to Ecosystems*. Berlin; Springer.

Kiessling, W., 2006. Geographic range and extinction risk: new lessons from ancient marine organisms. *Programme with Abstracts, Palaeogeography and Palaeobiogeography (Biodiversity in Space and Time) Conference, Cambridge*: 9.

King, C., 1983. Cainozoic micropalaeontological biostratigraphy of the North Sea. *Institute of Geological Sciences Report*, **82**/7.

Kjennerud, T. & Gillmore, G.K., 2003. Integrated Palaeogene palaeobathymetry of the northern North Sea. *Petroleum Geoscience*, **9**: 125–32.

Klemme, H.D. & Ulmishek, G.F., 1991. Effective petroleum source rocks of the world:

stratigraphic distribution and controlling depositional factors. *American Association of Petroleum Geologists Bulletin*, **75**(12): 1809–51.

Knight, R. & Mantoura, R.C.F., 1985. Chloroplast and carotenoid pigments in Foraminifera and their symbiotic Algae: analysis by high performance liquid chromatography. *Marine Ecology Progtress Series*, **23**: 241–9.

Knox, R.W.O'B., Corfield, R.M. & Dunay, R.E. (eds.), 1996. *Correlation of the Early Palaeogene in North-West Europe*. London; The Geological Society (Special Publication, No. 101).

Kogbe, C.A. & Me'hes, K., 1986. Micropaleontology and biostratigraphy of the coastal basins of West Africa. *Journal of African Earth Sciences*, **5**(1): 1–100.

Kohl, B., 1985. Early Pliocene benthic foraminifers from the Salinas basin, southeastern Mexico. *Bulletin of American Paleontology*, **88**(322).

Koho, K., 2008. Benthic Foraminifera: ecological indicators of past and present oceanic environments – a glance at the modern assemblages from Portuguese submarine canyons. *Geologi*, **60**: 161–6.

Koho, K.A., Kouwenhoven, T.J., de Stigter, H.C. & van der Zwaan, G.J., 2007. Benthic Foraminifera in the Nazare canyon, Portuguese continetal margin: sedimentary environments and disturbance. *Marine Micropaleontology*, **66**: 27–51.

Koho, K.A., Langezaal, A.M., van Lith, Y.A. et al., 2008. The influence of a simulated diatom bloom on deep-sea benthic Foraminifera and the activity of bacteria: a mesocosm study. *Deep-Sea Research I*, **5**: 696–719.

Koho, K.A., Pina-Ochoa, E., Geslin, E. & Risgaard-Petersen, N., 2010. Denitrification and life strategy of Globobulimina turgida under anoxic conditions: a laboratory study. *Abstracts, Forams2010 (International Symposium on Foraminifera), Bonn*: 123.

Komiya, T., Hirata, T., Kitajima, K. et al., 2008. Evolution of the composition of seawater though geologic time, and its influence on the evolution of life. *Gondwana Research*, **14**: 159–74.

Korchagin, O.A., Kuznetsova, K.I. & Bragin, N.Yu., 2003. Find of early planktonic foraminifers in the Triassic of the Crimea. *Doklady Earth Sciences*, **390**(4): 482–6.

Kotthoff, U., Katz, M.E. & McCarthy, F.M.G., 2010. Foraminifer- and palynology-based reconstructions of site–shoreline distance, sea-level, and ecosystems: new data from the New Jersey shallow shelf (IODP Expedition 313). *Abstracts, Forams2010 (International Symposium on Foraminifera), Bonn*: 124.

Koutsoukos, E.A.M., 1982. Geohistoria e paleoecologia das bacias marginais de Florianopolis e Santos. *Anais do XXXII Congresso Brasiliero de Geologia, Salvador, Bahia*, **5**: 2369–82.

Koutsoukos, E.A.M., 1984. Evolucao paleoecologica do Albiano ao Maestrichtiano na area noreste da bacia de Campos, Brasil, com base em foraminiferos. *Anais do XXXIII Congresso Brasiliero de Geologia, Rio de Janeiro*, **2**: 685–98.

Koutsoukos, E.A.M., 1985a. A area noroeste da bacia de Campos, Brasil do Mesocretaceo ao Neocretaceo: evolucao paleoambiental e paleogeografica de estudos de foraminferos. *IX Congreso Brasiliero de Paleontologia, Fortaleza, Ceara*.

Koutsoukos, E.A.M., 1985b. Distribucao paleobatimetrica de foraminiferos bentonicos do Cenozoico: margem continental Atlantica. *Trabahos Apresentados do VIII Congresso Brasiliero de Paleontologia, Rio de Janeiro*, **2**: 355–70.

Koutsoukos, E.A.M., 1985c. Paleobatimetria da margina continental do Brasil durante o Albiano. *IX Congreso Brasiliero de Paleontologia, Fortaleza, Ceara*.

Koutsoukos, E.A.M. (ed.), 2005. *Applied Stratigraphy*. Dordrecht; Springer.

Koutsoukos, E.A.M., Leary, P.N. & Hart, M.B., 1989. *Favusella* Michael (1972): evidence of ecophenotypic adaptation of a planktonic foraminifer to shallow-water carbonate facies during the mid-Cretaceous. *Journal of Foraminiferal Research*, **19**(4): 324–36.

Koutsoukos, E.A.M., Leary, P.N. & Hart, M.B., 1990. Latest Cenomanian–earliest Turonian low-oxygen tolerant benthonic Foraminifera: a case-study from the Sergipe basin (N.E. Brazil) and the western Anglo-Paris basin (southern England). *Palaeogeography, Palaeoclimatology, Palaeoecology*, **77**: 145–77.

Koutsoukos, E.A.M., Mello, M.R., de Azambuja Filho, N.C. et al., 1991. The Upper Aptian–Albian succession of the Sergipe basin, Brazil: and integrated paleoenvironmental assessment. *American Association of Petroleum Geologists Bulletin*, **75**(3): 479–98.

Koutsoukos, E.A.M. & Merrick, K.A., 1986. Foraminiferal paleoenvironments from the Barremian to Maestrichtian of Trinidad, West Indies. *Transactions of the 1st*

Geological Conference of the Geological Society of Trinidad and Tobago, 1985: 85–101.

Kouyoumontzakis, G., 1987. Le role des foraminiferes benthiques dans l'identification de l'etage bathyal sur le plateau continental congolais. *Geologie Mediteranenne*, **14**(1): 15–23.

Kovacs, J.S., 2005. Depth gradient proxies: palaeoecology versus sedimentology. Case study from the Turea Group deposits of the Paleogene Transylvanian Basin. *Acta Palaeontologica Romaniae*, **5**: 259–76.

Krainer, K. & Vachard, D., 2011. The Lower Triassic Werfen formation of the Karawanken mountains (southern Austria) and its disaster survivor microfossils, with emphasis on *Postcladella* n. gen. (Foraminifera, Milioliata, Cornuspirida). *Revue de Micropaleontologie*, **54**: 59–85.

Krutak, P.R., 1975. Environmental variation in living and total populations of Holocene Foraminifera and Ostracoda, coastal Mississippi, USA. *American Association of Petroleum Geologists Bulletin*, **59**(1): 140–60.

Ksiazkiewicz, M., 1975. Bathymetry of the Carpathian flysch basin. *Acta Geological Polonica*, **25**(3): 309–67.

Kucera, M., Aurahs, R. & Auch, A., 2010. Molecular phylogeny of planktonic Foraminifera. *Abstracts, Forams2010 (International Symposium on Foraminifera), Bonn*: 126.

Kuhnt, T., Friedrich, O., Herrle, J.O. & Schmiedl, G., 2010. Relationship between pore-density within benthic foraminifer and oxygen concentration. *Abstracts, Forams2010 (International Symposium on Foraminifera), Bonn*: 126–7.

Kumar, 1984. Middle Eocene–Early Oligocene biofacies and palaeoecology of the northern part of the Cambay basin. *Proceedings of the Xth Indian Colloquium on Micropalaeontology & Stratigraphy*: 289–98.

Kupper, I., 1960. Miogypsinen aus British West Afrika (Cameroon). *Science Reports, Tohoku University, Sendai, Second Series (Geology), Special Publication*, **4**: 56–69.

Kureshy, A.A., 1977a. The Cretaceous larger foraminiferal biostratigraphy of Pakistan. *Journal of the Geological Society of India*, **18**(12): 662–7.

Kureshy, A.A., 1977b. The Cretaceous planktonic foraminiferal biostratigraphy of Pakistan. *Palaeontological Society of Japan, Special Papers*, **21**: 223–31.

Kureshy, A.A., 1978a. The biostratigraphic correlation of sedimentary basins of West Pakistan. *Annales des Mines et de la Geologie*, **28**(II): 327–36.

Kureshy, A.A., 1978b. Tertiary larger foraminiferal zones of Pakistan. *Revista Espanola de Micropaleontologia*, **X**(3): 467–83.

Kureshy, A.A., 1978c. The Tertiary planktonic foraminiferal zones of Pakistan. *Revista Espanola de Micropaleontologia*, **IX**(2): 203–19.

Kureshy, A.A., 1980. Paleobiogeography of Cretaceous larger Foraminifera of Pakistan and the Caribbean region and their bearing on continental drift. *Cretaceous Research*, **1**: 93–100.

Kureshy, A.A., 1984. Neogene planktonic and larger foraminiferal datum planes of Pakistan and its calibration to paleomagnetic and radiometric time scale. *Revista Espanola de Micropaleontologia*, **XVI**: 399–414.

Kuroyanagi, A., Kawahata, H., Suzuki, A. *et al.*, 2009. Impacts of ocean acidification on large benthic Foraminifera: results from laboratory experiments. *Marine Micropaleontology*, **73**: 190–5.

Kuroyanagi, A., Tsuchiya, M., Kawahata, H. & Kitazato, H., 2008. The occurrence of two genotypes of the planktonic foraminifer *Globigerinoides ruber* (white) and paleoenvironmental implications. *Marine Micropaleontology*, **68**(3–4): 236–43.

Kuznetsova, K.I., Grigelis, A.A., Adjamian, J. *et al.*, 1996. *Zonal Stratigraphy and Foraminifera of the Tethyan Jurassic (Eastern Mediterranean)*. Amsterdam; Gordon & Breach.

Labune, C., Romero-Ramirez, A., Amoroux, J.M. *et al.*, 2012, Comparison of ecological quality indices based on benthic macrofauna and sediment profile images: a case study along an organic enrichment gradient off the Rhone river. *Ecological Indicators*, **12**(1): 133–42.

Lagoe, M.B., Gary, A., Zeller, S. *et al.*, 1994. Bathymetric zonation and relationships to sea-bed environments of modern benthic foraminiferal biofacies, northwestern Gulf of Mexico slope. *American Asociation of Petroleum Geologists Bulletin*, **78**(9): 1464 (abstract).

Lamb, J.L., 1964. The geology and paleontology of the Rio Aragua surface section, Serrania del Interior, State of Monagas, Venezuela. *Boletin Informativo Asociacion Venezolano de Geologia, Mineria y Petroleo, Caracas*, **7**(4): 111–23.

Langdon, G.S. & Malecek, S.J., 1987. Seismic stratigraphic study of two carbonate sequences, eastern Saudi Arabia. *American*

*Association of Petroleum Geologists Bulletin*, **71**(4): 403–18.

Langer, M.R. (ed.), 1995. Forams '94. Selected papers from an international symposium held at the University of California at Berkeley. *Marine Micropaleontology*, **26**(1–4).

Langer, M.R., 1999. Origin of Foraminifera: conflicting molecular and paleontological data? *Marine Micropaleontology*, **38**(1): 1–6.

Langer, M.R. & Hottinger, L., 2000. Biogeography of selected 'larger' Foraminifera. *Micropaleontology*, **46**(Suppl. 1): 105–26.

Langer, M.R. & Leppig, U., 2000. Molecular phylogenetic status of *Ammonia catesbyana* (d'Orbigny, 1839), and intertidal foraminifer from the North Sea. *Neues Jahrbuch für Geologie und Paläontologie*, **2000**(9): 545–56.

Langer, M.R. & Lipps, J.H., 2006. Assembly and persistence of Foraminifera in introduced mangroves on Moorea, French Polynesia. *Micropaleontology*, **52**(4): 343–55.

Langer, M.R., Lipps, J.H. & Werner, E.P., 1993. Molecular paleobiology of protists: amplification and direct sequencing of foraminiferal DNA. *Micropaleontology*, **39**(1): 63–8.

Langer, M.R., Silk, M.T.B. & Lipps, J.H., 1997. Global ocean carbonate and carbon dioxide production: the role of reef Foraminifera. *Journal of Foraminiferal Research*, **27**(4): 271–7.

Lankester, E.R. (ed.), 1903. *A Treatise on Zoology. Part 1, Introduction and Protozoa*. London; Adam and Charles Black.

Lankford, R.R., 1959. Distribution and ecology of Foraminifera from east Mississippi delta margin. *American Association of Petroleum Geologists Bulletin*, **43**: 2068–99.

Larkin, K.E. & Gooday, A.J., 2004. Soft-shelled monothalamous Foraminifera are abundant at an intertidal site on the south coast of England. *Journal of Micropalaeontology*, **23**(2): 135–7.

Larkin, K.E. & Gooday, A.J., 2009. Foraminiferal faunal responses to monsoon-driven changes in organic matter and oxygen availability at 140 m and 300 m water depth in the NE Arabian Sea. *Deep-Sea Research II*, **56** (6–7): 403–21.

Latif, M.A., 1961. The use of pelagic Foraminifera in the subdivision of the Paleocene–Eocene of the Rakhi Nala, West Pakistan. *Geological Bulletin, Punjab University*, **1**: 31–46.

Laudon, R.C., 1996. *Principles of Development Geology*. Engelwood Cliffs, NJ; Prentice Hall.

Launder, J.B. & Thompson, J.M.T. (eds.), 2009. *Geo-engineering Climate Change*. Cambridge; Cambridge University Press.

le Calvez, J., 1938. Recherches sur les foraminiferes. I. Developpement et reproduction. *Archives de Zoologie Experimentale et Generale*, **80**: 163–333.

xle Calvez, J., 1946. Place de la reduction chromatique et alternance de phases nucleaires dans le cycle des foraminiferes. *Comptes Rendus de l'Academie des Sciences, Paris*, **222**: 612–4.

xle Calvez, J., 1950. Recherches sur les foraminiferes. II. Place de le meiose et sexualite. *Archives de Zoologie Experimentale et Generale*, **87**: 211–43.

xle Calvez, Y., de Klasz, I. & Brun, L., 1971. Quelques foraminiferes de l'Afrique occidentale. *Revista Espanola de Micropaleontologia*, **3**(3): 305–26.

xle Calvez, Y., de Klasz, I. & Brun, L., 1974. Nouvelle contribution a la connaisance des microfaunes du Gabon. *Revista Espanola de Micropaleontologia*, **6**(3): 381–400.

xle Furgey, A. & St Jean, J., Jr., 1976. Foraminifera in brackish-water ponds designed for waste control and aquaculture studies in North Carolina. *Journal of Foraminiferal Research*, **6**(4): 274–94.

xle Nindre, Y.M., Manivit, J., Manivit, H. & Vaslet, D., 1990. Stratigraphie sequentielle du Jurassique et du Cretace en Arabie Saoudite. *Bulletin de la Societe Geologique de France*, **6**(6): 1025–34.

Leadbeater, S.C. & Riding, R. (eds.), 1986. *Biomineralization in Lower Plants and Animals*. Oxford; Systematics Association (Special Volume, No. 30).

Lear, C.H. & Rosenthal, Y., 2006. Benthic foraminiferal Li/Ca: insights into Cenozoic seawater carbonate saturation state. *Geology*, **34**(11): 985–8.

Leckie, R.M., 1987. Paleoecology of mid-Cretaceous planktonic Foraminifera: a comparison of open ocean and epicontinental sea assemblages. *Micropaleontology*, **33**(2): 164–76.

Leckie, R.M., Bralower, T.J. & Cashman, R., 2002. Oceanic Anoxic Events and plankton evolution: biotic response to tectonic forcing during the mid-Cretaceous. *Paleoceanography*, **17**(3): 13–1 to 13–29.

Lecointre, G. & le Guyader, H., 2006. *The Tree of Life: A Phylogenetic Classification*. Cambridge, Cambridge, MA; Harvard University Press.

Lee, J.J., 2006. Algal symbiosis in larger Foraminifera. *Symbiosis*, **42**: 63–75.

Lee, J.J. & Anderson, O.R. (eds.), 1991. *Biology of Foraminifera*. London; Academic Press.

Lee, J.J., Faber, W.W., & Lee, R.E. 1991. Granular reticular digestion: a possible preadaption to benthic foraminiferal symbiosis? *Symbiosis*, **10**: 47–61.

Lee, J.J., Freudenthal, H.D., Muller, W.A. *et al.*, 1963. Growth and physiology of Foraminifera in the laboratory: Part 3 – initial studies on *Rosalina floridana* (Cushman). *Micropaleontology*, **9**: 449–66.

Lee, J.J. & Hallock, P. (eds.), 2000. Advances in the biology of Foraminifera. *Micropaleontology*, **46** (Supplement 1).

Lee, J.J., Huttner, J. & Bovee, E.C. (eds.), 2002. *An Illustrated Guide to the Protozoa, Second Edition*. Lawrence Kansas; Society of Protozoologists/Allen Press.

Lee, J.J., McEnery, M., Pierce, S. *et al.*, 1966. Tracer experiments in feeding littoral Foraminifera. *Journal of Protozoology*, **13**: 659–70.

Lee, J.J. & Muller, W.A., 1967. Growth rates of Foraminifera in gnotobiotic culture. *Journal of Protozoology*, **13**(23): 86.

Lee, J.J., Muller, W.A., Stone, R.J. *et al.*, 1969. Standing crop of Foraminifera in sublittoral epiphytic communities of a Long Island salt marsh. *Marine Biology*, **4**: 44–61.

Lee, J.J. & Pierce, S., 1963. Growth and physiology of Foraminifera in the laboratory: Part 4 – monoxenic culture of an allogromiid with notes on its morphology. *Journal of Protozoology*, **10**: 401–11.

Lee, J.J., Pierce, S., Tentchoff, M. & McLaughlin, J.J., 1961. Growth and physiology of Foraminifera in the laboratory: Part 1 – collection and maintenance. *Micropaleontology*, **7**: 461–6.

Lee, J.J. & Zucker, W., 1969. Algal flagellate symbiosis in the foraminifer *Archaias*. *Journal of Protozoology*, **16**: 71–81.

Lehmann, C., Goff, J. & Jones, R.W., 2004. Jurassic carbonates and evaporites of the Middle East: a new look at an old play. *Abstracts, American Association of Petroleum Geologists International Conference and Exhibition, Cancun, Mexico*.

Lehmann, C., Goff, J. & Jones, R.W., 2006a. A detailed Jurassic sequence stratigraphic framework for the formation of dolomites along the Gotnia shelf margin. *Abstracts, Seventh Middle East Geosciences Conference and Exhibition, Manama, Bahrain, 2006 (GEO2006)*.

Lehmann, C., Goff, J. & Jones, R.W., 2006b. Facies distribution and dolomitization along the shelf margin of the Jurassic Gotnia basin. *Abstracts, American Association of Petroleum Geologists Annual Convention, Houston*.

Lehmann, C., Goff, J. & Jones, R.W., 2006c. Jurassic sequences of the northern Arabian Plate. *Abstracts, American Association of Petroleum Geologists International Conference and Exhibition, Perth, Australia*.

Lehner, P., Meijer, B. & Kooper, H., 1985. Mesozoic source rocks of the Middle East. *Proceedings of the Seminar on Source and Habitat of Petroleum in the Arab Countries, Kuwait, 1984*: 73–118.

Leiter, C. & Altenbach, A.V., 2010. Benthic Foraminifera from the diatomaceous mud belt off Namibia: characteristic species for severe anoxia. *Palaeontologia Electronica*, **13**(2).

Leorri, E., Cearreta, A. & Horton, B.P., 2008. A Foraminifera-based transfer function as a tool for sea-level reconstructions in the southern Bay of Biscay. *Geobios*, **41**: 787–97.

Leroy, D.O. & Hodgkinson, K.A., 1975. Benthonic Foraminifera and some Pteropoda from a deep-water dredge sample, northern Gulf of Mexico. *Micropaleontology*, **21**(4): 420–47.

Leroy, D.O. & Levinson, S.A., 1974. A deep-water Pleistocene microfossil assemblage from a well in the north west Gulf of Mexico. *Micropaleontology*, **20**(1): 1–37.

Less, G., 1987. Paleontology and stratigraphy of the European Orthophragmininae. *Geologica Hungarica, Series Paleontologica*, **51**.

Less, G. & Kovacs, L.O., 1995. Age-estimates by European Paleogene Orthophragmininae using numerical evolutionary correlation. *Geobios*, **29**: 261–85.

Lesser, M. (ed.), 2010. *Advances in Marine Biology, Vol. 58*. London; Academic Press.

Lesslar, P., 1987. Computer-assisted interpretation of depositional palaeoenvironments based on Foraminifera. *Geological Society of Malaysia Bulletin*, **21**: 103–19.

Leturmy, P. & Robin, C.(eds.), 2010. *Tectonic and Stratigraphic Evolution of Zagros and Maran during the Mesozoic–Cenozoic*. London; The Geological Society (Special Publication, No. 330).

Leutenegger, S., 1977. Reproduction cycles of larger Foraminifera and depth distribution of generations. *Utrecht Micropaleontological Bulletin*, **15**: 26–34.

Leutze, W.P., 1972. Stratigraphic utility of some Miocene and younger arenaceous Foraminifera. *Transactions of the Gulf Coast Association of Geological Societies*, **22**: 147–55.

Levandowsky, M. & Hutner, S.H. (eds.), 1980. *Biochemistry and Physiology of Protozoa*.

*Second Edition.* New York, NY; Academic Press.

Levin, L.A., Childers, S.E. & Smith, C.R., 1991. Epibenthic agglutinating foraminiferans in the Santa Catalina basin and their response to disturbance. *Deep-Sea Research*, **38**(4): 465–83.

Levin, L.A., Ekau, W., Gooday, A.J. *et al.*, 2009. Effects of natural and human-induced hypoxia on coastal benthos. *Biogeosciences*, **6**(10): 2063–98.

Levy, A., Mathieu, R., Poignant, A. *et al.*, 1988. Les Soritidae et les Peneroplidae dans le biofacies de la plate-forme des Bahamas. *Revue de Paleobiologie, Volume Special*, **2**: 833–41.

Licari, L.N., Schumacher, S., Wenzhofer, F. *et al.*, 2003. Communities and microhabitats of living benthic Foraminifera from the tropical East Atlantic: impact of different productivity regimes. *Journal of Foraminiferal Research*, **33**(1): 10–31.

Lightman, B. (ed.), 2004. *Thoemmes Dictionary of Nineteenth Century British Scientists.* Bristol; Thoemmes Press.

Lindsay, J.F., Holliday, D.W. & Hulbert, A.G., 1991. Sequence stratigraphy and the evolution of the Ganges–Brahmaputra delta complex. *American Association of Petroleum Geologists Bulletin*, **75**(7): 1233–54.

Linke, P. & Lutze, G.F., 1993. Microhabitat preferences of benthic Foraminifera: a static concept or a dynamic adaptation to optimise food acquisition? *Marine Micropaleontology*, **20**: 215–34.

Lipps, J.H., 1976. Feeding strategies and test function in Foraminifera. *Maritime Sediments Special Publication*, **1**: 100–10.

Lipps, J.H. (ed.), 1993. *Fossil Prokaryotes and Protists.* Oxford; Blackwell.

Lipps, J.H., Berger, W.H., Buzas, M.A. *et al.*, 1979. *Foraminiferal Ecology and Paleoecology.* Society of Economic Paleontologists and Mineralogists (Notes, Short Course No.6, Houston, Texas).

Lipps, J.H. & Finger, K.L., 2010. How many Foraminifera are there? *Abstracts, Forams2010 (International Symposium on Foraminifera), Bonn*: 132.

Lipps, J.H., Finger, K.L. & Walker, S.E., 2011. What should we call the Foraminifera? *Journal of Foraminiferal Research*, **41**(4): 309–13.

Lipps, J.H. & Goldstein, S.T. (eds.), 2006. Systematics and evolution of protists: fossils, morphology and molecules. *Anuario de Instituto de Geosciencias, UFRJ*, **29**(1): 1–212.

Lipps, J.H. & Valentine, J.W., 1970. The role of Foraminifera in the trophic structure of marine communities. *Lethaia*, **3**: 279–86.

Lister, J.J, 1895. Contributions to the life history of the Foraminifera. *Philosophical Transactions of the Royal Society of London, Series B*, **186**: 401–53.

Lister, J.J, 1905. On the dimorphism of the English species of *Nummulites* and the size of the megalosphere in relation to that of the microspheric and megalospheric tests of this genus of Foraminifera. *Proceedings of the Royal Society*, **B76**: 298–319.

Lister, B., 1906. The life history of the Foraminifera. *Transactions, British Asssociation for the Advancement of Science*, **D**: 583–96.

Liu, X. & Galloway, W.E., 1997. Quantitative determination of Tertiary sediment supply to the North Sea basin. *American Association of Petroleum Geologists Bulletin*, **81**(9): 1482–510.

Lob, C. & Mutterlose, J., 2012. The onset of anoxic conditions in the early Berraman of the Boreal realm evidenced by benthic Foraminifera. *Revue de Micropaleontologie*, **55**(3): 113–26.

Lobegeier, M.K. & Sen Gupta, B.K., 2008. Foraminifera of hydrocarbon seeps, Gulf of Mexico. *Journal of Foraminiferal Research*, **38**(2): 93–116.

Loeblich, A. & Tappan, H., 1964. *Treatise on Invertebrate Paleontology, Part C, Protista 2, Sarcodina, chiefly 'Thecamoebians' and Foraminiferida.* ; Lawrence, KS; University of Kansas Press.

Loeblich, A. & Tappan, H., 1987. *Foraminiferal Genera and their Classification.* New York; Van Nostrand Reinhold.

Lohmann, G.P., 1978. Abyssal benthonic Foraminifera as hydrographic indicators in the western South Atlantic Ocean. *Journal of Foraminiferal Research*, **8**(1): 6–34.

Lomando, A.J. & Harris, P.M. (compilers), 1988. *Giant Oil and Gas Fields.* Tulsa, OK; Society of Economic Paleontologists and Mineralogists (Core Workshop Volume).

Longet, D., Archibald, J.M., Keeling, P.J. & Pawlowski, J., 2003. Foraminifera and Cercozoa share a common origin according to RNA polymerase II phylogenies. *International Journal of Systematic and Evolutionary Microbiology*, **53**: 1735–9.

Longoria, J.F., 1984. Cretaceous biochronology from the Gulf of Mexico region based on planktonic microfossils. *Micropaleontology*, **30**(3): 225–42.

Longoria, J.F. & Gamper, M.A., 1975. The classification and evolution of Cretaceous planktonic Foraminifera, part I: the superfamily Hedbergelloidea. *Revista Espanola de Micropaleontologia, Numero Especial*, **1975**: 61–96.

Longoria, J.F. & Gamper, M.A. (eds.), 1998. *International Symposium on Foraminifera (Forams '98), Monterrey, Mexico, Proceedings and Abstracts with Programs*. Monterrey; Socieded Mexicana de Paleontologia.

Loubere, P. (ed.), 1995. Generation of the paleoceanographic signal. *Journal of Foraminiferal Research*, **25**(1).

Loubere, P., 1997. Benthic foraminiferal assemblage formation, organic carbon flux, and oxygen concentrations on the outer continental shelf and slope. *Journal of Foraminiferal Research*, **27**(2): 93–100.

Loubere, P., Gary, A. & Lagoe, M., 1993. Sea bed geochemistry and benthic foraminiferal bathymetric zones on the slope of the northwest Gulf of Mexico. *Palaios*, **8**: 439–49.

Loucks, R.G. & Sarg, J.F. (eds.), 1993. *Carbonate Sequence Stratigraphy: Recent Developments and Applications*. Tulsa, OK; American Association of Petroleum Geologists (Memoir, No. 57).

Loutfi, G. & Jaber, A.A., 1970. Geology of the upper Albian–Campanian succession in the Kuwait–Saudi Arabia neutral zone, offshore area. *Proceedings of the Seventh Arab Petroleum Congress, Paper*, **62**(B-3).

Lovell, B., 2006. Climate change: conflict of observational science, theory and politics: discussion. *American Association of Petroleum Geologists Bulletin*, **90**(3): 405–7.

Lovell, B., 2009. *Challenged by Carbon: The Oil Industry and Climate Change*. Cambridge; Cambridge University Press.

Lowe, J.J. & Walker, M.J.C., 1997. *Reconstructing Quaternary Environments. Second Edition*. Harlow; Longman.

Lowemark, L., Konstantinou, K.I. & Steinke, S., 2008. Bias in foraminiferal multispecies reconstructions of paleohydrographic conditions caused by foraminiferal abundance variations and bioturbational mixing: a model approach. *Marine Geology*, **256**: 101–6.

Lundquist, J.J., Culver, S.J. & Stanley, D.J., 1997. Foraminiferal and lithological indicators of depositional processes in Wilmington and South Heyes submarine canyons, US Atlantic continental slope. *Journal of Foraminiferal Research*, **27**(3): 209–31.

Luning, S., Kolonic, S., Belhadj, E.M. *et al.*, 2004. Integrated depositional model for the Cenomanian–Turonian organic-rich strata in north Africa. *Earth-Science Reviews*, **64**: 51–117.

Lutze, G.F. & Coulbourn, W.T., 1984. Recent benthic Foraminifera from the continental margin of northwest Africa: community structure and distribution. *Marine Micropaleontology*, **8**: 361–401.

Lys, M., 1980. Donnees nouvelles sur la stratigraphie des formations Paleozioques de la plateforme Arabe dans la region d'Hazro (Turquie). *Comptes Rendus de l'Academie des Sciences, Paris*, **D291**: 917–20.

Macellari, C.E. & de Vries, T.J., 1987. Late Cretaceous upwelling and anoxic - sedimentation in northwestern South America. *Palaeogeography, Palaeoclimatology, Palaeoecology*, **9**: 279–92.

Macaulay, C.I., Beckett, D., Braithwaite, K. *et al.*, 2001. Constraints on diagenesis and reservoir quality in the fratured Hasdrubal field, offshore Tunisia. *Journal of Petroleum Geology*, **24**(1): 55–78.

Mackensen, A., Grobe, H., Kuhn, G. & Futterer, D.K., 1990. Benthic foraminiferal assemblages from the eastern Weddell Sea between 68 and 73degS: distribution, ecology and fossilisation potential. *Marine icropaleontology*, **16**(3/4): 241–83.

MacGregor, D., 2010. Understanding African and Brazilian margin climate, topography and drainage systems, implications for predicting deepwater reservoirs and source rock burial history. *Search and Discovery Article*, **10270**.

MacGregor, D.S., Moody, R.T.J. & Clark-Lowes, D.D. (eds.), 1998. *Petroleum Geology of North Africa*. London; The Geological Society (Special Publication, No. 132).

MacQueen, R.W. & Leckie, D.A. (eds.), 1992. *Foreland Basins and Thrust Belts*. Tulsa, OK; American Association of Petroleum Geologists (Memoir, No. 55).

Madof, A.S., Christie-Blick, N. & Anders, M.H., 2009. Stratigraphic controls on a salt-withdrawal intraslope minibasin, north-central Green Canyon, Gulf of Mexico: implications for misinterpreting sea level change. *American Association of Petroleum Geologists Bulletin*, **93**(4): 535–61.

Magniez-Jannin, F. & Jacquin, T., 1988. Foraminfera and sedimentary sequences: towards a better understanding of the Cretaceous anoxic environments in the South Atlantic. *Revue de Paleobiologie, Volume Special*, **2**: 297–307.

Magno, M.C., Bergamin, L., Finoia, M.G. *et al.*, 2012. Correlation between textural charactertistics of marine sediments and benthic Foraminifera in highly anthropogenically-altered coastal areas. *Marine Geology*, **315**–318: 143–61.

Magoon, L.B. & Dow, W.G. (eds.), 1994. *The Petroleum System: From Source to Trap*. Tulsa, OK; American Association of Petroleum Geologists (Memoir, No. 60).

Mahanti, G., Patra, S., Banerjee, A. *et al.*, 2006. Outcrop study in south Cambay basin, India. *The Leading Edge*, July, 2006: 869–71.

Mahmud, S.A. & Sheikh, S.A., 2012. Reservoir potential of Lower Nari sandstones (Early Oligocene) in southern Indus basin and Indus offshore. *Search and Discovery Article*, **50582**.

Majid, A.H. & Veizer, J., 1986. Deposition and chemical diagenesis of Tertiary carbonates, Kirkuk oil field, Iraq. *American Association of Petroleum Geologists Bulletin*, **70**(7): 898–913.

Mamo, B., Strotz, L. & Dominey-Howes, D., 2009. Tsunami sediments and their foraminiferal assemblages. *Earth-Science Reviews*, **96**: 263–78.

Mancini, E.A., Aharon, P., Goddard, D.A. *et al.*, 2012. Basin analysis and petroleum system characterization and modelling, interior salt basins, central and eastern Gulf of Mexico. *Search and Discovery Article*, **10396**.

Mancini, E.A., Obid, J., Badali, M. *et al.*, 2008. Sequence-stratigraphic analysis of Jurassic and Cretaceous strata and petroleum exploration in the central and eastern Gulf coastal plain, United States. *American Association of Petroleum Geologists Bulletin*, **92**(12): 1655–86.

Mancini, E.A. & Puckett, T.M., 2005. Jurassic and Cretaceous Transgressive–Regressive (T–R) cycles, northern Gulf of Mexico, USA. *Stratigraphy*, **2**(1): 37–48.

Mancini, E.A., Puckett, T.M. & Tew, B.H., 1996. Integrated biostratigraphic and sequence stratigraphic framework for Upper Cretaceous strata of the eastern Gulf coastal plain, USA. *Cretaceous Research*, **17**: 645–69.

Mancini, E.A. & Tew, B.H., 1991. Relationships of Paleogene stage and planktonic foraminiferal zone boundaries to lithostratigraphic and allostratigraphic contacts in the eastern Gulf coastal plain. *Journal of Foraminiferal Research*, **21**(1): 48–66.

Mann, K.O. & Lane, H.R., 1995. *Graphic Correlation*. Tulsa, OK; Society of Economic Paleontologists and Mineralogists (Special Publication, No. 53).

Mann, P. (ed.), 1999. *Caribbean Basins*. Amsterdam; Elsevier (Sedimentary Basins of the World Series, No. 4).

Mann, U. & Stein, R., 1997. Organic facies variations, source rock potential and sea level changes in Cretaceous black shales of the Quebrada Ocal, upper Magdalena valley, Colombia. *American Association of Petroleum Geologists Bulletin*, **81**(4): 556–76.

Mann, P. & Escalona, A. (eds.), 2010. Tectonics, basinal framework, and petroleum systems of eastern Venezuela, the Leeward Islands, Trinidad and Tobago, and offshore areas. *Marine and Petroleum Geology (Special Issue)*, **28**(1): 1–278.

Mansurbeg, H., de Ros, L.F., Morad, S. *et al.*, 2012. Meteoric water diagenesis in Late Cretaceous canyon-fill turbidite reservoirs from the Espirito Santo Basin, eastern Brazil. *Marine and Petroleum Geology*, **37**: 7–26.

Margreth, S., Gennari, G., Ruggeberg, A. *et al.*, 2011. Growth and demise of cold-water coral ecosyetems on mud volcanoes in the west Alboran Sea: the messages from the planktonic and benthic Foraminifera. *Marine Geology*, **282**: 26–39.

Margreth, S., Ruggeberg, A. & Spezzaferri, S., 2009. Benthic Foraminifera as bioindicator for cold-water coral reef ecosystems along the Irish margin. *Deep-Sea Research I*, **56**: 2216–34.

Marshall, P.R., 2010. Geosteering with minimal LWD capability: biosteering during underbalanced, coiled tubing drilling operations for gas in the Middle East and the development of the Stratsteer concept. *Abstracts, EAGE Geosteering & Well Placement Workshop, Dubai*.

Marshall, P.R., Burchette, T. & Ali, K.S., 2007. Integrated subsurface geology and biosteering: a case study from the Sajaa field, Sharjah, UAE. *Abstracts, Integrated Petroleum Technology Conference, Dubai*.

Margulis, L., Corliss, J.O., Melkonian, M. *et al.*, 1990. *Handbook of Protoctista*. Boston, MA; Jones and Bartlett.

Mariner, N., Morhange, C., BouDagher-Fadel, M. *et al.*, 2005. Geoarchaeology of Tyre's ancient northern harbour, Phoenicia. *Journal of the Archaeological Society*, **32**(9): 1302–27.

Marquez, L., 2005. Foraminiferal fauna recovered after the Late Permian extinctions in Iberia and the westernmost Tethys area. *Palaeogeography, Palaeoclimatology, Palaeoecology*, **229**: 137–57.

Martin, P.A., Lea, D.W., Mashiotta, T.A. *et al.*, 1997. Glacial–interglacial variation in mean ocean Sr? *EOS (Transactions of the American Geophysical Union)*, **78**(46): F388.

Martin, R.A., Nesbitt, E.A. & Campbell, K.A., 2010. The effects of anaerobic methane oxidation on benthonic foraminiferal assemblages and stable isotopes on the Hikurangi margin of eastern New Zealand. *Marine Geology*, **272**: 270–84.

Martin, R.E. (ed.), 2000. *Environmental Micropaleontology*. Dordrecht; Kluwer Academic/Plenum Press.

Martinez, A.R., 1989. *Venezuelan Oil: Development and Chronology*. London; Elsevier Applied Science.

Martinez, J.I., 2003. The paleoecology of Late Cretaceous upwelling events from the upper Magdalena basin, Colombia. *Palaios*, **18**: 305–20.

Martinez-Dias, C., Wagner, R.H., Prinz, C.F.W. & Granados, L.F. (eds.), 1983. *The Carboniferous of the World, I: China, Korea, Japan and South-East Asia*. Madrid; Instituto Tecnologico Geominero de Espana, Madrid (International Union of Geological Sciences Publication).

Martinez-Dias, C., Wagner, R.H., Prinz, C.F.W. & Granados, L.F. (eds.), 1985. *The Carboniferous of the World, II: Australia, Indian Subcontinent, South Africa, South America & North Africa*. Madrid; Instituto Tecnologico Geominero de Espana, Madrid (International Union of Geological Sciences Publication).

Martinez-Dias, C., Wagner, R.H., Prinz, C.F.W. & Granados, L.F. (eds.), 1996. *The Carboniferous of the World, III: The Former USSR, Mongolia, Middle Eastern Platform, Afghanistan and Iran*. Madrid/Leiden; Instituto Tecnologico Geominero de Espana, Madrid/Nationaal Natuurhistorisch Museum, Leiden (International Union of Geological Sciences Publication).

Martinsen, O.J. & Dreyer, T. (eds.), 2001. *Sedimentary Environments Offshore Norway: Palaeozoic to Recent*. Amsterdam; Elsevier (NPF Special Publication, No. 10).

Massey, A.C., Gehrels, W.R., Charman, D.J. & White, S.V., 2006. An intertidal foraminifera-based transfer function for reconstructing Holocene sea-level change in southwest England. *Journal of Foraminiferal Research*, **36**(3): 215–32.

Masters, G.M. & Ela, W.P. (eds.), 2008. *Introduction to Environmental Engineering and Science. Third Edition*. Upper Saddle River, NJ; Pearson Education International.

Matera, N.J. & Lee, J.J., 1972. Environmental factors affecting the standing crop of Foraminifera in sublittoral and psammolittoral communities of a Long Island salt marsh. *Marine Biology*, **14**: 89–103.

Mateu, G., 1974. Foraminiferos recientes de la Isla Menorca (Baleares) y su applicacion como indicadores biologicos de contaminacion litoral. *Boletin de la Sociedad de Historia Natural de Baleares*, **19**: 89–112.

Mateu-Vicens, G., Pomar, L. & Ferrandez-Canadell, C., 2011. Nummulitic banks in the upper Lutetian 'Buil level', Ainsa basin, south central Pyrenean zone): the impact of internal waves. *Sedimentology* **59**(2): 527–52 (doi: 10.1111.j.1365–3091.2011.01263.x).

Matteucci, R., Carboni, M.G. & Pignatti, J.S. (eds.), 1994. Studies on ecology and paleoecology of benthic communities. *Bollettino della Societa Paleontologica Italiana, Vol. Spec.*, **2**.

Matthew, C. (ed.), 2004. *The New Dictionary of National Biography*. Oxford; Oxford University Press.

Maurer, F. & Rettori, R., 2002. Middle Triassic Foraminifera from the Seceda core (Dolomites, northern Italy). *Rivista Italiana di Paleontologia e Stratigrafia*, **108**(3): 391–8.

May, P.R., 1991. The eastern Mediterranean Mesozoic basin: evolution and oil habitat. *American Association of Petroleum Geologists Bulletin*, **75**(7): 1215–32.

McCrea, J.M., 1950. On the isotope chemistry of carbonates and a paleotemperature scale. *Journal of Chemical Physics*, **18**: 849–57.

McDonnell, A., Loucks, R.G. & Galloway, W.E., 2008. Paleocene to Eocene deep-water slope canyons, western Gulf of Mexico: further insights for the provenance of deep-water Wilcox group plays. *American Association of Petroleum Geologists Bulletin*, **92**(9): 1169–89.

McGann, M., Vrijenhoek, R.C., Johnson, S. *et al.*, 2010. Foraminiferal response to whale falls in the northeastern Pacific Ocean. *Abstracts, Forams2010 (International Symposium on Foraminifera), Bonn*: 140.

McGowran, B., 2005. *Biostratigraphy: Principles and Practice*. Cambridge; Cambridge University Press.

McGuire, M.D., Koepnick, R.B., Markello, J.R. *et al.*, 1993. Importance of sequence stratigraphic concepts in developments of reservoir architecture in Upper Jurassic grainstones, Hadriya and Hanifa reservoirs, Saudi Arabia. *Society of Petroleum Reservoirs*, **25578**: 489–99.

McGuire, W.J., Griffiths, D.R., Hancock, P.I. & Stewart, I.S. (eds.), 2000. *The Archaeology of Geological Catastrophes*. London; The Geological Society (Special Publication, No. 171).

McIlroy, D., Green, O.R. & Brasier, M.D., 2001. Palaeobiology and evolution of the earliest agglutinated Foraminifera: *Platysolenites, Spirosolenites* and related forms. *Lethaia*, **34**: 13–29.

McLaughlin, P.I., Brett, C.E., McLaughlin, S.L.T. & Cornell, S.R., 2004. High-resolution sequence stratigraphy of a mixed carbonate–siliciclastic cratonic ramp (Upper Ordovician, Kentucky-Ohio, USA): insights into the relative influence of eustasy and tectonics through analysis of facies gradients. *Palaeogeography, Palsaeoclimatology, Palaeoecology*, **210**(2–4): 267–94.

McNeil, D.H., Issler, D.R. & Snowdon, L.R., 1996. Colour alteration, thermal maturity and burial diagenesis in fossil foraminifers. *Geological Survey of Canada Bulletin*, **499**: 1–34.

Meisling, K.E., Cobbold, P.R. & Mount, V.S., 2001. Segmentation of an obliquely rifted margin, Campos and Santos basins, southeastern Brazil. *American Association of Petroleum Geologists Bulletin*, **85**(11): 1903–24.

Meisterfeld, R., Holzmann, M. & Pawlowski, J., 2001. Morphological and molecular characterization of a new terrestrial allogromiid species: *Edaphoallogromia australica gen. et sp. nov.* (Foraminifera) from northern Queensland (Australia). *Protist*, **152**: 185–92.

Melguen, M., Bolli, H.M., Ryan, W.B.F. *et al.*, 1975. Facies evolution and carbonate dissolution cycles in sediments from basins and continental margins of the eastern South Atlantic since Early Cretaceous. *IX Congres International de Sedimentologie, Nice*: 43–50.

Mello, M.R. & Katz, B.J. (eds.), 2000. *Petroleum Systems of South Atlantic Margins*. Tulsa, OK; American Association of Petroleum Geologists (Memoir, No. 73).

Menzies, R.J., George, R.Y. & Rowe, G.T., 1973. *Abyssal Environment and Ecology of the World Oceans*. New York; Wiley Interscience.

Metwalli, M.H., Philip, G. & Moussly, M.M., 1974. Petroleum-bearing formations in northeastern Syria and northern Iraq. *American Association of Petroleum Geologists Bulletin*, **58**(9): 1781–96.

Meulenkamp, J.E. (ed.), 1983. Reconstruction of marine paleoenvironments. *Utrecht Micropalaeontological Bulletin*, **30**.

Miall, A.D., 1986. Eustatic sea-level changes interpreted from seismic stratigraphy: a critique of the methodology with particular reference to the North Sea Jurassic. *American Association of Petroleum Geologists Bulletin*, **70**: 131–7.

Miall, A.D., 1990. *Principles of Sedimentary Basin Analysis*. New York, NY; Springer Verlag.

Miall, A.D., 1992. Exxon global cycle chart: an event for every occasion? *Geology*, **20**: 787–90.

Miall, A.D. (ed.), 2008. *Sedimentary Basins of the World: The Sedimentary Basins of the United States and Canada*. Amsterdam; Elsevier.

Miall, A.D., 2010. *The Geology of Stratigraphic Sequences*. Amsterdam; Springer Verlag.

Middlemiss, F.A., Rawson, P.F. & Newall, G. (eds.), 1971. Faunal provinces in space and time. *Geological Journal, Special Issue*, **4**.

Milliken, K.L., 2010. Agglutinated foraminifers as agents of textural coarsening in muddy sediments. *Abstracts, Forams2010 (International Symposium on Foraminifera), Bonn*: 143.

Milliman, J.D. & Farnsworth, K.L., 2011. *River Discharge to the Coastal Ocean: A Global Synthesis*. Cambridge; Cambridge University Press.

Milner, P.A., 1998. Source rock distribution and thermal maturity in the southern Arabian peninsula. *GeoArabia*, **3**(3): 339–56.

Ministry of Energy and Energy Industries, 2009. *Trinidad and Tobago: Celebrating a Century of Commercial Oil Production*. Trinidad and Tobago; Ministry of Energy and Energy Industries.

Mishra, J., Singh, G., Paul, R. *et al.*, 2012. Facies analysis of Early Miocene Bombay formation in Panna–Bassein–Heera area, Mumbai offshore basin. *Search and Discovery Article*, **50584**.

Mishra, P.K., 1996. Study of Miogypsinidae and associated planktonics from Cauvery, Krishna-Godavari and Andaman basins of India. *Geoscience Journal*, **47**(2): 123–51.

Mitchell, D.J.W., Allen, R.B., Salama, W. & Abouzakm, A., 1992. Tectonostratigraphic framework and hydrocarbon potential of the Red Sea. *Journal of Petroleum Geology*, **15**(2): 187–210.

Moguilevsky, A. & Whatley, R., 1996. *Microfossils and Oceanic Environments*. Aberystwyth; University of Wales, Aberystwyth Press.

Mohan, M., 1982. Palaeocene stratigraphy of western India. *The Palaeontological Society of India, Special Publication*, **1**: 21–36.

Mohinuddin, S.K., Satyanarayana, K. & Rao, G.N., 1991. Cretaceous sedimentation in the subsurface of Krishna-Godavari basin. *Journal of the Geological Society of India*, **41**: 533–9.

Mohriak, W.U., Danforth, A., Post, P.J. *et al.*, (eds.), 2012. *Conjugate Divergent Margins*. London; The Geological Society (Special Publication, No. 369).

Mojtahid, M., Jorissen, F., Durrieu, J. *et al.*, 2006. Benthic Foraminifera as bio-indicators of drill cuttings disposal in tropical East Atlantic outer shelf environments. *Marine Micropaleontology*, **61**: 58–75.

Mojtahid, M., Jorissen, F., Lansard, B. & Fontanier, C., 2010. Microhabitat selection of benthic Foraminifera in sediments off the Rhone river mouth (NW Mediterranean). *Journal of Foraminiveral Research*, **40**(3): 231–46.

Mojtahid, M., Jorissen, F., Lansard, B. *et al.*, 2009. Spatial distribution of live benthic Foraminifera in the Rhone prodelta: faunal response to a continental-marine organic matter gradient. *Marine Micropaleontology*, **70**: 177–200.

Mojtahid, M., Jorissen, F. & Pearson, T.H., 2008. Comparison of benthic foraminiferal and macrofaunal response to organic pollution in the Firth of Clyde (Scotland). *Marine Pollution Bulletin*, **56**: 42–76.

Mojtahid, M., Zubkov, M.V., Hartmann, M. & Gooday, A.J., 2011. Grazing of intertidal benthic Foraminifera on bacteria: assessment using pulse-chase radiotracing. *Journal of Experimental Marine Biology and Ecology*, **399**(1): 25–34.

Moodley, L. & Hess, C., 1992. Tolerance of infaunal benthic Foramiinfera for low and high oxygen concventrations. *The Biological Bulletin*, **183**: 94–8.

Moodley, L., van der Zwaan, G.J., Rutten, G.M.W. *et al.*, 1998. Subsurface activity of benthic Foraminifera in relation to porewater oxygen content: laboratory experiments. *Marine Micropaleontology*, **34**(1/2): 91–106.

Moran-Zenteno, D., 1994. *Geology of the Mexican Republic*. Tulsa, OK; American Association of Petroleum Geologists (Studies in Geology, No. 30).

Morard, R., Quillevere, F., Douady, C.J. *et al.*, 2010. The allopatric distribution of *Globoconella inflata* cryptic species and its potential use for monitoring past movements of the Antarctic subpolar front. *Abstracts, Forams2010 (International Symposium on Foraminifera), Bonn*: 144.

Morard, R., Quillevere, F., Escarguel, G. *et al.*, 2009. Morphological recognition of cryptic species in the planktonic foraminifer *Orbulina*

*universa*. *Marine Micropaleontology*, **71**: 148–65.

Moreira, J.L.P., Madeira, C.V., Gil, J.A. & Machado, M.A.P., 2007. Bacia de Santos. *Boletim de Geociencias, Petrobras, Rio de Janeiro*, **15**(2): 531–49.

Morigi, C., Jorissen, F.J., Gervais, A. *et al.*, 2001. Benthic foraminiferal faunas in surface sediments off NW Africa: relationship with organic flux to the ocean floor. *Journal of Foraminiferal Research*, **31**(4): 350–68.

Morigi, C., Sabbatini, A., Vitale, G. *et al.*, 2012. Foraminiferal biodiversity associated with cold-water coral carbonate mounds and open slope of SE Rockall Bank (Irish continental margin – NE Atlantic). *Deep Sea Research I*, **59**: 54–71.

Morigi, C., Vitale, G., Pancotti, I. *et al.*, 2010. Benthic Foraminifera associated with deep-water corals in the Rockall Bank (NE Atlantic): meso and microscale contrasts and comparison of stained and dead assemblages. *Abstracts, Forams2010 (International Symposium on Foraminifera), Bonn*: 144–5.

Morley Davies, A., 1971. *Tertiary Faunas. Second Edition (revised by F.E. Eames)*. London; George Allen & Unwin.

Morris, P. & Therivel, R. (eds.), 2001. *Methods of Environmental Impact Assessment. Second Edition*. London; Spon Press.

Morse, J.W., Andersson, A.J. & Mackenzie, F.T., 2006. Initial responses of carbonate-rich shelf sediments to rising atmospheric $pCO_2$ and 'ocean acidification': role of high Mg calcites. *Geochimica et Cosmochimica Acta*, **70**: 5814–30.

Morton, A.C. (ed.), 1992. *Geology of the Brent Group*. London; The Geological Society (Special Publication, No. 61).

Morton, R.A. & Suter, J.R., 1996. Sequence stratigraphy and composition of Late Quaternary shelf-margin deltas, northern Gulf of Mexico. *American Association of Petroleum Geologists Bulletin*, **80**(4): 505–30.

Mossadegh, Z.K., Haig, D.W., Allan, T. *et al.*, 2009. Salinity changes during Late Oligocene to Early Miocene Asmari formation deposition, Zagros mountains, Iran. *Palaeogeography, Palaeoclimatology, Palaeoecology*, **272**: 17–36.

Moullade, M. & Nairn, A.E.M. (eds.), 1978. *Phanerozoic Geology of the World, II: The Mesozoic, A*. Amsterdam; Elsevier.

Moullade, M., Peybernes, B., Rey, J. & Saint-Marc, P., 1985. Biostratigraphic interest and paleobiogeographic distribution of Early and Mid-Cretaceous Mesogean Orbitolinidae

(Foraminiferida). *Journal of Foraminiferal Research*, **15**(3): 149–58.

Mouty, M. & Saint-Marc, P., 1982. Le Cretace moyen du Massif Alaouite (nord-ouest Syrie). *Cahiers de Micropaleontologie*, **3**: 55–69.

Mucadam, R.M., 2010. Using 3-D X-ray micro-computed tomography (X-ray MCT) to analyze the building blocks, design and probable mechanical function of a foraminiferal shell. *Abstracts, Forams2010 (International Symposium on Foraminifera), Bonn*: 145.

Mudge, D.C. & Bujak, J.P., 1996. Palaeocene biostratigraphy and sequence stratigraphy of the UK central North Sea. *Marine and Petroleum Geology*, **13**(3): 295–312.

Mukherjee, A., Fryer, A.E. & Thomas, W.A., 2009. Geologic, geomorphic and hydrologic framework and evolution of the Bengal basin, India and Bangladesh. *Journal of Asian Earth Sciences*, **34**: 27–44.

Muller, W.A., 1975. Competition for food and other niche-related studies of three species of salt-marsh Foraminifera. *Marine Biology*, **31**: 339–51.

Mullineaux, L.S. & Lohmann, G.P., 1981. Late Quaternary stagnations and recirculation of the eastern Mediterranean: changes in the deep water recorded by fossil benthic Foraminifera. *Journal of Foraminiferal Research*, **11**(1): 20–39.

Murray, G.E., Rahman, A.U., & Yarborough, H., 1985. Introduction to the habitat of petroleum, northern Gulf (of Mexico) coastal province. *GCSSEPM Foundation Fourth Annual Research Conference Proceedings*: 1–24.

Murray, J.W., 1971. *An Atlas of British Recent Foraminiferids*. London; Heinemann Educational Books.

Murray, J.W., 1973. *Distribution and Ecology of Living Benthic foraminiferids*. London; Heinemann Educational Books.

Murray, J.W., 1976. A method of determining proximity of marginal seas to an ocean. *Marine Geology*, **22**: 103–19.

Murray, J.W., 1982. Benthic Foraminifera:the validity of living, dead or total assemblages for the interpretation of palaeoecology. *Journal of Micropalaeontology*, **1**: 137–40.

Murray, J.W., 1991. *Ecology and Palaeoecology of Benthic Foraminifera*. Harlow; Longman Scientific & Technical.

Murray, J.W., 2000. The enigma of the continued use of total assemblages in ecological studies of benthic foraminifera. *Journal of Foraminiferal Research*, **30**(3): 244–5.

Murray, J.W., 2001. The niche of benthic Foraminifera, critical thresholds and proxies. *Marine Micropaleontology*, **41**(1–2): 1–8.

Murray, J.W., 2006. *Ecology and Applications of Benthic Foraminifera*. Cambridge; Cambridge University Press.

Murray, J.W., 2007. Biodiversity of living benthic Foraminifera: How many species are there? *Marine Micropalaeontology*, **64**(3–4): 163–176.

Murray, J.W., 2012. Unravelling the life history of '*Polystomella crispa*': the roles of Lister, Jepps and Myers. *Journal of Micropalaeontology*, **31**(2): 121–9.

Murray, J.W., 2013. Living benthic Foraminifera: biogeographical distributions and the significance of rare morphospecies. *Journal of Micropalaeontology*, **32**(1): 1–58 (doi:10.1144/jmpaleo2012–010).

Murray, J.W. & Alve, E., 2011. The distribution of agglutinated Foraminifera in NW European seas: baseline data for the interpretation of fossil assemblages. *Palaeontologia Electronica*, **14**(2).

Murray, J.W., Alve, E. & Jones, R.W., 2011. A new look at modern agglutinated foraminiferal morphogroups: their value in palaeoecological interpretation. *Palaeogeography, Palaeoclimatology, Palaeoecology*, **309**: 229–41.

Murray, J.W. & Bowser, S.S., 2000. Mortality, protoplasm decay rate, and reliability of staining techniques to recognise 'living' foraminifera: a review. *Journal of Foraminiferal Research*, **30**(1): 66–70.

Murris, R.J., 1980a. Middle East: stratigraphic evolution and oil habitat. *American Association of Petroleum Geologists Bulletin*, **64**(5): 597–618.

Murris, R.J., 1980b. Middle East: stratigraphic evolution and oil habitat. *Geologie en Mijnbouw*, **60**: 467–86.

Mutti, E. & Carminatti, M., 2012. Deep-water sands of the Brazilian offshore basins. *Search and Discovery Article*, **30219**.

Myers, E.H., 1935a. Culture methods for the marine Foraminifera of the littoral zone. *Transactions of the American Microscopical Society*, **54**: 264–7.

Myers, E.H., 1935b. The life history of *Patellina corrugata* Williamson, a foraminifera. *Bulletin of Scripps Institution of Oceanography, Technical Series*, **3**(15): 355–92.

Myers, E.H., 1935c. Morphogenesis of the test and biological significanmce of dimorphism in the foraminifer *Patellina corrugata* Williamson. *Bulletin of Scripps Institution of Oceanography, Technical Series*, **3**(16): 393–404.

Myers, E.H., 1936. The life-cycle of *Spirillina vivipara* Ehrenberg, with notes on morphogenesis, systematics and distribution of the Foraminifera. *Journal of the Royal Microscopical Society*, **56**: 120–46.

Myers, E.H., 1938. The present state of our knowledge concerning the life cycle of the Foraminifera. *Proceedings of the National Academy of Sciences*, **24**: 10–17.

Myers, E.H., 1942a. A quantitative study of the productivity of the Foraminifera in the sea. *Proceedings of the American Philosophical Society*, **85**(4): 325–41.

Myers, E.H., 1942b. Biological evidence as to the rate at which tests of Foraminifera are contributed to marine sediments. *Journal of Paleontology*, **16**: 397–8.

Myers, E.H., 1942c. Life activities of Foraminifera in relation to marine ecology. *Proceedings of the American Philosophical Society*, **86**(3): 439–58.

Myers, E.H., 1943. Biology, ecology and morphogenesis of a pelagic foraminifer. *Stanford University Publications, Biology Series*, **9**: 5–30.

Nagappa, Y., 1959. Foraminiferal biostratigraphy of the Cretaceous–Eocene succession in the India–Pakistan–Burma region. *Micropalaeontology*, **5**(2): 145–92.

Nagendra, R., Kamalak Kana, B.V., Sen, G. *et al.*, 2010. Sequence surfaces and paleobathymetric trends in Albian to Maastrichtian sediments of Arilayur area, Cauvery basin, India. *Marine and Petroleum Geology*, **28**(4): 895–905.

Nagy, J., 1985a. Jurassic foraminiferal facies in the Statfjord area, northern North Sea – I. *Journal of Petroleum Geology*, **8**(3): 273–95.

Nagy, J., 1985b. Jurassic foraminiferal facies in the Statfjord area, northern North Sea – II. *Journal of Petroleum Geology*, **8**(4): 389–404.

Nagy, J., 1992. Environmental significance of foraminiferal morphogroups in Jurassic North Sea deltas. *Palaeogeography, Palaeoclimatology, Palaeoecology*, **95**: 111–34.

Nagy, J., 2005. Delta-influenced foraminiferal facies and sequence stratigraphy of Paleocene deposits in Spitsbergen. *Palaeogeography, Palaeoclimatology, Palaeoecology*, **222**: 161–79.

Nagy, J. & Alve, E., 1987. Temporal changes in foraminiferal faunas and impact of pollution in Dandebukta, Oslo Fjord. *Marine Micropaleontology*, **12**: 109–28.

Nagy, J., Dypvik, H. & Bjaerke, T., 1984. Sedimentological and paleontological analyses of Jurassic North Sea deposits from deltaic environments. *Journal of Petroleum Geology*, **7**(2): 169–88.

Nagy, J., Finstad, E.K., Dypvik, H. & Bremer, M.G.A., 2001. Response of foraminiferal facies to transgressive–regressive cycles in the Callovian of northeast Scotland. *Journal of Foraminiferal Research*, **31**(4): 324–49.

Nagy, J., Hess, S. & Alve, E., 2010. Environmental significance of foraminiferal assemblages dominated by small-sized *Ammodiscus* and *Trochammina* in Triassic and Jurassic delta-influenced deposits. *Earth-Science Reviews*, **99**: 31–49.

Nagy, J., Hess, S., Dypvik, H. & Bjaerke, T., 2011. Marine shelf to paralic biofacies of Upper Triassic to Lower Jurassic deposits in Spitsbergen. *Palaeogeography, Palaeoclimatology, Palaeoecology*, **300**: 138–51.

Nagy, J. & Johansen, H.O., 1991. Delta-influenced foraminiferal assemblages from the Jurassic (Toarcian–Bajocian) of the northern North Sea. *Micropaleontology*, **37**(1): 1–40.

Nagy, J., Reolid, R. & Rodriguez-Tovar, F.J., 2009. Foraminiferal morphogroups in dysoxic shelf deposits from the Jurassic of Spitsbergen. *Polar Research*, **28**: 214–21.

Nagy, J. & Seidenkrantz, M.-S., 2003. Jurassic marginal marine deposits on Anholt, Denmark. *Micropaleontology*, **49**(1): 27–46.

Nairn, A.E.M. & Stehli, F.G. (eds.), 1973. *The Ocean Basins and Margins, Volume I: The South Atlantic*. New York; Plenum Press.

Nanda, A.C. & Mathur, A.K. (co-conveners), 2000. *Neogene Sequences of India*. Geological Society of India.

Natland, M.L., 1963. Presidential address: paleoecology and turbidites. *Journal of Paleontology*, **37**(4): 946–51.

Natural History Museum, Basel, 1996. *Treatise on the Geology of Trinidad – Detailed Geological Maps and Sections*. Basel; Natural History Museum.

Nazri Ramli, M. & Ho Kiam Fui, 1984. Depositional environments and diagenesis of the F6 reef complex, central Luconia province, offshore Sarawak, Malaysia. *Proceedings of the ASCOPE/CCOP Workshop on Hydrocarbon Occurrences in Carbonate Rocks*: 269–92.

Neal, J.E., Stein, J.A. & Gamber, J.H., 1994. Graphic correlation and sequence stratigraphy in the Palaeogene of NW Europe. *Journal of Micropalaeontology*, **11**: 55–80.

Neale, J.W. & Brasier, M.D. (eds.), 1981. *Microfossils from Recent and Fossil Shelf Seas*. Chichester; Ellis Horwood.

Nederbragt, A.J., 1990. Late Cretaceous biostratigraphy and development of Heterohelicidae (planktonic Foraminifera). *Micropaleontology*, **32**(4): 329–72.

Neelin, J.D., 2010. *Climate Change and Climate Modeling*. Cambridge; Cambridge University Press.

Negi, A.S., Sahu, S.K., Thomas, P.D. *et al.*, 2006. Fusing geologic knowledge and seismic in searching for subtle hydrocarbon traps in India's Cambay basin. *The Leading Edge*, July 2006: 872–80.

Nelson, C.H. & Damuth, J.E., 2003. Myths of turbidite system control: insights provided by modern turbidite studies. *Abstracts, International Conference on Deep-Water Processes in Modern and Ancient Environments, Barcelona*: 32.

Nevle, R.J. & Bird, D.K., 2008. Effects of synpandemic fire reduction and reforestation in the tropical Americas on atmospheric $CO_2$ during the European conquest. *Palaeogeography, Palaeoclimatology, Palaeoecology*, **264**: 25–38.

Nguyen, T.M.P., Petrizzio, M.R. & Speijer, R., 2009. Experimental dissolution of a fossil foraminiferal assemblage (Paleocene–Eocene Thermal Maximum, Dababiya, Egypt): implications for paleoenvironmental reconstructions. *Marine Micropaleontology*, **73**: 241–58.

Nielsen, K.S., Schroder-Adams, C.J., Leckie, D.A. *et al.*, 2008. Turonian to Santonian paleoenvironmental changes in the Cretaceous Western Interior Sea: the Carlile and Niobrara formations in southern Alberta and southwestern Saskatchewan, Canada. *Palaeogeography, Palaeoclimatology, Palaeoecology*, **270**: 64–91.

Nigam, R., 2005. Addressing environmental issues through Foraminifera: case studies from the Arabian Sea. *Journal of the Palaeontological Society of India*, **50**(2): 23–56.

Nigam, R., Mazumder, A., Henriques, P.J. & Saraswat, R., 2007. Benthic Foraminifera as proxy for oxygen-depleted conditions off the central west coast of India. *Journal of the Geological Society of India*, **70**: 1047–54.

Nikravesh, M., Aminzadeh, F. & Zadeh, L.A. (eds.), 2003. *Soft Computing and Intelligent Data Analysis in Oil Exploration*. Amsterdam, Elsevier (Developments in Petroleum Science, No. 51).

Nixon, F.C., Reinhardt, E.G. & Rothaus, R., 2009. Foraminifera and tidal notches: dating neotectonic events at Korphos, Greece. *Marine Geology*, **257**(1–4): 41–53.

Nomaki, H., Heinz, P., Hemleben, C. & Kitazato, H., 2005. Behavior and response of deep-sea benthic Foraminifera to freshly supplied organic matter: a laboratory feeding experiment in microcosm environments. *Journal of Foraminiferal Research*, **35**(2): 103–13.

Nomaki, H., Ogawa, N.O., Takano, Y. *et al.*, 2011. Differing utilization of glucose and algal particulate organic matter by the deep-sea benthic organisms of Sagami Bay, Japan. *Marine Ecology Progress Series*, **431**: 11–24.

Nummedal, D., Bartov, Y., Sarg, R. & Boak, J., 2009. Oil shale stratigraphy: a global perspective. *Search and Discovery Article*, **30083**.

Nunn, L.L., 1986. Foraminiferal biostratigraphy and paleoecology of the Wilcox group (Paleocene–Eocene) in central and southern Louisiana. *Transactions – Gulf Coast Association of Geological Societies*, **XXXVI**: 511–6.

Oertli, H.J. (ed.), 1984. *Benthos '83*. Pau; Centre de Recherches Exploration-Production Elf-Aquitaine (Memoire, No. 6).

Ogg, J.G., Ogg, G. & Gradstein, F.M., 2008. *The Concise Geologic Time Scale*. Cambridge; Cambridge University Press.

Ojeda, H.A.O., 1982. Structural framework, stratigraphy and evolution of Brazilian margin basins. *American Association of Petroleum Geologists Bulletin*, **66**(6): 732–49.

Olausson, E. & Cato, I. (eds.), 1980. *Chemistry and Biochemistry of Estuaries*. Cambridge; John Wiley.

Oliveira-Silva, P., Barbosa, C.F., Machado de Almeira, C. *et al.*, 2012. Sedimentary geochemistry and foraminiferal assemblages in coral reef assessment of Abrolhos, southwest Atlantic. *Marine Micropaleontology*, **94–95**: 14–24.

Olson, H.C. & Leckie, R.M., (eds.), 2003. *Micropaleontologic Proxies for Sea-Level Change and Stratigraphic Discontinuities*. Tulsa, OK; Society of Economic Paleontologists and Mineralogists (Special Publication, No. 75).

Olsson, I., 1976. Distribution and ecology of the foraminiferan *Ammotium cassis* (Parker) in some Swedish estuaries. *Zoon*, **4**: 137–47.

Olsson, R.K., Hemleben, C., Berggren, W.A. & Huber, B.T. (eds.), 1999. *Atlas of Paleocene Planktonic Foraminifera*. Smithsonian Contributions to Paleobiology (No. 85).

Olsson, R.K. & Nyong, E.E., 1984. A paleoslope model for Campanian–lower Maestrichtian Foraminifera of New Jersey and Delaware. *Journal of Foraminiferal Research*, **14**(1): 50–68.

Orndorff, A.L. & Culver, S.J., 1998. Foraminifera of the Early Miocene upper part of the

Anahuac formation from a well in Vermilion Parish, Louisiana, U.S.A. *Journal of Foraminiferal Research*, **28**(4): 286–305.

Orr, J.C., Fabry, V.J., Aumont, O. *et al.*, 2005. Anthropogenic ocean acidification over the twenty-first century and its impact on calcifying organisms. *Nature*, **437**: 681–6.

Ortiz, S., Alegret, L., Payros, A. *et al.*, 2011. Distribution patterns of benthic Foraminifera across the Ypresian–Lutetian Gorrondatze section, northern Spain: response to sedimentary disturbance. *Marine Micropaleontology*, **78**: 1–13.

Osterman, L.E., 2003. Benthic foraminifers from the continental shelf and slope of the Gulf of Mexico: an indicator of shelf hypoxia. *Estuarine, Coastal and Shelf Science*, **58**(1): 17–35.

Osterman, L.E., Erlandsen, M. & Castenson, E.D., 2001. Benthic foraminiferal census data from surface sediment samples, western Gulf of Mexico Louisiana and Texas continental shelf and slope). *US Geological Survey Open File Report*, 01–182.

Osterman, L.E., Kelly, W.S. & Ricardo, J.P., 2008. Benthic foraminiferal census data from Louisiana continental shelf areas, Gulf of Mexico. *US Geological Survey Open File Report*, 2008–1348.

Osterman, L.E., Parrish, K. & Caplan, J., 2004. Benthic foraminiferal census data from Gulf of Mexico cores (Texas and Louisiana continental shelf). *US Geological Survey Open File Report*, 2004–1209.

Osterman, L.E., Poore, R.Z., Sarzenski, P.W. & Turner, R.E., 2005. Reconstructing a 180-year record of natural and anthropogenic induced hypoxia from the sediments of the LA continental shelf. *Geology*, **33**: 329–32.

Osterman, L.E., Poore, R.Z., Swarzenski, P.W. *et al.*, 2009. The 20th century development and expansion of Louisiana shelf hypoxia, Gulf of Mexico. *Geo-Marine Letters* (doi: 10.1007/s00367–009–0158–2).

Oti, M.N. & Postma, G. (eds.), 1995. *Geology of Deltas*. Rotterdam; Balkema.

Pacht, J.A., Sheriff, R.E. & Perkins, B.F. (eds.), 1996. *Stratigraphic Analysis Utilizing Advanced Geophysical, Drilling and Borehole Technology for Petroleum Exploration and Production*. Gulf Coast Section – Society of Economic Paleontologists and Mineralogists (GCS-SEPM) Foundation (Proceedings of the 17th Annual Research Conference).

Pande, D.K., Singh, R.R. & Chandra, K., 2008. Source rocks in deep water depositional systems of east and west coasts of India. *Search and Discovery Article*, **10169**.

Pandey, J., Azmi, R.J., Bhandari, A. & Dave, A. (eds.), 1996. *Contributions, XVth Indian Colloquium on Micropalaeontology and Stratigraphy, Dehra Dun*.

Pandey, J. & Choubey, M.S., 1984. Middle–Late Eocene palaeoecology and sediment distribution pattern in the south Cambay basin. *Proceedings of the Xth Indian Colloquium on Micropalaeontology & Stratigraphy*: 339–50.

Pandey, S.C., Pande, A. & Sharma, B.K., 2000. Source organics and depositional environment in Ratnagiri and Mukta blocks, Bombay offshore basin, India. *Proceedings, Fifth International Conference & Exhibition, Petroleum Geochemistry & Exploration in the Afro-Asian Region, New Delhi*: 325–9.

Panieri, G. & Camerlanghi, A., 2010. Benthic Foraminifera as proxy of methane emissions in the marine environment. *Abstracts, Forams2010 (International Symposium on Foraminifera), Bonn*: 154.

Panieri, G., Camerlanghi, A., Conti, S. *et al.*, 2009. Methane seepages recorded in benthic Foraminifera from Miocene seep carbonates, northern Apennines, (Italy). *Palaeogeography, Palaeoclimatology, Palaeoecology*, **284**: 271–82.

Panieri, G. & Sen Gupta, B.K., 2008. Benthic Foraminifera of the Blake Ridge hydrate mound, Western North Atlantic Ocean. *Marine Micropaleontology*, **66**(2): 91–102.

Papazzoni, C.A., 2010. Taphonomic index and transportation in nummulite banks and in nummulitic limestones. *Abstracts, Forams2010 (International Symposium on Foraminifera), Bonn*: 154–5.

Parente, M., Frijia, D., di Lucia, M. *et al.*, 2008. Stepwise extinction of larger Foraminifera at the Cenomanian–Turonian boundary: a shallow water perspective on nutrient fluctuations during Oceanic Anoxic Event 3 (Buonarelli Event). *Geology*, **36**(9): 715–8.

Parker, J.R. (ed.), 1993. *Petroleum Geology of Northwest Europe: Proceedings of the 4th Conference*. London; The Geological Society.

Passey, Q.R., Bohacs, K.M., Esch, W.L. *et al.*, 2010. From oil-prone source-rock to gas-producing shale reservoir – geologic and petrophysical characterization of unconventional shale-gas reservoirs. *SPE*, **131350**.

Patterson, R.T., Hutchinson, I., Guilbault, J.-P. & Clague, J.J., 2000. A comparison of the vertical zonation of diatom, foraminifera, and macrophyte assemblages in a coastal marsh: implications for greater sea-level resolution. *Micropaleontology*, **46**(3): 229–44.

Patterson, R.T., McKillop, W.B., Kroker, S. *et al.*, 1997. Evidence for rapid avian-mediated foraminiferal colonization of Lake Winnipegosis, Manitoba during the Holocene hypsithermal. *Journal of Paleolimnology*, **18**: 131–43.

Pawlowski, J., 2000. Introduction to the molecular systematics of Foraminifera. *Micropaleontology*, **46** (Suppl., 1): 1–12.

Pawlowski, J., Bolivar, I., Fahrni, J. & Zaninetti, L., 1994a. Taxonomic identification of Foraminifera using ribosomal DNA sequences. *Micropaleontology*, **40**(4): 373–7.

Pawlowski, J., Bolivar, I., Fahrni, J.F. *et al.*, 1997. Extreme differences in rates of molecular evolution of Foraminifera revealed by comparison of ribosomal DNA sequences and the fossil record. *Molecular Biology and Evolution*, **14**(5): 498–505.

Pawlowski, J., Bolivar, I., Fahrni, J.F. *et al.*, 1999a. Molecular evidence that *Reticulomyxa filosa* is a freshwater naked foraminifer. *Journal of Eukaryotic Microbiology*, **46** (6): 612–7.

Pawlowski, J., Bolivar, I., Fahrni, J.F. *et al.*, 1999b. Naked foraminiferans revealed. *Nature*, **399**: 27.

Pawlowski, J., Bolivar, I., Guiard-Maffia, J. & Gouy, M., 1994b. Phylogenetic position of Foraminifera inferred from LSU rRNA gene sequences. *Molecular Biology and Evolution*, **11**(6): 929–38.

Pawlowski, J., Fahrni, J.F. & Bowser, S.S., 2002a. Phylogenetic analysis and genetic diversity of *Notodendrodes hyalinosphaira*. *Journal of Foraminiferal Research*, **32**: 173–6.

Pawlowski, J., Fahrni, J.F., Brykczynska, U. *et al.*, 2002b. Molecular data reveal high taxonomic diversity of allogromiid Foraminifera in Explorers Cove (McMurdo Sound, Antarctica). *Polar Biology*, **25**: 96–105.

Pawlowski, J. & Gooday, A.J., 2009. Precambrian biota: protistan origin of trace fossils? *Current Biology*, **19**(1): R28–30.

Pawlowski, J., Gooday, A.J. & Cornelius, N., 2003a. Diversity of deep-sea benthic Foraminifera: molecular and morphological aspects. *Berichte zur Polar-und Meeresforschung*, **470**: 51–6.

Pawlowski, J. & Holzmann, M., 2002. Molecular phylogeny of Foraminifera: a review. *European Journal of Protistology*, **38**: 1–10.

Pawlowski, J. & Holzmann, M., 2008. Diversity and geographic distribution of benthic foraminifera: a molecular perspective. *Biodiversity and Conservation*, **17**: 317328.

Pawlowski, J., Holzmann, M., Berney, C. *et al.*, 2002c. Phylogeny of allogromiid Foraminifera inferred from SSU rRNA gene sequences. *Journal of Foraminiferal Research*, **32**: 334–43.

Pawlowski, J., Holzmann, M., Berney, C. *et al.*, 2003b. The evolution of early Foraminifera. *Proceedings of the National Academy of Sciences*, **100**(20), 11494–8.

Pawlowski, J., Holzmann, M., Fahrni, J. & Richardson, S.L., 2003c. Small subunit ribosomal DNA suggests that the xenophorean *Syringammina corbicula* is a foraminiferan. *Journal of Eukaryotic Microbiology*, **50**: 483–7.

Pawlowski, J., Lecroq, B. & Lejzerowicz, F., 2010. Hidden diversity of early Foraminifera unveiled by environmental DNA surveys. *Abstracts, Forams2010 (International Symposium on Foraminifera), Bonn*: 156.

Payne, S.N.J., Jones, R.W., Lowe, S., & Milner, P.S., 2003. The role and value of biosteering in reservoir exploitation. *Abstracts, AAPG International Conference, Barcelona*.

Payros, A., Bernaola, G., Orue-Etxebarria, X. *et al.*, 2007. Reassessment of the Early–Middle Eocene biomagnetochronollogy based on evidence from the Gorrondatxe section (Basque country, western Pyrenees). *Lethaia*, **40**: 183–95.

Payton, C.E. (ed.), 1977. *Seismic Stratigraphy: Applications to Hydrocarbon Exploration*. Tulsa, OK; American Association of Petroleum Geologists (Memoir, No. 26).

Pearce, C.R., Coe, A.L. & Cohen, A.S., 2010. Seawater redox variations during the deposition of the Kimmeridge Clay formation, United Kingdom (Uppwe Jurassic): evidence from molybdenum isotopes and trace metal ratios. *Paleoceanography*, **25**: PA4213.

Pearson, E.A. (ed.), 1960. *Waste Disposal in the Marine Environment*. New York; Pergamon Press.

Pearson, P.N., 1993. A lineage phylogeny for the Paleogene planktonic Foraminifera. *Micropaleontology*, **39**(3): 193–232.

Pearson, P.N., Olsson, R.K., Huber, B.T. *et al.*, 2006. *Atlas of Eocene Planktonic Foraminifera*. Cushman Foundation for Foraminiferal Research (Special Publication, No. 41).

Pearson, P.N. & Palmer, M.R., 2000. Atmospheric carbon dioxide concentrations over the past 60 million years. *Nature*, **406** (6797): 695–9.

Peres, W.E., 1993. Shelf-fed turbidite system model and its application to the Oligocene deposits of the Campos basin, Brazil. *American Association of Petroleum Geologists Bulletin*, **77**(1): 81–101.

Perez-Asensio, J.N., Aguiree, J., Schmiedl, G. & Civis, J., 2012. Messinian paleoenvironmental evolution of the lower Guadalquivir Basin (SW Spain) based on benthic foraminifera. *Palaeogeography, Palaeoclimatology, Palaeoecology*, **326**–328: 135–51.

Perez-Cruz, L.L. & Machain-Castillo, M.L., 1990. Benthic Foraminifera of the Oxygen Minimum Zone, continental shelf of the Gulf of Tehuantepec, Mexico. *Journal of Foraminiferal Research*, **20**(4): 312–25.

Perkins, B.F. & Martin, G.B. (eds.), 1985. *Habitat of Oil and Gas in the Gulf Coast*. Gulf Coast Section – Society of Economic Paleontologists and Mineralogists (GCS-SEPM) Foundation (Proceedings of the 4thAnnual Research Conference).

Perry, C.T., Berkeley, A. & Smithers, S.G., 2008. Microfacies characteristics of a tropical, mangrove-fringed shoreline, Cleveland Bay, Queensland, Australia: sedimentary and taphonomic controls on mangrove facies development. *Journal of Sedimentary Research*, **78**: 77–97.

Pessagno, E.A., Jr., 1969. *Upper Cretaceous Stratigraphy of the Western Gulf Coast Area of Mexico, Texas and Arkansas*. Boulder, CO; The Geological Society of America (Memoir, No. 111).

Petters, S.W., 1982. Central West African Cretaceous–Tertiary benthic Foraminifera and stratigraphy. *Palaeontographica, Abteilung A*, **179**: 1–104.

Pettitt, P. & White, M., 2012. *The British Palaeolithic*. New York; Routledge.

Pflum, C.E. & Frerichs, W.E., 1976. *Gulf of Mexico Deep-Water Foraminifers*. Cushman Foundation for Foraminiferal Research (Special Publication, No. 14).

Phleger, F.B., 1955. Ecology of Foraminifera in south-eastern Mississippi delta area. *Bulletin of the American Association of Petroleum Geologists*, **39**(5): 712–52.

Phleger, F.B., 1960. *Ecology and Distribution of Recent Foraminifera*. Baltimore, MD; The Johns Hopkins Press.

Phleger, F.B., 1963. Patterns of living marsh Foraminifera in south Texas coastal lagoons. *Boletin de la Sociedad Geologica Mexicana*, **28**(1): 1–44.

Phleger, F.B. & Parker, F.L., 1951. *Ecology of Foraminifera, northwest Gulf of Mexico*. Boulder, CO; Geological Society of America (Memoir, No. 46).

Phleger, F.B. & Soutar, A., 1973. Production of benthic Foraminifera in three east Pacific oxygen mimima. *Micropalaeontology*, **19**(1): 110–5.

Pianka, E.R., 1970. On r- and K- selection. *American Naturalist*, **104**: 592–7.

Pickering, K.T. & Owen, L.A., 1994. *An Introduction to Global Environmental Issues*. New York; Routledge.

Picou, E.B., Jr., Perkins, B.F., Rosen, N.C. & Nault, M.J. (eds.), 1999. *Gulf of Mexico Basin Biostratigraphic Index Fossils – A Geoscientists Guide. Foraminifers and Nannofossils, Parts I & II: Oligocene through Holocene Foraminifers*. Gulf Coast Section – Society of Economic Paleontologists and Mineralogists (GCS-SEPM) Foundation (Proceedings of the 18th Annual Research Conference).

Pignatti, J., Benedetti, A., di Carlo, M. *et al.*, 2010. Fathoming the diversity of Foraminifera: how many species are there? *Abstracts, Forams2010 (International Symposium on Foraminifera), Bonn*: 160.

Pilarczyk, J.E. & Reinhardt, E.G., 2011. Testing foraminiferal taphonomy as a tsunami indicator in a shallow arid system lagoon: Sur, Sultanate of Oman. *Marine Geology* (doi:10.1016/j.margeo.2011.12.002).

Pillet, L., de Vargas, C. & Pawlowski, J., 2010. Molecular identification of sequestered diatom chloroplasts and kleptoplastidy in Foraminifera. *Protist*, **162**: 394–404.

Pina-Ochoa, E., Hogslund, S., Geslin, E. *et al.*, 2010. Widespread occurrence of nitrate storage and denitrification among Foraminifera. *Abstracts, Forams 2010 (International Symposium on Foraminifera), Bonn*: 160.

Pindell, J.L. & Perkins, B.F. (eds.), 1992. *Mesozoic and Early Cenozoic Development of the Gulf of Mexico and Caribbean Region: A Context for Hydrocarbon Exploration*. Gulf Coast Section – Society of Economic Paleontologists and Mineralogists (GCS-SEPM) Foundation (Proceedings of the 13th Annual Research Conference).

Pipperr, M., 2011. Characterisation of Ottnangian (middle Burdigalian) palaeoenvironments in the North Alpine Foreland Basin using benthic Foraminifera: a review of the Upper Marine Molasse of southern Germany. *Marine Micropaleontology*, **79**: 80–99.

Pipperr, M. & Reichenbacher, B., 2010. Foraminifera from the borehole Altdorf (SE Germany): proxies for Ottnangian (Early Miocene) palaeoenvironments of the Central Paratethys. *Palaeogeography, Palaeoclimatology, Palaeoecology*, **289**: 62–80.

Pitman, J.K., Steinshouser, D. & Lewan, M.D., 2004. Petroleum generation and migration in the Mesopotamian basin and Zagros fold belt of Iraq: results from a basin-modeling study. *GeoArabia*, **9**(4): 41–72.

Pittet, B., van Buchem, F.S.P., Hillgartner, H., et al., 2002. Ecological succession, palaeoenvironmental change and depositional sequences of Barremian–Aptian shallow-water carbonates of northern Oman. *Sedimentology*, **49**: 555–81.

Pitts, M. & Roberts, M., 1997. *Fairweather Eden*. London; Century.

Platon, E., Sen Gupta, B.K., Rabalais, N.N. & Turner, R.E., 2005. Effect of seasonal hypoxia on the benthic foraminiferal community of the Louisiana inner continental shelf: the 20th century record. *Marine Micropaleontology*, **54**: 263–83.

Platon, E. & Sikora, P., 2005. StrataPlot: a new graphic correlation tool. *Abstracts, Geologic Problem Solving with Microfossils Conference, Rice University, Houston:* 59–60.

Poag, C.W. (ed.), 1981. *Ecologic Atlas of Benthic Foraminifera from the Gulf of Mexico*. Woods Hole, MA; Marine Science International.

Poag, C.W., 1984. Distribution and ecology of deep-water benthic Foraminifera in the Gulf of Mexico. *Palaeogeography, Palaeoclimatology, Palaeoecology*, **48**: 25–37.

Poag, C.W., 1985. *Benthic Foraminifera as Indicators of Potential Petroleum Sources*. Gulf Coast Section – Society of Economic Paleontologists and Mineralogists (GCS-SEPM) Foundation (Proceedings of the 4th Annual Research Conference).

Poag, C.W. & Tresslar, R.C., 1981. Living Foraminifera of West Flower Garden Bank, northernmost coral reef in the Gulf of Mexico. *Micropaleontology*, **27**(1): 31–70.

Poirier, C., Goubert, E. & Chaumillon, E., 2010. Foraminiferal record of climate and land use changes during the last centuries in Marennes–Oleron Bay (Atlantic coast of France). *Abstracts, Forams2010 (International Symposium on Foraminifera), Bonn*: 164.

Posamentier, H.W., Summerhayes, C.P., Haq, B.U. & Allen, G.P. (eds.), 1993. *Sequence Stratigraphy and Facies Associations*. Oxford; Blackwell (International Association of Sedimentologists Special Publication, No. 18).

Post, P., Olson, D., Lyons, V. et al. (eds.), 2004. *Salt–Sediment Interactions and Hydrocarbon Prospectivity: Concepts, Applications and Case Studies for the 21st Century*. Gulf Coast Section – Society of Economic Paleontologists and Mineralogists (GCS-SEPM) Foundation (Proceedings of the 24th Annual Research Conference).

Post, P., Rosen, N., Olson, D. et al. (eds.), 2005. *Petroleum Systems of Divergent Continental Margin Basins*. Gulf Coast Section – Society of Economic Paleontologists and Mineralogists (GCS-SEPM) Foundation (Proceedings of the 25th Annual Research Conference).

Postuma, J.A., 1971. *Manual of Planktonic Foraminifera*. Amsterdam; Elsevier.

Powell, A.J. & Riding, J.B. (eds.), 2005. *Recent Developments in Applied Biostratigraphy*. London; The Micropalaeontological Society (Special Publication).

Powers, R.W., 1968. Arabie Saoudite. *Lexique Stratigraphique Internationale*, **III**(10b1): 1–107.

Pratson, L.F. & Ryan, W.B.F., 1994. Pliocene to Recent infilling and subsidence of intraslope basins offshore Louisiana. *American Association of Petroleum Geologists Bulletin*, **78**(10): 1483–506.

Premoli-Silva, I. & Bolli, H.M., 1973. Late Cretaceous to Eocene planktonic Foraminifera and stratigraphy of Leg 15 sites in the Caribbean Sea. *Initial Reports of the Deep Sea Drilling Project*, **15**: 499–547.

Price, G.D., Vowles-Sheridan, N. & Anderson, M.W., 2008. Lower Jurassic mud volcanoes and methane, Kilve, Somerset, UK. *Proceedings of the Geologists' Association*, **119**: 193–201.

Prokoph, A., Rampini, M.E. & El-Bilali, H., 2004. Periodic components in the diversity of calcareous plankton and geological events over the past 230 Myr. *Palaeogeography, Palaeoclimatology, Palaeoecology*, **207**: 105–25.

Pryor, F., 2003. *Britain BC*. London; Harper Perennial.

Pujos-Lamy, A., 1984. Foraminiferes benthiques et bathymetrie: le Cenozoique du Golfe de Gascogne. *Palaeogeography, Palaeoclimatology, Palaeoecology*, **48**: 39–60.

Pulham, A., Pearce, J. & Boyd, T., 1991. Sedimentology and biostratigraphy of cores in Miocene deposits, offshore Louisiana, Gulf of Mexico. *Transactions, Gulf Coast Association of Geological Societies*, **XLI**: 556.

Punyasena, S.W., Jaramillo, C., de la Parra, F. & Yuelin Du, 2012. Probabilistic correlation of single stratigraphic samples: a generalized approach for biostratigraphic data. *American*

*Association of Petroleum Geologists Bulletin*, **96**(2): 235–44.

Quallington, A., 2011. Determining the paleogeographic evolution and source to sink relationships of Indian offshore sedimentary basins. *Search and Discovery Article*, **30159**.

Quillevere, F., Morard, R., de Vargas, C. *et al.*, 2010. Truncorotalia truncatulinoides and the value of direct morpho-genetic comparisons in planktonic Foraminifera. *Abstracts, Forams2010 (International Symposium on Foraminifera), Bonn*: 166–7.

Qing Li, Jiasheng Wang, Jianwen Chen & Qing Wei, 2010. Stable carbon isotopes of benthic foraminifers from IODP Expedition 311 as possible indicators of episodic methane seep events in a gas hydrate ecosystem. *Palaios*, **25**: 676–81.

Quadri, V.U.N. & Quadri, S.M.J.G., 1998. Pakistan has unventured regions, untested plays. *Oil & Gas Journal*, 5th January: 62–5.

Quadri, V.U.N. & Shuaib, S.M., 1986. Hydrocarbon prospects of southern Indus basin, Pakistan. *American Association of Petroleum Geologists Bulletin*, **70**(6): 730–47.

Quillmann, U., Marchitto, T.M., Jennings, A.E. *et al.*, 2012. Cooling and freshening at 8.2 ka on the NW Iceland shelf recorded in paired del18O and Mg/Ca measurements of the benthic foraminifer *Cibicides lobatulus*. *Quaternary Research*, **78**: 528–39.

Quinn, P.S. & Day, P.M., 2007. Ceramic micropalaeontology: the analysis of microfossils in ancient ceramics. *Journal of Micropalaeontology*, **26**(2): 159–68.

Quinterno, P.J. & Gardner, J.V., 1987. Benthic foraminifers on the continental shelf and upper slope, Russian River area, northern California. *Journal of Foraminiferal Research*, **17**(2): 132–52.

Rabalais, N.N. & Turner, R.E. (eds.), 2001. *Coastal Hypoxia: Consequences for Living Resources and Ecosystems*. Washington, DC; American Geophysical Union.

Raban, A., Reinhardt, E.G., McGrath, M. & Hodge, N., 1999. The underwater excavations 1993–1994. *Journal of Roman Archaeology, Supplementary Series*: 152–68.

Racey, A., 1995. Lithostratigraphy and larger foraminiferal (nummulitid) biostratigraphy of the Tertiary of northern Oman. *Micropaleontology*, **41**(Suppl.): 1–123.

Racey, A., 2001. A review of Eocene nummulite accumulations: structure, formation and reservoir potential. *Journal of Petroleum Geology*, **24**(1): 79–100.

Racey, A., Bailey, H.W., Beckett, D. *et al.*, 2001. The petroleum geology of the Early Eocene El Garia formation, Hasdrubal field, offshore Tunisia. *Journal of Petroleum Geology*, **24**(1): 29–53.

Rahaghi, A., 1980. *Tertiary Foraminiferal Assemblage of Qum-Kashan, Sabzewar and Jahrum Areas*. Tehran; National Iranian Oil Company (Geological Laboratories Publication No. 8).

Rahaghi, A., 1983. *Stratigraphy and Faunal Assemblage of Paleocene–Lower Eocene in Iran*. Tehran; National Iranian Oil Company (Geological Laboratories Publication No. 10).

Rahaghi, A., 1992. Remarks on the genera *Sirtina, Vanderbeekia, Iranites* and *Neumanites* (Foraminifera) from Upper Cretaceous of Iran with suggestion of a new subfamily Neumanitinae. *Revista Espanola de Micropaleontologia*, **24**(2): 119–29.

Rajshekar, C., 1995. Foraminifera from the Bagh group, Narmada basin, India. *Journal of the Geological Society of India*, **46**: 413–28.

Rajshekar, C. & Atpalkar, S., 1995. Foraminifera from the Nodular Limestone, Bilthana, Gujurat: stratigraphic significance. *Journal of the Geological Society of India*, **45**: 585–93.

Raju, A.T.R., 1968. Geological evolution of Assam and Cambay Tertiary basins of India. *American Association of Petroleum Geologists Bulletin*, **52**(12): 2422–37.

Raju, D.S.N., 1974. Studies of Indian Miogypsinidae. *Utrecht Micropaleontological Bulletin*, **9**.

Raju, D.S.N., 2008. High-resolution bio-, chrono-, and bio-sequence stratigraphy and sea level change: global vs. Indian record. *ONGC Bulletin*, **43**(1): 10–43.

Raju, D.S.N., 2011. Integrated bio-chrono-tectonostratigraphic and paleoenvironmental framework for hydrocarbon exploration in east coast basins of India. *Extended Abstracts, 2nd South Asian Geosciences Conference and Exhibition (GEOIndia), New Delhi*.

Raju, D.S.N., Bhandari, A. & Ramesh, P., 1999. Relative sea-level fluctuations during Cretaceous and Cenozoic in India. *Bulletin of the Oil and Natural Gas Corporation*, **36**(1): 185–201.

Raju, D.S.N., Guha, D.K., Bedi, T.S. *et al.*, 1970. Microfauna, biostratigraphy and paleoecology of the Middle Eocene to Oligocene sediments in western India. *Publication of the Centre of Advanced Study in Geology, Punjab University, Chindigarh*, **7**: 155–78.

Raju, D.S.N., Peters, J., Shanker, R. & Kumar, G., 2005. *An Overview of Litho-, Bio-, Chrono- and Sequence Stratigraphy and Sea-Level Changes of Indian Sedimentary Basins*. Dehra Dun; Association of Petroleum Geologists (Special Publication, No. 1).

Ramsay, A.T. (ed.), 1977. *Oceanic Micropaleontology*. London; Academic Press.

Ranaweera, K., Bains, S. & Joseph, D., 2009(a). Analysis of image-based classification of foraminiferal tests. *Marine Micropaleontology*, **72**: 60–5.

Ranaweera, K., Harrison, A.P., Bains, S. & Joseph, D., 2009(b). Feasibility of computer-aided identification of foraminiferal tests. *Marine Micropaleontology*, **72**: 66–75.

Rao, G.N., 2001. Sedimentation, stratigraphy and petroleum potential of Krishna-Godavari basin, east coast of India. *American Association of Petroleum Geologists Bulletin*, **85** (9): 1623–43.

Rao, G.N., Manmohan, N. & Prasad, P., 1996. A study of Paleocene depositional environments in Krishna-Godavari basin: an integrated approach. *Indian Journal of Petroleum Geology*, **5**(2): 1–20.

Rashid, S.H., 2010. Biofacies and palaeoenvironments of the Oxfordian Hanifa formation, Saudi Arabia. *Abstracts, Forams2010 (International Symposium on Foraminifera), Bonn*: 167–8.

Ratcliffe, K.T. & Zaitlin, B.A. (eds.), 2010. *Application of Modern Stratigraphic Techniques: Theory and Case Histories*. Tulsa, OK; Society of Economic Paleontologists and Mineralogists (Special Publication, No. 94).

Rathburn, A.E., Levin, L.A., Held, Z. & Lohmann, K.C., 2000. Benthic Foraminifera associated with cold methane seeps on the northern California margin: ecology and stable isotopic composition. *Marine Micropaleontology*, **38**(3–4): 247–66.

Rathburn, A.E., Martin, J.B., Waggoner, J.D. et al., 2010. Distribution patterns and isotopic signatures of benthic Foraminifera from methane seep clam beds: using Rose Bengal and Celltracker Green to evaluate stable isotopic disequlibrium. *Abstracts, Forams2010 (International Symposium on Foraminifera), Bonn*: 168.

Rathburn, A.E., Perez, M.E. & Lange, C.B., 2001. Benthic–pelagic coupling in the southern California Bight: relationships between sinking organic material, diatoms and benthic Foraminifera. *Marine Micropaleontology*, **43**(3–4): 261–72.

Rathore, S.S., 2007. *Exploring Exploration*. Dehra Dun; Association of Petroleum Geologists (Bulletin, No. 1).

Raup, D.M. & Sepkoski, J.J., Jr., 1982. Mass extinctions in the marine fossil record. *Science*, **215**: 1501–3.

Rawson, P.F. & Riley, L.A., 1982. Latest Jurassic–Early Cretaceous events and the 'late Cimmerian' unconformity in the North Sea area. *American Association of Petroleum Geologists Bulletin*, **66**: 2628–48.

Raza, H.A., Ali, S.M. & Ahmed, R., 1990. Petroleum geology of Kirthar sub-basin and part of Kutch basin. *Pakistan Journal of Hydrocarbon Research*, **2**(1): 29–73.

Raza Khan, M.S., Sharma, A.K., Sahota, S.K. & Mathur, M., 2000. Generation and hydrocarbon entrapment within Gondwanan sediments of the Mandapeta area, Krishna-Godavari basin, India. *Organic Geochemistry*, **31**: 1495-15-7.

Reddy, A.N., Satyanarayana, K., Bhaktavatsala, K.V. et al., 2005. Sequence stratigraphy and depositional process of Miocene sediments in KD structure, deepwaters of Krishna-Godavari basin, India. *Journal of the Geological Society of India*, **66**: 42–58.

Redmond, C.D., 1964a. The foraminiferal family Pfenderinidae in the Jurassic of Saudi Arabia. *Micropaleontology*, **10**(2): 251–63.

Redmond, C.D., 1964b. Lituolid Foraminifera from the Jurassic and Cretaceous of Saudi Arabia. *Micropaleontology*, **10**(4): 405–14.

Redmond, C.D., 1965. Three new genera of Foraminifera from the Jurassic of Saudi Arabia. *Micropaleontology*, **11**(2): 133–40.

Reeckmann, A., & Friedmann, G.M. (eds.), 1982. *Exploration for Carbonate Petroleum Reservoirs*. John Wiley.

Reichel, M., 1937. Etude sur les Alveolines. *Memoires Suisses de Paleontologie*, **57**: 1–147.

Reicherter, K., Michetti, A.M. & Silva, P.G. (eds.), 2009. *Palaeoseismology: Historical and Prehistorical Records of Ground Effects for Seismic Hazard Assessment*. London; The Geological Society (Special Publication, No. 316).

Reinhardt, E.G., 1999. Foraminifera from the inner harbour (area I14). *Journal of Roman Archaeology, Supplementary Series*: 346–9.

Reinhardt, E.G., Easton, N. & Patterson, R.T., 1996. Foraminiferal evidence of late Holocene sea-level change on Amerindian site distribution at Montague Harbour, British Columbia. *Géographie Physique et Quaternaire*, **50**(1): 35–6.

Reinhardt, E.G., Fitton, R. J. & Schwarcz, H. P., 2003. Isotopic (Sr, O, C) indicators of salinity and taphonomy in marginal marine systems. *Journal of Foraminiferal Research*, **33**(3): 262–72.

Reinhardt, E.G., Goodman, B.E., Boyce, J.I. et al., 2006. The tsunami of December 13, 115 A.D. and the destruction of Herod the Great's harbor at Caesarea Maritima, Israel. *Geology*, **34**(12): 1061–4.

Reinhardt, E.G. & Patterson, R.T., 1999. Foraminiferal analysis of three stratigraphic sections from the Inner Harbor at Caesarea Maritima. *Journal of Roman Archaeology, Supplementary Series*: 252–61.

Reinhardt, E.G., Patterson, R.T., Blenkinsop, J. & Raban, A., 1998a. Paleoenvironmental evolution of the inner basin of the ancient harbor at Caesarea Maritima, Israel: foraminiferal and Sr isotopic evidence. *Revue de Paleobiologie*, **17**(1): 1–21.

Reinhardt, E.G., Patterson, R.T. & Schröder-Adams, C.J., 1994. Geoarchaeology of the ancient harbor site of Caesarea Maritima, Israel: evidence from sedimentology and paleoecology of benthic Foraminifera. *Journal of Foraminiferal Research*, **24**(1): 37–48.

Reinhardt, E.G. & Raban, A., 1999. Catastrophic destruction of Herod the Great's harbor at Caesarea Maritima, Israel: geoarchaeological evidence. *Geology*, **27**(9): 811–4.

Reinhardt, E.G., Stanley, D. & Patterson, R.T., 1998b. Strontium isotopic-paleontological method as a high-resolution paleosalinity tool for lagoonal environments. *Geology*, **26** (11): 1003–6.

Reinhardt, E.G., Stanley, D.J. & Schwarcz, H., 2001. Human-induced desalinization of Manzala lagoon, Nile delta, Egypt: evidence from isotopic analysis of benthic invertebrates. *Journal of Coastal Research*, **17**(2): 431–442.

Reiss, Z. & Hottinger, L., 1984. *The Gulf of Aqaba: Ecological Micropalaeontology.* New York; Springer Verlag.

Remin, Z., Dubicka, Z., Kozlowska, A. & Kuchta, B., 2012. A new method of rock disintegration and foraminiferal extraction with the use of liquid nitrogen [lN2]. Do conventional methods lead to biased paleoecological and paleoenvironmental interpretations? *Marine Micropaleontology*, **86**–87: 11–4.

Renema, W., 2002. Larger Foraminifera as marine environmental indicators. *Scripta Geologica*, **124**.

Renema, W. (ed.), 2007. *Biogeography, Time and Place: Distributions, Barriers and Islands.* Dordrecht; Springer.

Renema, W., Bellwood, D.R., Braga, J.C. et al., 2008. Hopping hotspots: global shifts in marine biodiversity. *Science*, **321**: 654–7.

Renz, H.H., 1962. Stratigraphy and paleontology of the type section of the Santa Anita group and the overlying Merecure group, Rio Querecual, northeastern Venezuela. *Boletin Informativo Asociacion Venezolano de Geologia, Mineria y Petroleo, Caracas*, **5**: 89–108.

Reolid, J., Nagy, J. & Rodriguez-Tovar, F., 2010. Ecostratigraphic trends of Jurassic agglutinated foraminiferal assemblages as a response to sea-level changes in shelf deposits of Svalbard (Norway). *Palaeogeography, Palaeoclimatology, Palaeoecology*, **293**: 184–96.

Reolid, M., Nagy, J., Rodriguez-Tovar, F.J. & Oloriz, F., 2008b. Foraminiferal assemblages as palaeoenvironmental bioindicators in Late Jurassic epicontinental platforms: relation with trophic conditions. *Acta Palaeontologica Polonica*, **53**(4): 705–22.

Reolid, M., Rodriguez-Tovar, F.J. & Nagy, J., 2012a. Ecological replacement of Valanginian agglutinated Foraminifera during a maximum flooding event in the Boreal realm (Spitsbergen). *Cretaceous Research*, **33**: 196–204.

Reolid, M., Rodriguez-Tovar, F.J., Nagy, J. & Oloriz, F., 2008a. Benthic foraminiferal morphogroups of mid to outer shelf environments of the Late Jurassic (Prebetic zone, southern Spain): characterisation of biofacies and environmental significance. *Palaeogeography, Palaeoclimatology, Palaeoecology*, **261**: 280–99.

Reolid, M., Sebane, A., Rodriguez-Tovar, F.J. & Marok, A., 2012b. Foraminiferal morphogroups as a tool to approach the Toarcian Anoxic Event in the western Saharan Atlas (Algeria). *Palaeogeography, Palaeoclimatology, Palaeoecology*, **323**–325: 87–99.

Repetski, J.E. (ed.), 1996. *Abstracts of Papers, Eleventh North American Paleontological Convention.* Washington, DC; Smithsonian Institution (The Paleontological Society, Special Publication, No.8).

Resig, J., 1992. *Parabolivina peruensis*, a new oxygen minimum foraminifer from the Peru margin. *Journal of Foraminiferal Research*, **22**(1): 30–3.

Requejo, A.G., Wielcowsky, C.C., Klosterman, M.J. & Sassen, R., 1994. Geochemical characterisation of lithofacies and organic facies in Cretaceous organic-rich rocks from Trinidad, east Venezuela basin. *Organic Geochemistry*, **22**(3–5): 441–59.

Reuter, M., Piller, W.E., Harzhauser, M. et al., 2010. The Quilon limestone, Kerala basin,

India: an archive for Miocene Indo-Pacific seagrass beds. *Lethaia*, **44**(1): 76–86 2010 (doi: 10.1111/j.1502-3931.2010.00226x).

Reyment, R.A. & Bengtson, P.(eds.), 1981. *Aspects of Mid-Cretaceous Regional Geology*. London; Academic Press.

Reymond, B.A. & Stampfli, G.M., 1996. Three-dimensional sequence stratigraphy and subtle stratigraphic traps associated with systems tracts: West Cameron region, offshore Louisiana, Gulf of Mexico. *Marine and Petroleum Geology*, **13**(1): 41–60.

Reyre, D. (ed.), 1966. *Bassins Sedimentaires du Littoral Africain, Premiere Partie: Littoral Atlantique*. Paris; Association des Services Geologiques Africains.

Rhodes, E.G. & Moslow, T.F. (eds.), 1992. *Marine Clastic Reservoirs: Examples and Analogues*. New York; Springer Verlag.

Richardson, E.S. & Rona, P.A., 1980. Global Eocene plate reorganization: implications for petroleum exploration. *UN/ESCAP CCOP/SOPAC Technical Bulletin*, **3**: 25–36.

Riche, P. & Prestat, B., 1980. Paleogeographie du Cretace Moyen du Proche- et Moyen- Orient et sa signification petroliere. *Proceedings of the Tenth World Petroleum Congress, Bucharest*: 57–75.

Ricketts, E.R., Kennett, J.P., Hill, T.M. & Barry, J.P., 2009. Effects of carbon dioxide sequestration on California margin deep-sea foraminiferal assemblages. *Marine Micropaleontology*, **72**: 165–75.

Riebesell, U., Zondervan, I., Rost, B. *et al.*, 2000. Reduced calcification of marine plankton in response to increase atmospheric $CO_2$. *Nature*, **407**: 364–7.

Rios-Netto, A.M., Antunes, I.L., Bentes, D. & Abreu, C.J., 2010. Foraminiferal analyses, sea-level fluctuations and depositional dynamics of a modern deep-water lobe complex in Campos basin. *Abstracts, Forams2010 (International Symposium on Foraminifera), Bonn*: 171.

Risgaard-Petersen, N., Langezaal, A.M., Ingvardsen, S. *et al.*, 2006. Evidence for complete denitrification in a benthic foraminifer. *Nature*, **443**(7): 93–6.

Robaszynski, F. & Caron, M. (co-ords.), 1979. Atlas de Foraminiferes planctoniques du Cretace Moyen. *Cahiers de Micropaleontologie*, **1**/2: 1–185 and 1–181.

Robaszynski, F., Caron, M., Dupuis, C. *et al.*, 1990. A tentative integrated stratigraphy of the Turonian of central Tunisia: formations, zones and sequential stratigraphy in the Kalaat Senan area. *Bulletins des Centres*

*Recherches Exploration-Production Elf-Aquitaine*, **14**(1): 213–384.

Robaszynski, F., Caron, M., Gonzalez, J.M. & Wonders, A.A.H., 1984. Atlas of Late Cretaceous planktonic Foraminifera. *Revue de Micropaleontologie*, **26**(3/4): 145–305.

Roberts, D.G. & Bally, A.W. (eds.), 2012. *Regional Geology and Tectonics*. Amsterdam; Elsevier.

Roberts, H.H., Rosen, N.C., Fillon, R.H. & Anderson, J.B. (eds.), 2003. *Shelf-Edge Deltas and Linked Downslope Petroleum Systems: Global Significanmce and Future Exploration Potential*. Gulf Coast Section, Society of Economic Paleontologists and Mineralogists Foundation (Proceedings of the 23rd Annual Research Conference).

Roberts, J.M., Wheeler, A., Freiwald, A. & Cairns, S. (eds.), 2009. *Cold-Water Corals: The Biology and Geology of Deep-Sea Coral Habitats*. Cambridge: Cambridge University Press.

Roberts, M.B. & Parfitt, S.A. (eds.), 1999. *Boxgrove: A Middle Pleistocene Hominid Site at Eartham Quarry, Boxgrove, West Sussex*. English Heritage (Archaeological Report, No. 17).

Robertson, A.H.F., Searle, M.P. & Ries, A.C. (eds.), 1990. *The Geology and Tectonics of the Oman Region*. London; The Geological Society (Special Publication, No. 49).

Robinson, G.S., 1970. Change in the bathymetric distribution of the genus *Cyclammina*. *Transactions of the Gulf Coast Association of Geological Societies*, **20**: 201–9.

Robinson, G.S. & Kohl, B., 1978. Computer-assisted paleoecologic analyses and application to petroleum exploration. *Transactions of the Gulf Coast Association of Geological Societies*, **28**: 433–47.

Robinson, M.M. & McBride, R.A., 2006. Benthic Foraminifera from a relict flood tidal delta along the Virginia/North Carolina Outer Banks. *Micropaleontology*, **52**(1): 67–80.

Rodrigues, K., 1993. The Naparima Hill–Cruse/Forest/Gros Morne petroleum system of Trinidad: a quantitative evaluation of petroleum generated. *Extended Abstract, American Association of Petroleum Geologists Annual Convention, New Orleans*.

Rodriguez-Pinto, A., Pueyo, E.L., Serra-Kiel, J. *et al.*, 2012. Lutetian magnetostratigraphic calibration of larger Foraminifera zonation (SBZ) in the southern Pyrenees: the Isuela section. *Palaeogeography, Palaeoclimatology, Palaeoecology*, **333**–4: 107–20.

Roehl, P.O. & Choquette, P.W. (eds.), 1985. *Carbonate Petroleum Reservoirs*. New York; Springer Verlag.

Rogerson, M., Kouwenhouven, T.J., van der Zwaan, G.J. *et al*., 2006. Benthic Foraminifera of a Miocene canyon and fan. *Marine Micropaleontology*, **60**: 295–318.

Rogl, F., 1998. Palaeogeographic considerations for Mediterranean and Paratethyan seaways (Oligocene to Miocene). *Annalen des Naturhistorischen Museums, Wien*, **99A**: 279–310.

Rogl, F. & Gradstein, F.M. (eds.), 1988. *Proceedings of the Second Workshop on Agglutinated Foraminifera (Vienna, 1986)*. Vienna; Geologisches Bundesanstalt (Abhandlungen, No. 41).

Rohling, E.J., Sprovieri, M., Cane, T. *et al*., 2004. Reconstructing past planktonic foraminiferal habitats using stable isotope data: a case history for Mediterranean sapropel S5. *Marine Micropaleontology*, **50**(1–2): 89–124.

Rohr, G.M., 1991. Paleogeographic maps. Maturin basin of E. Venezuela and Trinidad. *Transactions of the Second Geological Conference of the Geological Society of Trinidad & Tobago:* 88–105.

Romero, J., Caus, E. & Rosell, J., 2002. A model for the palaeoenvironmental distribution of larger Foraminifera based on late Middle Eocene deposits on the margin of the South Pyrenean basin (NE Spain). *Palaeogeography, Palaeoclimatology, Palaeoecology*, **179**: 43–56.

Rosen, N.C., Breard, S.Q., Engelhardt-Moore, N. *et al*. (eds.), 2001. *Gulf of Mexico Basin Biostratigraphic Index Fossils – A Geoscientists Guide. Foraminifers and Nannofossils, Part III: Paleocene through Eocene Foraminifers*. Gulf Coast Section – Society of Economic Paleontologists and Mineralogists (GCS-SEPM) Foundation (Special Publication).

Rosen, R.N. & Hill, W.A., 1990. Biostratigraphic application to Pliocene–Miocene sequence stratigraphy of the western and central Gulf of Mexico and its integration to lithostratigraphy. *Transactions, Gulf Coast Association of Geological Societies*, **XL**: 737–43.

Rosen, R.N. & Rosen, N.C. (eds.), 2008. *Biostratigraphy and Chronostratigraphy of the Subsurface Cretaceous of the Gulf Coast Basin, Emphasizing the Woodbine–Tuscaloosa*. Gulf Coast Section – Society of Economic Paleontologists and Mineralogists (GCS-SEPM) Foundation (Special Publication, No. 3).

Rosoff, D.B. & Corliss, B.H., 1992. An analysis of Recent deep-sea benthic foraminiferal morphotypes from the Norwegian and Greenland Seas. *Palaeogeography, Palaeoclimatology, Palaeoecology*, **91**: 13–20.

Ross, C.A., 1967. Development of fusulinid (Foraminiferida) faunal realms. *Journal of Paleontology*, **41**(6): 1341–54.

Ross, C.A. & Haman, D. (eds.), 1987. *Timing and Depositional History of Eustatic Sequences: Constraints on Seismic Stratigraphy*. Cushman Foundation for Foraminiferal Research (Special Publication, No. 24).

Ross, C.A. & Ross, J.R.P., 1995. Foraminiferal zonation of Late Paleozoic depositional sequences. *Marine Micropaleontology*, **26**: 469–78.

x?ch.7Ross, C.A., Ross, J. & Brenckle, P.L. (eds.), 1997. *Late Paleozoic Foraminifera: their Biostratigraphy, Evolution and Paleoecology, and the Mid-Carboniferous Boundary*. Cushman Foundation for Foraminiferal Research (Special Publication, No. 36).

Ross, D.J.K. & Bustin, R.M., 2009. The importance of shale composition and pore structure upon gas storage potential of shale gas reservoirs. *Marine and Petroleum Geology*, **26**: 916–27.

Rossi, V. & Horton, B.P., 2009. The application of a subtidal Foraminifera-based transfer function to reconstruct Holocene paleobathymetry of the Po delta, northern Adriatic Sea. *Journal of Foraminiefral Research*, **39**(3): 180–90.

Rossi, V & Vaiani, S.C., 2008. Benthic foraminiferal evidence of sediment supply changes and fluvial drainage reorganization in Holocene deposits of the Po delta, Italy. *Marine Micropaleontology*, **69**(2): 106–18.

Rothaus, R., Reinhardt, E.G. & Noller, J., 2004. Regional considerations of coastline change, tsunami damage and recovery along the southern coast of the Bay of Izmit (the Kocaeli (Turkey) earthquake of 17 August 1999). *Natural Hazards*, **31**(1): 233–52.

Rothe, N., Gooday, A.J., Cedhagen, T. & Hughes, J.A., 2011. Biodiversity and distribution of the genus *Gromia* (Protista, Rhizaria) in the deep Weddell Sea (Southern Ocean). *Polar Biology*, **34**: 69–81.

Rothschild, L.J. & Lister, A.M. (eds.), 2003. *Evolution on Planet Earth: The Impact of the Physical Environment. Second Edition*. London; Academic Press.

Rottger, R., Kruger, R. & de Ruk, S., 1990. Trimorphism in Foraminifera (Protozoa): verification of an old hypothesis. *European Journal of Protistology*, **25**: 226–8.

Roure, F. (ed.), 1994. *Peri-Tethyan Platforms*. Paris; Editions Technip.

Rouvillois, A., 1967. Observations morphologiques, sedimentaires et ecologiques sur la plage de la ville Ger, dans l'estuaire de la Rance. *Cahiers Oceanographiques*, **19**: 375–89.

Rouvillois, A., 1972a. Biocoenese et taphocoenese de foraminiferes sur la plateau continental Atlantique au large de l'Ile d'Yeu. *Cahiers de Micropaleontologie, Ser. 3*, **1**: 1–10.

Rouvillois, A., 1972b. Influence du barrage de l'usine mare-motrice sur la morphologie, l'ecologie et la biocoenese de la plage de la ville Ger dans l'estuaire de la Rance. *Quatrieme Congres International de la Mer*, 115–23.

Rowe, G.T. & Pariente, V. (eds.), 1992. *Deep-Sea Food Chains and the Global Carbon Cycling*. Dordrecht; Kluwer Academic Publishers.

Roychoudhury, S.C. & Deshpande, S.V., 1982. Regional distribution of carbonate facies, Bombay offshore region, India. *American Association of Petroleum Geologists Bulletin*, **66**(10): 1483–96.

Ruban, D.A., 2011. Palaeozoic mass extinctions and foraminifers: a new insight? *Revue de Micropalaeontologie*, **54**: 237–8.

Ruckheim, S., Bornemann, A. & Mutterlose, J., 2006. Planktonic Foraminifera from the mid-Cretaceous (Barremian–Early Albian) of the North Sea basin: palaeoecological and palaeoceanographic implications. *Marine Micropaleontology*, **58**: 83–102.

Ruddiman, W.F., 2003. The Anthropogenic era began thousands of years ago. *Climate Change*, **61**: 261–93.

Ruddiman, W.F., 2005. *Plows, Plagiues and Petroleum: How Humans Took Control of Climate*. Princeton, NJ; Princeton University Press.

Ruddiman, W. F & Sarnthein, M., 1986. Paleoclimatic linkage between high and low latitudes. *Nature*, **322**: 211–2.

Ruggeberg, A., Dullo, C., Dorschel, B. & Hebbeln, D., 2007. Environmental changes and growth history of a cold-water carbonate mound (Propellor mound, Porcupine seabight). *International Journal of Earth Sciences*, **269**: 569–74.

Ruiz, C., Matthews, S., Goff, J. *et al.*, 2004. Mid-Miocene to Recent tectonostratigraphic evolution of the northwestern Dezful embayment, southwest Iran. *GeoArabia*, **9**(1): 121 (abstract).

Sabbatini, A., Bonatto, S., Gooday, A.J. *et al.*, 2010. Modern benthic foraminifers at Northern shallow sites of Adriatic Sea and soft-walled, monothalamous taxa: a brief overview. *Micropaleontology*, **56**(3–4): 359–76.

Sabbatini, A., Morigi, C., Negri, A. & Gooday, A.J., 2002. Soft-shelled benthic Foraminifera from a hadal site (7800m water depth) in the Atacama Trench (SE Pacific): preliminary observations. *Journal of Micropalaeontology*, **21**(2): 131–5.

Sabbatini, A., Morigi, C., Negri, A. & Gooday, A.J., 2007. Distribution and biodiversity of stained monothalamous Foraminifera from Tempelfjord, Svalbard. *Journal of Foraminiferal Research*, **37**(2): 93–106.

Sadek, A., 1992. *Geology of the Arab World*. Cairo; Cairo University.

Sadooni, F.N., 1993. Stratigraphic sequence, microfacies and petroleum prospects of the Yamama formation, Lower Cretaceous, southern Iraq. *American Association of Petroleum Geologists Bulletin*, **77**(11): 1971–88.

Sadooni, F.N. & Alsharhan, A.S., 2003. Stratigraphy, microfacies and petroleum potential of the Mauddud formation (Albian–Cenomanian) in the Arabian Gulf basin. *American Association of Petroleum Geologists Bulletin*, **87**(10): 1653–80.

Sahni, A. (ed.), 1996. *Cretaceous Stratigraphy and Palaeoenvironments*. Bangalore; Geological Society of India.

Saidova, K.M., 1981. *On an Up-To-Date System of Supraspecific Taxonomy of Cenozoic Benthonic Foraminifera*. Moscow; Institut Okeanologii P.P. Shirshova, Akademiya Nauk SSSR. In Russian.

Saint-Marc, P., 1982. Distribution paleoecologique et paleobiogeographique des grands foraminiferes du Cenomanien. *Revista Espanola de Micropaleontologia*, **14**: 247–62.

Saint-Marc, P., 1992. Biogeographic and bathymetric distribution of benthic Foraminifera in Paleocene El Haria formation, Tunisia. *Journal of African Earth Sciences*, **15**(3/4): 473–87.

Salami, M.B., 1976. Biology of *Trochammina* cf. *quadriloba* Hoglund (1947), an agglutinating foraminifer. *Journal of Foraminiferal Research*, **6**(2): 142–53.

Salem, M.J. & Belaid, M.N. (eds.), 1991. *The Geology of Libya*. Amsterdam; Elsevier.

Salem, M.J. & Oun, K.M. (eds.), 2003. *The Geology of Northwest Libya*. Malta; Gutenberg Press.

Salgueiro, E., Voelker, A., Abrantes, F. *et al.*, 2008. Planktonic Foraminifera from modern sediments reflect upwelling patterns off Iberia: insights from a regional transfer function. *Marine Micropaleontology*, **66**(3–4): 136–64.

Samanta, B.K., 1969. Taxonomy and stratigraphy of the Indian species of *Discocyclina*. *Geological Magazine*, **106**(2): 115–29.

Samanta, B.K., 1972. Planktonic foraminiferal biostratigraphy of the early Tertiary of the Rakhi Nala section, Suklaiman range, West Pakistan. *Journal of the Geological Society of India*, **13**(4): 317–28.

Samanta, B.K., 1973. Planktonic Foraminifera from the Paleocene–Eocene succession in the Rakhi Nala, Sulaiman range, Pakistan. *Bulletin of the British Museum (Natural History), Geology*, **22**(6): 421–82.

Samanta, B.K. (ed.), 1985. *Proceedings of the XI Indian Colloquium on Micropalaeontology & Stratigraphy, Part 1, Microfauna*. Calcutta; Geological, Mining & Metallurgical Society of India.

Sampo, M., 1969. *Microfacies and microfossils of the Zagros area, southwestern Iran*. Leiden; E.J. Brill.

Sams, R.H., 1995. Interpreted sequence stratigraphy of the Los Jabillos, Areo and (subsurface) Naricual formations, northern Monagas area, eastern Venezualan basin. *Boletin de la Sociedad Venezolana de Geologia*, **20**(1–2): 30–40.

Sancetta, C., Imbrie, J., Kipp, N.G. *et al.*, 1972. Climatic record in North Atlantic deep-sea core V23–82: comparison of the last and present interglacials based on quantitative time series. *Quaternary Research*, **2**: 363–7.

Sander, N.J., 1962. Apercu paleontologique et stratigraphique du Paleogene d'Arabie Saoudite. *Revue de Micropaleontologie*, **5**(1): 3–40.

Sander, N.J., 2012. Paleontologic and stratigraphic overview of the Paleogene in eastern Saudi Arabia. *Carnets de Geologie*, **2012**/14.

Sandon, H., 1932. *The Food of Protozoa*. Cairo; Faculty of Science, Egyptian University (Publication, No. 1).

Sanyal, A., Hemming, H.G., Broecker, W.S. *et al.*, 1996. Oceanic pH control on the boron isotopic composition of Foraminifera: evidence from culture experiments. *Paleoceanography*, **11**(5): 513–7.

Saraswati, P.K., 2010. Stable isotopic composition of modern larger Foraminifera and its connotation as palaeobiological proxy. *Abstracts, Forams2010 (International Symposium on Foraminifera), Bonn*: 175.

Saraswati, P.K., 2012. Early Palaeogene vertebrate-bearing deposits in Cambay and Kutch basins, India: a review of foraminiferal biostratigraphy and strontium isotope stratigraphy. *Abstracts, 4th International Geologica Belgica meeting*.

Sarma, M., 2012. Petroleum system modelling and risk analysis, Cambay basin, India. *Search and Discovery Article*, **10403**.

Sartorio, D. & Venturini, S., 1988. *Southern Tethyan Biofacies*. Milan; AGIP.

Sastri, V.V., Sinha, R.N., Singh, G. & Murti, K.V.S., 1973. Stratigraphy and tectonics of sedimentary basins of east coast of peninsular India. *American Association of Petroleum Geologists Bulletin*, **57**(4): 65–8.

Saunders, J.B., 1957. Trochamminidae and certain Lituolidae (Foraminifera) from the Recent brackish-water sediments of Trinidad, British West Indies. *Smithsonian Miscellaneous Contributions*, **134**(5): 1–16.

Saunders, J.B., 1958. Recent Foraminifera of mangrove swamps and river estuaries and their fossil counterparts in Trinidad. *Micropaleontology*, **4**(1): 79–92.

Saunders, J.B. (ed.), 1968. *Transactions of the Fourth Caribbean Geological Conference, Port-of-Spain, Trinidad & Tobago, 1965*. Port of Spain; Caribbean Printers.

Saunders, J.B. & Bolli, H.M., 1985. Trinidad's contributions to world biostratigraphy. *Transactions of the Fourth Latin American Geological Congress, Trinidad & Tobago*, **1979**: 781–95.

Scarparo Cunha, A.A. & Koutsoukos, E.A.M., 1998. Calcareous nannofossils and planktonic foraminifers in the Upper Aptian of the Sergipe basin, northeastern Brazil: palaeoecological inferences. *Palaeogeography, Palaeoclimatology, Palaeoecology*, **142**: 175–84.

Schafer, C.T., 1970. Studies of benthonic Foraminifera in Restigouche estuary, I: faunal distribution patterns near pollution sources. *Maritime Sediments*, **6**: 121–34.

Schafer, C.T., 1973. Distribution of Foraminifera near pollution sources in Chaleur Bay. *Water, Air & Soil Pollution*, **2**: 219–33.

Schafer, C.T., 1982. Foraminiferal recolonization of an offshore dump site in Chaleur Bay, New Brunswick, Canada. *Journal of Foraminiferal Research*, **12**(4): 317–26.

Schafer, C.T. & Cole, F.E., 1974. Distributions of benthonic Foraminifera: their use in delimiting local nearshore environments. *Geological Survey of Canada Paper*, **74**–30: 103–8.

Schafer, C.T., Collins, E.S. & Smith, J.N., 1991. Relationships of Foraminifera and thecamoebian distributions to sediments contaminated by pulp mill effluent: Saguenay Fiord, Quebec, Canada. *Marine Micropaleontology*, **17**(3/4): 255–83.

Schafer, C.T., Wagner, F.J.E. & Ferguson, C., 1975. Occurrence of Foraminifera, molluscs and ostracods adjacent to the industrialized shoreline of Canso Strait, Nova Scotia. *Water, Air & Soil Pollution*, **5**: 79–96.

Schaub, H., 1951. Stratigraphie und Palaontologie des Schlieren-flysches, mit besonderer Berucksichtigung der Palaocaenen und Untereocaenen Nummuliten und Assilinen. *Schweizerische Palaontolgische Abhandlungen*, **68**: 1–222.

Schaub, H., 1981. Nummulites et assilines de la Tethys Paleogene: Taxinomie, phylogenie et biostratigraphie. *Memoires Suisses de Paleontologie*, **104**–6: 1–236.

Schaudinn, P., 1895. Uber den Dimorphismus bei Foraminiferen. *Sitzungsberichte der Gesellschaft naturforschender Freunde zu Berlin*, **5**: 87–97.

Schaudinn, P., 1911. *Fritz Schaudinns Arbeiten*. Hamburg and Leipzig; Verlag Leopold Moss.

Scheibner, C., Reijmer, J.J.G., Marzouk, A.M. et al., 2003. From platform to basin: the evolution of a Paleoceen carbonate margin (eastern desert, Egypt). *International Journal of Earth Sciences*, **92**: 624–40.

Scheibner, C. & Speijer, R.J., 2009. Recalibration of the Tethyan shallow benthic zonation across the Paleocene–Eocene boundary: the Egyptian record. *Geologica Acta*, **7**(1–2): 195–214.

Scheibnerova, V., 1970. Foraminifera and their Mesozoic biogeoprovinces. *Record of the Geological Survey of New South Wales*, **13** (3): 1356–74.

Scheibnerova, V., 1978. Depth habitats of Cretaceous Foraminifera with special reference to globotruncanids. *Annales des Mines et de la Geologie*, **28**(2): 147–51.

Schenck, H.G., 1940. Applied paleontology. *American Association of Petroleum Geologists Bulletin*, **24**(1): 1752–78.

Scheuth, J.D. & Frank, T.D., 2008. Reef Foraminifera as bioindicators of coral reef health: Low Isles Reef, northern Great Barrier Reef, Australia. *Journal of Foraminiferal Research*, **38**(1): 11–22.

Schlee, J.S. (ed.), 1984. *Interregional Unconformities and Hydrocarbon Exploration*. Tulsa, OK; American Association of Petroleum Geologists (Memoir, No. 36).

Schluter, T., 2006. *Geological Atlas of Africa (With Notes on Stratigraphy, Tectonics, Economic Geology, Hazards and Geosites of Each Country)*. Berlin; Springer.

Schmidt, D.N., Thierstein, H.R., Bollmann, J. & Schiebel, R., 2004. Abiotic forcing of plankton evolution in the Cenzoic. *Science*, **303**: 207–10.

Schmiedl, G., de Bovee, F., Buscail, R. et al., 2000. Trophic control on benthic foraminiferal abundance and microhabitat in the bathyal Gulf of Lions, western Mediterranean Sea. *Marine Micropaleontology*, **40**: 167–88.

Schnitker, D., 1993. Ecostratigraphy of Plio-Pleistocene benthic foraminifers in ODP hole 625B and four Eureka holes from the Gulf of Mexico. *Micropaleontology*, **39**(4): 404–18.

Schofield, K., Rosen, N.C., Pfeiffer, D. & Johnson, S. (eds.), 2008. *Answering the Challenges for Production from Deep-Water Reservoirs: Analogiues and Case Studies to Aid a New Generation*. Gulf Coast Section – Society of Economic Paleontologists and Mineralogists (GCS-SEPM) Foundation (Proceedings of the 28th Annual Research Conference).

Schonfeld, J., 2001. Benthic Foraminifera and pore-water oxygen profiles: a re-assessment of species boundary conditions at the western Iberian margin. *Journal of Foraminiferal Research*, **31**(2): 86–107.

Schonfeld, J., 2012. History and development of methods in Recent benthic foraminiferal studies. *Journal of Micropalaeontology*, **31**: 53–72.

Schonfeld, J., Alve, E., Geslin, E. et al., 2012. The FOBIMO (FOraminiferal Bio-MOnitoring) initiative – towards a standardised protocol for soft-bottom benthic foraminiferal monitoring studies. *Marine Micropaleontology*, **94**–95: 1–13.

Schonfeld, J., Dullo, W.C., Pfannkuche, O. et al., 2011. Recent benthic foraminiferal assemblages from cold-water coral mounds in the Porcupine Seabight. *Facies*, **57**(2): 187–213.

Schopf, T.J.M. (ed.), 1972. *Models in Paleobiology*. San Francisco, CA; Freeman, Cooper and Co.

Schroder, R., 1975. General evolutionary trends in Orbitolinas. *Revista Espanola de Micropaontologia, Numero Especial*: 117–28.

Schroder, R. & Darmoian, S.A., 1977. *Gyroconulina columellifera* n.gen., n.sp., a complex ataxophragmiid Foraminifera from the Aqra limestone (Maastrichtian) of northern Iraq. *Bollettino della Societa Paleontologica Italiana*, **16**(1): 117–23.

Schroder, R. & Neumann, M. (co-ords.), 1985. Grandes foraminferes du Cretace moyen de la region Mediterraneenne. *Geobios, Memoire Special*, **7**: 1–161.

Schroder, T., 1992. A palynological zonation of the North Sea basin. *Journal of Micropalaeontology*, **11**(2): 113–26.

Schulz, H.D. & Zabel, M. (eds.), 2000. *Marine Geochemistry*. Berlin; Springer.

Schumacher, D. & Perkins, B.F. (eds.), 1990. *Gulf Coast Oils and Gases: Their Characteristics, Origin, Distribution, and Exploration and Production Significance*. Gulf Coast Section – Society of Economic Paleontologists and Mineralogists (GCS-SEPM) Foundation (Proceedings of the 9th Annual Research Conference).

Schumacher, S., Jorrisen, F.J., Dissard, D. *et al.*, 2007. Live (Rose Bengal stained) and dead benthic Foraminifera from the Oxygen Minimum Zone of the Pakistan continental margin (Arabian Sea). *Marine Micropaleontology*, **62**: 45–73.

Schumacher, M., Jorissen, F.J., Mackensen, A. *et al.*, 2010. Ontogenetic effects on stable isotopes in tests of live (Rose Bengal stained) benthic Foraminifera from the Pakistan continental margin. *Marine Micropaleontology*, **76**: 92–103.

Schweizer, M., Jorissen, F. & Geslin, E., 2011. Contributions of molecular phylogenetics to foraminiferal taxonomy: general overview and example of *Pseudoeponides falsobeccarii* Rouvillois, 1974. *Comptes Rendus Palevol*, **10**: 95–105.

Schweizer, M., Pawlowski, J., Duijnstee, I.A.P. *et al.*, 2005. Molecular phylogeny of the foraminiferal genus *Uvigerina* based on ribosomal DNA sequences. *Marine Micropaleontology*, **57**(3–4): 51–67.

Schweizer, M., Pawlowski, J., Kouwenhoven, T.J. *et al.*, 2008. Molecular phylogeny of Rotaliida (Foraminifera) based on complete small subunit rDNA sequences. *Marine Micropaleontology*, **66**(3–4): 233–46.

Schweizer, M., Pawlowski, J., Kouwenhoven, T.J. & van der Zwaan, B., 2009. Molecular phylogeny of common cibicidids and related Rotaliida (Foraminifera) based on small subunit rDNA sequences. *Journal of Foraminiferal Research*, **39**(4): 300–15.

Scott, D.B. & Lipps, J.H., 1995. Environmental applications of foraminiferal studies. *Journal of Foraminiferal Research*, **25**(3).

Scott, D.B. & Medioli, F.S., 1980. *Quantitative Studies of Marsh Foraminiferal Distributions in Nova Scotia: Implications for Sea-Level Research*. Cushman Foundation for Foraminiferal Research (Special Publication, No. 17).

Scott, D.B., Medioli, F. & Braund, R., 2003. Foraminifera from the Cambrian of Nova Scotia: the oldest multichambered foraminifera. *Micropaleontology*, **49**(2): 109–26.

Scott, D.B., Medioli, F.S. & Schafer, C.T., 2001. *Monitoring in Coastal Environments using Foraminifera and Thecamoebian Indicators*. Cambridge; Cambridge University Press.

Scott, D.B., Suter, J.R. & Kosters, E.C., 1991. Marsh Foraminifera and arcellaceans of the lower Mississippi delta: controls on spatial distributions. *Micropaleontology*, **37**(4): 373–92.

Scott, D.B., Tobin, R., Williamson, M. *et al.*, 2005. Pollution monitoring in two North American estuaries: historical reconstructions using benthic Foraminifera. *Journal of Foraminiferal Research*, **35**(1): 65–82.

Scott, G.H., Bishop, S. & Burt, B.J., 1990. *Guide to some Neogene Globorotaliids (Foraminiferida) from New Zealand*. Lower Hutt, New Zealand; New Zealand Geological Survey (Palaeontological Bulletin, No. 6).

Scott, R.W., 1990. *Models and Stratigraphy of Mid-Cretaceous Reef Communities, Gulf of Mexico*. Tulsa, OK; Society of Economic Paleontologists and Mineralogists/Society for Sedimentary Geology (Concepts in Sedimentology and Paleontology Series).

Scott, R.W., 2002. Upper Albian benthic foraminifers new in west Texas. *Journal of Foraminiferal Research*, **32**(1): 43–50.

Seears, H., 2011. *Biogeography and Phylogenetics of the Planktonic Foraminifera*. University of Nottingham (unpublished doctoral thesis).

Seibold, I. & Seibold, E., 1981. Offshore and lagoonal benthic Foraminifera near Cochin (south-west India): distribution, transport, ecological aspects. *Neues Jahrbuch fur Geologie und Palaontologie, Abhandlungen*, **162**: 1–56.

Seiglie, G.A., 1968. Foraminiferal assemblages as indicators of high organic carbon content in sediments and of polluted waters. *American Association of Petroleum Geologists Bulletin*, **52**: 2231–41.

Seiglie, G.A., 1971. A preliminary note on the relationships between foraminifers and pollution in two Puerto Rican bays. *Caribbean Journal of Science*, **11**: 93–8.

Seiglie, G.A., 1974. Foraminifers of Mayaguez and Anaso Bays and surroundings, part 4: relationships of foraminifers and pollution in Mayaguez Bay. *Caribbean Journal of Science*, **14**(1/2): 1–68.

Seiglie, G.A., 1975. Foraminifers of Guayanilla Bay and their use as environmental indicators. *Revista Espanola de Micropaleontologia*, **7**: 453–87.

Seiglie, G.A. & Baker, M.B., 1983. Some West African Cenozoic agglutinated foraminifers with inner structures: taxonomy, age and evolution. *Micropaleontology*, **29**: 391–403.

Seiglie, G.A. & Frost, S.H., 1979. Significance of middle Tertiary large Foraminfera common to West Africa and Caribbean. *American Association of Petroleum Geologists Bulletin*, **63**: 525 (abstract).

Seiglie, G.A., Fleisher, R.L. & Baker, M.B., 1986. Alveovalvulinidae n.fam., and Neogene diversification of agglutinated foraminifers with inner structure. *Micropaleontology*, **32**(2): 169–81.

Sejrup, H.P., Birks, H.J.B., Kristensen, D.K. & Madsen, H., 2004. Benthonic foraminiferal distributions and quantitative transfer functions for the northwest European continental margin. *Marine Micropaleontology*, **53**: 197–226.

Selley, R.C. (ed.), 1997. *African Basins*. Amsterdam; Elsevier (Sedimentary Basins of the World Series, No. 3).

Selley, R.C., 1998. *Elements of Petroleum Geology, Second Edition*. Toronto; Academic Press.

Semensatto, D.L., Jr. & Dias-Brito, D., 2007. Alternative saline solutions to float foraminiferal tests. *Journal of Foraminiferal Research*, **37**(3): 265–9.

Semensatto, D.L., Jr., Funo, R.H.F., Dias-Brito, D. & Coelho, C., Jr., 2009. Foraminiferal ecological zonation along a Brazilian mangrove transect: diversity, morphotypes and the influence of subaerial exposure time. *Revue de Micropaleontologie*, **52**: 67–74.

Semikhatov, M.A. & Chumakov, N.M., 2004. *Climate in the Epochs of Major Biospheric Transformations*. Moscow; Nauka.

Sen Gupta, B.K., 1977. The distribution of modern benthic Foraminifera on continental shelves of the world's ocean. *Indian Journal of Earth Sciences*, **4**(1): 60–83.

Sen Gupta, B.K. (ed.), 1999. *Modern Foraminifera*. Dordrecht; Kluwer Academic Publishers.

Sen Gupta, B.K. & Aharon, P., 1994. Benthic Foraminifera of bathyal hydrocarbon vents of the Gulf of Mexico: initial reports on communities and stable isotopes. *Geo-Marine Letters*, **14**(2–3): 88–96.

Sen Gupta, B.K. & Kilbourne, R.T., 1976. Depth distribution of benthic Foraminifera on the Georgia continental shelf. *Maritime Sediments Special Publication*, **1**: 25–38.

Sen Gupta, B.K., Lee, R.F. & May, M.S., III, 1981. Upwelling and an unusual assemblage of benthic Foraminifera on the northern Florida continental slope. *Journal of Paleontology*, **55**(4): 853–7.

Sen Gupta, B.K. & Machain-Castillo, M.L., 1993. Benthic Foraminifera in oxygen-poor habitats. *Marine Micropaleontology*, **20**: 183–201.

Sen Gupta, B.K. & Smith, L.E., 2010. Modern benthic Foraminifera of the Gulf of Mexico: a census report. *Journal of Foraminiferal Research*, **40**(3): 247–65.

Sen Gupta, B.K., Smith, L.E. & Lobegeier, M.K., 2007. Attachment of Foraminifera to vestimentiferan tubeworms at cold seeps: refuge from seafloor hypoxia and sulphide toxicity. *Marine Micropaleontology*, **62**(1): 1–6.

Sen Gupta, B.K., Turner, R.E. & Rabalais, N.N., 1996. Seasonal oxygen depletion in continental shelf waters of Louisiana: historical record of benthic foraminifers. *Geology*, **24**: 227–30.

Sepkoski, J.J., Jr., 1981. A factor analytical description of the Phanerozoic marine fossil record. *Paleobiology*, **7**: 36–53.

Septfontaine, M., 1988. Towards an evolutionary classification of Jurassic lituolids (Foraminifera) in carbonate platform environment. *Revue de Paleobiologie, Volume Spécial*, **2**: 229–56.

Septfontaine, M., Arnaud-Vaneau, A., Bassoullet, J.-P. *et al.*, 1991. Les foraminiferes imperfores des plates-formes carbonatees jurassiques: etat des connaisances et perspectives d'avenir. *Bulletin de la Societe Vaudoise des Sciences Naturelles*, **80**(3): 255–77.

Seranne, M. & Anka, Z., 2005. South Atlantic continental margins of Africa: a comparison of the tectonic vs. climate interplay on the evolution of equatorial west Africa and SW Africa margins. *Journal of African Earth Sciences*, **43**: 283–300.

Sergeeva, N.G., Anikeeva, O.V. & Gooday, A.J., 2005. The monothalamous foraminiferan *Tinogullmia* in the Black Sea. *Journal of Micropalaeontology*, **24**(2): 191–2.

Sergeeva, N.G., Anikeeva, O.V. & Gooday, A.J., 2010. Soft-shelled, monothalamous Foraminifera from the oxic/anoxic interface (NW Black Sea). *Micropaleontology*, **56**(3–4): 393–407.

Serra-Kiel, J., Hottinger, L., Caus, E. *et al.*, 1998. Larger foraminiferal biostratigraphy of the Tethyan Paleocene and Eocene. *Bulletin de la Societe Geologique de France*, **169**(2): 281–99.

Setty, M.G.A.P., 1976. The relative sensitivity of benthonic Foraminifera in the polluted marine environment of Cola Bay, Goa. *Proceedings of the Sixth Indian Colloquium on Micropalaeontology and Stratigraphy, Banaras*: 225–34.

Setty, M.G.A.P., 1982. Pollution effects monitoring with Foraminifera as indices in the Thana Creek, Bombay area. *Journal of Envrionmental Studies*, **18**: 205–9.

Setty, M.G.A.P., 1984. Benthic foraminiferal bioconeses in the estuarine regions of Goa. *Rivista Italiana di Paleontologia e Stratigrafia*, **89**: 437–45.

Setty, M.G.A.P., & Nigam, R., 1984. Benthonic Foraminifera as pollution indices in the marine environment of the west coast of India. *Rivista Italiana di Paleontologia e Stratigrafia*, **89**(3): 421–36.

Setudehnia, A., 1972. Iran, Part II: Southwest Iran. *Lexique Stratigraphique International*, **III**(9b).

Severin, K.P., 1983. Test morphology of benthic Foraminifer as a discriminator of biofacies. *Marine Micropaleontology*, **8**: 65–76.

Seyrafian, A., 2000. Microfacies and depositional environments of the Asmari formation, at Dehdez area (a correlation across central Zagros basin). *Carbonates and Evaporites*, **15**(2): 121–9.

Seyrafian, A. & Hamedani, A., 1998. Microfacies and depositional environment of the upper Asmari limestone (Burdigalian), north-central Zagros basin, Iran. *Neues Jahrbuch fur Geologie und Palaontologie, Abhandlungen*, **210**(2): 129–41.

Shackleton, N.J., Berger, A. & Pettier, W.R., 1990. An alternative astronomical calibration of the lower Pleistocene timescale based on ODP Site 677. *Philosophical Transactions of the Royal Society of Edinburgh, Earth Science*, **81**: 251–61.

Shackleton, N.J. & Opdyke, N.D., 1973. Oxygen isotope and paleomagnetic stratigraphy of equatorial Pacific core V28–238. *Quaternary Research*, **3**: 39–55.

Shaffer, B.L., 1990. The nature and significance of condensed sections in Gulf Coast late Neogene sequence stratigraphy. *Transactions, Gulf Coast Association of Geological Societies*, **XL**: 767–76.

Shakib, S.S., 1987. Age, biozonation and stratigraphy of Kazhdumi formation of SW Iran. *Rivista Italiana di Paleontologia e Stratigrafia*, **93**(2): 201–24.

Shakib, S.S., 1990. The biostratigraphical aspects of Gadvan formation (Barremian–Aptian) of southwest Iran. *Rivista Italiana di Paleontologia e Stratigrafia*, **96**(1): 111–32.

Shan Yu, Saint-Marc, P., Thonnat, M. & Berthold, M., 1996. Feasibility study of automatic identification of planktonic Foraminifera by computer vision. *Journal of Foraminiferal Research*, **26**(2): 113–23.

Shanmugam, G., Bloch, R.B., Mitchell, S.M. *et al.*, 1995. Basin-floor fans in the North Sea: sequence stratigraphic models vs. sedimentary facies. *American Association of Petroleum Geologists Bulletin*, **79**(4): 477–512.

Sharifi, A.R., Croudace, I.U. & Austin, R.L., 1991. Benthic foraminiferids as pollution indicators in Southampton Water, southern England. *Journal of Micropalaeontology*, **10**(1): 109–13.

Sharland, P.R., Archer, R., Casey, D.M. *et al.*, 2001. *Arabian Plate Sequence Stratigraphy*. Manama, Bahrain; Gulf PetroLink (GeoArabia Special Publication, No. 2).

Sharland, P.R., Casey, D.M., Davies, R.B. *et al.*, 2004. Arabian plate sequence stratigraphy: revisions to SP2. *GeoArabia*, **9**(1): 199–214.

Shaw, A.B., 1964. *Time in Stratigraphy*. New York; McGraw-Hill.

Shennan, I. & Andrews, J. (eds.), 2000. *Holocene Land–Ocean Interaction and Environmental Change around the North Sea*. London; The Geological Society (Special Publication, No. 166).

Shuaib, S.M., 1982. Geology and hydrocarbon potential of offshore Indus basin, Pakistan. *American Association of Petroleum Geologists Bulletin*, **66**: 940–6.

Siddiqui, N.K., 2004. Sui Main limestone: regional geology and the analysis of original pressures of a closed-system reservoir in Pakistan. *American Association of Petroleum Geologists Bulletin*, **88**(7): 1007–35.

Simmons, M.D. (ed.), 1994. *Micropalaeontology and Hydrocarbon Exploration in the Middle East*. London; Chapman & Hall.

Simmons, M.D., Preobrazhensky, M.B. & Bugrova, I.J., 1992. Biostratigraphic characterisation of carbonate sequences and systems tracts: examples from the Early Cretaceous of the Middle east and Turkmenia. *Abstracts, International Symposium on Mesozoic and Cenozoic Sequence Stratigraphy of European Basins, Dijon*: 290–1.

Simmons, M.D., Sharland, P.R., Casey, D.M. *et al.*, 2007. Arabian Plate sequence stratigraphy: potential implications for global chronostratigraphy. *GeoArabia*, **12**(4): 101–30.

Simo, J.A., Scott, R.W. & Masse, J.-P. (eds.), 1993. *Cretaceous Carbonate Platforms*. Tulsa, OK; American Association of Petroleum Geologists (Memoir, No. 56).

Singh, K., Aswal, H.S. & Swamy, S.N., 2008. Sequence biostratigraphy and hydrocarbon source potential of Mesozoic sediments,

Kutch basin, India. *GEO India Convention & Exhibition, New Delhi*: 1–6.

Singh, N.P., 1984. Addition to the Tertiary biostratigraphy of Jaisalmer basin. *Petroleum Asia Journal*, April 1984: 106–28.

Sinha, A.K., Prabhakar, V., Sharma, B.L. *et al.*, 2012. Source rock evaluation and petroleum system modelling in part of Bengal, basin, India. *Search and Discovery Article*, **50573**.

Sinha, D.K. (ed.), 2007. *Micropaleontology*. Sinibaldi; Narosa Publishing House.

Sintubin, M., Stewart, I.S., Niemi, T.M. & Altunel, E. (eds.), 2010. *Ancient Earthquakes*. Boulder, CO; The Geological Society of America (Special Paper, No. 471).

Sissingh, W., 1977. Biostratigraphy of Cretaceous calcareous nannoplankton. Geologie en Mijnbouw, **56**: 37–65.

Skelton, P., 2003. *The Cretaceous World*. Cambridge; Cambridge University Press.

Skinner, H.C., 1966. Modern palaeoecological techniques: an evaluation of the role of paleoecology on Gulf Coast exploration. *Transactions of the Gulf Coast Association of Geological Societies*, **16**: 59–79.

Skinner, H.C. & Steinkraus, W.E., 1972. *Gulf Coast Stratigraphic Correlation Methods with an Atlas and Catalogue of Principal Index Foraminifera*. New Orleans, LA; Louisiana Heritage Press.

Slatt, R.M., Rosen, N.C., Bowman, M. *et al.* (eds.), 2006. *Reservoir Characterization: Integrating Technology and Business Practices*. Gulf Coast Section – Society of Economic Paleontologists and Mineralogists (GCS-SEPM) Foundation (Proceedings of the 26th Annual Research Conference).

Sliter, W.V. (ed.), 1980. *Studies in Marine Micropaleontology and Paleoecology: A Memorial Volume to Orville L. Bandy*. Ithaca, New York; Cushman Foundation for Foraminiferal Research (Special Publication, No. 19).

Sliter, W.V., 1989. Biostratigraphic zonation for Cretaceous planktonic foraminifers examined in thin section. *Journal of Foraminiferal Research*, **19**(1): 1–19.

Sliter, W.V. & Baker, R.A., 1972. Cretaceous bathymetric distribution of benthic foraminifers. *Journal of Foraminiferal Research*, **2**(4): 167–83.

Sliter, W.V., Be, A.W.H. & Berger, W.H. (eds.), 1975. *Dissolution of Deep-Sea Carbonates*. Cushman Foundation for Foraminiferal Research (Special Publication, No. 13).

Sloss, L.L., 1963. Sequences in the cratonic interior of North America. *Geological Society of America Bulletin*, **74**: 93–114.

Smart, C.W. & Gooday, A.J., 1997. Recent benthic Foraminifera in the abyssal northeast Atlantic Ocean: relation to phytodetrital inputs. *Journal of Foraminiferal Research*, **27**(2): 85–92.

Smart, C.W. & Gooday, A.J., 2006. Benthic foraminiferal trends in relation to an organic enrichment gradient on the continental slope (850 m water depth) off North Carolina (USA). *Journal of Foraminiferal Research*, **36**(1): 34–43.

Smart, C.W. & Ramsay, A.T.S., 1995. Benthic foraminiferal evidence for the existence of an Early Miocene oxygen-depleted water mass. *Journal of the Geological Society*, **152**: 735–8.

Smith, A.B., Simmons, M.D. & Racey, A., 1990. Cenomanian echinoids, larger Foraminifera and calcareous Algae from the Natih formation, central Oman mountains. *Cretaceous Research*, **11**: 29–69.

Smith, C.J., Collins, L.S., Jaramillo, C. & Quiroz, L.I., 2010. Marine paleoenvironments of Miocene–Pliocene formations of north-central Falcon State, Venezuela. *Journal of Foraminiferal Research*, **40**(3): 266–82.

Smout, A.H., 1954. *Lower Tertiary Foraminifera of the Qatar Peninsula*. London; British Museum (Natural History).

Smout, A.H., 1956. Three new Cretaceous genera of Foraminifera related to the Ceratobuliminidae. *Micropaleontology*, **2**(4): 335–48.

Southall, K.E., Gehrels, W.R. & Hayward, B.W., 2006. Foraminifera in a New Zealand salt marsh and their suitability as sea-level indicators. *Marine Micropaleontology*, **60**: 167–79.

Southward, A.J., Tyler, P.A., Young, C.M. & Fuiman, L.A. (eds.), 2003. *Advances in Marine Biology, Vol. 46*. London; Academic Press.

Speijer, R.P., Kouwenhoven, T.J. & Nguyen, T.M.P., 2010a. Opportunities and pitfalls of the use of planktonic/benthic ratios in paleodepth reconstructions. *Abstracts, Forams2010 (International Symposium on Foraminifera), Bonn*: 186.

Speijer, R.P., van Loo, D., Cnudde, V. & Jacobs, P., 2010b. Use of high-resolution X-ray CT in biometric and phylogenetic studies of foraminifera. *Abstracts, Forams2010 (International Symposium on Foraminifera), Bonn*: 187.

Spencer, A.M. (ed.), 1986. *Habitat of Hydrocarbons on the Norwegian Continental Shelf*. Geilo, Norwegian Petroleum Society.

Spencer, R.S., 1988. Quantified intraspecific variation of common benthic Foraminifera from the northwest Gulf of Mexico: a potential paleobathymetric indicator. *Journal of Foraminiferal Research*, **22**(3): 274–92.

Spencer, R.S., 1996. A method for improving the precision of paleobathymetric estimates: an example from the northwest Gulf of Mexico. *Marine Micropaleontology*, **28**(3/4): 263–82.

Spindler, M. & Hemleben, C., 1980. Symbionts in planktonic Foraminifera (Protozoa). *Endocytobiology, Endosymbiosis and Cell Biology*, **1**: 133–40.

Spitz, K. & Trudinger, J., 2009. *Mining and the Environment*. Boca Raton, FL; CRC Press.

Stainforth, R.M., Lamb, J.L., Luterbacher, H. *et al.*, 1975. Cenozoic planktonic foraminiferal zonation and characteristics of index taxa. *University of Kansas Paleontological Contributions*, **62**: 1–425.

Stampfli, G., Zaninetti, L., Brönnimann, P. *et al.*, 1976. Trias de l'Elburz oriental, Iran: stratigraphie, sedimentologie, micropaleontology. *Rivista Italiana di Paleontologia*, **82**(3): 467–500.

Stanley, D.J., 1960. Stratigraphy and Foraminifera of lower Tertiary Vidono shale, near Puerto La Cruz, Barcelona. *American Association of Petroleum Geologists Bulletin*, **44**: 616–27.

Stanley, S.M., 2006. Influence of seawater chemistry on biomineralization throughout Phanerozoic time: paleontological end experimental evidence. *Palaeogeography, Palaeoclimatology, Palaeoecology*, **232**: 214–36.

Steel, R.J., Felt, F.L., Johannessen, E.P. & Mathieu, C. (eds.), 1995. *Sequence Stratigraphy of the Northwest European Margin*. Norwegian Petroleum Foundation (Special Publication, No. 5).

Stefanelli, S. & Kapotondi, L., 2008. Foraminiferal response to the deposition of insolation cycle 90 sapropel in different Mediterranean areas. *Journal of Micropalaeontology*, **27**: 45–61.

Steinhoff, I., Cicero, A.D., Koepke, K. *et al.*, 2011. Understanding the regional Haynesville and Bossier Shale depositional systems in east Texas and northern Louisiana: an integrated structural/stratigraphic approach. *Search and Discovery Article*, **50379**.

Steininger, F.F., Senes, J., Kleeman, K. & Rogl, F. (eds.), 1985. *Neogene of the Mediterranean Tethys and Paratethys@ Statigraphic Correlation Tables and Sediment Distribution Maps*. Vienna; University of Vienna Press.

Stewart, I.A. 2000. *The Molecular Evolution of Planktonic Foraminifera and its Implications for the Fossil Record*. University of Edinburgh (unpublished doctoral thesis).

Stewart, I.A., Darling, K.F., Kroon, D. *et al.*, 2001. Genotypic variability on subarctic Atlantic planktonic Foraminifera. *Marine Micropaleontology*, **43**(1–2): 43–53.

Stewart, J.R., Aspinall, S., Beech, M. *et al.*, 2011. Biotically constrained palaeoenvironmental conditions of a mid-Holocene intertidal lagoon on the southern shore of the Arabian Gulf: evidence associated with a whale skeleton from Musaffah, Abu Dhabi, UAE. *Quaternary Science Reviews*, **30**(25–6): 3675–90.

Stouff, V., Lesourd, M. & Debenay, J.P., 1999. Laboratory observations on asexual reproduction (schizogony) and ontogeny of *Ammonia tepida* with comments on the life cycle. *Journal of Foraminiferal Research* **29** (1): 75–84.

Strauss, J., Grossman, E.L., Carlin, J.A. & Dellapenna, T.M., 2012. 100 years of benthic foraminiferal history on the inner Texas shelf inferred from faunas and stable isotopes: preliminary results from two cores. *Continnetal Shelf Research*, **38**: 89–97.

Stringer, C., 2006. *Homo Britannicus: The Incredible Story of Human Life in Britain*. Allen Lane.

Strogen, P., Somerville, I.D. & Jones, G.L. (eds.), 1996. *Recent Advances in Lower Carbonferous Stratigraphy*. London; the Geological Society (Special Publication, No. 107).

Suhr, S.B., Pond, D.W., Gooday, A.J. & Smith, C.R., 2003. Selective feeding by benthic Foraminifera on phytodetritus on the western Antarctic Peninsula shelf: evidence from fatty acid biomarker analysis. *Marine Ecology – Progress Series*, **262**: 153–62.

Summa, L.L., Goodman, E.D., Richardson, M. *et al.*, 2003. Hydrocarbon systems of northeastern Venezuela: plate through molecular scale analysis of the genesis and evolution of the eastern Venezuelan basin. *Marine and Petroleum Geology*, **20**(3–4): 323–49.

Summerhayes, C.P., Prell, W.L. & Emeis, K.C. (eds.), 1992. *Upwelling Systems: Evolution since the Early Miocene*. London; The Geological Society (Special Publication, No. 64).

Summerhayes, C.P. & Shackleton, N.J., 1986. *North Atlantic Palaeoceanography*. London; The Geological Society (Special Publication, No. 21).

Sun, S.Q., 1992. Aragonite dissolution from hypersaline seawater: a hypothesis. *Sedimentary Geology*, **77**: 249–57.

Swamy, S.N. & Kapoor, P.N. (eds.), 2002. *Proceedings of the 1st Conference and*

*Exhibition on Strategic Challenges and Paradigm Shift in Hydrocarbon Exploration with Special Reference to Frontier Basins.* Association of Petroleum Geologists.

Szabo, F. & Kheradpir, A., 1978. Permian and Triassic stratigraphy, Zagros basin, southwest Iran. *Journal of Petroleum Geology,* **1**(2): 57–82.

Szarek, R., Nomaki, H. & Kitazato, H., 2007. Living deep-sea benthic Foraminifera from the warm and oxygen-depleted environment of the Sulu Sea. *Deep-Sea Research II,* **54**: 145–76.

Szinger, B., Toth, E., Gorog, A. & Viszkok, J., 2010. Advantages and limits of micro-CT application in Foraminifera studies. *Abstracts, Forams2010 (International Symposium on Foraminifera), Bonn*: 190.

Takayanagi, Y. & Saito, T. (eds.), 1992. *Studies in Benthic Foraminifera (Proceedings of the Fourth International Symposium on Benthic Foraminifera, Sendai, 1990 (Benthos '90)).* Tokyo; Tokai University Press.

Talib, A. & Gaur, K.N., 2008. Foraminiferal composition and age of the Chari formation, Jumara dome, Kutch. *Current Science,* **95** (3): 367–73.

Talib, A., Gaur, K.N. & Bhalla, S.N., 2007. Callovian–Oxfordian boundary in Kutch mainland, India: a foraminiferal approach. *Revue de Paleobiologie,* **26**(2): 625–30.

Tankard, A.J., Soruco, R.S. & Welsink, H.J., 1995. *Petroleum Basins of South America.* Tulsa, OK; American Association of Petroleum Geologists (Memoir, No. 62).

Tappan, H. & Loeblich, A.R., Jr., 1988. Foraminiferal evolution, diversification, and extinction. *Journal of Paleontology,* **62**(5): 695–714.

Tasker, A., Wilkinson, I.P., Williams, M. *et al.,* 2009. Using microfossils to provenance the materials for Roman mosaics: examples from Leicester and Silchester. *Abstracts, Palaeontological Association Annual Meeting, Birmingham*: 76.

Taylor, P.D. (ed.), 1995. *Field Geology of the British Jurassic.* London; The Geological Society.

Tendal, O.S., 1979. Aspects of the Biology of Komokiacea and Xenophyophoria. *Sarsia,* **64**: 13–7.

Tendal, O.S. & Hessler, R.R., 1977. An introduction to the biology and systematics of the Komokiacea (Textulariina, Foraminiferida). *Galathea Report,* **14**: 165–94.

Terken, J.M.J., 1999. The Natih petroleum system of north Oman. *GeoArabia,* **4**(2): 157–80

Termier, H., Termier, G. & Vachard, D., 1978. Biostratigraphie des terains du Maroc central depuis le Givetien jusqu'au Namurien inferieur. *Annales des Mnes et de la Geologie,* **28**: 39–63.

Tevesz, M.J.S. & McCall, P.L. (eds.), 1983. *Biotic Interactions in Recent and Fossil Benthic Communities.* New York; Plenum.

Thomas, A.N., 1952a. The Asmari limestone of south-west Iran. *Report of the Eighteenth Session, International Geological Congress, Great Britain, 1948,* **VI**: 35–44.

Thomas, A.N., 1952b. Facies variations in the Asmari limestone. *Report of the Eighteenth Session, International Geological Congress, Great Britain, 1948,* **X**: 74–82.

Thomas, F.C., Gradstein, F.M. & Griffiths, C.M., 1988. Bibliography and index of quantitative stratigraphy. *Committee on Quantitative Stratigraphy Special Publication,* **1**: 1–58.

Thompson, d'A.W., 1917. *On Growth and Form.* Cambridge; Cambridge University Press.

Tibert, N.E. & Leckie, R.M., 2004. High-resolution estuarine sea level cycles from the Late Cretaceous: amplitude constraints using agglutinated Foraminifera. *Journal of Foraminiferal Research,* **34**(2): 130–43.

Tipsword, H.L., Setzer, F.M. & Smith, F.L., Jr., 1966. Interpretation of depositional environment in Gulf Coast petroleum exploration from paleoecology and related stratigraphy. *Transactions of the Gulf Coast Association of Geological Societies,* **16**: 119–30.

Todd, R. & Brönnimann, P., 1957. *Recent Foraminifera and Thecamoebina from the Eastern Gulf of Paria, Trinidad.* Cushman Foundation for Foraminiferal Research (Special Publication, No. 3).

Todo, Y., Kitazato, H., Hashimoto, J. & Gooday, A.J., 2005. Simple Foraminifera flourish at the ocean's deepest point. *Science,* **307** (5710), 689.

Toland, C., Peebles, R.G. & Walkden, G.M., 1993. Upper Jurassic and basal Cretaceous outcrop sequence stratigraphy of Wadi Hagil, Ras Al Khaimah. *Society of Petroleum Engineers, Paper,* **25581**.

Toriyama, R., 1984. Summary of the fusuline faunas in Thailand and Malaysia. *Geology and Palaeontology of South-East Asia,* **25**: 137–46.

Toriyama, R., Hamada, T., Igo, H. *et al.,* 1975. The Carboniferous and Permian systems in Thailand and Malaysia. *Geology and Palaeontology of South-East Asia,* **15**: 39–76.

Torres, M.E., Martin, R.A., Klinkhammer, G.P. & Nesbitt, E.A., 2010. Post depositional alteration of foraminiferal shells in cold seep settings: new insights from flow-through time-resolved analyses of biogenic and inorganic seep carbonates. *Earth and Planetary Science Letters*, **299**: 10–22.

Toth, E., Gorog, A., Lecuyer, C. *et al.*, 2010. Palaeoeonvironmental reconstruction of the Sarmatian (Middle Miocene) Central Paratethys based on palaeontological and geochemical analyses of Foraminifera, ostracods, gastropods and rodents. *Geological Magazine*, **147**(2): 299–314.

Tozer, R., Akhter, S., Bennett, J. *et al.*, 2008. Frontier exploration of the deepwater Indus fan, Pakistan. *Abstracts, GeoIndia 2008, New Delhi*.

Tribovillard, N.-P., Stephan, J.-F., Manivit, H. *et al.*, 1991. Cretaceous black shales of Venezuelan Andes: preliminary results on stratigraphy and paleoenvironmental interpretation. *Palaeogeography, Palaeoclimatology, Palaeoecology*, **81**: 313–21.

Trifonova, E. & Vaptzarova, A., 1988. Paleoenvironments and foraminifers at the end of the Early Triassic and the beginning of the Middle Triassic epochs in Bulgaria. *Revue de Paleobiologie, Volume Special*, **2**: 161–6.

Tronchetti, G., 1981. Foraminiferes Cretaces dans les basins oust-Africains: queqlques representants de la superfamille Bolivinitacea. *Cahiers de Micropaleontologie*, **2**: 31–51.

Truett, J.C. & Johnson, S.R. (eds.), 2000. *The Natural History of an Arctic Oil Field: Development and the Biota*. San Diego, CA; Academic Press.

Truffleman, N.J., Hallock, P., Hine, P.C. & Peebles, M.W., 1991. Distribution of foraminferal tests in sediments of Serrannilla Bank, Nicaraguan Rise, southwestern Caribbean. *Journal of Foraminiferal Research*, **21**(1): 39–47.

Tsuchiya, M., Kitazato, H. & Pawlowski, J., 2003. Analysis of internal transcribed spacer of ribosomal DNA reveals cryptic speciation in *Planoglabratella opercularis*. *Journal of Foraminiferal Research*, **33**(4): 285–93.

Tsuchiya, M., Tazume, M. & Kitazato, H., 2008. Molecular characterization of the non-costate morphotypes of buliminid foraminifers based on internal transcribed region of ribosomal DNA (ITS rDNA) sequence data. *Marine Micropaleontology*, **69**(2): 212–24.

Tsujimoto, A., Yasuhara, M., Nomura, R. *et al.*, 2008. Development of modern benthic ecosystems in eutrophic coastal oceans: the foraminiferal record over the last 200 years, Osaka Bay, Japan. *Marine Micropaleontology*, **69**: 225–39.

Tucker, M.E., 1991. Sequence stratigraphy of carbonate–evaporite basins: models and application to the Upper Permian (Zechstein) of northeast England and adjoining North Sea. *Journal of the Geological Society*, **148**: 1019–36.

Tunnell, J.W., Jr., Felder, D.L. & Earle, S.A. (eds.), 2009. *Gulf of Mexico: Origin, Waters and Biota*. College Station, TX; Texas A & M University Press.

Tyson, R.V. & Pearson, T.H. (eds.), 1991. *Modern and Ancient Continental Shelf Anoxia*. London; The Geological Society (Special Publication, No. 58).

Tyson, R.V., Wilson, R.L. & Downie, C., 1979. A stratified water column environmental model for the Kimmeridge Clay. *Nature*, **277**: 377–80.

Tyszka, J., 1994. Response of Middle Jurassic benthic foraminiferal morphogroups to dysoxic/anoxic conditions in the Pieniny Klippen basin, Polish Carpathians. *Palaeogeography, Palaeoclimatology, Palaeoecology*, **110**: 55–81.

Tyszka, J., Jach, R. & Bubik, M., 2010. A new vent-related foraminifer from the lower Toarcian black claystone of the Tatra Mountains, Poland. *Acta Palaeontologica Polonica*, **55** (2): 333–42.

Tyszka, J., Oliwkiewicz-Miklasinska, M., Gedl, P. & Kaminski, M.A., 2005. *Methods and Applications in Micropalaeontology*. Krakow; Instytut Nauk Geologicznych, Polska Akamemia Nauk (Studia Geologica Polonica, No. 124).

Ufkes, E., Jansen, J.H.F. & Brummer, G.-J.A., 1998. Living planktonic Foraminifera in the eastern South Atlantic during spring: indicators of water masses, upwelling and the Congo (Zaire) river plume. *Marine Micropaleontology*, **33**(1/2): 27–54.

Underhill, J.R. (ed.), 1998. *The Development, Evolution and Petroleum Geology of the Wessex Basin*. London; The Geological Society (Special Publication, No. 133).

Urey, H.C., 1947. The thermodynamic properties of isotopic substances. *Journal of the Chemical Society*, **1**: 562–81.

Vachard, D., Gailot, J., Vaslet, D. & le Nindre, Y.-M., 2005. Foraminifers and algae from the Khuff formation (late Middle Permian–Early Triassic) of central Saudi Arabia. *GeoArabia*, **10**(4): 137–86.

Vachard, D., Hauser, M., Martini, R. *et al.*, 2002. Middle Permian (Midian) foraminiferal assemblages from the Batain plain (eastern Oman): their significance to Neotethyan paleogeography. *Journal of Foraminiferal Research*, **32**(2): 155–72.

Vachard, D., Massa, D. & Strank, A., 1993. Le Carbonifere du sondage A1–37 (Cyrenaique, Libye): analyse biostratigraphique, conséquences paleogeographiques. *Revue de Micropaleontologie*, **36**(2): 165–86.

Vachard, D., Pille, L. & Gaillot, J., 2010. Palaeozoic Foraminifera: systematics, palaoeecology and responses to the global changes. *Revue de Micropalaeontologie*, **53**(4): 209–54.

Vahrenkamp, V.C., 1996. Chemostratigraphy on the Lower Cretaceous Shu'aiba formation: a del13C reference profile for the Aptian stage from the southern Neo-Tethys Ocean. *American Association of Petroleum Geologists Bulletin*, **80**(5): 647–62.

Vahrenkamp, V.C., 2010. Chemostratigraphy on the Lower Cretaceous Shu'aiba Formation: a del13C Reference Profile for the Aptian Stage from the Southern Neo-Tethys Ocean. Manama, Bahrain; Gulf PetroLink (GeoArabia Special Publication, No. 4(1)), 107–37.

Vahrenkamp, V., Al Katheeri, F., van Laer, P. et al., 2012. Re-evaluation of the Late Jurassic stratigraphy of Abu Dhabi: a different tack on a carbonate–evaporite system and its implication for exploration plays. *Search and Discovery Article*, **50652**.

Vaidyanadhan, R. & Ramakrishnan, M., 2008. *Geology of India*. Bangalore; Geological Society of India.

van Andel, T. & Postma, H., 1954. Recent sediments of the Gulf of Paria. *Verhandelingen der Koninklijke Nederlandsche Akademie van Wetenschappen, Afdeeling Natuurkunde, Eerste Reeks*, **XX**(5): 1–247.

van Bellen, R.C., 1956. The stratigraphy of the 'Main Limestone' of the Kirkuk, Bai Hassan and Qarah Chauq Dagh structures in north Iraq. *Journal of the Institute of Petroleum*, **42**(293): 233–63.

van Buchem, F.S.P., Gerdes, K.D. & Esteban, M. (eds.), 2010. *Mesozoic and Cenozoic Carbonate Systems of the Mediterranean and the Middle East*. London; The Geological Society (Special Publication, No. 329).

van Buchem, F.S.P., Pittet, B., Hillgartner, H. et al., 2002a. High-resolution sequence stratigraphic architecture of Barremian/Apian carbonate systems in northern Oman and the United Arab Emirates (Kharaib and Shu'aiba formations). *GeoArabia*, **7**(3): 461–500.

van Buchem, F.S.P., Razin, P., Homewood, P.W. et al., 1996. High-resolution sequence stratigraphy of the Natif formation (Cenomanian/Turonian) in northern Oman: distribution of source rocks and reservoir facies. *GeoArabia*, **1**(1): 65–91.

van Buchem, F.S.P., Razin, P., Homewood, P.W. et al., 2002b. Stratigraphic organization of carbonate ramps and organic-rich intrashelf basins: Natih formation (Middle Cretaceous) of northern Oman. *American Association of Petroleum Geologists Bulletin*, **86**(1): 21–53.

van den Akker, T.J.H.A., Kaminski, M.A., Gradstein, F.M. & Wood, J., 2000. Campanian to Palaeocene biostratigraphy and palaeoenvironments in the Foula sub-basin, west of the Shetland Islands, UK. *Journal of Micropalaeontology*, **19**: 23–43.

van der Zwaan, G.J., Duijnstee, I.A.P., den Dulk, M. et al., 1999. Benthic foraminifers: proxies or problems? A review of paleoecological concepts. *Earth-Science Reviews*, **46**: 213–36.

van der Zwaan, G.J., Jorisen, F.J. & de Stigter, H.C., 1990. The depth dependency of planktonic/benthic foraminiferal ratios: constraints and applications. *Marine Geology*, **95**: 1–16.

van Dover, C.L., 2000. *The Ecology of Deep-Sea Hydrothermal Vents*. Princeton, NJ; Princeton University Press.

van Gorsel, J.T., 1988. Biostratigraphy in Indonesia: methods, pitfalls and new directions. *Proceedings of the Indonesian Petroleum Association Seventeenth Annual Convention*: 275–300.

van Hinsbergen, D.J.J., Kouwenhouven, T.J. & van der Zwaan, G.J., 2005. Paleobathymetry in the backstripping procedure: correction for oxygenation effects on depth estimates. *Palaeogeography, Palaeoclimatology, Palaeoecology*, **221**: 245–65.

van Hinte, J.E. (ed.), 1966. *Proceedings of the Second West African Micropalaeontological Colloquium (Ibadan, 1965)*. Leiden; E.J. Brill.

van Hinte, J.E., 1972. The Cretaceous time scale and planktonic foraminiferal zones. *Proceedings, Koninklijke Nederlandse Akademie van Wetenschappen, Ser. B*, **75**(1): 1–8.

van Hoey, G., Borja, A., Birchenough, S. et al., 2010. The use of benthic indicators in Europe from the Water Framework Directive to the Marine Strategy Framework Directive. *Marine Pollution Bulletin*, **60**: 2187–96.

van Marle, L.J., 1988. Bathymetric distribution of benthic Foraminifera on the Australian–Irian Jaya continental margin, eastern Indonesia. *Marine Micropaleontology*, **13**: 97–152.

van Marle, L.J., 1991. *Eastern Indonesian Late Cenozoic Smaller Benthic Foraminifera*.

Amsterdam; Koninklijke Nederlandse Akademie van Wetenschappen (Afdeeling Natuurkunde, Eerste Reeks, No. 34).

van Leeuwen, R.J.W., 1989. Sea-floor distribution and late Quaternary patterns of planktonic and benthic foraminifers in the Angola basin. *Utrecht Micropaleontological Bulletin*, **38**.

van Morkhoven, F.P.C.M., Berggren, W.A. & Edwards, A.S., 1986. *Cenozoic Cosmopolitan Deep-Water Benthic Foraminifera*. Pau, France; Elf-Aquitaine (Bulletin des Centres de Recherches Exploration-Production, Mem. No. 11).

van Vessem, E.J., 1978. Study of Lepidocyclinidae from south-east Asia, particularly from Java and Borneo. *Utrecht Micropaleontological Bulletin*, **19**: 1–163.

Vaslet, D., le Nindre, Y.-M., Vachard, D. *et al.*, 2005. The Permian–Triassic Khuff formation of central Saudi Arabia. *GeoArabia*, **10**(4): 77–132.

Veeken, P.C.H., 1997. The Cenozoic fill of the North Sea basin (UK sector, 56–62degN), a seismic stratigraphic study with emphasis on Paleogene massflow deposits. *Geologie en Mijnbouw*, **75**: 317–40.

Vella, P., 1964. Foraminifera and other fossils from late Tertiary deep-water coral thickets, Wairapa, New Zealand. *Journal of Paleontology*, **38**: 916–28.

Venec-Peyre, M.T., 1981. Les foraminiferes et la pollution: etude de la microfaune de la Cape du Dourduff (embrochure de la riviere de Morlaix). *Cahiers de Biologie Marine*, **22**: 25–33.

Venec-Peyre, M.-T., Lipps, J.H., Weber, M. & Bartolini, A., 2010. Amoco Cadiz oil spill and Foraminifera thirty years later. *Abstracts, Forams2010 (International Symposium on Foraminifera), Bonn*: 193.

Vennin, E., van Buchem, F.S.P., Joseph, P. *et al.*, 2003. A 3D outcrop model for Ypresian nummulitic carbonate reservoirs: Jebel Ousselat, northern Tunisia. *Petroleum Geoscience*, **9**: 145–61.

Verdenius, J.G., van Hinte, J.E. & Fortuin, A.R. (eds.), 1983. *Proceedings of the First Workshop on Arenaceous Foraminifera*. Trondheim; IKU (Continental Shelf Institute).

Viana, A.R. & Rebesco, M. (eds.), 2007. *Economic and Palaeoceanographic Significance of Contourite Deposits*. London; The Geological Society (Special Publication, No. 276).

Vilela, C.G., Batista, D.S., Neto, J.A.B. & Ghiselli, R.O., Jr., 2011. Benthic Foraminifera distribution in a tourist lagoon in Rio de Janeiro, Brazil: a response to anthropogenic impacts. *Marine Pollution Bulletin*, **62**: 2055–74.

Vilela, C.G. & Maslin, M., 1997. Benthic and planktonic foraminifers and stable isotope analysis of mass-flow sediments in the Amazon fan. *Proceedings of the Ocean Drilling Program, Scientfic Results*, **155**: 335–51.

Villamil, T., Arango, C., Weimer, P. *et al.*, 1998. Biostratigraphic techniques for analyzing benthic biofacies, stratigraphic condensation, and key surface identification, Pliocene and Pleistocene sediments, northern Green Canyon and Ewing Bank (offshore Louisiana), northern Gulf of Mexico. *American Association of Petroleum Geologists Bulletin*, **82**: 961–85.

Vincent, E., Killingley, J.S. & Berger, W.H., 1981. Stable isotope composition of benthic Foraminifera from the equatorial Pacific. *Nature*, **289**(5799): 639–43.

Vining, B.A. & Pickering, S.C. (eds.), 2010. *Petroleum Geology: From Mature Basins to New Frontiers – Proceedings of the 7th Petroleum Geology Conference*. London; The Geological Society (Petroleum Geology Conference Series).

Vinken, R. (co-ord.), 1988. The north-west European Tertiary basin: results of the International Geologcal Correlation Programme project No. 124. *Geologisches Jahrbuch, Reihe A*, **100**.

Vinogradov, A.P., 1953. *The Elemental Chemical Composition of Marine Organisms*. New Haven, CT; Yale University Press (Sears Foundation for Marine Research Memoir, No. 2).

Vismara-Schilling, A. & Coulbourn, W.T., 1991. Benthic foraminiferal thanatofacies associated with Late Pleistocene to Holocene anoxic events in the eastern Mediterranean Sea. *Journal of Foraminiferal Research*, **21**(2): 103–25.

von Koenigswald, G.H.R., Emeis, J.D., Buning, W.L. & Wagner, C.W. (eds.), 1963. *Evolutionary Trends in Foraminifera*. Amsterdam; Elsevier.

Wade, B.S., Pearson, P.N., Berggren, W.A. & Palike, H., 2011. Review and revision of Cenozoic tropical planktonic foraminiferal; biostratigraphy and calibration against the geomagnetic polarity and astronomical time scale. *Earth-Science Reviews*, **104**: 111–42.

Wade, C.M., Darling, K.F., Kroon, D. & Leigh Brown, A. J., 1996. Early evolutionary origin of the planktonic Foraminifera inferred from small subunit rDNA sequence comparisons. *Journal of Molecular Evolution*, **43**(6): 672–7.

Wakefield, M.I., Cook, R.J., Jackson, H. & Thompson, P., 2001. Interpreting biostratigraphical data using fuzzy logic: the identification of regional mudstones within the Fleming field, UK North Sea. *Journal of Petroleum Geology*, **24**(4): 417–40.

Wakefield, M.I. & Monteil, E., 2002. Biosequence stratigraphical and palaeoenvironmental findings from the Cretaceous through Tertiary succession, central Indus basin, Pakistan. *Journal of Micropalaeontology*, **21**: 115–30.

Walker, G. & King, D., 2008. *The Hot Topic: How to Tackle Global Warming and Still Keep the Lights On*. London; Bloomsbury.

Walker, K.R. & Bambach, R.K., 1974. Feeding by benthic invertebrates: classification and terminology for paleo-ecological analysis. *Lethaia*, **7**(1): 67–78.

Walker, M., 2005. *Quaternary Dating Methods*. Chichester; John Wiley.

Wandrey, C.J., 2004. Bombay geologic province Eocene to Miocene composite total petroleum system, India. *US Geological Survey Bulletin*, **2208**-F.

Wandrey, C.J., Law, B.E. & Shah, H.A., 2004. Sembar Goru/Ghazij composite total petroleum system, Indus and Sulaiman-Kirthar geologic provinces, Pakistan and India. *US Geological Survey Bulletin*, **2208**-C.

Wang, P. & Murray, J.W., 1983. The use of Foraminifera as indicators of tidal effects in estuarine deposits. *Marine Geology*, **51**: 239–50.

Warraich, M.Y. & Nishi, H., 2003. Eocene planktonic foraminiferal biostratigraphy of the Sulaiman range, Indus basin, Pakistan. *Journal of Foraminiferal Research*, **33**(3): 219–36.

Warraich, M.Y., Ogasawara, K. & Nishi, H., 2000. Late Paleocene to Early Eocene planktonic foraminiferal biostratigraphy of the Dungan formation, Sulaiman range, central Pakistan. *Paleontological Research*, **4**(4): 275–301.

Warwick, P.D., Johnson, E.A. & Khan, I.H., 1998. Collision-induced tectonism along the northwestern margin of the Indian subcontinent as recorded in the Upper Paleocene to Middle Eocene strata of central Pakistan (Kirthar and Sulaiman ranges). *Palaeogeography, Palaeoclimatology, Palaeoecology*, **142**: 201–16.

Watkins, J. (ed.), 1982. *Studies in Continental Margin Geology*. Tulsa, OK; American Association of PetroleumGeologists (Memoir, No. 34).

Watkins, J.G., 1961. Foraminiferal ecology around the Orange County, California ocean sewage outfall. *Micropaleontology*, **7**: 199–206.

Watkinson, M.P., Hart, M.B. & Johsi, A., 2007. Cretaceous tectonostratigraphy and the development of the Cauvery basin, southeast India. *Petroleum Geoscience*, **13**: 181–91.

Watson, P. & Nairn, E., 2005. Stratigraphy to Seismic (StS): a technique to provide digital biostratigraphic information fopr the seismic interpreter and its impact on geological problem solving. *Abstracts, Geologic Problem Solving with Microfossils Conference, Rice University, Houston*: 59–60.

Weeks, L.G. (ed.), 1958. *Habitat of Oil*. Tulsa, OK; American Association of Petroleum Geologists.

Wefer, G., Suess, E., Balzer, W. *et al.*, 1982. Fluxes of biogenic components from sediment trap deployment in circumpolar waters of the Drake Passage: *Nature*, **299**: 145–7.

Weimer, P., 1990. Sequence stratigraphy, facies geometries and depositional history of the Mississippi fan, Gulf of Mexico. *American Association of Petroleum Geologists Bulletin*, **74**(4): 425–53.

Weimer, P., Bouma, A., & Perkins, B.F. (eds.), 1994. *Submarine Fans and Turbidite Systems: Sequence Stratigraphy, Reservoir Architecture and Production Chjaracteristics, Gulf of Mexico and International*. Gulf Coast Section – Society of Economic Paleontologists and Mineralogists (GCS-SEPM) Foundation (Proceedings of the 15th Annual Research Conference).

Weimer, P. & Link, M.H. (eds.), 1991. *Seismic Facies and Sedimentary Processes of Submarine Fans and Turbidite Systems*. New York; Springer Verlag.

Weimer, P. & Posamentier, H.W. (eds.), 1993. *Siliciclastic Sequence Stratigraphy: Recent Developments and Applications*. Tulsa, OK; American Association of Petroleum Geologists (Memoir, No. 58).

Weiner, A.K.M., Aurahs, R. & Kucera, M., 2010. Molecular phylogeny of *Hastigerina pelagica* and *Hastigerinella digitata* and their phylogeography in the Mediterranean Sea. *Abstracts, Forams2010 (International Symposium on Foraminifera), Bonn*: 199.

Weinholz, W. & Lutze, G.F., 1989. The *Stilostomella* extinction. *Proceedings of the Ocean Drilling Program, Scientific Results*, **108**: 113–7.

Wernli, R., 1988. Les protoglobigérines (foraminifères) du Toarcien et de l'Aalénien du

Domuz Dag (Taurus Occidental, Turquie). *Ecologae Geologicae Heletiae*, **81**: 661–8.

Wescott, W.A., Krebs, W.N., Sikora, P.J. *et al.*, 1998. Modern applications of biostratigraphy in exploration and production. *The Leading Edge*, September 1998: 1204–10.

Wescott, W.A. & Boucher, P.J., 2000. Imaging submarine channels in the western Nile Delta and interpreting their paleohydraulic characteristics from 3-D seismic. *The Leading Edge*, June 2000: 580–91.

Wheeler, H.E., 1958. Base level, lithosphere surface and time-stratigraphy. *Geological Society of America Bulletin*, **75**: 599–610.

Whidden, K.J. & Jones, R.W., 2012. Correlation of early Paleogene global diversity patterns of Large Benthic Foraminifera with Paleocene and Eocene climatic events. *Palaios*, **27**: 235–51.

Whidden, K.J., Jones, R.W. & Afifi, T., 2009. Abundance and diversity patterns of nummulite species in the Paleogene of Libya as indicators of climatic and tectonic changes. *Abstracts, European Association of Geoscientists and Engineers (EAGE) 4th North African/Mediterranean Petroleum & Geosciences Conference & Exhibition, Tunis*.

White, M.R., 1992. On species identification in the foraminiferal genus Alveolina (Late Palaeocene–Middle Eocene). *Journal of Foraminiferal Research*, **22**(1): 52–70.

White, P., Siddall, R., Underwood, C. & BouDagher-Fadel, M., 2012. The geological provenance of coloured carbonate mosaic materials used at Fishbourne. *Journal of Roman Archaeology Supplementary Series*, **91**: 111–34.

Whittaker, J.E., 1988. *Benthic Cenozoic Foraminifera from Ecuador*. London; British Museum (Natural History).

Whittaker, J.E., Banner, F.T. & Jones, R.W., 1998. *Key Mesozoic Benthic Foraminifera of the Middle East*. London; The Natural History Museum.

Whittaker, J.E. & Hart, M.B. (eds.), 2010. *Micropalaeontology, Sedimentary Environments and Stratigraphy: a Tribute to Dennis Curry (1912–2001)*. London; The Micropalaeontological Society (Special Publication, No. 4).

Whittaker, J.E., Horne, D.J. & Jones, R.W., 2003. Micropalaeontology in the service of archaeology: advances in Quaternary biostratigraphy and palaeoenvironmental analysis using Foraminifera and ostracods. *Abstracts, The Micropalaeontological Society Annual General Meeting Presentations*.

Whybrow, P.J. & Hill, A. (eds.), 1999. *Fossil Vertebrates of Arabia*. New Haven, CT; Yale University Press.

Widmark, J.G.V., 1995. Multiple deep-water sources and trophic regimes in the latest Cretaceous deep sea: evidence from benthic Foraminifera. *Marine Micropaleontology*, **26**: 361–84.

Widmark, J.G.V. & Speijer, R.P., 1997. Benthic foraminiferal faunas and trophic regimes at the terminal Cretaceous Tethyan seafloor. *Palaois*, **12**: 354–71.

Wiedmann, J. (ed.), 1979. *Aspekte der Kreide Europas*. International Union of Geological Sciences (Series A, No. 6).

Wilgus, C.K., Hastings, B.S., Posamentier, H. *et al.*, 1988. *Sea-Level Changes: An Integrated Approach*. Tulsa, OK; Society of Economic Paleontologists and Mineralogists (Special Publication, No. 42).

Wilkinson, I.P., 2011. Foraminiferal biozones and their relationship to the lithostratigraphy of the Chalk group of southern Engand. *Proceedings of the Geologists' Association*, **122**: 842–9.

Wilkinson, I.P., Tasker, A., Gouldwel, A. *et al.*, 2010. Micropalaeontology reveals the source of building materials for a defensive earthwork (?English Civil War) at Wallingford Castle, Oxfordshire. *Journal of Micropalaeontology*, **29**: 87–92.

Wilkinson, K. & Stevens, C., 2003. *Environmental Archaeology: Approaches, Techniques & Applications*. Stroud; Tempus.

Williams, H.F.L., 1999. Foraminiferal distributions in tidal marshes bordering the Strait of Juan de Fuca: implications for paleoseismicity studies. *Journal of Foraminiferal Research*, **29**(3): 196–208.

Williams, M., Dunkerley, D., de Deckker, P. *et al.*, 2003. *Quaternary Environments. Second Edition*. London; Arnold.

Williams, M., Haywood, A.M., Gregory, F.J. & Schmidt, D.N. (eds.), 2007. *Deep-Time Perspectives on Climate Change*. London; The Geological Society (for The Micropalaeontological Society).

Williams, M.D., 1958. Stratigraphy of the Lower Indus basin, West Pakistan. *Proceedings, Fifth World Petroleum Congress*: 377–94.

Williams, R.A., Robinson, M.C., Fernandez, E.G. & Mitchum, R.M., Jr., 2001. Cotton Valley/ Bossier of east Texas: sequence stratigraphy recreates the depositional history. *Transactions, Gulf Coast Association of Geological Societies*, **LI**: 379–88.

Williams, R.G. & Follows, M.J., 2011. *Ocean Dynamics and the Carbon Cycle*. Cambridge; Cambridge University Press.

Williamson, M.A., 1987. A quantitative foraminiferal biozonation of the Late Jurassic and Early Cretaceous of the east Newfoundland basin. *Micropaleontology*, **33**(1): 37–65.

Wilson, A.O., 1981. Jurassic Arab-C and D carbonate petroleum reservoirs, Qatif field, Saudi Arabia. *Society of Petroleum Engineers Paper*, **9594**: 171–7.

Wilson, A.O., 1985. Jurassic source rocks in the western Arabian Gulf area. *Proceedings of the Seminar on Source and Habitat of Petroleum in the Arab Countries, Kuwait*, **1984**: 501–20.

Wilson, B., 2003. Foraminiferal and paleodepths in a section of the Early to Middle Miocene Brasso formation, central Trinidad. *Caribbean Journal of Science*, **39**(2): 209–14.

Wilson, B., 2004. Benthonic foraminiferal palaeoecology across a transgressive–regressive cycle in the Brasso formation (Early–Middle Miocene) of central Trinidad. *Caribbean Journal of Science*, **40**(1): 126–38.

Wilson, B., 2006. Trouble in paradise? A comparison of 1953 and 2005 benthonic foraminiferal seafloor assemblages at the Ibis field, offshore eastern Trinidad, West Indies. *Journal of Micropaleontology*, **25**: 157–64.

Wilson, B., 2007. Benthonic foraminiferal palaeoecology of the Brasso formation (*Globorotalia fohsi lobata* and *Globorotaluia fohsi robusta* [N12–N13] zones), Trinidad, West Indies: a transect through an Oxygen Minimum Zone. *Journal of South American Earth Sciences*, **23**: 91–8.

Wilson, B., 2008a. Benthonic foraminiferal palaeoecology indicates an Oxygen Minimum Zone and an allochthonous, inner neritic assemblage in the Brasso formation (Middle Miocene) at St Fabien quarry, Trinidad, West Indies. *Caribbean Journal of Science*, **44**(2): 228–35.

Wilson, B., 2008b. Using SHEBI (SHE analysis for Biozone Identification): to proceed from the top down or the bottom up? A discussion using two Miocene foraminiferal successions from Trinidad, West Indies. *Palaios*, **23**: 636–44.

Wilson, B., 2010. A lagoonal interlude with occasional hypersalinity in the deposition of the Early–Middle Miocene Brasso formation of Trinidad. *Journal of South American Earth Sciences*, **29**: 254–61.

Wilson, B., 2011. Alpha and beta diversities of Late Quaternary bathyal benthonic foraminiferal communities in the NE Caribbean Sea. *Journal of Foraminiferal Research*, **41**(1): 33–40.

Wilson, B. & Costelloe, A., 2011a. Abundance biozone boundary types and characteristics determined using beta diversity: an example using Pleistocene benthonic Foraminifera in DSDP Hole 148, eastern Caribean Sea. *Palaios*, **26**: 152–9.

Wilson, B. & Costelloe, A., 2011b. Benthonic foraminiferal paleoecology of the Pleistocene in DSDP Hole 148, Aves Ridge, eastern Caribean Sea. *Journal of Foraminiferal Research*, **41**(4): 363–70.

Wilson, B. & Horton, B.P., 2012. Determining carrying capacity from foraminiferal time-series studies. *Journal of Micropalaeontology*, **31**: 111–9.

Wilson, B., Jones, B. & Birjue, K., 2010. Paleoenvironmental interpretations based on foraminiferal abundance biozones, Mayo limestone, Trinidad, West Indies, including alpha and beta diversities. *Palaios*, **25**: 158–66.

Wilson, B., Miller, K., Thomas, A.-L. *et al.*, 2008. Foraminifera in the mangal at the Caroni Swamp, Trinidad: diversity, population structure and relation to sea level. *Journal of Foraminiferal Research*, **38**(2): 127–36.

Wilson, B., Orchard, K. & Phillip, J., 2012. SHE analysis for Biozone Identification among foraminiferal sediment assemblages on reefs and in associated sediment around St Kitts, eastern Caribbean Sea, and its environmental significance. *Marine Micropaleontology*, **82**–83: 38–45.

Wilson, R.C.L., Drury, S.A. & Chapman, J.L., 2000. *The Great Ice Age: Climate Change and Life*. New York; Routledge.

Wing, S.L., Gingerich, P.D., Schmitz, B. & Thomas, E., 2003. *Causes and Consequences of Gloally Warm Climates in the Early Paleogene*. Boulder, CO; The Geological Society of America (Special Papers, No. 369).

Winter, F.W., 1907. Zur Kenntnis der Thalamophoren. I. Untersuchung über *Peneroplis pertusus* (Forskal). *Archiv der Protistenkunde*, **10**: 1–113.

Wisshak, M. & Tapanila, L. (eds.), 2008. *Current Developments in Bioerosion*. Berlin; Springer.

Wollenburg, J. & Tiedemann, R., 2010. Initial results on methane-seepage-emulating culture experiments on barophilic deep-sea foraminifera. *Abstracts, Forams2010 (International Symposium on Foraminifera), Bonn*: 201.

Wonders, A.A.H., 1980. Middle and Late Cretaceous planktonic Foraminifera of the western Mediterranean area. *Utrecht Micropaleontological Bulletin*, **24**: 1–158.

Wood, C., 2003. *Environmental Impact Assessment: A Comparative Review*. Second Edition. Harlow; Pearson/Prentice Hall.

Wood, L., 2000. Chronostratigraphy and tectonostratigraphy of the Columbus basin, offshore eastern Trinidad. *American Association of Petroleum Geologists Bulletin*, **84**(12): 1905–28.

Woodroffe, S.A., Horton, B.P., Larcombe, P. & Whittaker, J.E., 2005. Intertidal mangrove Foraminifera from the central Great Barrier Reef, Australia: implications for sea-level reconstruction. *Journal of Foraminiferal Research*, **35**(3): 259–70.

Woodland, A.W. (ed.), 1975. *Petroleum and the Continental Shelf of North-West Europe*. New York; John Wiley.

Woodside, P.R., 1981. Petroleum geology of Trinidad and Tobago. *Oil and Gas Journal*, 28th September, 1981: 314–87.

Woulds, C., Cowie, G.L., Levin, L.A. *et al.*, 2007. Oxygen as a control on seafloor biological communities and their roles in sedimentary carbon cycling. *Limnology and Oceanography*, **52**(4): 1698–1709.

Wray, C.G., Langer, M. R., DeSalle, R. *et al.*, 1995. Origin of the Foraminifera. *Proceedings of the National Academy of Sciences, USA*, **92**: 141–5.

Wright, R.I., 1968. Miliolidae (foraminiferos) recientes del estuario del Rio Quequen Grande (Provincia de Buenos Aires). *Revista del Museo Argentino ed Ciencias Naturales 'Bernardino Rivadavia', Hidrobiologia*, **2**: 225–56.

Wright, T., 2001. *Bolivinoides* saves Ipswich from flooding disaster! *Newsletter of Micropalaeontology*, **65**: 17.

Xiang Rong, Liu Fang, Chen Zhong *et al.*, 2010. Recent progress in cold seep benthic Foraminifera. *Advances in Earth Science*, **25**(2): 193–202.

Xue, L. & Galloway, W.E., 1995. High-resolution depositional framework of the Paleocene Middle Wilcox strata, Texas coastal plain. *American Association of Petroleum Geologists Bulletin*, **79**(2): 205–30.

Xuefeng Yu, Weijian Zhou, Yongyang Huang *et al.*, 2010. Peat records of human impact on the atmosphere in north-west China during the late Neolithic and Bronze Ages. *Palaeogeography, Palaeoclimatology, Palaeoecology*, **286**: 17–22.

Yamasaki, M., Sasaki, A., Oda, M. & Domitsu, H., 2008. Western equatorial Pacific planktonic foraminiferal fluxes and assemblages during a La Nina year (1999). *Marine Micropaleontology*, **66**(3–4): 304–19.

Yang-Logan, L.C., Tveit, R., Bailey, H.W. & Gallagher, L.T., 1996. Biosteering: a biostratigraphic application to horizontal drilling in the Eldfisk field, Norwegian North Sea. *Transactions of the 1995 AAPG Mid-Continent Section Meeting*.

Yanko, V. & Flexer, A., 1991. Foraminiferal benthonic assemblages as indicators of pollution: example of the north-western shelf of the Black Sea. *Proceedings of the Third Annual Symposium on the Mediterranean of Israel*.

Yanko, V.V. & Flexer, A., 1992. Microfauna as possible indicator of hydrocarbon seepages: method for oil–gas reconnaissance. *Transactions of the Israeli Geological Society Meeting*, March 1992: 169–70.

Yanko, V., Flexer, A., Kress, N. *et al.*, 1992. Response of benthic Foraminifera to various pollution sopurces: implications for pollution monitoring. *Journal of Foraminiferal Research*, **24**(1): 1–17.

Yanko, V. & Kronfeld, J., 1992. Low and high magnesian calcitic tests of benthic Foraminifera chemically mirror morphological deformations. *Proceedings of the Fourth International Conference on Paleoceanography, Kiel*: 308.

Yanko, V. & Kronfeld, J., 1993. Trace metal pollution affects the carbonate chemistry of benthic foraminiferal shells. *Proceedings of the 24th Annual Meeting of the Israel Society for Ecology and Environental Quality Studies, Tel Aviv*.

Yanko, V., Kronfeld, J. & Flexer, A., 1994. Response of benthic Foraminifera to various pollution sources: implications for pollution monitoring. *Journal of Foraminiferal Research*, **24**(1): 1–17.

Young Hoon Chung, Su-Yeong Yang & Jae Woo Kim, 2012. Numerical simulation of deep biogenic gas play northeastern Bay of Bengal, offshore northwest Myanmar. *Search and Discovery Article*, **50562**.

Yu, J. & Elderfield, H., 2007. Benthic foraminiferal B/Ca ratios reflect deep water carbonate saturation state. *Earth and Planetary Science Letters*, **258**(1–2): 73–86.

Zaigham, N.A. & Mallick, K.A., 2000. Prospect of hydrocarbon associated with fossil rift structures of the southern Indus basin, Pakistan. *American Association of Petroleum Geologists Bulletin*, **84**(11): 1833–48.

Zakrevskaya, E., Stupin, S. & Bugrova, E., 2009. Biostratigraphy of larger Foraminifera in the Eocene (upper Ypresian–lower Bartonian) sequences of the southern slope of the western Caucasus (Russia, NE Black Sea): correlation with regional and standard planktonic foraminiferal zones. *Geologica Acta*, **7**(1–2): 259–79.

Zaninetti, L., 1979. L'etude des foraminiferes des mangroves actuelles: reflexion sur les objectifs at sur l'etat des connaisances. *Archives des Sciences*, **32**(2): 151–61.

Zaninetti, L., Bronnimann, P., Huber, H. & Moshtagian, A., 1978. Microfacies et microfaunes du Permien au Jurassique au Kuh-e Gahkum, sud-Zagros, Iran. *Rivista Italiana di Paleontologia*, **84**(4): 865–96.

Zapata, E., Padron, V., Madrid, I. *et al.*, 2003. Biostratigraphic, sedimentologic and chemostratigraphic study of the La Luna formation (late Turonian–Campanian) in the San Miguel and Las Hernandez sections, western Venezuela. *Palaios*, **18**: 367–77.

Zellers, S.D. & Gary, A.C., 2007. Unmixing foraminiferal assemblages: polytopic vector analysis applied to Yakataga formation sequences in the offshore Gulf of Alaska. *Palaios*, **22**: 47–59.

Zhang, J., Gilbert, D., Gooday, A.J. *et al.*, 2010. Natural and human-induced hypoxia and consequences for coastal areas: synthesis and future development. *Biogeosciences*, **7**: 1443–67.

Zhang, J., Miller, K.G. & Berggren, W.A., 1993. Neogene planktonic foraminiferal biostratigraphy of the northeastern Gulf of Mexico. *Micropaleontology*, **39**(4): 299–326.

Ziegler, K., Turner, P. & Daines, S.E. (eds.), 1997. *Petroleum Geology of the Southern North Sea: Future Potential*. London; The Geological Society (Special Publication, No. 123).

Ziegler, M.A., 2001. Late Permian to Holocene paleofacies evolution of the Arabian plate and its hydrocarbon occurrences. *GeoArabia*, **6**(3): 445–503.

Zivkovic, S. & Glumac, B., 2007. Paleoenvironmental reconstruction of the Middle Eocene Trieste-Pazin basin (Croatia) from benthic foraminiferal assemblages. *Micropaleontology*, **53**(4): 285–310.

Zou, F., Slatt, R., Bastidas, R. & Ramirez, B., 2012. Integrated outcrop reservoir characterization, modelling and simulation of the Jackfork group at the Baumgartner Quarry area, western Arkansas: implications to Gulf of Mexico deep-water exploration and production. *American Association of Petroleum Geologists Bulletin*, **96**(8): 1429–48.

Zutshi, P.L. & Panwar, M.S., 1997. *Geology of Petroliferous Basins of India*. Dehra Dun; Oil and Natural Gas Corporation.

# Index

374

For EU product safety concerns, contact us at Calle de José Abascal, 56–1°, 28003 Madrid, Spain or eugpsr@cambridge.org.

www.ingramcontent.com/pod-product-compliance
Ingram Content Group UK Ltd.
Pitfield, Milton Keynes, MK11 3LW, UK
UKHW051510240426
470322UK00008B/113